Experimental Methods for Membrane Applications in Desalination and Water Treatment

Experimental Methods for Membrane Applications in Desalination and Water Treatment

SERGIO G. SALINAS-RODRÍGUEZ

LOREEN O. VILLACORTE

PUBLISHING

Published by: **IWA Publishing**
Unit 104 – 105, Export Building
1 Clove Crescent
London E14 2BA, UK
Telephone: +44 (0)20 7654 5500
Fax: +44 (0)20 7654 5555
Email: publications@iwap.co.uk
Web: www.iwapublishing.com

First published 2024

British Library Cataloguing in Publication Data
A CIP catalogue record for this book is available from the British Library

Library of Congress Cataloguing in Publication Data
A catalogue record for this book is available from the Library of Congress

Reference:
Salinas Rodriguez SG, Villacorte LO (2024) Experimental Methods for Membrane Applications in Desalination and Water Treatment, 1st edn IWA Publishing, London. doi: 10.2166/9781789062977

Cover design: Hans Emeis
Graphic design: Hans Emeis

ISBN 9781789062960 (Hardback)
ISBN 9781789062977 (eBook)
ISBN 9781789062984 (ePub)

Foreword

Experimental Methods for Membrane Applications in Desalination and Water Treatment

Water

Few other substances are so abundant on our beautiful planet that they would be able to cover it in a layer more than three kilometers thick. Seen from space, our planet is blue and white from water.

Still, water scarcity is a grave and global issue, because water is often not in the right place, at the right time, and in the right quality. This book is dedicated to the last issue, water quality. More specifically, the focus is on experimental membrane processes in water treatment (this, of course, you will have picked up from the book title). Membrane processes in water treatment are literally as old as life itself, but still a vibrant experimental field, as will be clear when you enjoy the book.

As a technology, membrane filtration is highly effective, proven to be able to mitigate the increasingly global challenges of water scarcity and limited access to clean water. Depending on membrane type, the filtration process can remove a wide range of water contaminants, making it uniquely suitable for purifying unconventional but abundant water sources such as seawater, highly polluted surface or groundwater, and various types of wastewater. As water scarcity impacts billions of people globally, thousands of membrane-based purification plants have been planned or installed in both developed and developing regions. This means that plant engineers and operators who have process and analytical knowledge of membrane technology are urgently needed. Researchers are also needed to further improve the sustainability and economic feasibility of the technology.

Unfortunately, knowledge on membrane processes is currently fragmented in various academic publications, most of which are not freely available to operators, engineers and researchers, particularly in developing countries. This book aims to address this critical issue by bringing it all together in a series of chapters written by some of the foremost experts in the field.

The Grundfos Foundation is proud to co-sponsor this book.

Poul Toft Frederiksen
Head of Programme, Research and Learning,
The Grundfos Foundation

Contributors

Chapter(s)

Aamer Ali, *PhD, MSc, Assistant Professor* 5
Center for Membrane Technology, Department of Chemistry and Bioscience, Aalborg University, Denmark

Adam C. Hambly, *PhD, Senior Researcher Water Technology & Processes* 12
Dep. of Environmental and Resource Engineering, Technical University of Denmark, Denmark

Alberto Tiraferri, *PhD, MSc, Full Professor of Applied Environmental Engineering* 4
Department of Environment, Land and Infrastructure Engineering, Politecnico di Torino, Italy

Almotasembellah Abushaban, *PhD, MSc, Assistant Professor in Water Desalination* 1, 7, 15
The Applied Chemistry and Engineering Research Center of Excellence, Mohammed VI Polytechnic University, Morocco

Barun Lal Karna, *ME (Research), Assistant Chief Engineer, Sofitel Sydney Darling Harbour* 11
Water Research Centre, School of Civil and Environmental Engineering, University of New South Wales, Australia

Cejna Anna Quist-Jensen, *PhD, MSc, Associate Professor of Membrane Technology* 5
Center for Membrane Technology, Department of Chemistry and Bioscience, Aalborg University, Denmark

Claus Hélix-Nielsen, *PhD, MSc, Professor DTU Sustain & President Danish Natural Sciences Academy* 4
Head of Dep. of Environmental and Resource Engineering, Technical University of Denmark, Denmark

Francisco Javier García Picazo, *MEng, Assistant Chemist* 19
Environmental Chemistry Services, City of San Diego, United States of America

Guillem Gilabert-Oriol, *PhD, MSc, Research and Development Leader* 2, 3
DuPont Water Solutions, Spain, Adjunct Professor in the Universitat Rovira i Virgili, Spain

Gustavo A. Fimbres Weihs, *PhD, Lead Research Fellow* 19
School of Chemical and Biomolecular Engineering, The University of Sydney, Australia

Helen Rutlidge, *PhD, BSc, Lecturer* 11
School of Chemical Engineering, University of New South Wales, Australia

Helga Calix Ponce, *MSc, Researcher* 13
Denmark

Irena Petrinic, *PhD, MSc, Associate Professor* 4
University of Maribor, Slovenia

Jan Frauholz, *MSc, Process Engineer* 4
Aquaporin A/S, Denmark & RWTH Aachen, Germany

Javier Rodriguez Gómez, *Laboratory supervisor* 18
Genesys and PWT brands, H_2O innovation, Spain

Jia Xin Tan, *BEng* 19
Faculty of Chemical and Process Engineering Technology,
Universiti Malaysia Pahang Al-Sultan Abdullah, Malaysia

Johannes S. Vrouwenvelder, *PhD, MSc, Professor of Environmental Science and Engineering,* 17
Director of Water Desalination and Reuse Center (WDRC)
Biological & Environmental Science & Engineering Division (BESE), King Abdullah University of Science and Technology (KAUST), Saudi Arabia
Delft University of Technology , Faculty of Applied Sciences, Department of Biotechnology , The Netherlands

Karima Bakkali, *BSc, Research Engineer* 15
Mohammed VI Polytechnic University, Morocco

Kathleen Foo, *MSc, PhD Research Fellow* 19
Faculty of Chemical and Process Engineering Technology,
Universiti Malaysia Pahang Al-Sultan Abdullah, Malaysia

Léonie Le Bouille, *MSc, Research Fellow* 15
IHE Delft Institute for Water Education / CIRSEE Suez / Delft University, The Netherlands / France

Loreen O. Villacorte, *PhD, MSc, Lead Water Treatment Specialist* 1, 13
Global Technology and Innovation, Grundfos Holding A/S, Denmark

Luca Fortunato, *PhD, MSc, Research Scientist* 17
Water Desalination and Reuse Center (WDRC), Biological & Environmental Science & Engineering Division (BESE), King Abdullah University of Science and Technology (KAUST), Saudi Arabia

Lucia Ruiz Haddad, *MSc, PhD Candidate* 14
Environmental Science and Engineering Program, Biological and Environmental Science and Engineering (BESE) Division, King Abdullah University of Science and Technology (KAUST), Thuwal, Kingdom of Saudi Arabia

Maria Salud Camilleri-Rumbau, *PhD, MSc, Researcher, R&D Project Manager* 4
Technology Centre of Catalonia - Fundació Eurecat, Spain / Aquaporin A/S, Denmark

Mohamed Chaker Necibi, *PhD, Associate Professor in Circular Economy* 15
International Water Research Institute (IWRI), Mohammed VI Polytechnic University, Morocco

Mohamed Fauzi Haroon, *PhD, Associate Director Analytical Science & Technology* 14
Moderna, United States of America

Mohammad Mahdi A. Shirazi, *PhD, MSc, Senior Postdoctoral Fellow* 5
Center for Membrane Technology, Department of Chemistry and Bioscience,
Aalborg University, Denmark

Mohaned Sousi, *PhD, MSc* 16
Water Supply, Sanitation and Environmental Engineering Department,
IHE Delft Institute for Water Education, The Netherlands

Morten Lykkegaard Christensen, *PhD, MSc, Associate Professor of Wastewater Treatment and Membrane Technology Department of Chemistry and Bioscience,* 2
Aalborg University, Denmark

Muhammad Ali, *PhD, Martin Naughton Assistant Professor in Environmental Microbiology* 14
Department of Civil, Structural & Environmental Engineering, Trinity College Dublin,
The University of Dublin, Ireland

Muhammad Nasir Mangal, *PhD, MSc* 10
Membrane specialist, Berghof Membranes, The Netherlands

Nuria Peña García, *PhD, MSc, Research Director* 9, 18
Director of Scientific Global Services for Genesys and PWT brands, H_2O innovation, Spain

Pascal E. Saikaly, *PhD, MSc, Professor of Environmental Science and Engineering,* 14
Chair of Environmental Science and Engineering Program, Biological and Environmental Science and Engineering (BESE) Division, King Abdullah University of Science and Technology (KAUST), Thuwal, Kingdom of Saudi Arabia

Pierre Le-Clech, *PhD, MSc, Associate Professor of Water and Wastewater treatment* 11
School of Chemical Engineering, University of New South Wales, Australia

Poul Toft Frederiksen, *PhD, MSc, Head of Programme Research and Learning* F
The Grundfos Foundation, Denmark

Pouyan Mirzaei Vishkaei, *MSc* 1
Researcher, Water Supply, Sanitation and Environmental Engineering Department,
IHE Delft Institute for Water Education, The Netherlands

Rita Kay Henderson, *PhD, MSc, Professor of Water Quality and Treatment* 11
School of Chemical Engineering, University of New South Wales, Australia

Sergio G. Salinas-Rodriguez, *PhD, MSc, Associate Professor of Water Supply Engineering* 1, 7, 8, 15
Water Supply, Sanitation and Environmental Engineering Department,
IHE Delft Institute for Water Education, The Netherlands

Steven J. Duranceau, *PhD, PE, Professor and Director* 6
Environmental Systems Engineering Institute Department of Civil, Environmental & Construction Engineering, College of Engineering and Computer Science,
University of Central Florida, United States of America

Urban J. Wünsch, *PhD, Postdoctoral Researcher* 12
National Institute of Aquatic Resources, Technical University of Denmark, Denmark

Vanida A. Salgado-Ismodes, *MSc, PhD research fellow* 7
Water Supply, Sanitation and Environmental Engineering Department,
IHE Delft Institute for Water Education, The Netherlands

Victor Augusto Yangali Quintanilla, *PhD, MSc, Lead Water Treatment Specialist* 4
Global Technology and Innovation, Grundfos Holding A/S, Denmark

Victoria Sanahuja-Embuena, *PhD, MSc, Chemical Engineer, Scientist* 4
Aquaporin A/S, Denmark

Wen Yew Lam, *BEng* 19
Faculty of Chemical and Process Engineering Technology,
Universiti Malaysia Pahang Al-Sultan Abdullah, Malaysia

Weng Fung Twong, *BEng* 19
Faculty of Chemical and Process Engineering Technology,
Universiti Malaysia Pahang Al-Sultan Abdullah, Malaysia

Xuan Tung Nguyen, *BSc, Project Manager* 4
Aquaporin Asia, Singapore

Yie Kai Chong, *BEng* 19
Faculty of Chemical and Process Engineering Technology,
Universiti Malaysia Pahang Al-Sultan Abdullah, Malaysia

Yuli Ekowati, *PhD, MSc, Postdoctoral Researcher* 1, 13
Global Technology and Innovation, Grundfos Holding A/S, Denmark

Yong Yeow Liang, *PhD, Senior Lecturer* 19
Faculty of Chemical and Process Engineering Technology, Centre for Research in Advanced Fluid and Processes, Universiti Malaysia Pahang Al-Sultan Abdullah, Kuantan, Pahang, Malaysia

About the editors

Sergio G. Salinas-Rodriguez is Associate Professor and desalination and water treatment technology professional at IHE Delft Institute for Water Education. He has a PhD in Desalination and Water Treatment from the Technical University of Delft, an MSc in Water Supply Engineering from UNESCO-IHE Institute for Water Education, a Master's in Irrigation and Drainage and a BSc in Civil Engineering from San Simon Major University. He also obtained the University Teaching Qualification in the Netherlands.

He has over 75 publications in books, chapters, international peer-reviewed journals and conference proceedings in the areas of seawater and brackish water desalination, water treatment, water reuse, and natural organic matter characterization.

Sergio is involved in teaching and curriculum development of the MSc Programme in Water and Sustainable Development at IHE Delft. His projects comprise capacity building, research and innovation (e.g., EU-MEDINA, EU-MIDES, EU-MAR2PROTECT). He has mentored more than 50 MSc students, co-promoted 4 PhD students, and currently supervises 2 PhD students. Sergio lectures and coordinates several courses on Desalination and membrane technology.

Loreen Ople Villacorte is Lead Water Treatment Specialist at Global Technology and Innovation in Grundfos Denmark. He has broad experience in research, conceptualization, development and validation of water treatment technologies including applications of traditional and emerging membrane technologies.

For the last 18 years, he has held various roles in the academia and the industry across three countries (Philippines, Netherlands and Denmark), primarily driving research and technology development projects to tackle water challenges in drinking water production and transport, wastewater treatment or reuse, oil-water separation and industrial cooling systems. Most of these projects were implemented through cross-functional collaborations and involves understanding the physics, biology and chemistry of water to enable development of effective treatment solutions. He has published >25 scientific articles and filed numerous patents in water treatment and desalination applications.

He is a civil engineer with a master's degree in water supply engineering at IHE-Delft and a doctoral degree in desalination and water treatment from the Technical University of Delft, IHE-Delft and Wetsus.

Contents

doi: 10.2166/9781789062977_0001

Chapter 1

Feedwater Quality Guidelines and Assessment Methods for Membrane-based Desalination

Loreen O. Villacorte, Grundfos, Denmark

Yuli Ekowati, Grundfos, Denmark

Almotasembellah Abushaban, UM6P, Morocco

Pouyan Mirzaei Vishkaei, IHE Delft, The Netherlands

Sergio G. Salinas-Rodriguez, IHE Delft, The Netherlands

The learning objectives of this chapter are the following:

- To review the existing feedwater quality guidelines for membrane-based desalination
- To present and discuss the existing and proposed methods for assessing fouling and scaling potential of feedwater.

1.1 INTRODUCTION

Amidst the global problem of dwindling freshwater water resources, desalination of unconventional but abundant water resources such as seawater and brackish water has grown rapidly over the last three decades. From a global operational capacity of ~7.5 million m^3/day in 1990 to ~115 million m^3/day in 2023, water desalination technologies have been the leading solution to address the growing municipal, agricultural and industrial demand for clean freshwater (Figure 1a and 1b). Furthermore, desalination technologies are generally applied for triple barrier wastewater reuse applications, which currently has a global installed capacity of >60 million m^3/year (Birch *et al.*, 2023).

The chapter is from the book Experimental Methods for Membrane Applications in Desalination and Water Treatment, Sergio G. Salinas-Rodriguez, Loreen O. Villacorte (Eds).

Seawater is the main water source for desalination globally, with the exception of North America, where the majority of applications is based on brackish water desalination. Japan, South Korea, Taiwan, and China desalinate seawater, brackish water, and wastewater effluent at relatively similar rates.

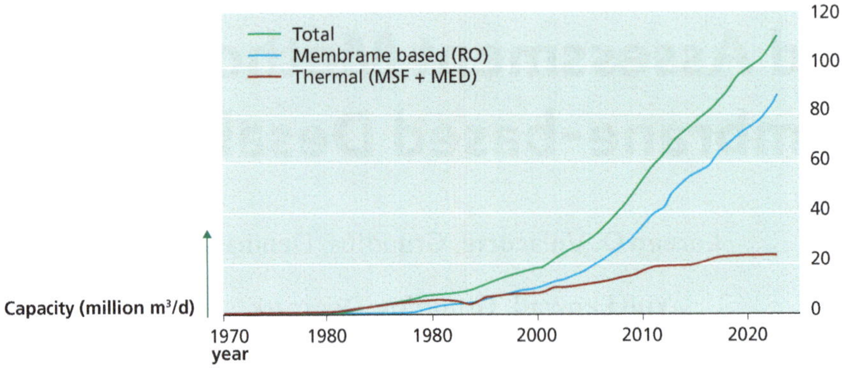

Figure 1a The growth of global desalination application in terms of online production capacity since 1970 (top) and current online seawater desalination by technology, region and end-user. Produced with information from (DesalData, 2023). MSF = Multi-stage flash distillation, MED = Multi-effect distillation, RO = Reverse osmosis

In terms of technology, membrane-based desalination using reverse osmosis (RO) dominates the application (~74% of global capacity). This is mainly driven by the significantly lower investment cost and energy requirements today are lower than thermal processes (e.g., MSF, MED). A large majority (>75%) of the desalinated water are used for supplying drinking water supply while about 20% are used in industries. Most of the Middle East countries rely on desalination for municipal use, while countries such as China, India, South Korea, Brazil, Taiwan, Chile, Indonesia use desalination to satisfy industrial demand.

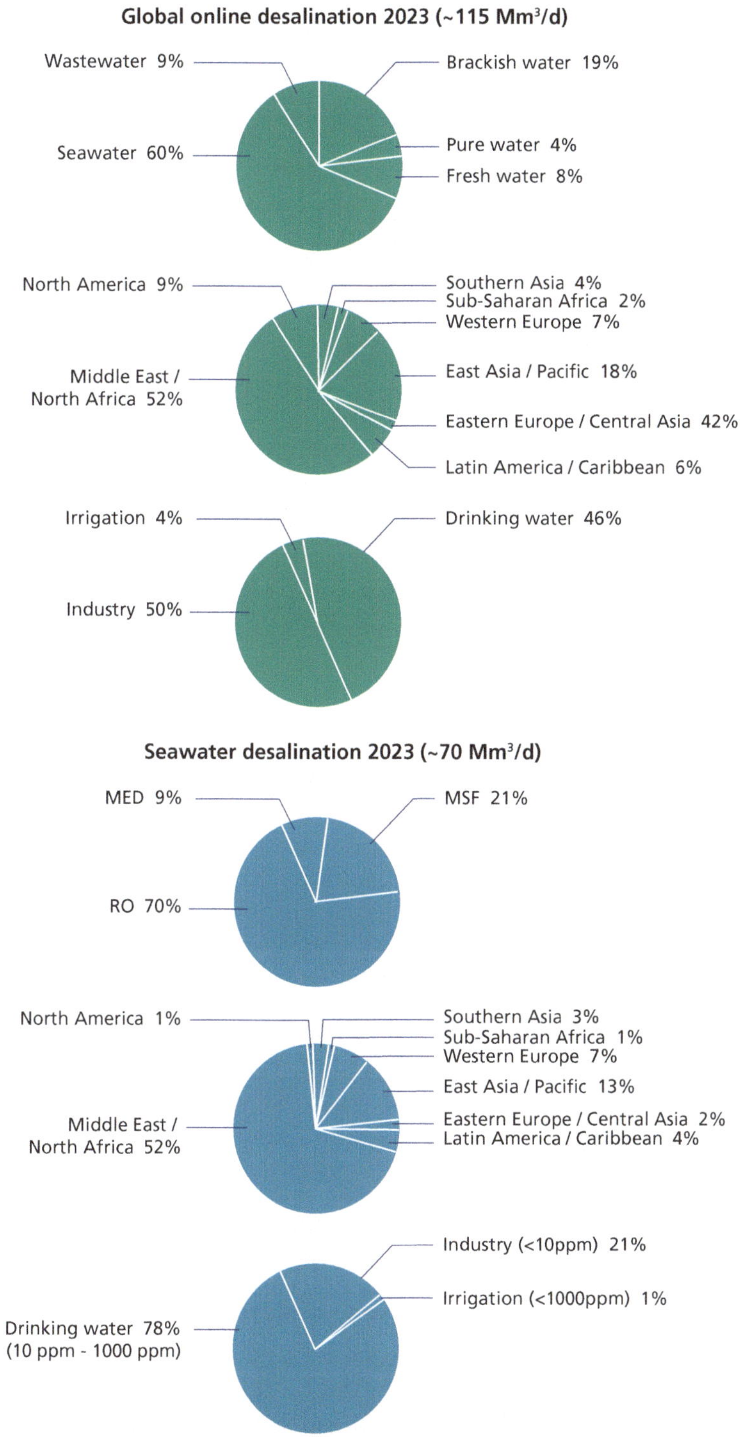

Figure 1b Current online global desalination (top 3) and online seawater desalination (bottom 3) by technology, region and end-user. Produced with information from (DesalData, 2023)

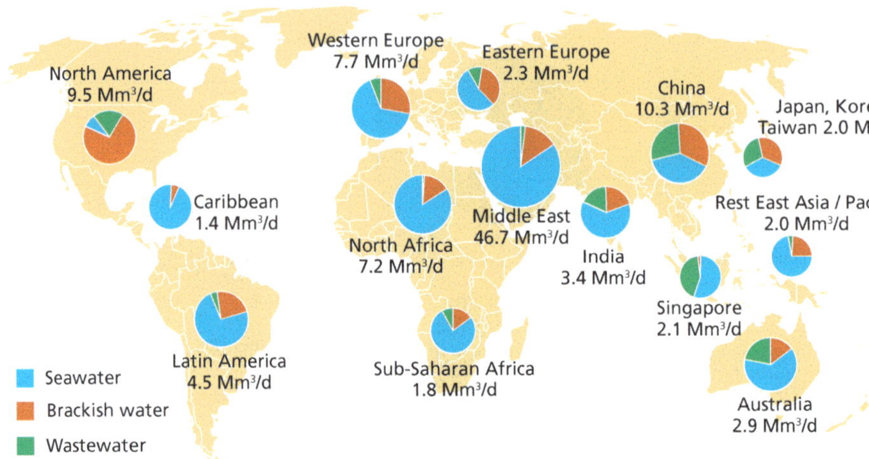

Figure 2 Feedwater sources and (online, in construction) production capacities of desalination plants in different geographic locations in 2023 (updated from Salinas Rodriguez and Schippers (2021) with information from DesalData (2023)).

The price per cubic metre of desalinated water has reduced significantly over the years due to more efficient membrane production, implementation of energy recovery devices, cost of engineering, etc, and a more competitive market. The specific energy consumption has already been reduced by at least 50% over the last 20 years and the overall carbon footprint of desalination could be reduced down further by switching to renewable energy sources (Birch *et al.*, 2023). On the downside, membrane fouling and scaling are the main 'Achilles heel' for the sustainable application of RO (Voutchkov, 2010, Salinas Rodriguez, 2021).

Fouling and scaling in membranes can lead to a variety of problems, such as the need for (frequent) chemical cleaning, reduction of production capacity, higher energy consumption, decrease in produced water quality, that it makes RO production facilities less reliable, and require more frequent membrane replacement (Dhakal *et al.*, 2020, Salinas Rodriguez *et al.*, 2021b). Fouling and scaling are broadly categorized into i) particulate/colloidal fouling due to suspended and colloidal matter, ii) inorganic fouling due to iron and manganese, iii) organic fouling due to organic compounds e.g., polymers, iv) biofouling due to growth of bacteria, and v) scaling due to deposition of sparingly soluble compounds.

During RO operation, membrane fouling and scaling may manifest in three ways, namely: i) increasing the differential pressure across the spacer in spiral wound elements due to 'clogging', resulting in potential membrane damage (such as telescoping, channelling, or squeezing); ii) increasing membrane resistance (or decreasing the normalized permeability) due to deposition and/or adsorption of materials on the membrane surface, resulting in higher required feed pressure to maintain capacity; and iii) increasing in normalized salt passage due to concentration polarization in the fouling layer, resulting in higher salinity in the product water.

Particulate and colloidal fouling are mostly well controlled by the pre-treatment systems (mostly media filtration or membrane filtration), but the occurrence of organic fouling and biofouling is still a major issue in RO membranes, and is the main reason for the need for frequent cleaning of the reverse osmosis membranes (Peña *et al.*, 2022).

To minimize the occurrence of membrane fouling/scaling in RO, pre-treatment of the feedwater is essential. Additionally, methods and tools can help significantly, by monitoring the performance of the pre-treatment with regards to fouling/scaling control and process optimization. Pre-treatment can take place in the form of media filters with or without coagulation, membrane filtration with or without inline coagulation (e.g., ultrafiltration), and dissolved air floatation in combination with the previous mentioned two options.

Along with the increase in the number of desalination plants (>22,800 plants in 2023), the capacity of newly installed plants has also increased significantly over time. A growing preference for extra-large (XL) plants (capacity >50,000 m^3/d) has been reported in recent years (Birch *et al.* (2023); Kurihara and Ito (2020). More XL seawater RO (SWRO) plants are expected in the future. This means reliable pre-treatment systems and monitoring tools will be essential for these XL plants, as Cleaning-in-Place (CIP) of membrane modules more than once per year is rather challenging. The design and operational settings of such pre-treatment systems will depend on the water quality and their temporal variations of the water source alongside the feedwater quality guidelines provided by the membrane supplier.

For a long time, the silt density index (SDI) has served as a sum 'king/ultimate' parameter for assessing RO feed water. DuPont (2020) for the first time introduced the MFI-0.45 in their RO feedwater guidelines. This is a major step forward due to the limitations of the SDI in assessing fouling in RO (Schippers *et al.*, 2014). In addition, the inclusion of parameters like AOC and BFR bring relevance for the monitoring of the biofouling potential of RO feedwater as several types of fouling take place simultaneously. Table 1 presents the recommended guideline values for RO feedwater by RO manufacturers and literature. The majority of RO membrane manufacturers are in agreement with the recommendations by DuPont although they main guideline is the SDI value less than 4-5 and preferable less than 3 for RO feedwater.

Table 1 Parameters and recommended guideline values for RO feed water

Parameter	Unit	DuPont (2023)	Other sources	Standard Methods
(a) Particulate fouling indicators				
SDI_{15}	%/min	<5 (target <3)	<5 (target <3) (Wilf and Klinko, 2016) < 3 (Badruzzaman *et al.*, 2019) < 4 (Voutchkov, 2010)	(ASTM D4189 - 07)
MFI-0.45	s/L^2	4 (target <1)		(ASTM D8002 - 15)
MFI-UF	s/L^2	-	<490 at 15 lmh (safe MFI*) (Salinas Rodríguez, 2011)	
Turbidity	NTU	< 1	< 0.5 (Badruzzaman *et al.*, 2019) < 0.1 (Voutchkov, 2010)	(ASTM D1889-00)
(b) Organic fouling indicators				
Oil and grease	mg/L	0.1	< 0.1 (Badruzzaman *et al.*, 2019) < 0.02 (Voutchkov, 2010)	(ASTM D7575-11)
TOC	mg-C/L	3	< 2 (Badruzzaman *et al.*, 2019) < 2 (target <0.5) (Voutchkov, 2010)	(ASTM D2579-93e1)
SUVA	L/mg-m		< 4 (USEPA, 2005)	
COD	mg/L	10		(ASTM D1252-06(2020))
(c) Biological fouling indicators				
AOC	μg/L Ac-C	10 (target <5)	<10 μg-C acetate/L (threshold for biofouling in freshwater) (van der Kooij *et al.*, 1982)	(NEN 6271:1995 nl)
BGP	μg-C/L	-	-	<70 (Abushaban, 2019)
BFR	pg-ATP/cm^2	5 (target <1)	< 1 (Vrouwenvelder and van der Kooij, 2001)	
PO_4-P	μg/L		0.3 μg P/L (Vrouwenvelder *et al.*, 2010)	
(d) Inorganic fouling and scaling indicators				
Ferrous iron	mg/L	4	< 2 (Badruzzaman *et al.*, 2019) < 2 (Voutchkov, 2010)	(ASTM D1068-15)
Ferric iron	mg/L	0.05	< 0.1 (Badruzzaman *et al.*, 2019) <0.05 (Voutchkov, 2010)	(ASTM D1068-15)
Manganese	mg/L	0.05	0.05 (Badruzzaman *et al.*, 2019) 0.02 (Voutchkov, 2010)	(ASTM D858-17)
Aluminium	mg/L	0.05		(ASTM D857-17)
Silica	mg/L		20 (Badruzzaman *et al.*, 2019)	(ASTM D859-16(2021)e1)
pH	-		4-11 (Voutchkov, 2010)	(ASTM D1293-12)
LSI (freshwater)	-	Concentrate LSI < 0 (if no antiscalant is added)		(ASTM D3739-19)

Parameter	Unit	DuPont (2023)	Other sources	Standard Methods
S&DSI (seawater)	-	Concentrate S&DSI < 0 (if no antiscalant is added)		(ASTM D4582-91(2001))
(e) Membrane material limits				
Temperature	°C		< 35 (Voutchkov, 2010)	
Free chlorine	mg/L	<0.1	< 0.1 (Badruzzaman *et al.*, 2019) < 0.01 (Voutchkov, 2010)	(ASTM D1253-14(2021)e1)
ORP	mV	<175-200		(ASTM D1498-14(2022)e11)

* Safe MFI is a value for RO feedwater that will yield a 1 bar pressure increase in a 6 months period.

In addition to the established feedwater quality parameters in Table 1, tens of thousands of scientific articles and patents were published over the past 30 years describing or applying new assessment tools/indices for evaluating the fouling/scaling potential of RO feedwater as well as to characterize the impact of specific feedwater components to RO operation (Figure 3). Some of these tools were also applied to optimize the design and operation of RO pre-treatment system, including MF/UF processes (see Chapter 2). The succeeding sections review the advantages as well as the challenges of applying these assessment tools in membrane-based desalination systems.

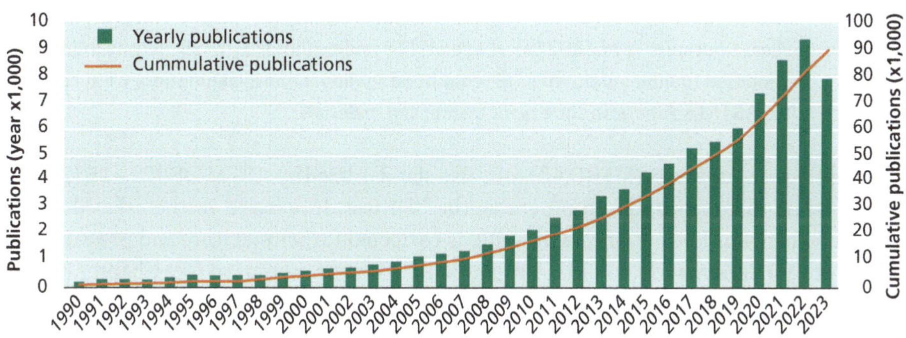

Figure 3 Number of scientific and patent publications related to fouling and scaling assessment in reverse osmosis process from 1990 and 2023 (November). Data generated through Google Scholar using the search string: "(fouling OR scaling) AND (characterization OR assessment OR potential OR indicator OR index) AND (reverse osmosis)".

1.2 PARTICULATE FOULING POTENTIAL

Fouling indices to measure the particulate fouling potential of RO feedwater have been in development since the 1960's (Figure 4). The oldest and most widely used index, the silt density index (SDI) has been standardised by ASTM D4189 - 14 (2014), is applied worldwide as it is simple to perform and with low-cost consumables (see Chapter 6).

However, increasingly the value of this test to predict the rate of fouling in RO systems due to particle deposition is being questioned. The limitations of the SDI test are well documented (Schippers and Verdouw, 1980, Nahrstedt and Camargo Schmale, 2008, Alhadidi *et al.*, 2011a, Alhadidi *et al.*, 2011b, Rachman *et al.*, 2013, Salinas Rodriguez *et al.*, 2019) and include no correction for feedwater temperature variation (higher SDI values at higher temperatures); the result is heavily dependent on the permeability of the test membrane filter; not applicable for testing high fouling feed water e.g., raw water – ASTM recommends that turbidity should be < 1 NTU; not applicable for testing UF system permeate, which is increasingly being used in desalination pre-treatment; no linear relation with colloidal/suspended matter; fouling potential of particles smaller than 0.45 μm are not taken into account; it is not based on any filtration mechanism.

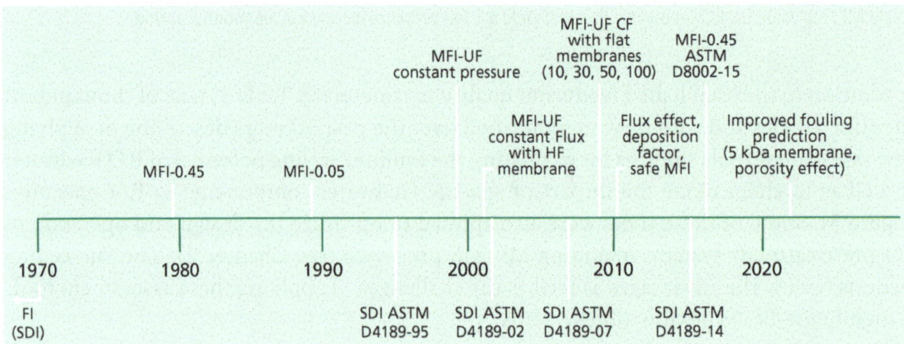

Figure 4 Historical development of fouling indices for particulate fouling assessment (adapted from Salinas Rodriguez *et al.* (2021a)). Legend: FI = fouling index, SDI = silt density index, MFI = modified fouling index, MFI-UF = modified fouling index ultrafiltration, CF = constant flux, ASTM = American society for testing and materials.

The modified fouling index (MFI-0.45) test, also standardized by ASTM D8002 - 15 (2015), uses the same materials and equipment as the SDI test. It is based in the cake filtration fouling mechanism. The obtained MFI value is corrected for temperature and pressure and shows a linear relation with colloidal/suspended matter concentration (see Chapter 7). The predicted rate of fouling turns out to be very low at a level of MFI = 1 s/L^2, which is in the range of SDI 1 to 3.

Based on the low sensitivity of MFI-0.45, the MFI-UF test with ultra-filtration membranes was developed to capture these smaller particles (see Chapter 8). The strong dependence of MFI on flux, means that to be able to predict accurately the potential of particulate fouling in RO systems, the MFI should preferably be measured at a flux similar to a RO system (15- 20 $L/m^2/h$) or extrapolated from higher fluxes. A theoretical 'safe MFI' was proposed assuming e.g., an allowable increase in NDP of 1 bar in 6 months (Salinas Rodríguez, 2011, Salinas Rodriguez *et al.*, 2019). The safe MFI calculated for a deposition factor $\Omega = 1$ (worst case) at a flux of 15 $L/m^2/h$ (average flux in RO), has been reported about 490 s/L^2. And could be used as a threshold value for assessing RO feed water quality. Good correlation with RO membrane fouling development was observed when applying the MFI-UF prediction model (Abunada *et al.*, 2023).

1.3 INORGANIC FOULING AND SCALING POTENTIAL

Inorganic fouling occurs in RO processes due to the deposition of insoluble or sparingly soluble inorganic compounds from the feedwater to the RO membrane. In many literatures, inorganic fouling also includes scaling, however, deposition of metal ions, such as iron, manganese and aluminium is also called as inorganic fouling. Iron, manganese and aluminium ions are present naturally in surface water and groundwater. Fouling due to these ions can be reduced by pre-treatment, such as aeration, oxidation, and filtration. Membrane manufacturers provide guidelines for feedwater quality compatible with the membrane which typically includes limits on iron, manganese, and aluminium concentrations (Table 1). Analytical methods to quantify critical inorganic components in the feedwater or to identify accumulation on the membrane surface are well established, which usually includes ICP and SEM-EDX. There is no single analytical method than can completely identify all inorganic foulants so it is important to get familiarized with these methods (see full list of tools in Chapter 9) when studying inorganic fouling.

Scaling on RO membranes is specifically caused by precipitation and accumulation of sparingly soluble salts usually as a consequence of up-concentrating these salts in the RO process, eventually exceeding their solubility limit. The type of scaling depends on the ion composition, temperature and pH of the feedwater. Scaling issues typically observed in RO processes are calcium carbonate, calcium sulphate, calcium phosphate, barium sulphate, and silica/metal silicates. Scaling can be avoided by limiting the RO recovery, acid dosing, and antiscalant dosing. The latter is the most preferable approach because it does not compromise the RO production and it is effective for different types of scaling with only low dose requirement. Pretreatment of feedwater by ion exchange or lime softening can also be applied to control scaling.

The scaling potential of feedwater in the RO system can be assessed based on scaling indices. The commonly used indices for RO application are saturation index (SI) and supersaturation ratio (Sr), Langelier saturation index (LSI), and Stiff-Davis stability index (S&DSI). SI and Sr are applicable for all types of scaling species while LSI and S&DSI are specifically used for calcium carbonate scaling. A positive value of the indices generally indicates supersaturation condition where precipitation can occur. Calculation of saturation indices are available in standardized ASTM methods (ASTM D4692-01, D3739-19, D4582-23) and in literatures (e.g., Mangal *et al.* (2021)). In practice, the scaling potential of RO feedwater are usually determined using the design software developed by antiscalant chemical suppliers and membrane manufacturers, such as WAVE (DuPont), IMSDesign (Hydranautics) and MembraneMaster MM5 (Genesys). A geochemical modelling program, PHREEQC, can also be used to calculate SI.

An emerging parameter known as induction time has been increasingly used to predict or study scaling in RO. Scaling in RO occurs when the lattice ions in supersaturated solution start to agglomerate and form nuclei or clusters. If the size of the cluster is below the critical size, then the formed crystal will re-dissolve into the solution and if the cluster size is above the critical size, then the crystal become stable and grow bigger. Induction time is defined as the time required for the supersaturated solution to form nuclei in the critical cluster size dimensions just before it becomes stable and starts to accelerate growth (He *et al.*, 1995,

Boerlage *et al.*, 2000, Boerlage, 2001). Induction time can be determined by measuring the change in pH, turbidity, conductivity, and specific ion concentration in the water over a period in a controlled lab environment (Waly, 2011, Mangal, 2023). Measuring induction time is potentially useful tool in developing new design, pretreatment and operational strategies to control scaling in RO but better understanding is needed to measure impact of other inorganics as well as organic components in the water on the induction time of specific salts.

1.4 ORGANIC FOULING POTENTIAL

The accumulation of organic matter in membrane systems can directly cause substantial decline in operational performance (e.g., permeability). Organic foulants are typically abundant in surface waters but are also present in ground water sources. They can originate from natural sources from human activities (e.g., sewage discharge) or from chemicals used in the pre-treatment processes. Identifying organic foulants and understanding their characteristics provides valuable insights on how to prevent organic fouling, by choosing the appropriate pre-treatment and operational and cleaning strategies. Currently available assessments tools for organic fouling can range from simple spectrophotometric methods to more advanced chromatographic techniques (see Figure 5).

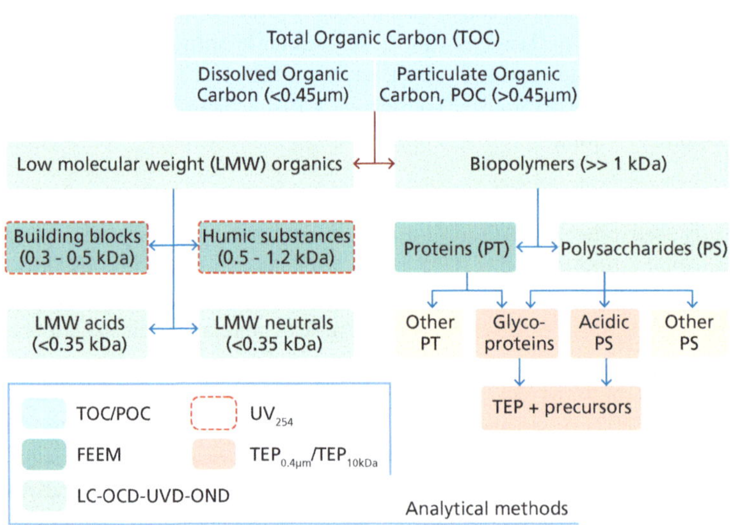

Figure 5 Overview of the different fractions of organic matter in the water and the applicable analytical methods to identify or quantify them (Villacorte *et al.*, 2021). LC-OCD-UVD-OND is liquid chromatography (LC) with inline detectors for organic carbon (OCD), UV absorbance at 254 nm (UVD) and organic nitrogen (OND); FEEM is fluorescence excitation-emission matrices; TEP refers to transparent exopolymer particles measured with a 0.4 μm or 10 kDa membrane.

1.4.1 Organic carbon

Traditionally, the presence of organic matter in RO feedwater is assessed by measuring the total organic carbon (TOC) or dissolved organic carbon (DOC). Routine TOC/ DOC measurements have been used for monitoring the bulk organic fouling potential of the feedwater. However, not all organic carbon in the water can be directly associated with organic fouling. So, in many cases, TOC/ DOC as such is not sufficient to assess the variations in organic fouling potential of the feedwater. According to Voutchkov (2010), if TOC concentration is below 0.5 mg/L then biofouling is unlikely; and above 2 mg/L, biofouling is very likely. Nevertheless, as the TOC is a bulk concentration value, it is very important to identify the fraction of the TOC responsible for bacterial growth. Cationic organic polymers have also been reported with negative fouling effects on RO membranes, as they may coprecipitate with negatively charged antiscalants and foul the membrane irreversibly (Ekowati *et al.*, 2014, Peña *et al.*, 2015, DuPont, 2023).

1.4.2 UV absorbance and fluorescence

Hydrophobic and aromatic organic compounds such as humic substances can be abundant in surface water sources. UV absorbance at 254 nm (UVA_{254}) is typically used as an indicator of their relative abundance. Specific UVA_{254} (SUVA), defined as the ratio between UVA_{254} and DOC, is a parameter shown to correlate with the aromatic contents and hydrophobicity of organic matter (Baghoth, 2012). Such measurement is simple, fast and can be measured routinely. UV absorbing aromatic and hydrophobic organics are typically removed by coagulation pretreatment more efficiently than the non-UV absorbing hydrophilic components (Matilainen *et al.*, 2011). Hence, a low SUVA after pretreatment does not necessarily mean low organic fouling potential because hydrophilic compounds may still remain in the RO feedwater. Moreover, humic substances were reported to be less problematic foulant than hydrophilic organic substances (Amy *et al.*, 2011). Therefore, it is recommended that UVA_{254} or SUVA is supplemented with other measurements when assessing organic fouling potential of RO feedwater.

An emerging technique to characterize organic foulants is by measuring differences in fluorescence spectra associated with specific organic compounds in a sample using spectrofluorometer (see Chapter 12). The technique can be applied on both liquid (e.g., feedwater) and solid (e.g., membrane surface) samples, and generates a three-dimensional excitation-emission matrix (EEM). The location of EEM peaks provides a qualitative indication of types of organic molecules present in water samples (Westerhoff *et al.*, 2001). For example, humic-like fluorescence peak could be clearly discriminated from protein-like peak in the EEM spectra. In general, fluorescence spectroscopy can be used as a rapid and sensitive method to characterize dissolved organic matter, but analysis is limited to organic components in the water that contains fluorophores. For instance, other organic matter components such as polysaccharides do not fluoresce and could not be analyzed using the EEM spectra. Therefore, FEEM is typically not a standalone method for organic fouling assessment.

1.4.3 LC-OCD

Liquid chromatography-organic carbon detection (LC-OCD) is an advanced method for fractionation and measurement of organic carbon by size-exclusion chromatography followed by inline analyses through multiple detectors (i.e., organic carbon, UV_{254} and organic nitrogen). While DOC measures organic carbon in bulk, LC-OCD fractionates DOC based on their molecular weight and hydrophobicity (see Chapter 11). LC-OCD measures the concentration of organic carbon fractions as biopolymers, humic substances, building blocks, low molecular weight acids, and low molecular weight neutrals (Huber *et al.*, 2011). The limit of detection of the method for each fraction is at ppb level and the method has been adapted for high salinity water (Amy *et al.*, 2011). LC-OCD method has been applied in many studies to characterize organic matter in surface water for assessment of fouling potential of the different fractions and their removal through the water treatment processes (Lozier *et al.*, 2009, Salinas Rodriguez *et al.*, 2009, Villacorte *et al.*, 2009, Simon *et al.*, 2013, Ho *et al.*, 2015, Shanmuganathan *et al.*, 2015, Jeong *et al.*, 2016, Yin *et al.*, 2019, Altmann *et al.*, 2023).

1.4.4 TEP

Transparent exopolymer particles (TEP) and their precursors can have a major role in organic and biological fouling in membrane filtration processes. Berman and Holenberg (2005) first proposed the potential role of TEP as a major initiator of biofilm leading to biofouling in reverse osmosis (RO) membranes. Consequently, various experimental studies investigated the role of TEP, on biofouling (Bar-Zeev *et al.*, 2012) and organic fouling (Villacorte *et al.*, 2021) in membrane systems. Several methods have been developed, adopted, modified, and demonstrated to quantify TEP and elucidate their role to membrane fouling (see Chapter 13). The TEP0.4μm and TEP10kDa methods has been successfully used to semi-quantitatively demonstrate the role of TEP on the operational performance of membrane processes (Villacorte *et al.*, 2015a). They have been applied as an indicator of fouling potential of RO feedwater and showed significant correlation with MFI-UF (Villacorte *et al.*, 2015b). So far, no study has successfully determined the threshold level of TEP in the feedwater at which membrane fouling will likely not occur. TEP methods still have some inherent limitations (Discart *et al.*, 2015, Bittar *et al.*, 2018, Li *et al.*, 2018), so it should be implemented with proper attention to the protocol used and by someone who is experienced with laboratory analytical techniques.

The quantification of algae in the RO feedwater source can act as an indication of the occurrence of algal blooms, which can generate organic foulants like TEP. Algae can be quantified directly through microscopic counting as cell density or indirectly through chlorophyll-a measurement. Standard chlorophyll-a methods are widely available (Arar and Collins, 1997; ASTM D3731-20, 2020; Lipps *et al.*, 2023c). A spike in algae concentrations can coincide with an increase of organic fouling mainly due to extracellular substances released by algae. However, a spike in algae density or chlorophyll-a concentration in the water does not necessarily result in high organic fouling because bloom-forming algal species can vary in shapes/sizes, specific chlorophyll-a concentration, TEP/EPS production and their characteristics that affects their removal in the pre-treatment process and their organic fouling potential to RO.

1.4.5 Oil and grease

One of the most detrimental types of organic foulants are oily compounds which can impact both the operation and integrity of the membrane units. Ideally, oil and grease should not exceed 0.1 mg/L in the feed water of a RO system because it can attach and accumulate on the membrane surface which may lead to irreversible organic fouling. Pre-treatment is necessary in the case of treating produced water from oil and gas extraction and industrial wastewater. The method to determine oil and grease in water consists of extraction by liquid/liquid extraction, solid phase extraction, or microwave extraction and measurement by gravimetric and infrared analysis. The standard method for determination of oil and grease in water through these various extraction methods and analyses can be found in Lipps *et al.* (2023b), USEPA (2010), ASTM D7066-04 (2010) and ASTM D7575-11 (2017). Generally, the infrared methods are more sensitive compared to gravimetric methods with detection limit of approximately 1 mg/L and even down to 0.1 mg/L when using tetrachloroethylene as the extraction solvent (Farmaki *et al.*, 2007).

1.5 BIOFOULING POTENTIAL

Biofouling occurs due to the growth of microorganisms on the membrane and feed spacer of the RO system. Biofouling is a common issue in most RO desalination plants (Peña *et al.*, 2022) and is often inevitable even when bacteria/microorganisms are completely removed through the pre-treatment system (i.e., using microfiltration or ultrafiltration system). If a single bacteria/ microorganism finds their way to reach the RO system, it can rapidly grow and form a biofilm layer on the membrane and/or feed spacer of the RO when nutrient concentrations in the feedwater are limited. Biofouling occurs often in plants with open water sources (e.g., sea, river, lake) as they typically contain higher concentrations of organics and other nutrients. Thus, biofouling of brackish water RO is less frequent than that of seawater RO system.

1.5.1 Bacterial growth potential

Given that RO biofouling is mainly due to bacterial growth on the membrane surface and feed spacer, a method to measure the bacterial growth potential (BGP) in the RO feed water and through the pre-treatment process was developed by Abushaban (2019). The full description of the protocol and optimization of each step are discussed in Chapter 15. The basic concept of the method is to measure the growth of constant number of indigenous bacteria due to the presence of any nutrients (C: N: P) in a seawater sample. The limit of detection (LoD) of the method is 10 μg-C (as glucose)/L which is low enough to measure the BGP in the BWRO feedwater. However, this LoD might not be low enough in some BWRO system where BGP can be much lower, especially after the pre-treatment process.
The correlation between BGP in the SWRO feed water and biofouling in selected SWRO membrane systems in Australia, Europe, and the Middle East was investigated (Abushaban *et al.*, 2019b, Abushaban *et al.*, 2020, Abushaban *et al.*, 2021). Results show that a higher BGP in the SWRO feedwater (100 - 950 μg-C/L) corresponds to a higher normalised pressure drop or higher CIP cleaning frequency in the RO system, a demonstration of the applicability of BGP as a biofouling indicator in RO systems. It was estimated that a BGP value of 70 μg-C/L in the SWRO feedwater requires once per year CIP frequency (Abushaban, 2019, Abushaban *et al.*, 2019a, Dhakal *et al.*, 2020, Salinas Rodriguez *et al.*, 2021b). Consequently, a safe level of BGP (below 70 μg-C/L as glucose equivalent) was preliminarily proposed to control biofouling in SWRO desalination plants.

The advantage of the BGP method over other methods is that it measures the growth of indigenous bacteria until when the biodegradable nutrients present in seawater sample is depleted. Moreover, the duration of the test is around 4-5 days which is relatively short compared to conventional assimilable organic carbon (12-14 days) and biomass production potential (15-28 days) methods. A test duration of that takes days can still be a practical limitation of the method, particularly, when the concentration of BGP varies significantly (hourly or daily) in the water source. However, it should be noted that biofilm formation usually takes couple of weeks to be grow on the membrane system. Thus, getting the results within a few days can still be considered as an early detection of biofouling and a corrective action can still be made to remove or control growth of microorganisms.

There are other limitations in the application of the BGP method. Firstly, the protocol itself is quite complex. The complexity is due to the requirement of carbon-free glassware of each step. Any introduction of carbon or organic matter during handling and measuring BGP will negatively affect the results (Abushaban, 2019). Secondly, the cost of measuring BGP is high as it needs a qualified and skilled technician to measure the sample for around two weeks (including preparation and measurements) and the reagents used to measure microbial adenosine triphosphate (ATP) are also expensive. The frequency of measuring the BGP in SWRO feed water and/or along the pre-treatment processes, depends on the water source quality and the expected variation in the quality from season/month/week/day to another. Finally, another limitation of BGP method is that the results are influenced by the salinity of the water which is normally not the case in many desalination plants where the salinity of the water source is somehow constant. If the salinity of the water source is changed, it means new calibration curves for both ATP and BGP should be established. The higher the salinity the lower ATP signal is expected and thus lower slope of BGP.

In general, BGP method is a promising assessment method to control biofouling in SWRO system. However, the method is rather complex and thus it is currently applied mainly for research studies and not yet on a routine basis.

1.5.2 Assimilable organic carbon

Assimilable organic carbon (AOC) method has been developed 4 decades ago, mainly to monitor growth potential in drinking water distribution systems. In the recent years, the method has been adapted for seawater application. The difference between BGP and AOC methods is that AOC uses a single strain of bacteria while BGP uses a mix of indigenous bacteria (Abushaban *et al.*, 2022). The use of single strain of bacteria enables standardization the method (use one conversion value for samples from different location). Further optimization of the inoculum used in the test led to substantial reduction the duration of the test to 1-2 days.

The AOC method is that the method is simpler to implement than the BGP method. However, it is considered less accurate than BGP as it represents bacterial growth of a single strain of bacteria. It was also reported that the difference in bacterial growth of a single strain of bacteria is typically at least 20% lower than the growth of indigenous bacteria in fresh water (Ross *et al.*, 2013).

Various methods of AOC have been developed over the years using different bacterial strains such as Vibrio fischeri and Vibrio harveyi (Weinrich *et al.*, 2011, Jeong *et al.*, 2013). The LoD of these two methods are 10 and 0.1 μg-C/L, respectively. However, the extremely low LoD (0.1 μg-C/L) reported is questionable as it was calculated after subtracting the AOC of the blank, which was >50 μg-C/L.

Weinrich *et al.* (2015) reported a good correlation between AOC and differential pressure increase and specific flux decline at the Tampa Bay pilot seawater desalination plant where the feedwater AOC concentrations measured were between 22 and 161 μg-C/L. Consequently, a preliminary threshold concentration of AOC (50 μg-C/L) was proposed using Vibrio harveyi bacteria in seawater (Weinrich *et al.*, 2015). So far, the reported AOC concentrations in SWRO feed water varies between 10 and 220 μg-C/L.

Overall, the AOC method is applicable to monitor biofouling potential along the SWRO pre-treatment process and in the SWRO feed water. However, the accuracy of the reported bacterial growth may not represent actual conditions in the RO system as only one single strain is used to mimic the bacterial growth.

1.5.3 Biodegradable dissolved organic carbon

Not all DOC is bioavailable or can be directly utilized by microorganisms. Biodegradable dissolved organic carbon (BDOC) represents the fraction of DOC that can be utilized by microorganisms. It is calculated by subtracting the initial DOC of the water sample from the final DOC at the end of incubation period. The incubation period in BDOC measurement can be varied, depending on the time required to reach a stable DOC. Servais *et al.* (1987) suggested 4 weeks of incubation period while shorter incubation periods were introduced in various application of BDOC method (Joret *et al.*, 1989, Frías *et al.*, 1992, Kadjeski *et al.*, 2020). BDOC includes a larger fraction of total organic carbon (10-30%) with LoD of 0.1-0.2 mg/L which is about 10 folds higher than LoD of BGP and AOC. In general, the BDOC method is time consuming for routine monitoring and less sensitive compared to AOC and BGP methods. Additionally, in membrane filtration, it is likely that a large portion of BDOC is retained on the membrane while still allowing the majority of AOC to pass through (Escobar *et al.*, 2000, Escobar and Randall, 2001).

1.5.4 Phosphate

Biofilm formation in membranes can be largely influenced by the C:N:P nutrient mass ratio in the water which is ideally 100:23:4.3. Based on this ratio, the requirement for phosphorus (P) is lower than other substrates (carbon and nitrogen), so a small change in P can lead to a significant change in microbial growth. Limiting P down to a low level can disrupt the nutrient balance, restricting bacterial growth in the water and reducing biofouling (van der Kooij *et al.*, 2007, Galjaard *et al.*, 2008, Jacobson *et al.*, 2009, Vrouwenvelder *et al.*, 2010, Kim *et al.*, 2014). Reliable analytical methods to measure phosphate down to sub-ppb level is critical to this strategy. Standard phosphate analytical methods are available such as the widely used phosphomolybdenum blue method (Lipps *et al.*, 2023a). The method has been applied and adapted in biofouling studies of water containing very low phosphate concentrations (Abushaban *et al.*, 2020, Javier *et al.*, 2020). Ultra-low phosphate concentration down to below 0.3 μg PO_4-P/L was reported to limit biofouling even in water with high concentration of organic carbon (Vrouwenvelder *et al.*, 2010). Some antiscalants (e.g., phosphonates) which are added to the feedwater to prevent scaling in RO, contain phosphate which has been found to cause biofouling (Vrouwenvelder *et al.*, 2000, Sweity *et al.*, 2013, Sweity *et al.*, 2015, Hasanin *et al.*, 2023). It is therefore recommended that the dosage and type of antiscalants be taken into consideration when applied as pretreatment for RO.

1.6 OUTLOOK AND OPPORTUNITIES

RO desalination of brackish and saline water sources is increasingly applied globally to solve water scarcity challenges in the utility and industry sector. Assessing the fouling and scaling potential of the feedwater source is highly critical when designing and operating the RO desalination plant including its pre-treatment units. For many years, RO membrane suppliers have provided specific guidelines of the ideal RO feedwater quality to minimize possible particulate, inorganic, organic and biological fouling, and scaling issues in the plant. However, some of these standard water quality parameters and indices have some limitations so alternative/supplemental assessment and characterization tools has been developed over the years for feedwater monitoring and experimental investigations.

Measuring the individual concentration of all potential foulants present in the RO feed water is sort of a mission impossible due to costs, duration and specialized laboratory facilities needed. Particulate fouling indices like the SDI or MFI-0.45 can be used in a daily basis for monitoring of water quality with onsite measurements. Other parameters like the MFI-UF constant flux are promising due to its sensitivity and ability to predict accurately the rate of RO membrane fouling.

Standardized inorganic fouling and scaling assessment methods are widely available for lab-based analyses. Online monitoring is currently a challenge and would be an important area for further development. The induction time concept for predicting when scaling can occur is a promising tool to developing RO plant design and control strategies to minimize scaling issues (e.g., Mangal *et al.* (2022)).

The presence of organic foulants can be routinely measured with offline/online TOC measurements. Routine chlorophyll-a measurement can be beneficial for feedwater sources that are prone to seasonal algal blooms. For in-depth investigations of organic fouling, more advanced or complex assessment methods (e.g., LC-OCD, FEEM, TEP) can be further considered.

Biological fouling is currently the leading cause of operational challenges in RO applications. Genomic tools (see Chapter 14) have been applied to identify bacterial communities often associated with biofilm development in RO. Promising methods for assessing biofouling potential of RO feedwater are either based on the bacterial growth capacity (see Chapters 15 and 16) or based on the concentration of specific limiting nutrient (phosphate) in the feedwater. However, these methods either requires days to weeks of incubation to generate results or have high sensitivity to contamination. Future developments should focus on overcoming either one of these limitations.

Ideally, assessment methods and indices should be made practical for onsite monitoring and low cost for consumables, enabling it for wide use. Testing conditions should be standardized/calibrated and should specify the method LOD and their applicability in fresh and saline water matrices. The need and opportunities for real time online monitoring should be explored further. In some cases, a few analyses per day is sufficient and in other cases once a day or once a week are considered sufficient.

Chemicals used in the treatment process may contribute to the membrane fouling development due to introduction of nutrients or organic foulants. Hence, it is recommended that operators also assess the fouling potential of such chemicals before applying it to the RO system.

Current RO feedwater quality guideline values recommended by membrane manufacturers may need to be verified under local conditions and modified/adjusted accordingly based on latest developments. On the other hand, feedwater guidelines for emerging desalination technologies such as forward osmosis (see Chapter 4) and membrane distillation (see Chapter 5) are currently non-existent. Future developments of these new technologies should also include understanding their propensity to fouling or scaling by applying current or new feedwater quality assessment methods.

1.7 ABBREVIATIONS AND SYMBOLS

AOC	Assimilable organic carbon
ATP	Adenosine tri-phosphate
BGP	Bacterial growth potential
BFR	Biofilm formation rate
BWRO	Brackish water reverse osmosis
CIP	Cleaning in place
COD	Chemical oxygen demand
DOC	Dissolved organic carbon
EEM	Excitation-emission matrix
LC-OCD	Liquid chromatography-organic carbon detection
LoD	Limit of detection
MED	Multi-effect distillation
MFI-0.45	Modified fouling index, constant pressure, 0.45 μm filter
MFI-UF	Modified fouling index, constant flux, 10 kDa or 100 kDa membrane filter
MSF	Multi stage flash distillation
RO	Reverse osmosis
SDI	Silt density index, %/min
S&DSI	Stiff and Davis saturation index
SWRO	Seawater reverse osmosis
TEP	Transparent exopolymer particles
TOC	Total organic carbon

NB. This chapter is currently being prepared for submission as a scientific article in a journal.

1.8 REFERENCES

Abunada M (2023). Prediction of Particulate Fouling in Reverse Osmosis Systems: MFI-UF Method Development and Application. PhD Dissertation, IHE Delft

Abunada M, Dhakal N, Gulrez R, Ajok P, Li Y, Abushaban A, Smit H, Moed D, Ghaffour N, Schippers JC, Kennedy MD (2023) Prediction of particulate fouling in full-scale reverse osmosis plants using the modified fouling index – ultrafiltration (MFI-UF) method. Desalination 553: 116478 DOI https://doi.org/10.1016/j.desal.2023.116478

Abushaban A (2019) Assessing Bacterial Growth Potential in Seawater Reverse Osmosis Pretreatment: Method Development and Applications CRC Press9781000034707.

Abushaban A, Salinas-Rodriguez SG, Dhakal N, Schippers JC, Kennedy MD (2019a) Assessing pretreatment and seawater reverse osmosis performance using an ATP-based bacterial growth potential method. Desalination 467: 210-218 DOI https://doi.org/10.1016/j.desal.2019.06.001

Abushaban A, Salinas-Rodriguez SG, Kapala M, Pastorelli D, Schippers JC, Mondal S, Goueli S, Kennedy MD (2020) Monitoring Biofouling Potential Using ATP-Based Bacterial Growth Potential in SWRO Pre-Treatment of a Full-Scale Plant. Membranes 10: 360

Abushaban A, Salinas-Rodriguez SG, Pastorelli D, Schippers JC, Mondal S, Goueli S, Kennedy MD (2021) Assessing Pretreatment Effectiveness for Particulate, Organic and Biological Fouling in a Full-Scale SWRO Desalination Plant. Membranes 11: 167

Abushaban A, Salinas-Rodriguez SG, Philibert M, Le Bouille L, Necibi MC, Chehbouni A (2022) Biofouling potential indicators to assess pretreatment and mitigate biofouling in SWRO membranes: A short review. Desalination 527: 115543 DOI https://doi.org/10.1016/j.desal.2021.115543

Abushaban A, Salinas-Rodriguez SG, Schippers JC, Kennedy MD (2019b) Assessing pretreatment and seawater reverse osmosis performance using an ATP-based bacterial growth potential method. Desalination 467: 210-218 DOI https://doi.org/10.1016/j.desal.2019.06.001

Alhadidi A, Blankert B, Kemperman AJB, Schippers JC, Wessling M, van der Meer WGJ (2011a) Effect of testing conditions and filtration mechanisms on SDI. Journal of Membrane Science 381: 142-151

Alhadidi A, Kemperman AJB, Schippers JC, Wessling M, van der Meer WGJ (2011b) The influence of membrane properties on the Silt Density Index. Journal of Membrane Science 384: 205-218

Altmann T, Rousseva A, Vrouwenvelder J, Shaw M, Das R (2023) Effectiveness of ceramic ultrafiltration as pretreatment for seawater reverse osmosis. Desalination 564: 116781 DOI https://doi.org/10.1016/j.desal.2023.116781

Amy GL, Salinas Rodríguez SG, Kennedy MD, Schippers JC, Rapenne S, Remize P-J, Barbe C, Manes C-LdO, West NJ, Lebaron P, Kooij Dvd, Veenendaal H, Schaule G, Petrowski K, Huber S, Sim LN, Ye Y, Chen V, Fane AG (2011) Water quality assessment tools. In: Drioli E, Criscuoli A, Macedonio F (eds) Membrane-Based Desalination - An Integrated Approach (MEDINA): 192 10.2166/9781780400914

Arar EJ, Collins GB (1997) Method 445.0: In vitro determination of chlorophyll a and pheophytin a in marine and freshwater algae by fluorescence United States Environmental Protection Agency, Office of Research and Development, National Exposure Research Laboratory, Cincinnati, pp. 1-22

ASTM D857-17 (2017) Standard Test Method for Aluminum in Water ASTM International, West Conshohocken, PA, pp. 5 https://www.astm.org/d0857-17.html.

ASTM D858-17 (2017) Standard Test Methods for Manganese in Water ASTM International, West Conshohocken, PA, pp. 9 https://www.astm.org/d0858-17.html.

ASTM D859-16(2021)e1 (2021) Standard Test Method for Silica in Water ASTM International, West Conshohocken, PA, pp. 5 https://www.astm.org/d0859-16r21e01.html.

ASTM D1068-15 (2016) Standard Test Methods for Iron in Water ASTM International, West Conshohocken, PA, pp. 13 https://www.astm.org/d1068-15.html.

ASTM D1252-06(2020) (2020) Standard Test Methods for Chemical Oxygen Demand (Dichromate Oxygen Demand) of Water ASTM International, West Conshohocken, PA https://www.astm.org/d1252-06r20.html.

ASTM D1253-14(2021)e1 (2021) Standard Test Method for Residual Chlorine in Water ASTM International, West Conshohocken, PA, pp. 6 https://www.astm.org/d1253-14r21e01.html.

ASTM D1293-12 (2018) Standard Test Methods for pH of Water ASTM International, West Conshohocken, PA, pp. 10 https://www.astm.org/d1293-12.html.

ASTM D1889-00 (2017) Standard Test Method for Turbidity of Water (Withdrawn 2007) ASTM International, West Conshohocken, PA https://www.astm.org/d1889-00.html.

ASTM D2579-93e1 (2021) Standard Test Method for Total Organic Carbon in Water (Withdrawn 2002) ASTM International, West Conshohocken, PA https://www.astm.org/d2579-93e01.html.

ASTM D3731-20 (2020) Standard Practices for Measurement of Chlorophyll Content of Algae in Surface Waters ASTM International, West Conshohocken, PA, pp. 5 www.astm.org.

ASTM D3739-19 (2019) Standard Practice for Calculation and Adjustment of the Langelier Saturation Index for Reverse Osmosis ASTM International, West Conshohocken, PA, pp. 4 https://www.astm.org/d3739-19.html.

ASTM D4189 - 07 (2007) Standard Test Method for Silt Density Index (SDI) of Water ASTM International, West Conshohocken, PA

ASTM D4189 - 14 (2014) Standard Test Method for Silt Density Index (SDI) of Water ASTM International, West Conshohocken, PA

ASTM D4582-91(2001) (2017) Standard Practice for Calculation and Adjustment of the Stiff and Davis Stability Index for Reverse Osmosis ASTM International, West Conshohocken, PA, pp. 4 https://www.astm.org/d4582-91r01.html.

ASTM D7066-04 (2010) Standard Test Method for dimer/trimer of chlorotrifluoroethylene (S-316) Recoverable Oil and Grease and Nonpolar Material by Infrared Determination ASTM International, West Conshohocken, PA, pp. 9 www.astm.org.

ASTM D7575-11 (2017) Standard Test Method for Solvent-Free Membrane Recoverable Oil and Grease by Infrared Determination ASTM International, West Conshohocken, PA, pp. 12 www.astm.org.

ASTM D8002 - 15 (2015) Standard Test Method for Modified Fouling Index (MFI-0.45) of Water ASTM International, West Conshohocken, PA www.astm.org.

Badruzzaman M, Voutchkov N, Weinrich L, Jacangelo JG (2019) Selection of pretreatment technologies for seawater reverse osmosis plants: A review. Desalination 449: 78-91 DOI https://doi.org/10.1016/j.desal.2018.10.006

Baghoth SA (2012) Characterizing natural organic matter in drinking water treatment processes and trains CRC Press/Balkema, Delft. 9781138000261.

Bar-Zeev E, Berman-Frank I, Girshevitz O, Berman T (2012) Revised paradigm of aquatic biofilm formation facilitated by microgel transparent exopolymer particles. Proceedings of the National Academy of Sciences 109: 9119-9124 DOI doi:10.1073/pnas.1203708109

Bassa FGM (2021). Evaluating surface water pre-treatment with the MFI-UF for assessing fouling in membrane systems. MSc, IHE Delft UWS-WSE/21.25.

Belila A, El-Chakhtoura J, Otaibi N, Muyzer G, Gonzalez-Gil G, Saikaly PE, van Loosdrecht MCM, Vrouwenvelder JS (2016) Bacterial community structure and variation in a full-scale seawater desalination plant for drinking water production. Water Research 94: 62-72 DOI https://doi.org/10.1016/j.watres.2016.02.039

Berman T, Holenberg M (2005) Don't fall foul of biofilm through high TEP levels. Filtration & Separation 42: 30-32 DOI https://doi.org/10.1016/S0015-1882(05)70517-6

Birch H, Pasture LdL, Gall M, Gasson C, Pankratz T, Quaresma M, Qureshi Z, Walker C (2023) Market focus deck Desalination and reuse October 2023 Media Analytics Ltd., Oxford, UK. 978-1-907467-67-7.

Bittar TB, Passow U, Hamaraty L, Bidle KD, Harvey EL (2018) An updated method for the calibration of transparent exopolymer particle measurements. Limnology and Oceanography: Methods 16: 621-628 DOI https://doi.org/10.1002/lom3.10268

Boerlage SFE (2001) Scaling and Particulate Fouling in Membrane Filtration Systems Swets&Zeitlinger Publishers, Lisse. 90 5809 242 9.

Boerlage SFE, Kennedy MD, Aniye MP, Abogrean EM, El-Hodali DEY, Tarawneh ZS, Schippers JC (2000a) Modified Fouling Index ultrafiltration to compare pretreatment processes of reverse osmosis feedwater. Desalination 131: 201-214

Boerlage SFE, Kennedy MD, Bremere I, Witkamp GJ, van der Hoek JP, Schippers JC (2000b) Stable barium sulphate supersaturation in reverse osmosis. Journal of Membrane Science 179: 53-68 DOI 10.1016/s0376-7388(00)00504-4

DesalData (2023) 36th Desalination Plants Inventory In: Global Water Intelligence (ed)

Dhakal N, Abushaban A, Mangal MN, Abunada M, Schippers JC, Kennedy M (2020) Membrane Fouling and Scaling in Reverse Osmosis. In: Sapalidis A (ed) Membrane Desalination: 325-344

Dhakal N, Salinas Rodriguez SG, Schippers JC, Kennedy MD (2015) Induction time measurements in two brackish water reverse osmosis plants for calcium carbonate precipitation. Desalination and Water Treatment 53: 285-293 DOI 10.1080/19443994.2014.903870

Discart V, Bilad MR, Vankelecom IFJ (2015) Critical Evaluation of the Determination Methods for Transparent Exopolymer Particles, Agents of Membrane Fouling. Critical Reviews in Environmental Science and Technology 45: 167-192 DOI 10.1080/10643389.2013.829982

DuPont (2020) FILMTEC™ Reverse Osmosis Membranes Technical Manual, Form No. 45-D01504-en, Rev. 3Water solutions, pp. 207

DuPont (2023) FILMTEC™ Reverse Osmosis Membranes Technical Manual, Form No. 45-D01504-en, Rev. 16Water solutions, pp. 188 https://www.dupont.com/content/dam/dupont/amer/us/en/water-solutions/public/documents/en/RO-NF-FilmTec-Manual-45-D01504-en.pdf.

Ekowati Y, Msuya M, Salinas Rodriguez SG, Veenendaal G, Schippers JC, Kennedy MD (2014) Synthetic organic polymer fouling in municipal wastewater reuse reverse osmosis. Journal of Water Reuse and Desalination 4: 125-136 DOI doi:10.2166/wrd.2014.046

Escobar IC, Hong S, Randall AA (2000) Removal of assimilable organic carbon and biodegradable dissolved organic carbon by reverse osmosis and nanofiltration membranes. Journal of Membrane Science 175: 1-17 DOI https://doi.org/10.1016/S0376-7388(00)00398-7

Escobar IC, Randall AA (2001) Assimilable organic carbon (AOC) and biodegradable dissolved organic carbon (BDOC):: complementary measurements. Water Research 35: 4444-4454 DOI https://doi.org/10.1016/S0043-1354(01)00173-7

Farmaki E, Kaloudis T, Dimitrou K, Thanasoulias N, Kousouris L, Tzoumerkas F (2007) Validation of a FT-IR method for the determination of oils and grease in water using tetrachloroethylene as the extraction solvent. Desalination 210: 52-60 DOI https://doi.org/10.1016/j.desal.2006.05.032

Frías J, Ribas F, Lucena F (1992) A method for the measurement of biodegradable organic carbon in waters. Water Research 26: 255-258 DOI https://doi.org/10.1016/0043-1354(92)90226-T

Galjaard G, Lampe M, Kamp PC (2008) (8) years of RO-experience at WTP Heemskerk biofouling aspects. Paper presented at the Membrane Technology Conference, Naples, Florida2008

Hasanin G, Mosquera AM, Emwas A-H, Altmann T, Das R, Buijs PJ, Vrouwenvelder JS, Gonzalez-Gil G (2023) The microbial growth potential of antiscalants used in seawater desalination. Water Research 233: 119802 DOI https://doi.org/10.1016/j.watres.2023.119802

He S, Oddo JE, Tomson MB (1995) The Nucleation Kinetics of Barium Sulfate in NaCl Solutions up to 6 m and 90 °C. Journal of Colloid and Interface Science 174: 319-326 DOI https://doi.org/10.1006/jcis.1995.1397

Ho JS, Ma Z, Qin J, Sim SH, Toh C-S (2015) Inline coagulation–ultrafiltration as the pretreatment for reverse osmosis brine treatment and recovery. Desalination 365: 242-249 DOI https://doi.org/10.1016/j.desal.2015.03.018

Huber SA, Balz A, Abert M, Pronk W (2011) Characterisation of aquatic humic and non-humic matter with size-exclusion chromatography – organic carbon detection – organic nitrogen detection (LC-OCD-OND). Water Research 45: 879-885 DOI http://dx.doi.org/10.1016/j.watres.2010.09.023

Jacobson JD, Kennedy MD, Amy G, Schippers JC (2009) Phosphate limitation in reverse osmosis: An option to control biofouling? Desalination and Water Treatment 5: 198-206 DOI 10.5004/dwt.2009.578

Javier L, Farhat NM, Desmond P, Linares RV, Bucs S, Kruithof JC, Vrouwenvelder JS (2020) Biofouling control by phosphorus limitation strongly depends on the assimilable organic carbon concentration. Water Research 183: 116051 DOI https://doi.org/10.1016/j.watres.2020.116051

Jeong S, Naidu G, Vigneswaran S, Ma CH, Rice SA (2013) A rapid bioluminescence-based test of assimilable organic carbon for seawater. Desalination 317: 160-165 DOI http://dx.doi.org/10.1016/j.desal.2013.03.005

Jeong S, Naidu G, Vollprecht R, Leiknes T, Vigneswaran S (2016) In-depth analyses of organic matters in a full-scale seawater desalination plant and an autopsy of reverse osmosis membrane. Separation and Purification Technology 162: 171-179 DOI https://doi.org/10.1016/j.seppur.2016.02.029

Joret C, Levi Y, Gilbert M (1989) The measurement of biodegradable organic carbon (BDOC): a tool in water treatment. Water Supply 7: 41-45

Kadjeski M, Fasching C, Xenopoulos MA (2020) Synchronous Biodegradability and Production of Dissolved Organic Matter in Two Streams of Varying Land Use. Frontiers in Microbiology 11 DOI 10.3389/fmicb.2020.568629

Khan MT, Busch M, Molina VG, Emwas A-H, Aubry C, Croue J-P (2014) How different is the composition of the fouling layer of wastewater reuse and seawater desalination RO membranes? Water Research 59: 271-282 DOI https://doi.org/10.1016/j.watres.2014.04.020

Kim C-M, Kim S-J, Kim LH, Shin MS, Yu H-W, Kim IS (2014) Effects of phosphate limitation in feed water on biofouling in forward osmosis (FO) process. Desalination 349: 51-59 DOI https://doi.org/10.1016/j.desal.2014.06.013

Kurihara M, Ito Y (2020) Sustainable Seawater Reverse Osmosis Desalination as Green Desalination in the 21st Century. Journal of Membrane Science and Research 6: 20-29 DOI 10.22079/jmsr.2019.109807.1272

Li X, Skillman L, Li D, Ela WP (2018) Comparison of Alcian blue and total carbohydrate assays for quantitation of transparent exopolymer particles (TEP) in biofouling studies. Water Research 133: 60-68 DOI https://doi.org/10.1016/j.watres.2017.12.021

Lipps WC, Braun-Howland EB, Baxter TE (2023a) 4500-P Phosphorus. In: American Public Health Association, American Water Works Association, Water Environment Federation, (ed) Standard Methods for the Examination of Water and Wastewater 24th Edition, 24th edn:

Lipps WC, Braun-Howland EB, Baxter TE (2023b) 5520 Oil and grease. In: American Public Health Association, American Water Works Association, Water Environment Federation, (ed) Standard Methods for the Examination of Water and Wastewater 24th Edition, 24th edn:

Lipps WC, Braun-Howland EB, Baxter TE (2023c) 10150 Chloropyll a. In: American Public Health Association, American Water Works Association, Water Environment Federation, (ed) Standard Methods for the Examination of Water and Wastewater 24th Edition, 24th edn:

Lozier JC, Bankston A, Beaty J, Garcia-Aleman J, Scharf R, Amy G, Salinas Rodríguez SG (2009) Use of advanced NOM characterization methods to trace the fate of organic contaminants from a membrane backwash recycle scheme. Paper presented at the Membrane Technology Conference, Memphis, Tennessee United States2009

Mangal MN (2023) Controlling scaling in groundwater reverse osmosis: minimizing antiscalant consumption Veenman, The Netherlands. 978-90-365-5588-3. 10.3990/1.9789036555890

Mangal N, Salinas Rodriguez SG, Yangali Quintanilla VA, Schippers JC, Kennedy MD (2021) Ch 08 - Scaling. In: Salinas Rodriguez SG, Schippers JC, Amy GL, Kim IS, Kennedy MD (eds) Seawater Reverse Osmosis Desalination: Assessment and Pre-treatment of Fouling and Scaling, 1st edn: 207-242 10.2166/9781780409863_0207

Matilainen A, Gjessing ET, Lahtinen T, Hed L, Bhatnagar A, Sillanpää M (2011) An overview of the methods used in the characterisation of natural organic matter (NOM) in relation to drinking water treatment. Chemosphere 83: 1431-1442 DOI https://doi.org/10.1016/j.chemosphere.2011.01.018

Nahrstedt A, Camargo Schmale J (2008) New insights into SDI and MFI measurements. Water Science and Technology: Water Supply 8: 401-412 DOI https://doi.org/10.2166/ws.2008.087

NEN 6271:1995 nl (1995) Bacteriological examination of water - Determination of the easily assimilable organic carbon (AOC) content NEN, Netherlands https://www.nen.nl/en/nen-6271-1995-nl-16389.

Peña N, del Vigo F, Salmerón O, Rodriguez J, Borrel A, Chesters SP (2015) Pre-coating of outside-inside capillary UF membranes with iron hydroxide particles to limit non-backwashable fouling during seawater algal blooms. Desalination and Water Treatment 55: 2973–2987 DOI 10.1080/19443994.2014.957950

Peña N, Rodriguez J, del Vigo F, Chesters SP (2022) 20 years of data from 500 seawater membrane autopsies. Paper presented at the IDA World Desalination Congress, Sydney, Australia, 9-13 October 2022

Rachman RM, Ghaffour N, Waly F, Amy GL (2013) Assessment of Silt Density Index (SDI) as fouling propensity parameter in Reverse Osmosis (RO) desalination systems. Desalination and Water Treatment 51: 1091-1103 DOI doi:10.1080/19443994.2012.699448

Ross PS, Hammes F, Dignum M, Magic-Knezev A, Hambsch B, Rietveld LC (2013) A comparative study of three different assimilable organic carbon (AOC) methods: results of a round-robin test. Water Supply 13: 1024-1033 DOI 10.2166/ws.2013.079

Salinas Rodriguez SG (2021). Fouling and scaling in seawater reverse osmosis desalination. The Source: 4https://www.thesourcemagazine.org/fouling-and-scaling-in-seawater-reverse-osmosis-desalination/

Salinas Rodríguez SG (2011) Particulate and organic matter fouling of SWRO systems: Characterization, modelling and applications CRC Press/Balkema, Delft. http://dx.doi.org/10.1201/b11609

Salinas Rodriguez SG, Boerlage SFE, Kenedy MD, Schippers JC (2021a) Ch 04 - Particulate fouling. In: Salinas Rodriguez SG, Schippers JC, Amy GL, Kim IS, Kennedy MD (eds) Seawater Reverse Osmosis Desalination: Assessment and Pre-treatment of Fouling and Scaling, 1st edn: 85-124 10.2166/9781780409863_0085

Salinas Rodriguez SG, Kennedy MD, Schippers JC, Amy GL (2009) Organic foulants in estuarine and bay sources for seawater reverse osmosis – Comparing pre-treatment processes with respect to foulant reductions. Desalination and Water Treatment 9: 155-164 DOI 10.5004/dwt.2009.766

Salinas Rodriguez SG, Schippers JC (2021) Ch 01 - Introduction to desalination. In: Salinas Rodriguez SG, Schippers JC, Amy GL, Kim IS, Kennedy MD (eds) Seawater Reverse Osmosis Desalination: Assessment and Pre-treatment of Fouling and Scaling, 1st edn: 26 10.2166/9781780409863_0001

Salinas Rodriguez SG, Schippers JC, Amy GL, Kim IS, Kennedy MD (2021b) Seawater Reverse Osmosis Desalination: Assessment and Pre-treatment of Fouling and Scaling, 1st edn IWA Publishing, London. 9781780409856. 10.2166/9781780409863

Salinas Rodriguez SG, Sithole N, Dhakal N, Olive M, Schippers JC, Kennedy MD (2019) Monitoring particulate fouling of North Sea water with SDI and new ASTM MFI0.45 test. Desalination 454: 10-19 DOI https://doi.org/10.1016/j.desal.2018.12.006

Schippers JC, Salinas Rodriguez SG, Boerlage SFE, Kennedy MD (2014). Why MFI is edging SDI as a fouling index. Desalination & Water Reuse: 28-32

Schippers JC, Verdouw J (1980) The modified fouling index, a method of determining the fouling characteristics of water. Desalination 32: 137-148

Servais P, Billen G, Hascoët M-C (1987) Determination of the biodegradable fraction of dissolved organic matter in waters. Water Research 21: 445-450 DOI https://doi.org/10.1016/0043-1354(87)90192-8

Shanmuganathan S, Nguyen TV, Jeong S, Kandasamy J, Vigneswaran S (2015) Submerged membrane – (GAC) adsorption hybrid system in reverse osmosis concentrate treatment. Separation and Purification Technology 146: 8-14 DOI https://doi.org/10.1016/j.seppur.2015.03.017

Simon FX, Penru Y, Guastalli AR, Esplugas S, Llorens J, Baig S (2013) NOM characterization by LC-OCD in a SWRO desalination line. Desalination and Water Treatment 51: 1776-1780 DOI 10.1080/19443994.2012.704693

Sweity A, Oren Y, Ronen Z, Herzberg M (2013) The influence of antiscalants on biofouling of RO membranes in seawater desalination. Water Research 47: 3389-3398 DOI https://doi.org/10.1016/j.watres.2013.03.042

Sweity A, Zere TR, David I, Bason S, Oren Y, Ronen Z, Herzberg M (2015) Side effects of antiscalants on biofouling of reverse osmosis membranes in brackish water desalination. Journal of Membrane Science 481: 172-187 DOI https://doi.org/10.1016/j.memsci.2015.02.003

USEPA (2005) Membrane Filtration Guidance Manual. EPA 815-R-06-009 USEPA, Online, pp. 35 https://nepis.epa.gov/Exe/ZyPDF.cgi/901V0500.PDF?Dockey=901V0500.PDF.

USEPA (2010) Method 1664 Revision B: n-hexane extractable material (HEM; oil and grease) and silica gel treated n-hexane extractable material (SGT-HEM; non-polar material) by extraction and gravimetry USEPA, Online, pp. 35 https://www.epa.gov/sites/default/files/2015-08/documents/method_1664b_2010.pdf.

van der Kooij D (1992) Assimilable organic carbon as an indicator of bacterial regrowth. J American water works association 84: Van der Kooij, D. 1992. Assimilable organic carbon as an indicator of bacterial regrowth. J. AWWA 1984(1992):1957-1965.

van der Kooij D, Hijnen W, Cornelissen E, Van Agtmaal S, Baas K, Galjaard G (2007) Elucidation of membrane biofouling processes using bioassays for assessing the microbial growth potential of feed water. Paper presented at the Membrane Technology Conference, Tampa Bay, Florida2007

van der Kooij D, Visser A, Hijnen WAM (1982) Determining the concentration of easily assimilable organic carbon in drinking water. Journal AWWA 74: 540-545 DOI https://doi.org/10.1002/j.1551-8833.1982.tb05000.x

Villacorte LO, Boerlage SFE, Dixon MB (2021) Ch 06 - Algal Blooms and RO desalination. In: Salinas Rodriguez SG, Schippers JC, Amy GL, Kim IS, Kennedy MD (eds) Seawater Reverse Osmosis Desalination: Assessment and Pre-treatment of Fouling and Scaling:

Villacorte LO, Ekowati Y, Calix-Ponce HN, Schippers JC, Amy GL, Kennedy MD (2015a) Improved method for measuring transparent exopolymer particles (TEP) and their precursors in fresh and saline water. Water Research 70: 300-312 DOI http://dx.doi.org/10.1016/j.watres.2014.12.012

Villacorte LO, Ekowati Y, Winters H, Amy G, Schippers JC, Kennedy MD (2015b) MF/UF rejection and fouling potential of algal organic matter from bloom-forming marine and freshwater algae. Desalination 367: 1-10 DOI https://doi.org/10.1016/j.desal.2015.03.027

Villacorte LO, Kennedy MD, Amy GL, Schippers JC (2009) The fate of Transparent Exopolymer Particles (TEP) in integrated membrane systems: Removal through pre-treatment processes and deposition on reverse osmosis membranes. Water Research 43: 5039-5052 DOI 10.1016/j.watres.2009.08.030

Voutchkov N (2010) Considerations for selection of seawater filtration pretreatment system. Desalination 261: 354-364 DOI https://doi.org/10.1016/j.desal.2010.07.002

Vrouwenvelder JS, Beyer F, Dahmani K, Hasan N, Galjaard G, Kruithof JC, Van Loosdrecht MCM (2010) Phosphate limitation to control biofouling. Water Research 44: 3454-3466 DOI https://doi.org/10.1016/j.watres.2010.03.026

Vrouwenvelder JS, Manolarakis SA, Veenendaal HR, van der Kooij D (2000) Biofouling potential of chemicals used for scale control in RO and NF membranes. Desalination 132: 1-10 DOI https://doi.org/10.1016/S0011-9164(00)00129-6

Vrouwenvelder JS, van der Kooij D (2001) Diagnosis, prediction and prevention of biofouling of NF and RO membranes. Desalination 139: 65-71 DOI https://doi.org/10.1016/S0011-9164(01)00295-8

Waly TKA (2011) Minimizing the use of chemicals to control scaling in SWRO - Improved prediction of the scaling potential of calcium carbonate CRC Press / Balkema, Leiden. 978-0-415-61578-5.

Waly TKA, Kennedy MD, Witkamp G-J, Amy G, Schippers JC (2012) The role of inorganic ions in the calcium carbonate scaling of seawater reverse osmosis systems. Desalination 284: 279-287 DOI http://dx.doi.org/10.1016/j.desal.2011.09.012

Weinrich L, LeChevallier M, Haas C (2015) Application of the bioluminescent saltwater assimilable organic carbon test as a tool for identifying and reducing reverse osmosis membrane fouling in desalination Water Reuse Research Foundation, Virginia, USA https://www.waterrf.org/research/projects/application-bioluminescent-saltwater-assimilable-organic-carbon-test-tool

Weinrich LA, Schneider OD, LeChevallier MW (2011) Bioluminescence-based method for measuring assimilable organic carbon in pretreatment water for reverse osmosis membrane desalination. Appl Environ Microbiol 77: 1148-1150 DOI 10.1128/aem.01829-10

Westerhoff P, Chen W, Esparza M (2001) Fluorescence Analysis of a Standard Fulvic Acid and Tertiary Treated Wastewater. Journal of Environmental Quality 30: 2037-2046 DOI https://doi.org/10.2134/jeq2001.2037

Wilf M, Klinko K (2016) Effect of new pretreatment methods and improved membrane performance on design of RO seawater systems Hydranautics, Online, pp. 1-17 https://membranes.com/wp-content/uploads/Documents/Technical-Papers/Application/SWRO/Effect-Of-New-Pretreatment-Methods-And-Improved-Membrane-Performance-On-Design-Of-RO-Seawater-Systems.pdf.

Yin W, Li X, Suwarno SR, Cornelissen ER, Chong TH (2019) Fouling behavior of isolated dissolved organic fractions from seawater in reverse osmosis (RO) desalination process. Water Research 159: 385-396 DOI https://doi.org/10.1016/j.watres.2019.05.038

Zhang M, Jiang S, Tanuwidjaja D, Voutchkov N, Hoek EMV, Cai B (2011) Composition and Variability of Biofouling Organisms in Seawater Reverse Osmosis Desalination Plants. Appl Environ Microbiol 77: 4390-4398 DOI doi:10.1128/AEM.00122-11

Part 1

Membrane processes

doi: 10.2166/9781789062977_0027

Chapter 2

Microfiltration and ultrafiltration

Morten Lykkegaard Christensen, Aalborg University, Denmark

Guillem Gilabert-Oriol, DuPont Water Solutions, Spain

The learning objectives of this chapter are the following:

- give an understanding of how microfiltration and ultrafiltration is used for water and wastewater treatment
- give knowledge of how membranes system is best operated to reduce membrane fouling and ensure high flux
- give an introduction to different cleaning options and how hydraulic cleaning methods are optimized
- give knowledge of how microfiltration and ultrafiltration system are designed and proper membrane is selected.

2.1 INTRODUCTION

Microfiltration and ultrafiltration are pressure-driven membrane filtration processes, where a transmembrane pressure (TMP) is used to press water through the membrane. They are operated at relatively low TMP compared with the other pressure-driven membrane processes i.e., nanofiltration and reverse osmosis (Figure 1). Micro- and ultrafiltration membranes are porous membranes and the separation based on sieving effects (size exclusion). This means that large particles or macromolecules are rejected by the membrane, whereas small molecules pass the membrane.

Microfiltration membrane typically have pores larger than 0.1 μm whereby bacteria and particles can be removed by the membrane (Figure 1). Often a nominal pore size is given from the manufacturer, but the performance of the membrane will also depend on pore size distribution and membrane materials. Microfiltration is often used as pretreatment prior to other membrane filtration processes such as RO, electrodialysis, and membrane distillation. Ultrafiltration membranes have pores sizes below 0.1 μm and retains macromolecules (Figure 1). Ultrafiltration membranes are used for removal of colloids, proteins, virus. Molecular weight cut-off (MWCO) is used to describe the membrane, instead of pore size. MWCO is the lowest molecular weight where more than 90% of the macromolecules are rejected by the membrane.

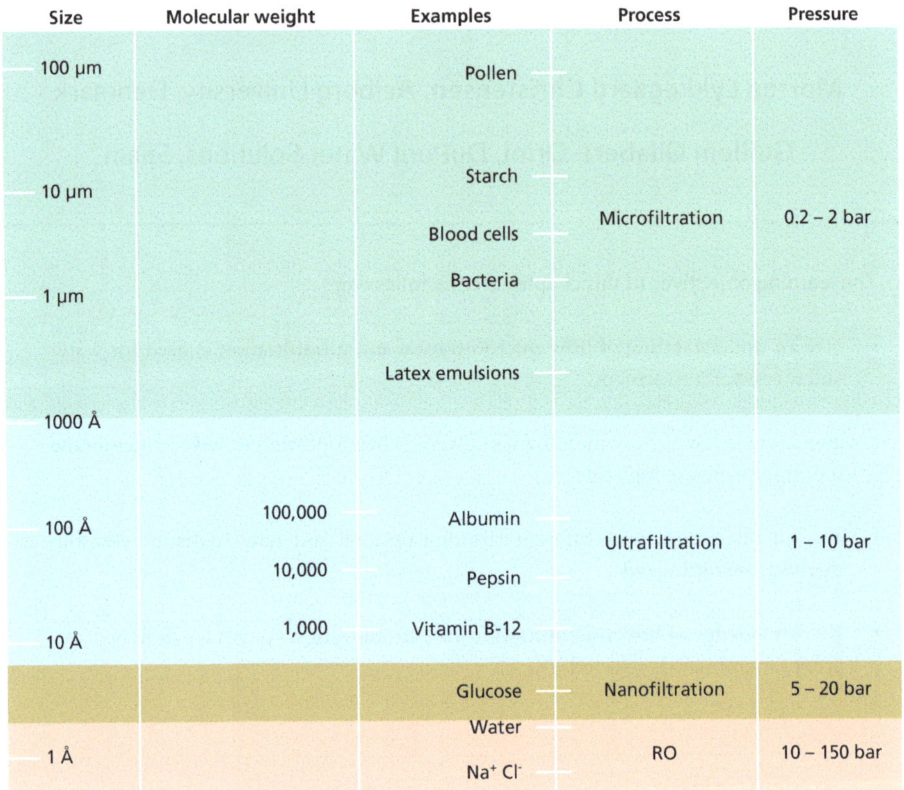

Figure 1 Pressure driven filtration processes

2.1.1 Advantages of ultrafiltration compared to conventional treatment

Micro- and ultrafiltration offers several advantages compared with conventional filtration methods. As an example, ultrafiltration used as a pretreatment step to treat water has experienced an impressive increase as a result of the continuous search for cost-effective technologies which enable a sustainable production of water (Chu *et al.*, 2009). Key benefits associated to the ultrafiltration technology versus conventional pretreatment are a low footprint, the ability to remove virus and bacteria and to significantly reduce colloids, suspended particles, turbidity and some total organic carbon. Even more importantly, the ability to reliably provide good quality filtrate water to the downstream reverse osmosis are the most remarkable benefits associated with this technology (Mourato *et al.*, 2003).

As another example Milwaukee, on the United States of America suffered back in 1993 on of the worst modern water-born epidemies due to Cryptosporidium microscopic parasite causing diarrhea. Cryptosporidium is known for being highly resistant to chlorination. This can pose a risk when conventional water treatment schemes based on sand filter and chlorination are used (Morris *et al.,* 2005).

Since then, ultrafiltration has emerged as a preferred technology to treat drinking water, thus replacing old water treatment schemes based on sand filters and chlorination. Ultrafiltration, as it is based on an absolute pore-size filtration, can eliminate Cryptosporidium or other chlorine-resistant organisms, as they cannot pass through the membrane (Pressdee, 2005). Additional benefits of ultrafiltration versus sand filters are summarized in the table 1.

Table 1 Summary of the benefits of ultrafiltration over sand filters (DuPont, 2022)

	Sand Filtration	**Ultrafiltration**
Pathogenic bacteria removal (coliforms)	≤2 log	≥ 4 log
Pathogenic virus removal (enteric)	≤ 3 log	≥ 4 log
Water effluent quality (Turbidity)	0.1 NTU	0.01 NTU
Organics removal (TOC)	5%	34%
Lower Silt Density Index	3.2	2.6
Footprint reduction	No	99%
Improve plant availably (reduce downtime)	95%	> 98%
Operate reverse osmosis at higher flux	< 14 L/m²h	> 14 L/m²h

Most of the advantages of ultrafiltration rely on the small pores for water transport, typically in the range of 20-30 nm. This allows the system to be significantly more compact, reducing the required footprint by up to 99% compared to a conventional sand filter. Additionally, thanks to this smaller pore size, it is very challenging for bacteria, parasites, protozoa and even viruses to pass through the membrane. This greatly improves the water effluent quality produced. When ultrafiltration is used as a pretreatment to the reverse osmosis, this means that the reverse osmosis system can operate at a higher flux, as particulate fouling risk is eliminated.

2.2 DESIGN AND OPTIMIZE MEMBRANE PROCESSES

Before setting up a micro- and ultrafiltration process, the following five points must be considered (Raghunath *et al.*, 2012)

1) Objective of the process and the success criteria
2) Pre-treatment of feed
3) Membrane selection
4) Module selection
5) Operation parameter optimization

It is usually not possible to follow the points chronologically, as e.g., a proper pre-treatment depends on choice of membrane, membrane module and operational parameters. In the text the objectives of the membrane will be discussed first, then membrane selection, operation, membrane module and finally pre-treatment.

2.3 OBJECTIVE OF THE FILTRATION PROCESS

Ultra- and microfiltration are used in many industries, at water facilities and wastewater plants. The objective of the process varies, and the relevant stream is not the same for all filtration processes but can be

1. Concentrate e.g., for recovery of macromolecules,
2. Filtrate e.g., for removal of pathogenic microorganism from drinking water
3. Both concentrate and filtrate e.g., for harvesting valuable product from waste stream by fractionation of macromolecules.

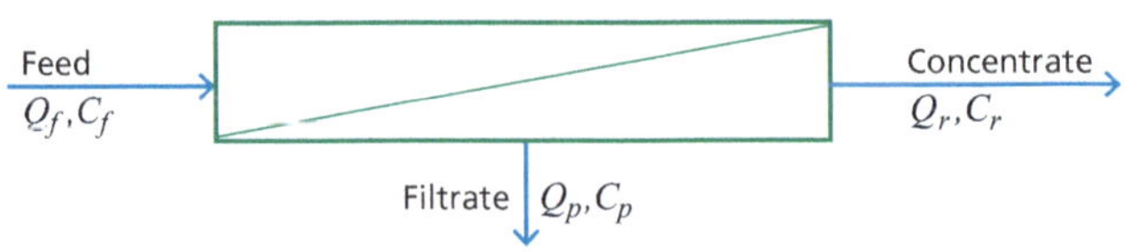

Figure 2 Membrane filtration process and mass balance.

This is important because it affect the optimal choice of membrane, modules, and operation parameters. In order to define the success criteria different key equations and parameters will be defined.

The optimal design of the membrane process depends on the flow and concentrations of the inlet, the wanted flow and concentration of the outlets, and the operation time of the membrane system (Figure 2). Q is the volumetric flow and C is the concentration of the compounds of interest. A key factor for membrane operations, is most particles or macromolecules (rejected materials) are concentrated in the concentrate stream, which can be found by defining a concentration factor.

$$CF = \frac{Q_f}{Q_r} \qquad \text{Eq. 1}$$

The concentration factor should be high and concentration factors up to 10 is typically set as the goal. The concentration factor gives a maximum value for how much the product can be concentrated and if the particles are fully rejected by the membrane, $C_r = CF \times C_f$.
Another key parameter is the yield (Y)

$$Y = \frac{Q_r C_r}{Q_f C_f} \qquad \text{Eq. 2}$$

i.e., fraction of the wanted product that are left in the concentrate. The yield should be high and ideally 1.

Filtrate may be the main product stream i.e., if the membrane is used for water treatment. *CF* must then be high as well because as much water as possible have to be recovered, but often the recovery is calculated instead, as it represents the water that is being produced compared to the water that is being fed into the ultrafiltration membrane

$$\%Recovery = \frac{Q_p}{Q_f} = 1 - \frac{1}{CF} \qquad \text{Eq. 3}$$

Another important parameter is availability. Availability represents the amount of time, in percentage, that the ultrafiltration is operating. Availability is defined as

$$\%Availability = \frac{t_{operation}}{t_{operation} + t_{stopped}} \qquad \text{Eq. 4}$$

This is an important factor, since as an example, if the membrane needs a lot of time to be properly cleaned, the overall ultrafiltration efficiency will be low.

$$\%Efficiency = \%Recovery \times \%Availability \qquad \text{Eq. 5}$$

Filtrate flux (J_w) across a membrane can be described as the filtrate flow (Q_p) normalized by the membrane's active area (A). Its units are typically expressed as liters/(m^2·h) or LMH, or as US gallons/(ft^2·day) or GFD.

$$J_w = \frac{Q_p}{A} \qquad \text{Eq. 6}$$

One useful parameter to calculate is the net flux (J_{net}). This represents the net flux a micro- and ultrafiltration installation is producing in a whole year. This takes into consideration the time installation is operating after discounting the time the plant is stopped, and the net water it produced after discounting the water that was produced but later consumed during, for example, cleanings. It can be calculated by multiplying the design flux by the efficiency

$$J_{net} = J_w \times \%Efficiency \qquad \text{Eq. 7}$$

This parameter is important since sometimes, in order to optimize the plant throughput production, it might be, for example, more beneficial to increase the design flux while decreasing a bit the recovery and availability, if this can help getting an overall higher net flux. The net flux can be used to calculate the required membrane area

2.4 MEMBRANE TYPES

The selection of the membrane is important for the filtration process, and different thing must be addressed. It is important to have a high flux through the membrane and a high rejection of particles and macromolecules. However, high rejection usually results in lower flux or higher energy demand, so often these two key parameters must be balanced depending on the propose of the filtration process. It is important to have a membrane with minimal risk of fouling or clogging, and a membrane with high lifetime, which depends on the composition of the membrane, surface properties and pore size.

Different types of membrane modules exist, such as flat sheet membranes (e.g., 1×1 m and 200 μm thick), spiral wound membranes, hollow fiber membranes and tubular membranes. Module length is typically around 1 m and the membrane area for one module can for large installation be up to 40-80 m^2.

Ultrafiltration and microfiltration can be operated either outside-in, or inside-out, depending on the type of features that is mainly desired. Another classification criteria depends on whether all the water that goes into the membrane is being treated, this is called dead-end filtration; or if a portion of the water is not being treated, it is called cross-flow filtration. Cross-flow filtration has the advantage of better managing fouling and lower energy consumption, but at the expense of a reduction on water recovery. Dead-end filtration has the advantage of having a higher water recovery and a smaller footprint, but fouling is more difficult to control.

Outside-in membranes are hollow fibers characterized by having the water to be treated outside the fiber, and the filtrated water collected inside the hollow fiber. These are typically made of Polyvinylidene fluoride (PVDF). They are usually operated in dead-end mode. Key features of this technology involve a higher resistant to chemicals, and the ability to perform air scouring during cleanings.

Inside-out membranes are hollow fibers characterized by having the water to be treated outside the fiber, and the filtrated water collected inside the hollow fiber. These are typically made of Poly(ether-Sulfone) (PES). They are usually operated in dead-end mode. Key feature of this technology is a higher membrane permeability.

Additionally, ceramic membranes are also used in the industry. Ceramic membranes are usually made of Aluminum, Silica, Titanium, and Zirconium oxides. They are typically operated in cross-flow filtration (Gruskevica and Mezule, 2021). One of the key challenges of ceramic membranes are its cost, and managing fouling properly. Key features of this technology are its thermal and chemical stability, as well as being able to operate at higher fluxes (Sondhi *et al.*, 2003).

Membranes system can also be classified as submerged and pressurized system. Submerged membranes have the advantage that they can be visually observed, and they usually operate at lower energy. Pressurized membranes have the advantage of operating at higher fluxes, a lower footprint (Nick, 2019) , but might suffer from higher fouling rates, as a possible result of higher compaction of the fouling layer. (Kim *et al.*, 2015)

The use of coagulation in microfiltration and ultrafiltration can be used as a way to reduce fouling rates and increase overall flux (Konieczny *et al.*, 2009). Coagulants are typically iron or aluminum based.

2.5 BASIC EQUATIONS

The filtrate flux (J_p) must be high and is a function of permeability and transmembrane pressure (TMP)

$$J_p = K \times TMP = \frac{TMP}{\eta \times R_m} \qquad \text{Eq. 8}$$

where K is the permeability (or more correctly the permeance) in m/(s·Pa), R_m is the hydraulic resistance of the membrane in m^{-1}, and η is the viscosity in Pa·s.

For filtration of pure water, the permeability is only dependent on membrane properties and can be calculated from membrane thickness, pore numbers and size.

$$K = \frac{\pi \sum_i (n_i d_i^4)\rho}{128 \times \eta \tau l} \qquad \text{Eq. 9}$$

Where n is number of pores per square meter, d_i is pore diameter, or hydraulic diameter for non-cylindrical pores (m), τ is tortuosity (-) and l is membrane thickness (m).

The ideal membrane is thin, to ensure high permeability, but thick enough to withstand the transmembrane pressure. To improve permeability but still ensure the mechanical properties asymmetric membrane or thin-film composite membrane are usually used (Figure 3). Composite membrane consists of a thin active layer, where the separation happens, and a support layer with low hydraulic resistance.

Figure 3 Anisotropic membranes

Notice that the number of pores per square meter is another important parameter and implicit given from the porosity of the membrane.
Besides flux, membrane rejection (R) is important and is defined as

$$R = 1 - \frac{C_p}{C_r} \qquad \text{Eq. 10}$$

where R is a number between 0 and 1, where all particles are rejected by the membrane if $R = 1$ whereas the particles can freely pass the membrane if $R = 0$

For some applications, it is important that some molecules pass the membrane and other are retained by the membrane. The selectivity (α) can then be calculated and used to compare different type of membranes

$$\alpha_{i/j} = \frac{1-R_i}{1-R_j} \text{ and } \alpha_{i/j} \geq 1 \qquad \text{Eq. 11}$$

Where *i* is the particle that should pass the membrane, and *j* is the particles that should be rejected by the membrane. Several particles or macromolecules are partly rejected by the membrane. The reason for this is that pores are not unisized, but a large pore size distribution are often observed (Figure 4).

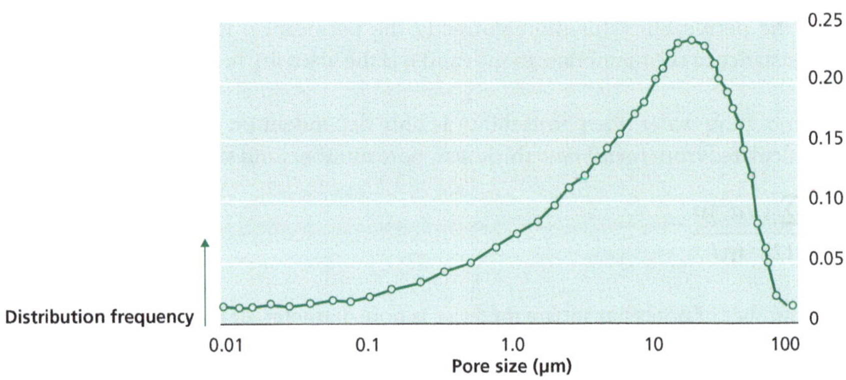

Figure 4 Pore size distribution microfiltration membranes (Data obtained from Liu *et al.*, 2018)

For water treatment and sterilization, it is important to ensure that bacteria cannot pass the membrane. Here the size of the largest pores is critical. A method to determine the size of the largest pores or ensure that all pores are below a given size is the bubble point method (breakthrough pressure). An alternative method is the water intrusion procedure, which is described elsewhere (see 'water intrusion procedure').

For wetted membranes, the air pressure must overcome the capillary pressure of the pores, before liquid can pass the pore (Figure 5).

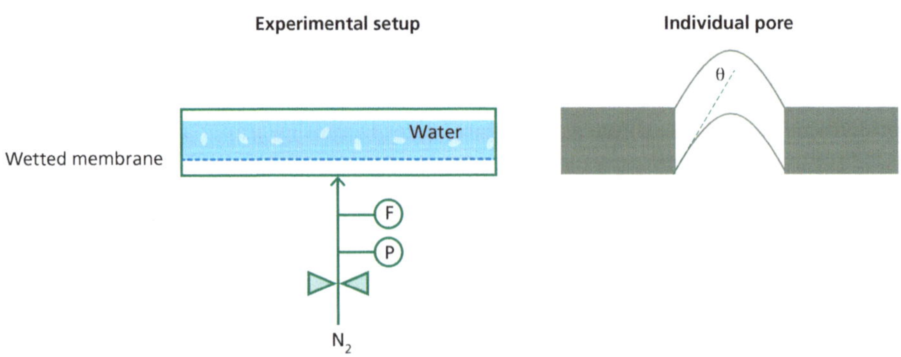

Figure 5 Bubble point method

The pressure required to overcome the capillary pressure is called breakthrough pressure (P^*)

$$P^* = \frac{4\gamma}{d} \qquad \text{Eq. 12}$$

where γ is the surface tension of the liquid. The surface tension for water is 0.072 N/m. For ultrafiltration membrane it is more complicated to measure pores size, instead cut-off values are often determined instead. MWCO can be determined by filtering mixtures of macromolecules and measuring rejection of the molecules (Figure 6). It is thereby possible to determine the size of the molecules where more than 90% of the molecules are retained by the membrane.

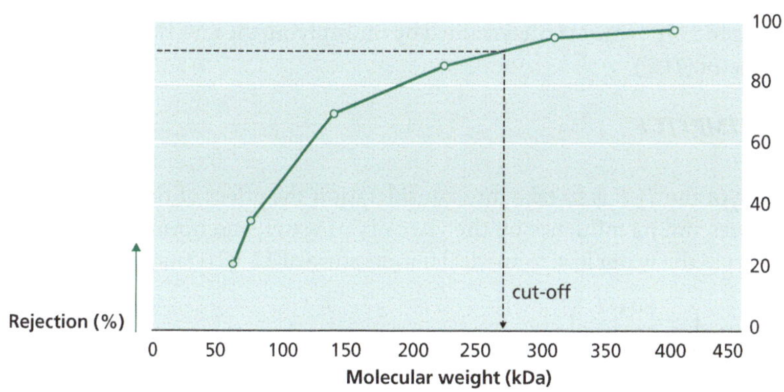

Figure 6 Molecular weight cut-off of PES/PSFNA membrane measured by using a mixture of PEG molecules (Data obtained from Wu *et al.*, 2018)

The selection of the membrane depends on feed composition and required quality of permeate or molecules that must be concentrated. The following parameters for the membrane must be considered

1. Selectivity (pore size and cut-off) so the membrane reject the particle that should be concentrated or removed
2. Permeability (high flux through the membrane at low pressure)
3. Thermal stability
4. Mechanical properties
5. Chemical stability (resistance for cleaning process)
6. Membrane composition and surface properties (to avoid adsorption of foulant)

2.6 NORMALIZATION

Filtrate flow permeating through a micro- and ultrafiltration membrane is heavily influenced by the water temperature. This happens as the process of water moving through convection across the membrane pores is facilitated by a higher temperature, as water then becomes less viscous, and less energy (TMP) is required for water to pass through the pores. On the other side, a lower water temperature increases the water viscosity and therefore more energy (TMP) is required for water to pass through the membrane pores.

Therefore, data normalization is of utmost importance in microfiltration and ultrafiltration membranes to properly assess the performance of a membrane-based installation. As an example, if data would not be normalized, one could observe an increase over time of raw TMP. This could induce the plant engineer to think that there might be irreversible fouling developing over time on the membrane system. If then this data would be operated, one could see that this TMP increase over time is due to a decrease in temperature over time as a result of moving to winter season. After normalizing this data considering water temperature, it could be observed that TMP evolution over time, thus dismissing any fouling issue.

The normalized *TMP* (*TMP**) is determined by multiplying the raw *TMP* by the temperature correction factor (*TCF*)

$$TMP^{*} = TMP \times TCF \qquad \text{Eq. 13}$$

The purpose of the *TCF* is to take into consideration the effect of the Temperature (T) in Celsius degrees and its influence on the viscosity of water, and normalizing its value from the temperature the water has, to an ideal temperature of 25 ºC (Daucik and Dooley, 2008)

$$TCF = \frac{10^{\left(\frac{247.8}{25+273.16-140}\right)}}{10^{\left(\frac{247.8}{T+273.16-140}\right)}} \qquad \text{Eq. 14}$$

It should be noticed that filtrate flux can also be normalized. This can be especially important if a system is operated at constant *TMP*, as then Flux will change over time. It might also be relevant in order to assess the stress that the ultrafiltration is operated, as the higher the flux, the more stressed the system will be, and the more difficult it would be to be able to operate the system under a sustainable way

$$TMP^{*} = TMP \times TCF \qquad \text{Eq. 15}$$

Special attention should be put in order to not normalize the filtrate flux twice, as normalize filtrate flux needs to be calculated using a non-normalized *TMP*, and then applying the *TCF* in the equation described above.

2.7 MEMBRANE FOULING

Membrane fouling is an inevitable phenomenon in membrane filtration. Membrane filtration occurs when particles are deposited on a membrane surface or in membrane pores in a process. Fouling reduces the permeability of the membrane and thereby increase the required pressure to keep the flux constant. Further, fouling may change the selectivity of the membrane. This may in some cases be beneficial i.e., it may result in better water treatment. However most often all type of fouling must be minimized, and different methods exists to reduce fouling build up or remove already formed fouling. Microfiltration is typically operated at low pressure (up to some bars) or flux. Too high flux increases the risk of accumulation of materials at the membrane surface. Thus, pressure must be kept low to avoid too high flux and thereby reduce fouling risk. Ultrafiltration membranes is operated at higher pressure to obtain same flux through membrane due to the smaller pores and higher hydraulic resistance. Micro- and ultrafiltration membrane are typically cleaned hydraulically or chemically. Cleaning reduces the operation time of the membrane whereby the availability of the membrane is reduced cf., Eq. (4).

Membrane fouling can be due to internal fouling such as adsorption of materials in the pores or pore blocking, and it can be external such as gel formation, precipitation of salts (scaling) and biofilm growth. The type of fouling is important for the optimal operational parameter, cleaning strategy and different method exists to analyses fouling (later chapters). Besides direct analysis of the fouling materials and the membrane, flux-pressure data can give some hint of the type of fouling and is useful to optimize operational parameters such as flux, pressure and cross-flow.

If flux-pressure data is used to quantify fouling, the resistance-in-series model is usually used to determine individual contributors to resistance (Géasan-Guiziou *et al.,* 1999) . One method is to separate fouling into reversible fouling (R_{rev}) or irreversible fouling (R_{irrev}). Reversible fouling can be removed by hydraulically cleaning whereas irreversible fouling cannot.

For fouled membrane, the water flux through the membrane can be described as

$$J_p = \frac{\Delta P}{\eta \sum_i R} \qquad \text{Eq. 16}$$

Where the resistance is the sum of individual contributors to the hydraulic resistance

$$\sum_i R = R_m + R_{irrev} + R_{rev} \qquad \text{Eq. 17}$$

Membrane resistance can be determined for a pristine membrane by filtering clean water (i.e., RO treated water)

$$R_m = \frac{\Delta P}{\eta J_{w,0}} \qquad \text{Eq. 18}$$

Determination of membrane resistance is critical to determine the other resistances

Irreversible fouling can be determined by filtering the suspension, clean the membrane and measure the water flux after cleaning.

$$R_{irrev} = \frac{\Delta P}{\eta J_{w,cleaned}} - R_m \qquad \text{Eq. 19}$$

Irreversible fouling is often ascribed to internal fouling i.e., pore blocking, and adsorption, and is problematic as it results in a permanent reduction of the membrane permeability. Further, it is difficult to avoid by changing operation condition of the filtration process. It depend on the type of membrane chosen i.e., pore size distribution and composition. Thus, if the irreversible resistance is too high, another membrane may be chosen, or feed must be pretreated prior to the filtration.

Reversible fouling usually increases continuously during filtration and can be calculated as

$$R_{rev} = \frac{\Delta P}{\eta J_w} - R_m - R_{ir} \qquad \text{Eq. 20}$$

The reversible fouling strongly depends on the operational conditions and will be discussed more deeply in next section.

2.8 SUSTAINABLE FLUX

Most filtration processes are operated at constant flux to have a stable output, i.e., a constant cross-flow (CF). The flux must be set so the energy cost is low, and the required membrane are is low. Often the system is operated so build-up of reversible fouling is low. Reversible fouling is a result of external fouling and strongly dependent on the transport of particles and molecules to the surface of the membrane. High flux increases the transport of material to the surface and thereby increase fouling build-up. At low pressures, material transport to the surface is low, and flux increase almost linearly with pressure also called the pressure-controlled region (Figure 7). At high pressure, the effect of increasing the pressure becomes less important for the flux. If pressure is increased, flux increase as well but decline to a lower steady state flux. Thus, performance of the membrane process can no longer be improved by increasing the pressure (pressure-independent region). Turbulence and high shear at the membrane surface can remove part of the external fouling layer and thereby reduce fouling build-up. A common method is to apply a high crossflow (up to 2-3 m/s) to reduce fouling built up and increase flux. Besides high crossflow shear can be induced by using turbulence promoters, rotating, or vibrating membrane or air scouring. If the concentration in the feed is low, dead-end, or semi-dead-end setup can be used to lower the energy consumption. At higher concentration feeds crossflow are required and for high viscous feed, rotating membrane are may be the most energy efficient solution.

A critical flux has been defined as the transition from the pressure-controlled region to the pressure independent region (Cleck *et al.*, 2003, Bacchin *et al.*, 2009). In the ideal case, the filtrate flux follows that of clean water until the critical flux, but often the filtrate flux is lower than the pure water flux due to irreversible fouling. The optimal flux is typical lower than the critical flux; a number around 75% of the critical flux can be used (Raghunath *et al* 2012).

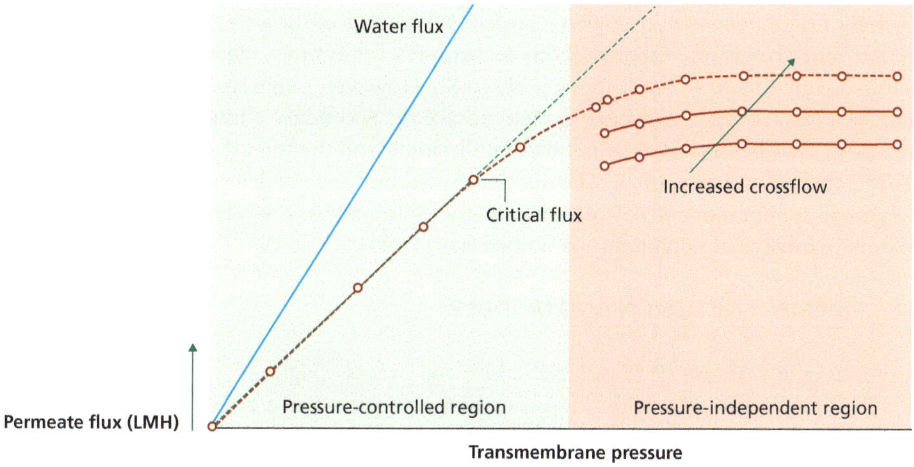

Figure 7 Flux-pressure curve and definition of critical flux

For long operation, membrane performance will usually decline also if the flux is lower than the critical flux. An alternative to the critical flux, is therefore the concept of sustainable flux. The sustainable flux is the maximum flux at which the fouling rate is acceptable and can be handled by hydraulic and chemical cleaning. At constant flux the fouling rate (*FR*) can be defined as

$$FR = \frac{\Delta TMP}{\Delta t} \qquad \text{Eq. 21}$$

By assuming that the transmembrane pressure increases linearly with time. Notice at high flux, a *TMP* jump may be observed where the fouling rate increases with time. At lower flux, it is usually reasonable to assume a linear relationship between *TMP* and time at constant flux. The sustainable flux can be determine using the flux step method, where the flux is gradually increased, and *FR* measured. The sustainable flux is then defined as the point where *FR* is exceeded a given threshold value for the fouling rate.

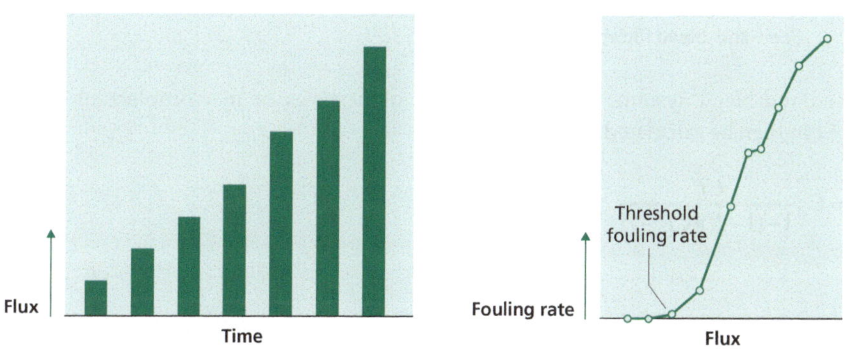

Figure 8 Step-flux test for determination of sustainable flux (Modified from Wang *et al.*, 2014)

Critical and sustainable flux is higher increased at higher cross-flow i.e., for microfiltration of lactalbumin suspension, the critical flux increase a factor of three, when the cross-flow was increased from 0.5 to 4-5 m/s (Vyas *et al.*, 2002). However, a high crossflow is costly; thus there is a balance with high flux vs. high cross-flow. Secondary effluent from wastewater treatment have been treated by membrane filtration, and the membrane have been tested at different flux and cross-flow. The optimal crossflow has been determined to 1.5 m/s for polymeric membranes and 4.5 m/s for ceramic membrane as the ceramic membranes was more expensive than polymeric one (Owen *et al.*, 1994).

2.9 MEMBRANE DESIGN AND MODULE

A typical membrane setup is the feed-and-bleed process (Figure 9). The system consists of a feed pump that ensure the transmembrane pressure and a constant flow of feed to the system. The recirculation pump is added to ensure a high crossflow at the membrane surface. The crossflow can be up to 2-3 m/s. At higher crossflow, the energy consumption increases rapidly. Due to the high crossflow, the pressure drop along the membrane module may decline significantly and in worst case, the transmembrane pressure becomes negative at the end of the module. The permeate flux is usually kept constant and thc transmembrane pressure regulated to ensure a constant flux.

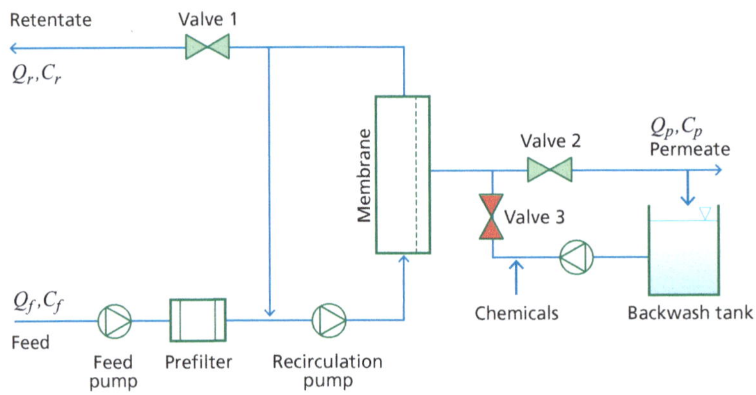

Figure 9 Feed-and-bleed filtration set-up.

For a feed-and-bleed system, the concentration of particles or macromolecules in the retentate can then be calculated as

$$C_r = C_f \frac{CF}{1-(1-R)(CF-1)} \quad \text{Eq. 22}$$

Where C_f and C_r is the concentration of particles in the feed and retentate, respectively. *CF* is the concentration factor and given as the ratio between the feed flow (Q_f) and the retentate flow (Q_r). If the particles are fully retained by the membrane, then $R = 1$ and $C_r = C_f \times CF$.

The yield can be calculated as

$$Y = \frac{1}{1-(1-R)(CF-1)} \qquad \text{Eq. 23}$$

2.10 PRETREATMENT

Pre-treatments are often required to prevent fouling, and is a necessary step for most membrane filtration process. Often bigger solids must be removed. In this case, macrofiltration is used. Its aim is to reduce big solids such as branches and leaves, that could otherwise potentially clog the ultrafiltration membranes and reduce its effectivity. Different methods exist, being the most typical screening, the use of hydro-cyclones, prefiltration using cartridge filters or multimedia filters. Suspended particles and colloidal particles can be problematic in filtration as such particles are difficult to remove from the membrane surface. Thus, it may be necessary to flocculate particles before the pre-filtration. Flocculation is usually done by adding salts (ferric, aluminum or poly-aluminum salts), polymers or combinations, whereby particles aggregates and can be removed by filters. As an example, coagulation and prefiltration have been used for treatment of raw water: a coagulation using iron coagulant ($FeCl_3$) with anionic polyelectrolyte in the first step and aluminum coagulant in the second one was used before raw water enters the pre-filtration (Sakola and Konierczny, 2004). The pre-treatment reduces the negative effect of membrane fouling, but also improved the quality of the treated water (Sakola and Konierczny, 2004)

2.11 CLEANINGS

Microfiltration and ultrafiltration is a typically a semi-batch process, specially for dead-end filtraiton applications. This is because it mainly handles particulate fouling, it gets fouled quickly. This means that the system needs to operate for 20 to 90 min depending on the water type and undergo a cleaning to restore its permeability. Optimizing cleanings remains thus of outmost importance to maximize the efficiency of the filtration system (Gilabert-Oriol, 2021). The main type of fouling that micro- and ultrafiltration membrane experience is detailed below. The most frequent one is particulate fouling. This is the most common and occurrent one. As the main task of an ultrafiltration membrane is to remove particles represented by total suspended solids (TSS) that give turbidity to the water. As water is filtrated through the membrane, it gets clogged by particles. Therefore, is it of utmost importance to perform a backwash to the membrane, so these particles can get detached from the membranes and its permeability can be restored.

A backwash (BW) typically consists of multiple steps. These steps can consist of an air scour, which blows air across the membranes, and shakes them to create abrasion and detach foulants accumulated on the top of the fibers. Later, a draining can be done to empty the module and remove the detached foulants. Then, a backwash top can be performed, together with an Air Scour. The main purpose of the backwash is to removed particles that are blocking the pores. Water with foulants is removed through the concentrate side of the module located at the top part of it. A backwash bottom follows the same approach, but water with foulants is evacuated through the bottom part of the module. It is not practical to perform an air scour in this step, as air is blown from the bottom of the element, and this would interfere with the aeration which is also blown from the bottom of the module. Finally, a forward flush is performed in order to provide a shear force and remove any remaining foulant from the top of the membrane. Key steps in this backwash process are performing a backwash top with and air scour, followed by a forward flush. By optimizing the duration of each one of these sub steps, as well as the frequency of backwashes, the overall efficiency of the whole ultrafiltration treatment can be greatly optimized (Gilabert-Oriol *et al.*, 2021).

A chemical enhanced backwash (CEB) consists of a typical Backwash sequence, where sodium hypochlorite (NaOCl) is dosed. As filtration cycles are done, bacteria start to grow and develop a biofilm. This biological fouling, or biofilm, is noted typically after 1 to 2 days of operation. It can be assessed as after certain number of backwashes, the TMP cannot be recovered to its initial values.

A cleaning-in-place (CIP) consists of a tailor-made chemical cleaning. It is typically performed every 3 to 4 months. As the membranes get cleaned on a regular basis, very specific foulant specific to the water type being treated starts to slowly build up over time. Typical CIPs performed are a caustic CIP, used to remove organics being accumulate on the membrane, where NaOH is used; and acid CIPs, used to remove scaling or metallic-based foulants, where HCl or oxalic acid is used.

2.11.1 Optimization of hydraulic cleaning

In order to remove fouling, hydraulic cleaning strategies can be used such as backwash, backflush or relaxation. Relaxation is done by closing the valve 1 and maybe valve 2 at the permeate side in Figure 9, whereby the transmembrane pressure drops to zero. Backwash or backflush are done by opening valve 3 and pump permeate through the membrane in Figure 9. Backwash and relaxation are typically done at regular intervals after 30-90 min and the duration is typical 1-3 min. Backflush are shorter cycles e.g., 30-60 s flush every 15 min. Regular hydraulic cleaning removes the reversible fouling. The efficiency of backwash, backflush or relaxation can be calculated as

$$h = \frac{TMP_f - TMP_0}{TMP_f - TMP_i} \qquad \text{Eq. 24}$$

Where TMP_i is the pressure at the beginning of the filtration cycle, TMP_f at the end of the filtration cycle and TMP_0 after hydraulic cleaning.

Another method to find the best strategy for hydraulic cleaning is to calculate the net flux (J_{net}) defined in Eq. (9). An example of an optimization of relaxation time for a membrane bioreactor shows that the optimal relaxation time is 3 min (Figure 10). Relaxation have been used in a membrane bioreactor setup used to treat municipal wastewater. For example for MBR systems with flat-sheet membrane, because backwash is not possible for these types of membranes as it will destroy the membranes. Thus, relaxation have been used as a gentle cleaning method to keep the flow high.

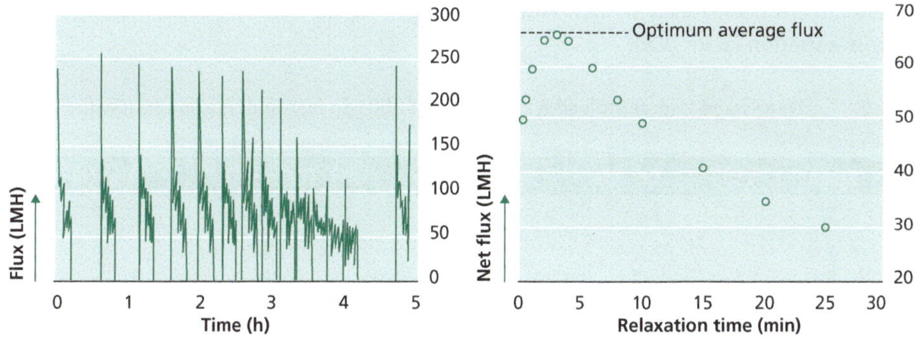

Figure 10 (a) Relaxation experiment at constant pressure during operation. (b) Net flux was calculated including both operation time and relaxation time in calculation. Optimum flux was obtained after 3 min. (Christensen *et al.*, 2016) The process was operated at constant transmembrane pressure.

At long term operations, the performance of the membrane will usually decline, and hydraulic cleaning is not sufficient. In these cases, it is necessary to use chemical to clean the membrane, one method is to add chemical to the backwash water known as chemically enhanced backwash (Figure 9), but the membrane can also be chemical cleaned from the feed side.

2.12 MEMBRANE CASCADES

Membrane system are designed to minimize the required membrane area. Membrane fouling are more pronounced at higher dry matter concentration of the feed. Further, in feed and-bleed system the concentration of feed near the membrane surface is almost the same as in the retentate. Thus, it will sometimes beneficial to operate more membranes in series. Modules in series reduce membrane area demand and lower energy consumption.

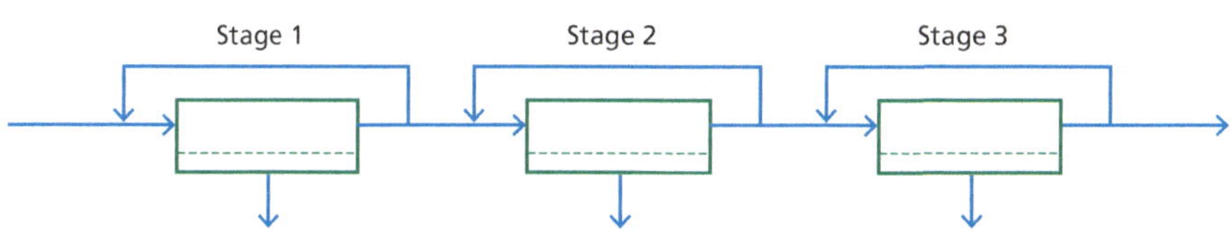

Figure 11 Feed-and-bleed in series to reduce required membrane area

An example is given for microfiltration of milk assuming the filtrate flux is given as

$$J_p = A + B\ln(CF) = 40\frac{L}{m^2h} - 14\frac{L}{m^2h}\ln(CF)$$

Where the flux decreases with increasing dry matter content in the feed and therefore with the concentration factor (CF).

Table 2 Three stage feed-and-bleed system to treat 5000 L/h feed and concentrate it with a factor of 5

Stage	CF	Q_p (L/h)	J_w (L/(m²h))	Area (m²)
1	1.5	1,667	34.3	48.0
2	2.5	1,333	27.2	49.1
3	5.0	1,000	17.5	57.2
Sum	**5.0**	**4,000**		**154.9**

Assuming a feed flow of 5000 L/h, rejection of SS is 1, and a required CF of 5, a single-stage feed-and-bleed system will result in a permeate flow of 4000 L/h and filtrate flux of 17.5 L/(m^{2}h). Hence the required membrane area is 229.0 m^2. This membrane area can be reduced to 154.9 m^2 for at three-stage feed-and bleed system (Table 2). This reduces the required membrane area with more than 30%.

2.13 SUMMARY

Microfiltration and ultrafiltration are used to treat water, as well as for concentration of dry matter or macromolecules. Membranes are selected based on their rejection, but it is also important to consider the risk for membrane fouling, and the chemical and mechanical stability of the membrane. Adsorption of molecules or pore blocking can sometimes be avoided by selecting an alternative membrane.

Most operation is done at constant operating flux, since most plants are designed to produce a certain amount of treated water per day. When designing the installation, it is important to set up the operate at a flux below the sustainable flux. This helps to avoid a too steep increase on trans-membrane pressure, and helps overall operating the microfiltration and ultrafiltration plants in a sustainable way. Higher permeate flux can be obtained with a high crossflow at the membrane surface, typical up to 2-3 m/s, but it also increases energy consumption and increase cost. Membrane processes can be done as multi-stage operation to reduce the required membrane area.

2.14 REFERENCES

Bacchin P. Aimar P., Field R.W. (2006) Critical and sustainable fluxes: Theory, experiments and applications. Journal of membrane science 281, 42-69.

Christensen ML., Bugge TV. Hede BH, Nierychlo M. Larsen P. Jørgensen M. K. (2016) Effects of relaxation time on fouling propensity in membrane bioreactors. Journal of Membrane Science. 504, 176-184

Chu R., J. Wei, M. Busch, Economic evaluation of UF+SWRO in seawater desalination. Chinese Desalination Association Conference (2009), Qing Dao, China.

Daucik K., R.B. Dooley, Revised Supplementary Release on Properties of Liquid Water at 0.1 MPa, The International Association for the Properties of Water and Steam, September 2008.

DuPont WaterApp, Pretreatment Advisor, Dupont (2022-09-09).

Gésan-Guiziou G., Boyaval E., Baufin G. (1999) Critical stability conditions in crossflow microfiltration of skimmed milk: transition to irreversible deposition. Journal of membrane science 158 211-222.

Gilabert-Oriol, G. (2021). Ultrafiltration Membrane Cleaning Processes. In Ultrafiltration Membrane Cleaning Processes. De Gruyter.

Gilabert Oriol, G., Hassan, M., Dewisme, J., Busch, M., & Garcia-Molina, V. (2013). High efficiency operation of pressurized ultrafiltration for seawater desalination based on advanced cleaning research. Industrial & Engineering Chemistry Research, 52(45), 15939-15945.

Gruskevica, K., & Mezule, L. (2021). Cleaning methods for ceramic ultrafiltration membranes affected by organic fouling. Membranes, 11(2), 131.

Kim, Y. J., Jung, J. W., & Lee, S. (2015). Comparison of fouling rates for pressurized and submerged ultrafiltration membranes. Desalination and Water Treatment, 54(13), 3610-3615.

Konieczny, K., Sakol, D., Płonka, J., Rajca, M., & Bodzek, M. (2009). Coagulation—ultrafiltration system for river water treatment. Desalination, 240(1-3), 151-159.

Le Cleck P., Jefferson B., Chang I.S. Judd S.J. (2003) Critical flux determination by the flux-step method in a submerged membrane bioreactor. Journal of membrane science 227, 81-93.

Liu, J., Huo, W., Zhang, X., Ren, B., Li, Y., Zhang, Z., Yang, J., (2018) Optimal design on the high-temperature mechanical properties of porous alumina ceramics based on fractal dimension analysis. Journal of Advanced Ceramic, 7, (2), 89-98.

Morris, Robert D., Elena N. Naumova, and Jeffrey K. Griffiths. 'Did Milwaukee experience waterborne cryptosporidiosis before the large documented outbreak in 1993?.' Epidemiology (1998): 264-270.

Mourato D., M. Singh, C. Painchaud, R. Arviv, Immersed membranes for desalination pre-treatment, International Desalination Association (IDA) World Congress, Bahamas, Paradise Island, Bahamas, 2003.

Nick, N., (2019) Pros And Cons Of Different Types Of Ultrafiltration Technology Configurations, Weter Online.

Owen G., Bandi M. Howell J.A. Churchouse S.J. (1994) Economic assessment of membrane processes for water and wastewater treatment. Journal of membrane science 102, 77-91.

Pressdee, J., Rezania, S., & Hill, C. (2005). Minneapolis Water Works' ultrafiltration plant gets off to a big start. Journal-American Water Works Association, 97(12), 56-63.

Raghunath B., Bin W,m Pattnaik P, Janssens J. (2012) Best practice for optimization and scale-up of microfiltration TFF processes. Bioprocessing Journal. 11 (1) 30-39.

Sakola C, Konieczny K (2004) Application of coagulation and conventional filtration in raw water pretreatment before microfiltration membranes. Desalination 162 61-67.

Sondhi, R., Bhave, R., & Jung, G. (2003). Applications and benefits of ceramic membranes. Membrane Technology, 2003(11), 5-8.

Vyas H.K. Bennett R.J. Marshall A.D. (2002) Performance of crossflow microfiltration during constant transmembrane pressure and constant flux operations. International dairy journal, 12, 473-479

Wang Z Field RW., Qu F., Han Y., Liang H Lia G., (2014) Use of threshold flux concept to aid selection of sustainable operating flux: A multi-scale study from laboratory to full scale. Separation and Purification Technology 123 (2014) 69–78.

Wu X., Xie Z., Wang H., Zhao C., Ng D., Zhang K (2018). Improved filtration performance and antifouling properties of polyethersulfone ultrafiltration membranes by blending with carboxylic acid functionalized polysulfone RSC ADV 2018, 8, 7774-7784.

doi: 10.2166/9781789062977_0047

Chapter 3

Reverse Osmosis and Nanofiltration

Guillem Gilabert-Oriol, DuPont Water Solutions, Spain

The learning objectives of this chapter are the following:

- give an understanding of the importance of reverse osmosis and nanofiltration membranes and how they are used for water treatment
- give an introduction to the different equations that govern reverse osmosis fundamentals and how to control, analyze and normalize a membrane installation
- give knowledge of how reverse osmosis and nanofiltration membranes are designed
- give knowledge on how membranes system is best operated to reduce membrane fouling and ensure system availability.

3.1 THE RISE OF REVERSE OSMOSIS

Water scarcity is being recognized as one of the main threats that mankind is facing globally (Fritzmann, *et al.*, 2007). Reverse Osmosis (RO) membrane technology has developed as a promising technology to address this problem, holding roughly 44% market share and growing among all the desalinating technologies (Valavala, *et al.*, 2011). This increased market adoption has been driven as materials have been improved and costs have dropped (Greenlee, *et al.*, 2009. This is especially relevant in arid regions such in the Middle East countries (ME), where population is located in arid and semi-arid regions, with a very limited rainfall, and where due to high ambient temperatures, evaporation contributes to a higher stress degree to the naturally available water sources. Moreover, water scarcity is aggravated by the population increase this region is exposed, as well as the economic development (Guo, *et al.*, 2000). All these factors, together with the favorable energy to product quality ratio that seawater reverse osmosis (SWRO) offers, has situated this technology as one key driver to sustain population living standards in ME countries (Carroll, *et al.*, 2000).

The first technologies used to desalinate seawater were using thermal processes where seawater is evaporated, and then the steam, which is free of salts, is recondensed to obtain fresh water. These thermal driven technologies that rely on distillation are multistage-flash distillation (MSF), multiple-effect distillation (MED), and vapor compression desalination processes (VCD). The main drawback of these methods is the significant amount of energy per cubic meter of water produced, compared to modern reverse osmosis based desalination. As Figure 1 shows, reverse osmosis (RO) desalination, specially when coupled with energy recovery devices (ERD) is 10 times more energy efficient than multistage flash desalination, and 4 times more efficient than vapor compression distillation (Kumar *et al.*, 2017, Kim *et al.*, 2019).

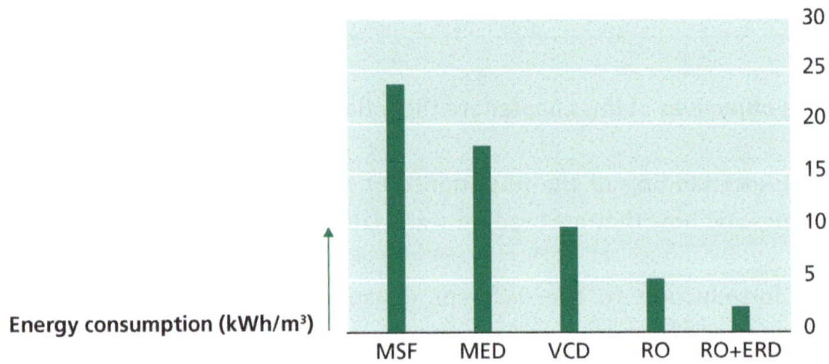

Figure 1 Energy consumption of different desalination technologies

One of the key aspects that has allowed reverse osmosis desalination to be so energy efficient is the introduction of energy recovery devices (Kadaj and Bosleman, 2018). These systems are like heat exchangers operation units, but instead of exchanging thermal energy, then they exchange pressure. This allows to recovery almost all the energy that was previously lost in the concentrate water stream of a seawater reverse osmosis system, and use it to pressurize the same volumetric flow in the feed of the reverse osmosis system. If we take seawater desalination as an example, this can reduce the energy expense in a seawater desalination plant by 55%. Also, a smaller high pressure pump is needed to pressurize the feed of a revere osmosis system, as 55% of the flow is already pressurized coming from the energy recovery device. It is worth mentioning that previously, the concentrate stream of a reverse osmosis system that could come pressurized up to 80 bar was discharge into the open atmosphere and all this energy was lost.

3.2 SUSTAINAIBLITY OF REVERSE OSMOSIS

Reverse osmosis membranes membrane technology offers a solution to achieve the sustainability development goals that the United Nations has set up for 2030. This has been stated and recognized by the United Nations (SDG, 2023). Thanks to the desalination technology, it is possible to fight against water scarcity, and obtain water of high quality.

If the energy is powered through renewable energies, such as through solar power or photovoltaics, it is possible to achieve drinking water of high quality. This allows to fight against climate change, as well as to provide unlimited amount of drinking water for the population, that it is not linked to whether it rains or note in the nature.

It is also important to make sure that desalination concentrate discharge is done properly, and that tis brine is properly managed through diffusors or mixing it with seawater, so its discharge does not affect marine species or the environment (Fernández-Torquemada *et al.*, 2019).

The use of brine can also be used as a resource, to get value out of the brine, and being able to extract valuable minerals such as sodium chloride, magnesium compounds, bromide, and rubidium, among others. This is important, as this allows reducing the cost of water of the desalination technologies, as well as preserving natural resources such as landscape and mountains from invasive mining extraction operations (Casas *et al.*, 2014).

Finally, some endeavors such as the Water Positive initiative, aims to take the sustainability impact of desalination and water reuse one step further. This is inspired by the carbon credits system, but for water credits. It aims to help those companies that aim to become water neutral in terms of its water footprint, so that they can compensate their water negative use, with those companies that are net producers of water, such as the desalination and water reuse installations. This can help driving awareness of the importance to reduce the water footprint, and helping preserve this valuable resource, as well as to help making sustainable water treatment processes more affordable.

3.3 UNDERSTANDING THE OSMOSIS PROCESS

In order to understand why reverse osmosis is called with this name, it is important to first understand what osmosis means. Osmosis is a natural process that only takes places when there is a semi- permeable membrane.

3.3.1 Semi-permeable membranes

A semi-permeable membranes is defined by letting a solvent like water pass through it, but not letting a solute like salt pass through the membrane. In order to better understand this process, it can be useful to imagine a simplified scenario, where only water is considered for a solvent, and only sodium chloride (NaCl) is considered for a solvent. When sodium chloride is dissolved in water, both sodium and chloride are separated in terms of Na^+ ions and Cl^- ions, following the equation detailed in Equation 1.

$$NaCl \rightarrow Na^+ + Cl^- \qquad \text{Eq. 1}$$

Although both atomic radius of water molecule (H_2O) and Na^+ ions and Cl^- ions are similar, and therefore it could be expected that both water and sodium and chloride ions can pass through the membrane in a similar rate, this is not the case. The reason for this phenomenon, is the solvation phenomena. In order to maintain electrical neutrality, when ions are dissolved in water, they become surrounded by water molecules. Since water molecules are polar, they tend to orient their mostly negative charge, with the sodium positively charged ions. The same happens for chloride negatively charged ions, as they get surrounded by the negatively charged side of the water molecules. The ultimate consequence for both chloride and sodium ions is that their effective sizes drastically increase, as a result of being surrounded by water molecules. This is the main reason why small molecular weight species that are charged are much better rejected from a solute like water. Therefore, the smaller a species is, and the more charged it is, the better it will be rejected by a reverse osmosis membrane.

The solvation effect is represented in Figure 2, where it can be seen the solvation effect of water molecules, represented by red (oxygen) and grey molecules (hydrogen) to sodium ions (blue) and chloride ions (green) as they pass through a space in the reverse osmosis membrane.

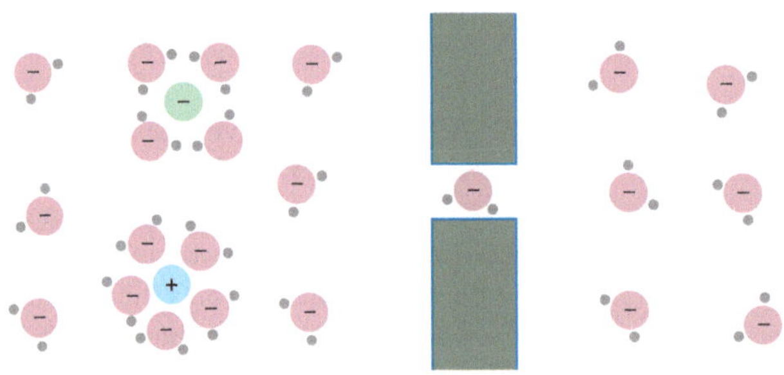

Figure 2 Solvation effect of water (red and grey molecules) to sodium ions (blue) and chloride ions (green) as they pass through a space in the reverse osmosis membrane

It should be noted that this "pore" drawn in this diagram, represent a space that is created in the polyamide chain. As polymers rotate as a result of being above 0º Kelvin temperature, small "pores" like this drawn in the diagram are created, where water can pass through it freely through a pressure driven convective flow (Wang *et al.*, 2023).

3.3.2 The reverse osmosis process

Once the concept of a semi-permeable membrane is defined, and it remains clear that it lets a solvent like water to pass, while a solute like sodium chloride cannot pass, the natural process of osmosis can be defined. As mentioned, the osmosis process only makes sense when a semi-permeable membrane is present. Osmosis is defined as a natural process when a specie that is at a higher concentration goes to dilute the other side of the membrane that has lower concentration of this species. This is typically the case of solutes like water. Water can move through a semi-permeable membrane, while salt cannot. Therefore, water with lower salt concentration like fresh water will always go to dilute the water with higher salt concentration like seawater, as this will have a lower water concentration. Since this natural phenomenon is not the goal of producing fresh drinking water, as it is not desired to get fresh water consumed. In order to achieve the opposite result, and be able to generate fresh drinking water from seawater, the reverse osmosis process was invented. This process consist into applying a pressure on the seawater that is enough in order to reverse this process, and being able to generate fresh water instead of consuming it.

Osmosis process happens in nature. One example are cherries. When it rains, rain droplets cover the cherry skin. The cherry skin acts like a semi permeable membrane, it lets the water to pass through, but not salts or sugars to escape. Therefore, after it rains, water travels from the outside of the skin, where the water concentration is higher, towards the inside of the cherry skin, where the water concentration is lower as a result of the fiber and fructose that cherries contain. The ultimate result is that as water from rain enters the inside of the cherry, it can eventually crack the cherry skin as its volume increases, and the cherry skin might not be able to expand properly before cracking to account for the increase in the cherries' volume.

In order to visualize this, two different solutions with the same volume each one. These solutions can be put in contact through a semi permeable membrane. The first solution is seawater, which it is assumed to have 30 g salt with 970 g of water. This represents a 3.1% salt concentration. The second solution is fresh water, which it is assumed to have 1 g of salt with 999 g of water. This represents a 0.1% salt concentration This set up can be observed in Figure 3. In this case, two solutions are separated. Fresh water will go to dilute seawater, while salt will not be able to pass.

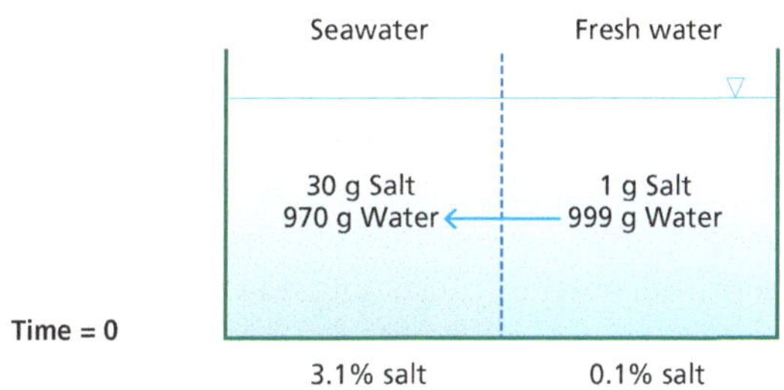

Figure 3 Initial experimental set up

As salt cannot pass, water from the fresh water side will go to dilute the seawater. Water will keep passing until both sides salt concentration are the same. This will happen after 935 g of water have passed from the fresh water side to the see water side. At this equilibrium point, concentrations in both sides will be the same at 1.6%. It is important to notice that salt mass in both sides is kept the same, but water mass has changed. This change in water volume, which increases in the seawater side but decreases in the fresh water side, leads to a difference in water height between the seawater and fresh water side. This difference in height is what is called osmotic pressure. This osmosis process be seen in Figure 4.

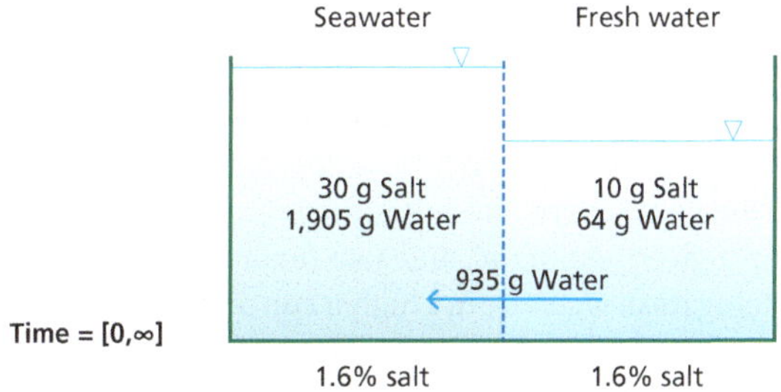

Figure 4 Osmosis process

If one aims to reverse this naturally occurring osmosis process, one needs to apply a pressure that at least is the same as this osmotic pressure. Once this pressure is applied, the water flow is reverses. If exactly the same pressure as the osmositc pressure is applied, it will be possible to reverse the 935 g of water flow from the seawater side to the fresh water side. This process can be observed in Figure 5.

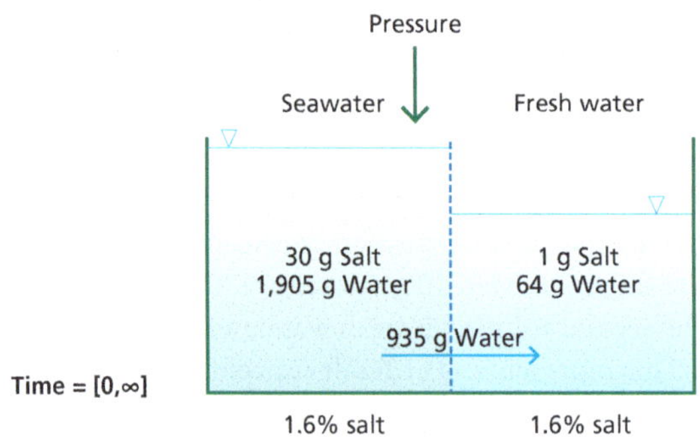

Figure 5 Reverse osmosis process

After applying the same pressure as the osmotic pressure, the equilibrium will again be reached, and the initial state will be created. Since typically the goal is to produce fresh water from seawater, and not to prevent osmosis to happen by reaching an equilibrium as the one depicted in Figure 6, a pressure greater than the osmotic pressure will be needed to be applied to produce more fresh water by consuming seawater.

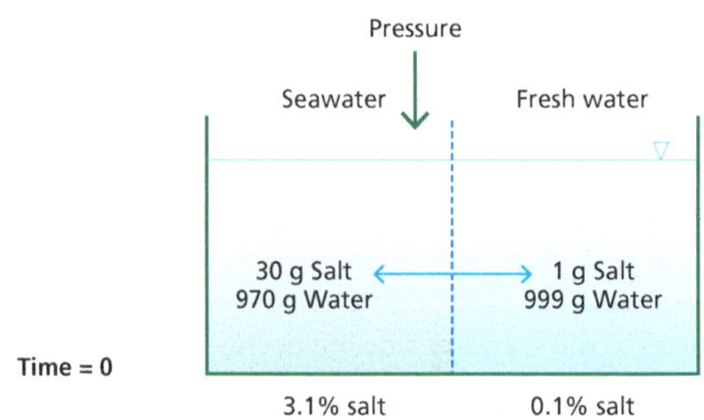

Figure 6 Equilibrium step

3.4 EQUATIONS

Reverse osmosis membranes are defined by a simple set of equations. These equations are used to control the flow of a solvent like water through the membrane, as well as the flow of solutes like salt through a membrane, and also finally to calculate the osmotic pressure needed to start producing fresh water from a higher concentration water. Also, the equations that are used to characterize a reverse osmosis system are presented.

3.4.1 Fundamental equations

Osmotic pressure is defined by the Greek letter pi (Π). Osmotic pressure can be calculating by multiplying the temperature (T) in Kelvin degrees, with the ideal gas constant (*R*) and the solute concentration (*C*) and the osmotic pressure coefficient (Φ). This formula is shown in Equation 2.

3.4.1.1 Osmotic pressure

The osmotic pressure coefficient represents how well a solute dissociates in water. For NaCl, which fully dissociates in sodium and chloride ions following Equation 1 it will be equal to 1. For other species that do not dissociate at all in water, its value will be equal to 0. Concentration is typically expressed in molar mas (mol/L). Ideal gas law is typically expressed as 0.08314 L·bar/K·mol. Temperature is expressed in kelvin degrees (K).

$$\pi = \varphi CRT \qquad \text{Eq. 2}$$

One easy way to remember the osmotic pressure equations is thinking how for diluted solutions, the osmotic pressure resembles the Van't Hoff equation for ideal gas laws. Van't Hoff equation is shown in Equation 3.

$$PV = nRT \qquad \text{Eq. 3}$$

Rearranging terms of Van't Hoff Equation 3 the same equation than osmotic pressure Equation 2 can be obtained, as shown in Equation 4. It is important however to highlight that the osmotic pressure dissociation coefficient needs to be factor in this equation.

$$P = \frac{n}{V}RT = CRT \qquad \text{Eq. 4}$$

The net driving pressure (*NDP*) represents how much of an energy driving for it exists across the membrane. This is obtained by discounting the osmotic pressure (Π) to the pressure that is applied to make a membrane permeate water (*P*). Pressure units are typically expressed in bar or psi. This formula is shown in Equation 5.

$$NDP = P - \pi \qquad \text{Eq. 5}$$

To better understand how to calculate the osmotic pressure, the following example can be studied. To calculate the osmotic pressure of a solution that has 2 g/L sodium chloride dissolved in water, Equation 2 can be used.

The first step is to transform the mass concentration to molar concentration. This is achieved in Equation 6.

$$\frac{2\,g\ NaCl}{L}\,\frac{1\ mol\ NaCl}{58.44\ g\ NaCl} = \frac{0.0342\ mol\ NaCl}{L} \qquad \text{Eq. 6}$$

Since sodium chloride fully dissociates in water following Equation 1, osmotic pressure dissociation coefficient can be assumed 1. Therefore it is possible to calculate the concentration of each individual ion dissolved in water as described in Equation 7.

$$0.0342\ mol\ NaCl \rightarrow 0.0342\ mol\ Na^{+} + 0.0342\ mol\ Cl^{-} \qquad \text{Eq. 7}$$

Finally, each contribution of the osmotic pressure needs to be calculated for each individual ion. This is achieved by using Equation 2 in each individual sodium and chloride ion. This can be seen in both Equation 8 and Equation 9 respectively, where it can be seen that both sodium and chloride ions have the same osmotic pressure individual contribution of 0.85 bar each.

$$P_{Na+} = \frac{0.0342\ mol\ Na^{+}}{L}\frac{0.08314\ L\ bar}{K\ mol}298K = 0.85\ \text{bar} \qquad \text{Eq. 8}$$

$$P_{Cl-} = \frac{0.0342\ mol\ Cl^{-}}{L}\frac{0.08314\ L\ bar}{K\ mol}298K = 0.85\ \text{bar} \qquad \text{Eq. 9}$$

Finally, tot total osmotic pressure of sodium chloride can be calculated adding each individual ion osmotic pressures. This is shown in Equation 10, where it can be seen that the total osmotic pressure of a 2 g/L sodium chloride solution equals 1.7 bar. A rule of thumb to quickly estimate the osmotic pressure is to divide the total dissolved solids by 100. This gives an approximation of the osmotic pressure in psi. To have it in bar, this resulting value needs to be multiplied by 14.5.

$$P_T = P_{Na}^{+} + P_{Cl}^{-} = 1.7\ \text{bar} \qquad \text{Eq. 10}$$

3.4.1.2 Water flux

Water flux across a membrane (F_w) is defined as the multiplication of the water permeability value, called *A*-value (*A*) with the net driving pressure, which is obtaining by subtracting the osmotic pressure gradient (∏) to the pressure gradient applied to the membrane (*P*). Flux of water is typically expressed in US gallons divided per square feet per day (gfd) or in liters divided per hour per square meter per day (LMH). *A*-value expresses membrane water permeability, and it is typically expressed in US gallons divided per square feet per day per psi (gfd/psi) or in liters divided per hour per square meter per bar (L/m^2·h·bar or simply LMH/bar). Pressure is typically expressed in psi or bar. This formula is shown in Equation 11.

$$F_w = A(P - \pi) \qquad \text{Eq. 11}$$

It is important to notice how this equation resembles the Darcy's law equation, which states that a flow across porous membrane is proportional to the pressure that is applied. Darcy's Law can be seen in Equation 12, where k represents the permeability coefficient, μ the dynamic viscosity, and L the membrane thickness. All these parameters can be incorporated into the A-value membrane permeability coefficient.

$$F_w = \frac{k}{\mu L} P \quad \text{Eq. 12}$$

This is the same equation that governs the transport ort of water across an ultrafiltration membrane. Therefore, it can be concluded that for a solvent like water, when it faces a semi-permeable membrane where it can pass freely through it, it acts as a pressure-driven convective flow filtration.

3.4.1.3 Salt transport

The flux of salt across a membrane (F_s) is described as the multiplication of the salt coefficient value, also referred as B-value, with the concentration gradient of solutes across the membrane. The flux of salt across a membrane is typically expressed in pounds divided per square feet per day (lbfd) or in grams divided per hour per square meter per day (GMH). B-value, or salt diffusion coefficient is typically expressed in US gallons divided per square feet per day (gfd) or in liters divided per hour per square meter ($L/m^2 \cdot h$ or simply LMH). Concentration is usually expressed in pounds divided by US gallon (lb/gal) or grams divided per liter (g/L). This formula is described in Equation 13.

$$F_s = BC \quad \text{Eq. 13}$$

It is important to notice how this equation resembles the Fick's law equation. Fick's law describes de diffusion transport of mass across a membrane. As salt cannot pass through the cavities that are created in a membrane, it needs to pass through diffusion. It can be observed how diffusion is noted as with a D, and this corresponds to the B-value diffusion coefficient. Concentration gradient stays the same. This formula is shown in Equation 12.

$$F_s = DC \quad \text{Eq. 14}$$

3.4.1.4 The difference between convective and concentration driven flows

Typically, a pressure driven convective flow is several orders of magnitude higher than the mass transfer coefficient that can be achieved through diffusion. This is why reverse osmosis membranes are able to separate so well a solute from a solvent. This mainly happens since for a semi-permeable membrane, a solvent like water can travel across the membranes just following a pressure gradient. This means that water sees 'pores' across the membrane. However, salt cannot pass through these 'pores', as because of solvation, dissociated species in water are too big to pass through these 'pores', and the only way they have to pass through a membranes is through diffusion.

The following examples illustrates the different order of magnitude difference between a convective flow like water, and a diffusion flow like sodium chloride across the membrane.

To calculate the flux of water across a membrane, it can be assumed a water permeability A-value of 4 LMH/bar, a 15 bar feed pressure, a 1 bar osmotic pressure.

$$F_w = A(P-\pi) = 4(15-1) = 56\,L/m^2h = 56{,}000\,gm^2h \qquad \text{Eq. 15}$$

To calculate the flux of salt across this same membrane, a salt diffusion coefficient of 0.2 LMH and a concentration of salts of 2 g/L can be assumed.

$$F_s = BC = 0.2\times 2 = 0.4g/m^2h \qquad \text{Eq. 16}$$

As it can be observed from this example, reverse osmosis membranes are really selective to water mass transport across the membrane when compared to salt mass transport. These serval order of magnitude different in mass transport clearly illustrate the difference between convective and diffusive flow.

3.4.2 System equations

A reverse osmosis system is mainly composed by a feed flow (Q_f), and then this feed flow gets divided between the filtrated flow that is treated with the membrane active layer, which is called the permeate flow (Q_p), and the concentrate flow (Q_c), which has all the rejected salts or spices that could not permeate the membrane. Concentrate flow is sometimes also referred as brine or retentate. Flows are usually specified with in cubic meters per hour or day (m^3/h or m^3/d), or in US gallons per day (gfd). Plant capacity represents the permeate flow a desalination plant can produces, and it is usually expressed in millions liters per day (MLD). 1,000 m^3/d equals 1 MLD. A simple reverse osmosis diagram can be found in Figure 7.

Figure 7 Reverse osmosis diagram

A reverse osmosis system is characterized by having close water mass balance. This means that all the water that is entering the reverse osmosis membrane (Q_f) needs to exit the reverse osmosis system through either the permeate (Q_f) or through the concentrate flow (Q_c). This formula is depicted in Equation 16.

$$Q_f = Q_p + Q_c \qquad \text{Eq. 17}$$

A reverse osmosis system is also characterizing by having a neutral salts mass balance. This means that all salts that are entering the system will be also exiting the reverse osmosis membrane by either the permeate of the concentrate. Individual salts concentration for the feed (C_f), permeate (C_p) and concentrate (C_c) are typically represented in g/L or in mg/L (ppm). This is represented in Equation 18.

$$Q_f C_f = Q_p C_p + Q_c C_c \qquad \text{Eq. 18}$$

Flux (F) represents a flow relative (Q) to the membrane active area (A) it is permeating. Flux is typically measured in cubic meters per hour or day per square meter (m^3/h or m^3/d) or in US gallons per day per square feet (US gall/(d·ft^2) or gfd). This formula is shown in Equation 19.

$$F = \frac{Q}{A} \qquad \text{Eq. 19}$$

Reverse osmosis system recovery (R) represents the process water yield, and is calculated dividing the permeate flow (Q_p) by the feed flow (Q_f). It is expressed as a percentage. This formula is shown in Equation 20.

$$R = \frac{Q_p}{Q_f} \qquad \text{Eq. 20}$$

Salt passage (SP) represented the percentage concentration of salt that is passing through the membrane compared to the initial salt concentration being treated. It is calculated by dividing the concentration of salt (C_p) in the permeate by the concentration of salt in the feed (C_f). This parameter is useful from a physics point of view as it lets directly comparing two different membrane performances. This formula is shown in Equation 21.

$$SP = \frac{C_p}{C_f} \qquad \text{Eq. 21}$$

Salt rejection (SR) represents the percentage on how much salt a membrane is rejecting. It is calculated by subtracting 1 minus salt passage (SP). This parameter is useful as it enables to quickly realize how much solute or salts are being rejected by a membrane system. However, in order to perform comparative evaluation of two different membranes performance, it is usually necessary to do any comparative evaluation using the salt passage parameter. This formula is shown in Equation 22.

$$SR = 1 - SP \qquad \text{Eq. 22}$$

Another factor that is calculated is the plant availability (Av). This represents the amount of time in percentage that the plant is in operation producing water (t_{op}) and therefore not stopped versus the total time of the time period being considered (t_T). This formula is shown in Equation 20.

$$Av = \frac{t_{op}}{t_T} \qquad \text{Eq. 23}$$

3.4.3 Factors affecting membrane performance

Several factors can affect membrane performance. The three factors that are usually most relevant are the effect that feed pressure increase, feed concentration increase, and feed temperature increase have on membrane performance. In order to understand how these factors changes, only three equations are required. These are the osmotic pressure equation, shown in Equation 5, the water transport equation, shown in Equation 11 and the salt transport equation, shown in Equation 13.

A summary table highlighting these interactions can be found in Table 1.

Table 1 Summary table of the effect of feed pressure, concentration and temperature on the salt rejection and flux of a reverse osmosis membrane

	Flux	Salt Rejection	
Pressure ↑	↑	↑	$F_w = A \cdot (P\uparrow - \pi)$ $F_s = B \cdot C$
Concentration ↑	↓	↓	$F_w = A \cdot (P - \pi\uparrow)$ $F_s = B \cdot C\uparrow$
Temperature ↑	↑	↓	$F_w = A\uparrow \cdot (P - \pi\uparrow)$ $F_s = B\uparrow\uparrow \cdot C$

3.4.3.1 Feed pressure

When feed pressure increases, water flux also increases as a result of an increase in the pressure. Salt passage across the membrane stays constant, but since more water is passing across the membrane, whet the water flow is divided by the same amount of salt, the final salt concentration in the permeate decreases. Therefore, salt rejection increases.

3.4.3.2 Feed concentration

As feed concentration increases, osmotic pressure also increases. This leads to a direct reduction in the water flux across the membrane, as there is less net driving pressure available for the membrane to permeate. Additionally, as concentration increases, the flux of salt directly increases. This leads to a decrease in water passing through the membrane, which is divided by a higher salt passing through the membrane, thus increasing the salt passage and decreasing the salt rejection.

3.4.3.3 Feed temperature

As feed water temperature increases, the water becomes less viscous. This means that with the same amount of energy, more water can permeate through the membrane. This eventually leads to improving the water permeability value (*A*-value) and therefore the permeate water flux. With regards to the salt rejection, an increase in temperatures improves much more the salt diffusion factor (*B*-value). Therefore, there is a higher increase in salt is passing through the membrane than water passing through the membrane, and as a result, salt rejection decreases.

3.4.3.4 Concentration polarization

Concentration polarization is the phenomenon in which as membrane removes water from the feed solution, solutes are pushed towards the boundary layer of the membrane. This leads to a decrease in performance in the reverse osmosis membrane system as, since the membrane is filtration, the membrane sees the concentration in the boundary layer and not in the feed solution. As this salt concentration is higher, this means that the membrane experiences a decline in flux and salt rejection due to this phenomenon (Sablani *et al.*, 2001). Concentration polarization can be minimizing by controlling the membrane recovery rate, as well as through membrane element design elements such as a feed spacer that is able to provide a proper mixing and therefore minimizes the accumulation of solutes in the membrane boundary layer.

3.5 REVERSE OSMOSIS MEMBRANES

Reverse osmosis and nanofiltration membranes are pressure-driven membrane filtration processes, where feed pressure bigger than the osmotic pressure is used to filtrate water through the membrane.

Commercial elements used in large industrial installations are standardized. They are usually referred as spiral-wound polyamide based membrane configured in a cylindrical shape with a typical diameter is 8-in (20 cm), with a typical length of 40-in (1 m). For smaller industrial application where the water capacity required is lower, elements with 4-in diameter and 40-in elements are also used. Finally, elements used in home drinking applications are less standardized, and their size in terms of diameter and length can vary depending on each manufacturer. Examples of these elements are 1.8-in and 2.5-in diameter elements with 12-in or 14-in length. An example of a DuPont FilmTec™ BW30 PRO-400 membrane can be found in Figure 8. In this example, feed flow will enter the membrane through the anti-telescoping device on the left. The permeate flow will be collected in each membrane leave, and finally collected through the inner permeate water tube. The permeate flow will be exiting the membrane through the permeate water channel located in the center of the membrane, leaving through the right of the membrane. The concentrate flow will also be exiting the membrane through the right part through the anti-telescoping device on the right part. It is useful to realize that the remaining feed water that exits the membrane is what it is called concentrate.

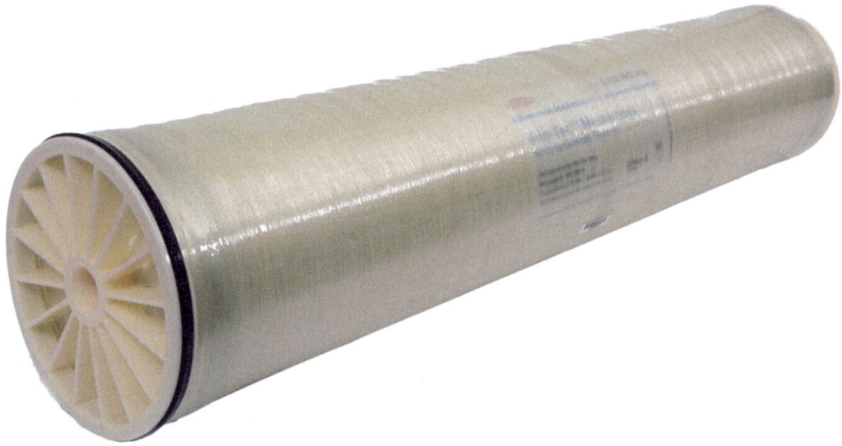

Figure 8 DuPont FilmTec™ BW30 PRO-400 membrane

A reverse osmosis membrane is typically composed by an active polyamide base layer that is around 0.2 μm thick. This polyamide membrane is also referred as the active layer, as is the one responsible from separating the salt solutes from the water solvent. As this membrane is so thin, in order to be able to precipitate it during the phase inversion process, a support layer typically consisting of polysulphone is used. This allows the proper precipitation of polyamide on the polysulphone. The polysulphone layer has a thickness of around 40 μm thick. As still this thickness is rather seen, in order to enhance its mechanical structure, this layer is put in a polyesther reinforcing layer, typically consisting of 120 μm thick layer. This three multi-layer structure is usually referred as thin-film (TFC) composite layer (DuPont, 2023). A schematic of this arrangement can be found in Figure 9.

Figure 9 Thin-film composite reverse osmosis membrane multi-layer composition

A scanning electron microscopy (SEM) image courtesy of DuPont FilmTec™ membranes depicting the main three layers in a reverse osmosis membrane can be found in Figure 10.

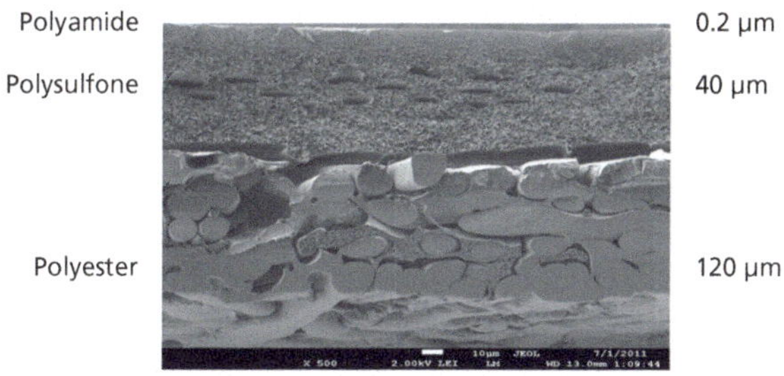

Figure 10 SEM image of a reverse osmosis membrane

Nanofiltration membranes are very similar to reverse osmosis membranes. Their main difference is that the active layer usually consists of a polypiperazine polymer. Typically, nanofiltration membranes are used when only certain solutes are needed to be separated, but not all of them. This allows to significant energy savings. Examples of their use are sulphate removing nanofiltration membranes. These membranes can let sodium chloride pass through their active layer, but they remove sulphates and other divalent ions. This is especially useful for oilfield applications, where seawater is used for injecting it into the oil wells. In this application, no sodium chloride needs to be removed. However, to prevent multiple problems, sulphate needs to be removed. By using nanofiltration membranes, the operating pressure can be reduced from 70 bar to 15 bar, therefore saving a lot of energy.

Typically a revers osmosis membrane spiral-wound element consists of multiple polyamide sheets that are rolled together. Each membrane sheet is separated from the one on its top by a feed spacer. Inside each membrane there is a permeate spacer. The role of the spacers is to provide mechanical integrity into the reverse osmosis element so that it can be properly folded. The feed spacer also plays a crucial role into minimizing the concentration polarization effect, as well as controlling biofouling and saving energy. Minimizing the dead spaces inside a membrane is important to prevent biological growth inside of a membrane. This schematic shown in Figure 11 shows the main parts of a spiral wound reverse osmosis element.

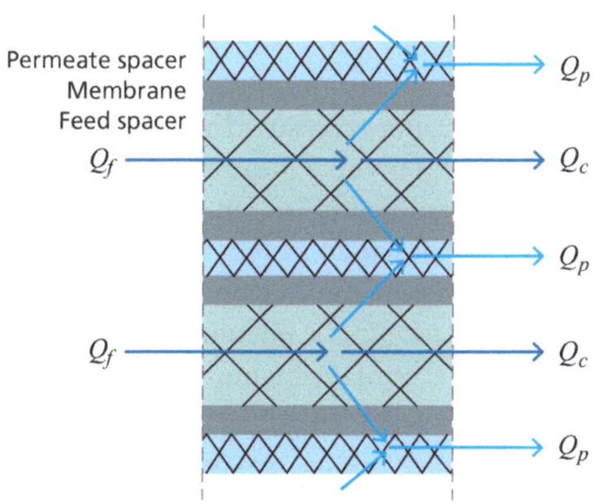

Figure 11 A spiral wound reverse osmosis elements with its parts

3.5.1 The significance of desalination

As of 2023, there are more than 21,000 desalination plants in the world. All these plants provide a total install capacity of newly created fresh water equivalent to 100,000,000 m^3/d (Climate ADAPT, 2023).

3.6 PERFORMANCE MONITORING

Typically, systems are designed to provide a constant yearly water production capacity. Therefore systems are designed at a constant flux rate. So, when systems suffer from fouling or changes in operating condictiones such as temperature decrease or salinity increase, typically their flux rate would decrease. To prevent this, more energy is used, so that the membrane system can compensate for the decrease in membrane permeability and be able to provide the same operating flux.

Therefore, it is of utmost importance to periodically monitor the performance of a reverse osmosis systems. There are three key parameters that need to be monitored in a reverse osmosis.

The first one is the energy consumption, monitored through the high pressure pump operating pressure. This is an important parameter because the energy consumption directly impacts the operating expenses (OPEX) of a reverse osmosis pump. Additionally, the high-pressure pump needs to have enough capacity to increase the pressure it is delivering to the membrane system. If the pump cannot deliver enough pressure, the whole desalination installation can start suffering from a decrease in the water it is delivering. This can lead to water shortages and even being outside the offset contract. This is why typically plants are designed in a conservative way. To properly size the high pressure pump, typically the lowest yearly temperature is used to size the pump, as the lowest water temperature will provide the highest pressure needed to sustain the targeted design installation permeate flow capacity.

The second parameter that is key to monitor a membrane system is the permeate water quality. Typically, the water conductivity is monitored. Conductivity is typically expressed in μS/cm. Conductivity is used to estimate the total dissolved solids (TDS) salinity of water. A good rule of thumb for low salinities water is that 2 μS/cm are equal to 1 mg/L (ppm) salinity. For seawater types, a good rule of thumb is that 1.4 μS/cm are equal to 1 mg/L (ppm). Sometimes, beside general salinity measured by conductivity, specific spices that are important are also specified. This can involve measuring specific targets such as alkalinity, boron, and pH, among others. Water quality is of utmost importance, because ultimately, water treatment plants are usually designed to provide a warrantied water quality that is typically limited to be below a certain limit. Therefore, water quality typically acts as the independent variable when designing a membrane system. Typically, a membrane installation is designed taking into account the highest water temperature thought the yearly temperature cycle. This is because at the highest temperature is when the water quality will be the worst, and therefore show the highest permeate water total suspended solids value. The third important parameter to monitor is the membrane pressure drop (*dP*). Pressure drop is a factor that is important to measure as it directly affects the energy consumption. However, this parameter is key as when pressure drop increases, this means that the membrane feed-concentrate channel is getting blocked. This can happen if the membrane is experiencing fouling, as when the membrane gets fouled, this means it is getting obstructed, and therefore it is more difficult for water to travel across the membrane. They key problem this issue presents, is as if the membrane gets too much blocked, the membrane can start to get mechanically damaged, and it can eventually lead to its irreversible damage and the membrane can stop working as intended. Therefore, when pressure drop starts to increase to higher values, it is a good habit to perform a cleaning-in-place (CIP), in order to try to recover the membrane performance to the initial pressure drop values. It should be noted that the more a membrane has higher pressure drop, the more difficult it becomes to clean the membrane and restore its performance to its initial values.

This is why it is so important to clean the membranes early, so that it is easier to recover the membrane performance. Because membranes systems typically suffer from fouling, especially in the areas of the planet where temperature is usually higher and there is therefore a higher water demand due to water scarcity issues, typically plants are designed with redundant trains. A train is a collection of pressure vessels. Each pressure vessel usually contains up to 7 membranes in series, and a train usually has dozens of pressure vessels. This means that for example, in a plant that has 11 reverse osmosis trains, 1 of this trains can be used to be put in operation when one train needs to be going through a chemical cleaning. This designs are called in the industry *N*-1 designs.

3.7 NORMALIZATION

Monitoring feed pressure, permeate conductivity and pressure drop is of utmost importance in order to be able to anticipate to possible problems that might arise from changes in operating conditions and in fouling. As it was explained previously, as a result of water temperature or water salinity changes, the energy consumption and permeate water quality can change. Therefore, it is of utmost importance to understand if these changes in water quality or energy consumptions are due to unexpected problems such as fouling, or they are normal and expected as a result of the physical principles that were previously explained. This is when membrane normalization comes into approach.

The key normalized parameters that are needed to be analyzed in a reverse osmosis membrane system are the normalized permeate flow, the normalized salt rejection, and the normalized pressure drop.

Normalization can be done using the equations listed below, or using computer assisted programs such as the FT-Norm PRO that DuPont offers.

3.7.1 Why normalization matters

The example detailed in Table 2 is a good example of why normalization matters. This is an example of a reverse osmosis plant where after 5 days of operation, the feed water temperature decreases from 25 ºC to 21 ºC. As it can be seen in the monitoring parameters, the plant is delivering a constant permeate production of 50 m^3 /h with a constant feed flow of 100 m^3 /h. This represents a 50% recovery. Feed salinity is also constant at 1,000 mg/L. However, it can be observed how feed pressure increases from 8 bar to 8.81 bar, and permeate quality decreases from 30 mg/L to 27 mg/L. Without normalizing the data, it would be challenging to understand if this increase in feed pressure and improvement in water quality is a result of expected thermodynamics, or it is a result of the membrane getting for example fouled. Normalization is the tool that allows to properly perform this assessment.

In this particular example, it can be seen that normalized permeate flow stays constant, while normalized salt rejection also stays constant. This means that the increase in feed pressure and improvement in water quality is not related to fouling, but do to they normal behavior of the membrane system. This happens as seen previously, when the temperature decreases, water quality improves as less salt passes through the membrane. Additionally, more energy is needed to pump the water through the membrane as its viscosity increases. If the normalized permeate flow would show, for example, a decrease over time, it could be suspected that fouling is responsible for this loss in membrane permeability. If normalized salt rejection would be, for example, decreasing, this could also indicate the likelihood of issues in the membrane system.

Table 2 Normalization example

Days	Feed flow (m^3/h)	Permeate flow (m^3/h)	Feed pressure (bar)	Feed TDS (mg/L)	Permeate TDS (mg/L)	Temperature (°C)	Normalized permeate flow (m^3/h)	Normalized salt rejection
0	100	50	8.00	1,000	30	25	50	97.0%
1	100	50	8.00	1,000	30	25	50	97.0%
2	100	50	8.19	1,000	29	24	50	97.0%
3	100	50	8.39	1,000	28	23	50	97.0%
4	100	50	8.59	1,000	27	22	50	97.0%
5	100	50	8.81	1,000	27	21	50	97.0%

3.7.2 Equations

Normalization can be done with operating software's such as DuPont's FT-Norm PRO software, or using the equations found in (DuPont, 2023).

3.7.2.1 Normalized permeate flow

To normalize the permeate flow (Q_{pn}) the following formula found in Equation 24 can be used. It must be noted that the subscript 0 references to the conditions being normalized to. This can typically be the first data when the system is stabilized, or even an ideal projection that the system is designed to operate at normal or ideal conditions. The other data without subscript refers to the data at the time instance being normalized to. Units are in bar for the pressure drop, and in m³ /h or m³ /day for the flows.

$$Q_{pn} = Q_p \frac{NDP_0}{NDP} \frac{TCF_0}{TCF} \qquad \text{Eq. 24}$$

Net driving pressure (*NDP*) is calculate taking the feed pressure, and subtracting half the pressure drop or differential pressure (*dP*) to the permeate pressure (P_p) and the average feed concentrate osmotic pressure (Π_{fc}). All pressures are measured in bar. This is shown in Equation 25. It should be noted that the osmotic pressure in the permeate has been omitted for simplification.

$$NDP = P_f - \frac{dP}{2} - P_p - \pi_{fc} \qquad \text{Eq. 25}$$

Pressure drop (*dP*) is calculated using Equation 26. To calculate it, the concentrate pressure (P_c) needs to be subtracted to the feed pressure (P_f).

$$dP = P_f - P_c \qquad \text{Eq. 26}$$

The average feed-concentrate osmotic pressure (Π_{fc}) can be calculated using Equation 27 (DuPont, 2021). It should be noted that Temperature (T) is in kelvin.

$$\pi_{fc} = 0.00265 \times C_{fc} \times T \qquad \text{Eq. 27}$$

The average feed-concentrate concentration (C_{fc}) can be calculated using Equation 28. It uses the feed concentration (C_f) and the membrane system recovery (*R*). Its units are in mg/L.

$$C_{fc} = C_f \frac{\ln \frac{1}{1-R}}{R} \qquad \text{Eq. 28}$$

To calculate the temperature correction factor, Equation 29 can be used. It should be noted that also Temperature (T) needs to be in kelvin. Also, the *k* parameter depends on water temperature. If temperature ≥ 25 ºC, *k* = 2640. If temperature ≤ 25 ºC *k* = 3020.

$$TCF = e^{k\left(\frac{1}{298} - \frac{1}{T}\right)} \qquad \text{Eq. 29}$$

3.7.2.2 Normalized salt rejection

In order to normalize the salt rejection, the following formula is used. This is displayed in Equation 30 (DuPont, 2021).

$$SP_n = SP\frac{Q_p \times TCF_0 \times C_{fc0} \times C_f}{Q_{p0} \times TCF \times C_{fc} \times C_{f0}} \qquad \text{Eq. 30}$$

3.7.2.3 Normalized pressure drop

Normalized pressure drop (dP_n) can be calculated using the formula displayed in Equation 31. To do this calculation, the standard pressure drop (dP) is used, together with the division of the baseline average feed-concentrate flow (Q_{fc0}) and the average feed-concentrate flow (Q_{fc}). Units are in bar for the pressure drop, and in m^3 /h or m^3 /day for the flows.

$$dP_n = dP\frac{C_{fc0}}{C_{fc}} \qquad \text{Eq. 31}$$

3.8 FOULING

Fouling is the phenomenon in which membranes get dirty. Fouling is one of the major challenges that is nowadays affecting the reverse osmosis industry (Kucera, 2011). The key problem that fouling offers is that it is difficult to foresee, and also it is difficult to deal with it once fouling starts to appear. It is sometimes also difficult to identify the type of fouling present.

Fouling is typically characterized because of the accumulation of unwanted spices that get deposited inside the membrane, leading to a decrease in its performance. Typically fouling leads to higher operating costs in membrane systems, as it can lead to higher energy consumption, lower water quality, lower water production, higher chemical costs because of the cleanings, and even an irreversible damage to the membrane. Fouling can be characterized into four major types, plus a fifth one that relates to membrane integrity failure.

3.8.1 Biofouling

The main type of fouling is biological fouling or simply biofouling. Biofouling happens as certain bacteria, that have evolved to form a biofilm when they found a solid surface, meet the membrane and the feed spacer. When this happens, these bacteria attach on the membrane, and they start to form a biofilm.

Biofouling is characterized by an exponential increase in the feed-concentrate pressure drop of the reverse osmosis membrane. It can lead to an increase in the energy consumption and if not deal properly, it can mechanically damage the reverse osmosis membrane. Biofouling is typically present in the first elements of a pressure vessel.

Once a biofilm is established on the membrane, it is difficult to remove. Useful strategies involve cleaning the membranes with caustic cleanings (CIP). These cleanings are most effective the higher the temperature and pH. If possible by the membrane characteristics, effective cleanings will involve pH around 13 at a temperature around 35 ºC using NaOH. It is important to have enough time for the chemicals to get soaked in the membranes, and then flush the soaked solution effectively. This combination of steps can be repeated multiple times until the soaking solution appears clean.

Other more preventive approaches involve starving the bacteria from growing. This can be achieved using pre-treatment technologies like the DuPont B-Free™ technology, that is able to eliminate nutrients before they reach the reverse osmosis system (Kucera, 2011). This system has been proved extremely effective in preventing biofouling development into reverse osmosis systems.

Other concepts that are practiced involve using non-oxidizing biocides, such as DBNPA or 2,2-dibromo-3-nitrilopropionamide (DBNPA). This biocide is described as non-oxidizing the polyamide active layer of the reverse osmosis membrane. Some considerations that need to be taken into account with this solution is its limited compatibility with drinking water application, and also the fact that bacteria can get used to the biocide, and at certain point, it can stop being effective and biofouling can develop again.

3.8.2 Organic fouling

The second type of fouling is the organic fouling. Organic fouling happens when organics molecules get accumulated on the membrane surface. It typically leads to decline in normalized permeate flow.

The main way to deal with organic fouling is through performing caustic based chemical cleanings (CIP). These cleanings are most effective the higher the temperature and pH. The same cleaning procedure as with biofouling can be followed.

3.8.3 Particulate fouling

Particulate fouling, or colloidal fouling, refers to the accumulation of particles in the membrane. This can be due to an inadequate pretreatment. This type of fouling leads to a rapid increase in the feed-concentrate pressure drop, as the feed spacer channel fills quickly with particles.

In order to clean particular fouling effectively, it might be necessary to perform a caustic based chemical cleanings (CIP). These cleanings are most effective the higher the temperature and the pH is. If possible by the membrane characteristics, effective cleanings will involve pH around 13 at a temperature around 35 ºC.

A good solution to deal with this type of fouling can involve upgrading the pretreatment from a conventional one to a membrane based one, like ultrafiltration. Another solution could be making sure there are no fiber breakages in the ultrafiltration part, and if there are, repairing them with glue and pins, or replacing the damage modules with new ones.

3.8.4 Scaling

Scaling is a type of fouling that typically occurs when non too soluble salts starts to precipitate on the membrane module. This can happen if water recovery is too high, or if temperature or water composition changes. A recommended way to prevent scaling is consulting a specialized anitscalant company. They have powerful software that simulate the operating conditions at the targets recoveries and temperatures, and recommend the best antiscalants and their concentration to used, based on the spices that have higher risk to precipitate as they have the risk to surpass its solubility limit. Scaling typically happens in the last elements of a pressure vessel, as there is where there is less water, and the water is more concentrated. So dissolved spices have a higher risk to precipitate. Scaling typically leads to an increase in pressure drop, as well as to a decrease in water quality or salt rejection. As salts start to precipitate on the membrane surface, this affects the concentration polarization, and increases the effective concentration of salts on the boundary layer. Scaling can also therefore lead to a decrease in normalized permeate flow, since the osmotic pressure is greatly increase, thus reducing the net driving pressure.

In order to clean a scaled membrane, it is important, if possible, to autopsy a membrane, to understand the type of scaling present. There are multiple companies that are specialized in offering this service. To clean the membrane, it is recommended to perform an acid chemical cleaning (CIP). If the membrane characteristics allows it, a pH of 1 at room temperature is effective. It is recommended to use HCl to prevent any further scaling cause by other species such as sulfuric acid. Sometimes, membrane can have some organic or biofouling too. Therefore, it is also recommended to always start with a caustic cleaning as previously described, and then follow the acid cleaning step.

3.8.5 Integrity failure

Integrity failure occurs when the membrane is suffering from a non-fouling related damage.

This can involve chemical oxidation of the membrane. This can happen if for example, sodium hypochlorite (NaOCl) from the ultrafiltration pre-treatment manages to reach the reverse osmosis membrane, as the polyamide can get damage and eventually eliminated. Symptoms involve an increase in the normalized salt passage, as since the membrane does not have an active layer, it stops separating salt from water. Once oxidation is detected, it is important to eliminate the source that is causing the chemical that is leading to oxidation to leach. A strategy to properly address the leaching of NaOCl in ultrafiltration membranes is described here (Gilabert-Oriol, 2021).

Other types of mechanical failures might involve o-ring failure. This can happen when the o-ring that is used to separate a membrane gets pinched, there is a by-pass of water from the feed side to the permeate side. This leads to an increase in the normalized salt passage. In order to fix this, a probing test needs to be done in each membrane connection inside a pressure vessel. A methodology to perform this test is described elsewhere (DuPont, 2021).

Compaction can happen when a membrane is operating at a too high pressure and temperature. Compaction can be reversible or irreversible, or a combination of both. Typicall effects of compaction involve a high increase in energy consumption, observed by a decline in normalized permeate flow. Compaction can lead to an improvement in water quality, as the membrane becomes more dense, it is more difficult for the salt to pass through it. When compaction is identified, it is important to either use compaction resistant reverse osmosis membranes, or to decrease, if possible, the target permeate flux, specially in periods of high temperature, with the aim to reduce the operating pressure.

3.9 REFERENCES

Carroll T., S. King, S. R. Gray, B. A. Bolto, N. A. Booker, The fouling of microfiltration membranes by NOM after coagulation treatment. Water Research 34.11 (2000) 2861-2868

Casas, S., Aladjem, C., Larrotcha, E., Gibert, O., Valderrama, C., & Cortina, J. L. (2014). Valorisation of Ca and Mg by products from mining and seawater desalination brines for water treatment applications. Journal of Chemical Technology & Biotechnology, 89(6), 872-883.

Climate ADAPT, consulted on August 19, 2023, https://climate-adapt.eea.europa.eu/en/metadata/adaptation-options/desalinisation

DuPont, FilmTec™ FT-Norm software, 2021

DuPont, FilmTec™ Reverse Osmosis Membranes Technical Manual, 2023.

Fernández-Torquemada Y., A. Carratalá, J. L. Sánchez Lizaso. Impact of brine on the marine environment and how it can be reduced. Desalination and Water Treatment 167, 27–37. (2019).

Fritzmann C., J. Löwenberg, T. Wintgens, T. Melin, State-of-the-art of reverse osmosis desalination, Desalination 216.1 (2007) 1-76

Gilabert-Oriol G. Ultrafiltration Membrane Cleaning Processes: Optimization in Seawater Desalination Plants. Walter de Gruyter GmbH & Co KG; 2021 Jun 8.

Greenlee L.F., D.F. Lawler, B.D. Freeman, B. Marrot, B., P. Moulin, P, Reverse osmosis desalination: water sources, technology, and today's challenges, Water research 43.9 (2009) 2317–2348

Guo W., H. H. Ngo,J. Li, A mini-review on membrane fouling, Bioresource technology 122 (2012), 27-34

Kadaj E., R. Bosleman, Energy recovery devices in membrane desalination processes. In Renewable energy powered desalination handbook (pp. 415-444). Butterworth-Heinemann. (2018)

Kim J., K. Park, D. R. Yang, S. Hong, A comprehensive review of energy consumption of seawater reverse osmosis desalination plants. Applied Energy, 254, (2019) 113652

Kucera J. Reverse Osmosis: Industrial applications and processes. Wiley; 2011.

Kumar M., T. Culp, Y. Shen, Water desalination: History, advances, and challenges. In Proc., Frontiers of Engineering: Reports on Leading-Edge Engineering from the 2016 Symp,55-132 (2017)

Massons G., G. Gilabert-Oriol, S. Arenas-Urrera, J. Pordomingo, J.C. González-Bauzá, E. Gasia, M. Slagt. Industrial scale pilot at Maspalomas I desalination plant demonstrates the efficiency of DuPont™ B-Free™ pretreatment–a new breakthrough solution against biofouling, Desalination and Water Treatment, 1-5, 2022

Sablani S.S., M.F.A. Goosen, R. Al-Belushi, M. Wilf, (2001). Concentration polarization in ultrafiltration and reverse osmosis: a critical review. Desalination, 141(3), pp.269-289.

Sustainable clean water through solar-powered desalination for water-scarce islands and coastal regions (SDG: 2, 3, 6, 8, 11, 12, 14), Section #SDGAction42477, https://sdgs.un.org/partnerships/sustainable-clean-water-through-solar-powered-desalination-water-scarce-islands-and, consulted on August 19, 2023

Valavala R., J. Sohn, J. Han, N. Her, Y. Yoon, Pretreatment in Reverse Osmosis Seawater Desalination: A Short Review, Environmental Engineering Research 16.4 (2011) 205-212

Wang L., J. He, M. Heiranian, , H. Fan,, L. Song, , Y. Li, M. Elimelech. (2023). Water transport in reverse osmosis membranes is governed by pore flow, not a solution-diffusion mechanism. Science Advances, 9(15), eadf8488.

doi: 10.2166/9781789062977_0071

Chapter 4

Forward Osmosis

Maria Salud Camilleri-Rumbau, Aquaporin, Denmark/Eurecat, Spain
Xuan Tung Nguyen, Aquaporin, Singapore
Victoria Sanahuja-Embuena, Aquaporin, Denmark
Jan Frauholz, Aquaporin, Denmark/RWTH Aachen, Germany
Victor Augusto Yangali Quintanilla, Grundfos, Denmark
Alberto Tiraferri, Politecnico di Torino, Italy
Irena Petrinic, University of Maribor, Slovenia
Claus Hélix-Nielsen, Technical University of Denmark, Denmark

The learning objectives of this chapter are the following:

- To define principles of forward osmosis
- To define and apply forward osmosis parameters for assessing performance
- To present and discuss the basic equations governing forward osmosis performance using typical experimental modes
- To understand the theoretical background of forward osmosis performance and performance prediction using modeling tools.

4.1 INTRODUCTION: PRINCIPLES OF FORWARD OSMOSIS

Forward osmosis (FO) is an osmotically driven membrane technique which allows the separation of water from a feed solution through a semi-permeable membrane using osmotic pressure gradient as a driving force. Although during the last decade there have been important advances in FO in terms of material development and processes, the few commercially available products and best practices for effluent processing makes the standardization of the FO applicability challenging.

In terms of materials used for fabrication of FO membranes, cellulose acetate (CTA) membranes and thin film composite (TFC) polyamide (PA) membranes are the most widely known, both being commercially available (Xiao *et al.*, 2017). New approaches using new materials for FO purposes have also been developed. To name a few, these are double-skinned membranes, membranes obtained by layer-by-layer techniques, mixed organic-inorganic membranes and aquaporin-based membranes, which consist of TFC PA with embedded aquaporins (Suwaileh *et al.*, 2020).

Despite FO being a developing technology, in the recent years, there has been an increasing interest for its use in industry. This opens countless opportunities for further developing membrane configurations that can be used in an industrial setting. In terms of applications, FO is a versatile membrane technique with a broad applicability spectrum within the water treatment sector (desalination, municipal wastewater, industrial wastewater, potable and non-potable water reuse, etc.), the management of process water (biorefineries, pharmaceutical processes, etc.) and food and beverage processes (concentration of flavours and aromas, juices, etc.).

Regarding membrane performance, FO relies on the osmotic pressure of the two solutions separated by the semi-permeable membrane. The pass of water is allowed due to osmotic gradient resulting in a concentrating process for the lower osmotic pressure solution (known as feed) and a dilution process for the higher osmotic pressure solution (known as draw solution). As for any other membrane-based process, water flux (J_w) and forward solute rejection (R) can be obtained from experimental data (see section 4.3.1); while a parameter known as reverse solute flux (J_s), unique for FO, allows to track the loss of draw solute into the feed solution.

During FO processing, concentration polarization can severely affect membrane performance in both feed and draw sides of the membrane. Both external concentration polarization (ECP) at the feed side of the membrane as well as internal concentration polarization (ICP) at the draw side of the membrane, play a detrimental role that cannot be overlooked when interpreting the FO process performance and when evaluating its applicability (see section 4.3.2. for more details). Concentration polarization can act in combination with fouling, scaling and/or a combination of both fouling mechanisms, to decrease the net driving force available for mass transport.

With this Chapter, the authors have put together the most relevant experimental practices in the FO field in order to guide researchers and engineers towards getting hands-on experiences in FO.

4.2 MATERIALS AND EXPERIMENTAL SET-UP

4.2.1 Membrane configurations

Membranes for FO processes and generally for liquid-liquid separations, can be found commercially either as flat-sheets or hollow tubes. The modules of these can be classified, similarly as done for other membrane processes, in four module types: plate-and-frame, spiral-wound, hollow fiber and tubular. The first two types are made of flat-sheet membranes, while the last two types are composed of hollow tube membranes.

Plate-and-frame modules are the simplest module configuration where membrane sheets are mounted in frames closely together. The feed solution to be treated passes alongside the sheets surface and it gets collected at the end as a concentrated retentate, while the permeate is collected in its own channel. Spiral-wound are membrane sheets rolled in alternating order together with turbulence promoting plastic grids, called spacers. The feed solution is introduced at one end of the module and flows axially on the active layer and feed spacer side of the membrane. The permeate is collected in the envelope and led to the permeate channel which is centrally located. Hollow fiber modules consist oppositely of tubes, packed closely together and placed inside a vessel. Here, the feed solution passes through the lumen of the hollow fiber, permeates through the membrane towards the shell side and exits the module. When their inner diameter rises to 5 mm or above, and the module packing density significantly decreases, the module is known as a tubular module.

There are intrinsic advantages and disadvantages of using each module configuration and their usage would depend on the intended process application. For example, plate-and-frame and tubular configurations are usually used with extremely high-particulate streams and/or high-viscosity solutions. The simplicity of the plate-and-frame configuration allows for high cross-flow velocities, reducing fouling by increasing longitudinal shear stress. Additionally, these membranes can be easily cleaned which increases the lifetime of the membrane. However, this module type is expensive to manufacture, and the packing density is low, which increases the membrane installation footprint considerably. Similarly, tubular modules can be more tolerant to fouling and clogging due to the large inner diameter of their hollow tube and the possibility of operating at high cross-flow velocities. However, it suffers from the same disadvantages as its counterpart, the plate-and-frame module due to a low packing density. Spiral-wound and hollow fiber configurations are thus the most commonly used module configurations for membrane-based liquid-liquid separation and FO processes, as they can cover a broad range of applications by balancing effectiveness and price.

4.2.2 Experimental modes

In general, there are three main experimental modes, regardless of process configuration being single-stage or multi-stage, that can be defined when working with FO membrane processes:

- Single pass mode
- Batch-batch mode
- Semi-batch mode

The main differences between these operational modes are related to how the feed and draw solutions are being processed within the membrane module. For example, in single pass mode, the feed and draw solutions enter the module and have a unidirectional contact area across the membrane, where the solutions are not recirculated. The feed recovery or concentration is achieved in one pass. Oppositely, in batch mode, both feed and draw solutions are recirculated to their respective holding tanks. This means that the feed solution gets increasingly concentrated in the feed tank, while the draw solution gets diluted with time. During semi-batch mode operations, one of the two solutions, either feed or draw solutions, are run in a batch mode while the other solution runs in single pass mode. Typically, the feed will run in a batch mode, allowing for continuous concentration during processing, while the draw solution will run in single pass to minimize loss in FO performance due to dilution of the draw solution.

The advantages of the operational modes above depend on the type of feed, on the process that needs to be undertaken, and on the objective of the FO system. For instance, during food valuable concentration processes, a semi-batch mode will be preferred, while a batch-batch mode might be preferred during concentration of secondary effluent by using sea water brine as a draw solution.

Test setup – Semi-batch mode (feed in batch mode vs. draw in single pass mode)

Figure 1 shows the schematic outline of a semi-batch FO setup. The draw solution will become diluted during the feed concentration process. The draw outlet can be discarded to the drain. The feed solution will be continuously recirculated to the FO membrane module and thus concentrated.

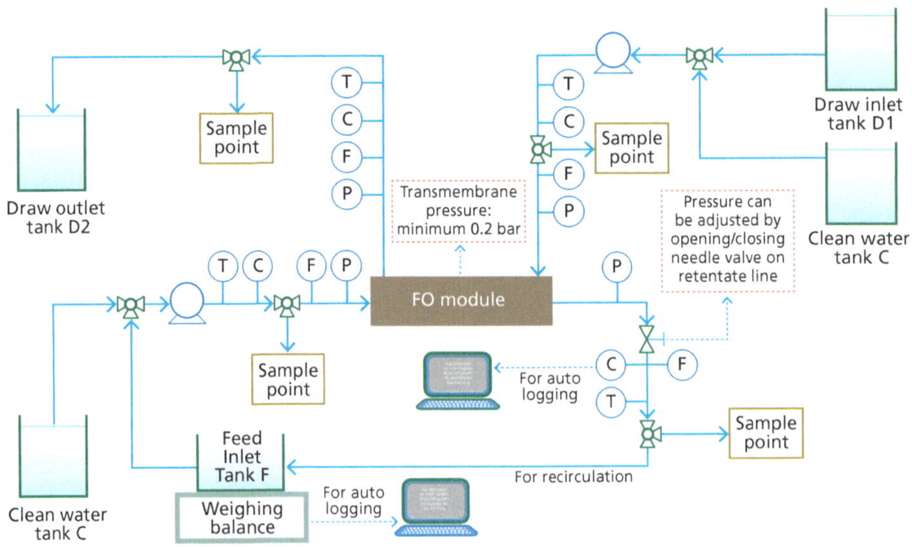

Figure 1 Setup for FO semi-batch operation

Recommendations for running an FO application test:

1. To select the type and strength of the draw solution, refer to section 4.2.3.
2. Start the feed pump to fill in the system. Afterwards, start the draw pump and adjust to the operating conditions as indicated by the FO membrane/FO system manufacturers, always ensuring that the system operating conditions are in agreement with the membrane manufacturer operating limits. Ensure removal of air in the system. The operating conditions can be modified during the application test. For example, the feed inflow can be increased to enhance shear on the membrane surface and delay fouling, while the draw inflow can be increased if flux (J_w) falls below 1 L/m^{2}h.
3. TMP must always be kept positive where possible. For this, it is crucial to monitor that the feed inlet and outlet pressure readings do not overcome the maximum TMP and inlet pressure specified by the membrane module manufacturer. See more details on the effects of negative TMP in section 4.3.3.
4. To be able to calculate compound mass balances, weight of samples taken from the feed inlet, feed outlet and draw outlet should be considered throughout the concentration process. This will allow monitoring of J_w and water recovery values accurately during the FO process.

For further recommendations on process parameters and constraints, refer to Table 1.

Table 1 Process parameters to be considered during FO application test

Process parameter	Process values - considerations
Recommended (and maximum) application flow rates, L/h	Refer to FO membrane/system manufacturer's recommendations.
Minimum feed outlet flow, L/h	Refer to FO membrane/system manufacturer's recommendations. The user needs to make sure that the feed outlet has a minimum flow to avoid module damage.
TMP, bar	Refer to FO membrane/system manufacturer's recommendations. Generally, it should be around or just above 0 bar.
Feed and draw inlet pressure, bar	Refer to manufacturer's recommendations on pressure tolerance for the specific FO product.
Recommended (and maximum) temperature, °C	Refer to manufacturer's recommendations. Generally, the FO module should be able to run feed and draw solutions at room temperature. Be aware that increases in operating temperature could affect FO performance, e.g., increased J_w and J_s. The module temperature tolerance should not be surpassed. Refer also to manufacturer's recommendations for CIP procedures.

Data collection during the application test

The data to be collected during the test is shown in Table 2. For detailed calculations, refer to equations on 'Typical parameters and phenomena' in section 4.3.1.

Table 2 Process parameters to be measured during the semi-batch application test and where to make the related readings

Purpose/calculated values	Process parameters	Where to measure
Water flux (J_w)	Flow/Weight	Feed bulk
TMP	Pressure	Feed inlet/outlet, draw inlet/outlet
Feed inlet pressure	Pressure	Feed inlet
Maintain stable temperature (T)	Temperature	Feed outlet (if measuring conductivity at feed outlet)
Osmotic pressure (indirect measurement)	Conductivity	Feed outlet
Ensure minimum feed outlet flow	Flow	Feed outlet

4.2.3. Draw solutions: properties, regeneration, types and selection criteria

The performance of FO applications depends on the draw solution, which provides the driving force for water permeation. An adequate choice of draw solute agent can maximize the water flux (J_w) and the water recovery of the system. In addition, the reverse solute flux (J_s) can be reduced and the regeneration costs can be lowered, which usually represents the largest operational costs in FO applications. Therefore, this subsection will give a short summary of the most important properties of draw agents, their influence on the membrane performance, available draw regeneration methods, and discuss advantages and disadvantages of different classes/types of draw agents. Eventually, guidelines regarding the selection of suitable draw solutes are given.

Main properties of a draw solution

Osmotic pressure

The driving force in forward osmosis processes is provided by the difference in osmotic pressure across the active layer of the membrane, as defined as follows:

$$\Pi = -\frac{RT}{V_m} \cdot \ln(a_w) = -\frac{RT}{V_m} \cdot \ln(\gamma_w \chi_w) \qquad \text{Eq. 1}$$

Where R is the gas constant, T the temperature, V_m the partial molar volume of water, a_w the water activity, γ_w the activity coefficient, and χ_w the mole fraction.

In practice, assumptions are made to simplify osmotic pressures estimation, often also due to unknown activity coefficients. In very dilute solutions the solvent activity coefficient can be assumed to be close to 1, resulting in the validity of the van't Hoff equation, as follows:

$$\Pi = c_m RT \qquad \text{Eq. 2}$$

where c_m is the osmotic concentration (osmolarity). The osmolarity is the molar concentration of osmotic active solutes. For a salt solution, such as NaCl which dissociates into two ions, the osmolarity equals twice the molarity assuming complete dissociation.

Even though the van't Hoff's equation is formally only valid for very dilute solutions, it often provides acceptable accuracy for salt-based draw solutions, especially if direct measurements of the osmolarity (e.g., freezing-point or vapor pressure osmometry) are available. In comparison, prediction of osmotic pressures of organics, polymers, or other draw solutes can lead to significant deviations.

Diffusivity and viscosity

In most technical applications, the draw solution is applied on the support layer side of the membrane leading inevitably to driving force losses due to concentration polarization effects (see section 4.3.2). Besides the properties of the membrane (structural parameter), the extent of ICP is mainly related to the draw solution concentration and diffusivity. The use of a higher draw concentration, hence nominal driving force, does not correlate with a linear increase in the water flux. A higher draw concentration leads to larger relative driving force losses due to stronger ICP related to the convective transport of draw solutes away from the active layer. Since the transport of draw solutes towards the active layer is diffusive only, the diffusivity of the draw agent significantly influences the extent of polarization effects. Diffusivity depends mainly on the solution's viscosity and the diffusion coefficient. Overall, draw solutions of low viscosity and high diffusion coefficient are the best choice when considering FO system productivity.

Regeneration of draw solutions

Continuous FO processes require a reconcentration of the diluted draw solution. Up to now, the regeneration process remains the bottleneck in the draw solution selection. Among the most studied draw regeneration methods are membrane-based processes, such as reverse osmosis (RO), nanofiltration (NF), and membrane distillation, as well as evaporative technologies and electrodialysis. While pressure-driven membrane processes are limited in achievable osmotic pressures due to allowable pressure, the evaporation of water is highly energy intensive. This has led to ideas for the implementation of FO within applications that do not require a draw regeneration. Important to mention are (so-called direct) FO processes using seawater or a concentrated fertilizer as draw solution, which may be discarded or beneficially used once diluted, to avoid costly regeneration. Furthermore, novel types of draw solutions have been designed to overcome existing challenges in draw regeneration. These so-called responsive draw solutions exploit drastic changes in physical and chemical properties of the draw agent provided by external stimuli, such as heat or pH, enabling a practical and efficient draw regeneration. Please note that the energy required for draw agent regeneration is always somewhat related to its target osmotic pressure due to thermodynamics considerations, thus simplicity of regeneration should not be confused with cost of regeneration and the two issues should be considered separately and simultaneously to design a feasible and effective FO system.

Types of draw solutes

In theory, any water-soluble component exhibiting an osmotic pressure can be used as a draw solute. Considering the above-described influence of draw properties on the process performance, small solutes of high osmotic pressures (high solubilities) are preferred. A variety of different draw solutes including salts, small organic molecules (e.g., sugars), volatile organic compounds, nanoparticles, polymers, or hydrogels have been investigated up to now.

Among the most studied and applied draw solutes are inorganic and organic salts offering the advantages of high osmotic pressures, low viscosities, high diffusivity, electrical charge, and low toxicity. As a result, salt-based draw solution can reach a high water flux, exhibit comparable low driving force losses, enable a hazard-free operation and simple regeneration by pressure-driven membrane processes such as reverse osmosis. Most salts are inexpensive, available as food grade quality, and their replenishment costs are low.

Due to the larger variety of salts, salt-based draw solutions can be selected with regard to the specific process requirements. For example, multivalent ions of higher molecular weights can be selected for food and beverage application to minimize the reverse solute flux into the product stream. Furthermore, studies indicate that the rejection of feed compounds can be enhanced by selecting an appropriate draw solute.

Worth mentioning is sodium chloride (NaCl), one of the most studied draw solutes. It is often used as a benchmark draw agent to evaluate membrane and process performance. Many membrane manufacturers and published articles use the specific reverse solute flux (J_s/J_w) of NaCl as key characteristic to account for the membrane's selectivity for water transport.

Besides the comparable high reverse solute flux of salts, their regeneration remains the bottleneck of salt-based draw solutions due to osmotic pressure limitation. Novel approaches such as osmotically-assisted reverse osmosis (OARO) may enable higher draw concentrations in the future (Peters and Hankins, 2020).

To exceed the osmotic pressure limitation of conventional draw solutes, a variety of responsive draw solutes which can switch solubility properties by external stimuli were studied. Among the most studied responsive draw solutions are thermo- and CO_2-responsive draw agents. Thermo-responsive draw solutions exploit temperature-dependent miscibility gaps between water and polymers or ionic liquids. By exceeding the lower critical solution temperature, the diluted draw solution separates into two phases whereof one phase is rich in draw agent and the other is water-rich. CO_2-responsive draw solutions undergo acid base reaction and are often amine-based. Amines can react reversibly with CO_2 and form bicarbonate salts which can be used as draw agents. Upon heating or purging with inert gas, the diluted amine bicarbonate draw solute decomposes into either a water-insoluble liquid or gaseous amines such as trimethylamine or ammonia. Disadvantages of responsive draw solutions are related to the costs, toxicity (amines), or low membrane performance due to severe ICP (polymers).

Selection of draw solutes

The right choice of draw agent depends on the specific application and feed stream, the target recovery, the process configuration, availability and costs of different energy forms (electrical, waste heat, ...), space requirements, commercial assessment, as well as further considerations. The following guidance can assist in selecting an adequate draw agent:

1. Osmotic pressure of the draw solution
   - Too low osmotic pressure of the draw solution induces low membrane performance, prolonging the process time or requiring larger membrane areas. Both may contribute to target compound losses from the feed solution as well as draw solute contamination of the feed solution.
   - Too high osmotic pressures can accelerate membrane fouling and scaling. Regeneration costs can increase due to dissipation of osmotic potential.
   - Rule of thumb: ratio in osmotic pressure between the draw solution and concentrate should not be below 1.1 - 1.2
2. Draw regeneration method:
   - The draw regeneration process is often limiting the choice of draw type and strength
   - Reverse osmosis:
     - RO is the most energy efficient process to regenerate draw solutions
     - Limited by hydraulic pressure (65 bar, high-pressure RO: 120 bar)
     - Osmotically-assisted reverse osmosis is not established but can overcome osmotic pressure limitations
   - Evaporators
     - Energy intense regeneration with no limitations regarding osmotic pressures
     - Corrosive draw solutes (e.g., chlorides) can drastically increase the CAPEX due to material requirements and should be avoided
   - Electrodialysis
     - High CAPEX
     - Limited to ionic draw solutes
     - Energy efficient at low osmotic pressure range
   - Membrane distillation
     - Energy intensive regeneration, but often low grade heat sources can be used
     - Not-yet-established technology due to currently low performance and specific current limitations related to module configurations, fouling/scaling, and long-term stability
   - Responsive draw recovery
     - Still under development, offering the potential to concentrate draw solution to high osmotic pressures
   - Energy intensive, but often enabling the utilization of low-grade heat sources
3. Applications:
   - Food and beverage
     - Only food-approved draw solutes are applicable (sugars, salts)
     - Multivalent ions and agent of higher molecular weight can reduce unwanted reverse solute flux
     - Target compound rejection can be increased by selecting draw solutes which are already present in the feed stream
   - Wastewater concentration
     - Lower concentration of draw agents may be beneficial to reduce fouling propensity
     - Target compound rejection can be increased by selecting draw solutes which are already present in the feed stream

4.3 EXPERIMENTAL METHODS

4.3.1. Typical parameters and phenomena

The most important process parameters are the water flux J_w, the reverse solute flux J_s, the specific reverse solute flux J_s/J_w, the recovery, and the rejection (both forward and reverse).

The **water flux** J_w is defined as the areal permeation rate of water as follows:

$$J_w = \frac{Q_{permeate}}{A_{membrane}} = \frac{Q_{feed} - Q_{concentrate}}{A_{membrane}} = \frac{Q_{draw\ out} - Q_{draw\ in}}{A_{membrane}} \qquad \text{Eq. 3}$$

Where Q is the flow rate and the active membrane area (*A*) is in the denominator. As seen above, J_w can be calculated based on the difference in feed in- and outlet flow rates as well as

based on the difference on the draw side. Its unit is L/m^{2}h. In batch operation, J_w can be determined by measuring the change in feed or draw weight under the assumption that only water permeates the membrane.

The **reverse solute flux** J_s is defined as follows:

$$J_s = \frac{m_{solute}}{A_{membrane}} \qquad \text{Eq. 4}$$

Where the mass flux of draw solute is in the numerator and the active membrane area (*A*) is in the denominator. The reverse solute flux is determined by measuring the draw solute concentration in the concentrate stream. Depending on the feed composition, an appropriate measurement of draw solute concentration, such as conductivity, ICP-OES, or HPLC, can be used. The reverse solute flux is usually given in g/m^{2}h.

The **specific reverse solute flux** J_s/J_w is defined as the ratio between J_s and J_w. It is a measure of the selectivity for water permeation over draw solute transport given in g/L.

The **transmembrane pressure (TMP)** is defined as the average hydraulic pressure between the feed side and the the draw side of the membrane, given as follows:

$$TMP = \frac{(p_{feed} + p_{feed\,out}) - (p_{draw\,in} + p_{draw\,out})}{2} \qquad \text{Eq. 5}$$

The **recovery (Rec)** defines the ratio of the volume of recovered water to the volume of feed solution. In single-pass operation the membrane recovery is defined by using the permeate and feed flow rates as follows:

$$\text{Rec}_{membrane} = \left(\frac{Q_{permeate}}{Q_{feed}}\right) \times 100\% = \left(1 - \frac{Q_{concentrate}}{Q_{feed}}\right) \cdot 100\% = \left(\frac{Q_{draw\ out} - Q_{draw\ in}}{Q_{feed}}\right) \cdot 100\% \qquad \text{Eq. 6}$$

In batch processes where the feed solution is constantly concentrated, the recovery is defined as follows:

$$\mathrm{Rec}(t) = \left(1 - \frac{V_{feed}(t)}{V_{feed}(t_0)}\right) \times 100\% \qquad \text{Eq. 7}$$

Assuming only water to permeate the membrane and a constant density of the feed solution $\rho_{feed(t)}$ allows calculating the recovery by weights instead of volumes.

In food and beverage processes, concentration factors (CF) are often used instead of recovery, where CF is:

$$CF = \frac{1}{1 - \frac{R}{100\%}} \qquad \text{Eq. 8}$$

The average **membrane forward rejection** *R* of a compound *i* (moving forward from the feed to the draw side) is commonly defined using the concentration ratio between permeate and feed. To take the concentration difference between incoming feed and outgoing concentrate stream into consideration, the average concentration on the feed side of the membrane is often used:

$$R_i = \left(1 - \frac{c_{i,permeate}}{\frac{c_{i,concentrate} + c_{i,feed}}{2}}\right) \times 100\% \qquad \text{Eq. 9}$$

In contrast to most other membrane applications, the permeate concentration cannot be directly measured due to its dilution by the draw. Therefore, its average must be calculated based on a mass balance of component *i* on the draw side.

Taking the draw flow rate into consideration leads to:

$$c_{i,permeate} \times Q_{permeate} = c_{i,draw\ out} \times Q_{draw\ out} - c_{i,draw\ in} \times Q_{draw\ in}$$

Inserting equation 1 (J_w) and rearranging leads to:

$$c_{i,permeate} = \frac{c_{i,draw\ out} \times Q_{draw\ out} - c_{i,draw\ in} \times Q_{draw\ in}}{J_w \times A_{membrane}} \qquad \text{Eq. 10}$$

While in (draw) batch operation the ingoing target solute concentration in the draw solution needs to be considered, in single-pass operation $C_{i,draw\ in}$ is in most cases negligible.

The achieved membrane forward rejection depends on the membrane type, operation conditions (e.g., flow rates of draw and feed solutions), the water flux, as well as the recovery, and it is different for different compounds.

It is important to note that the above membrane rejection calculations consider the observed rejection and not the real compound rejection, as it is calculated considering:

1) The total mass that has passed through the membrane during the entire pass and the average J_w (or entire time in batch mode) and not the mass that is passing across the membrane in each location along the module (or at any given time in batch mode);
2) The feed or draw bulk concentration and therefore not considering the higher compound concentrations reached at the active layer interface due to the polarization phenomena. Larger molecules are often better rejected than smaller ones. In addition, uncharged organic molecules show lower rejection than charged molecules due to missing electrostatic repulsion. Even during batch concentration processes the rejection may change significantly as shown in the case of urea (Figure 2).

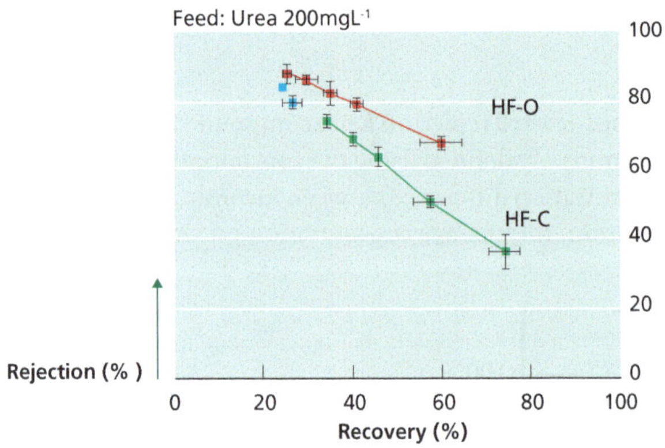

Figure 2 Urea rejection variation with recovery rate for HF–C (chlorinated membranes) and HF–O (non-modified membranes). Adapted from Sanahuja-Embuena *et al.* (2019).

4.3.2 FO process design constraints and considerations

To design a specific FO process and experimental setup, users are strongly advised to refer to the manufacturer's FO module datasheet to understand the operating limits of the given modules. It is also recommended that the user reads any other documentation provided by the FO manufacturer.

Concentration polarization (ECP/ICP)

As seen in Figure 3, concentration polarization occurs on both sides of the membrane due to the permeation of water concentrating the feed solution while diluting the draw solution. A distinction is made between external concentration polarization (ECP) on the active layer side of the membrane and polarization effects in the membrane's support layer referred to as internal concentration polarization (ICP). Depending on the membrane orientation, i.e. FO mode (where the feed solution is in contact with the active layer) or PRO mode (where feed solution is in contact with the membrane support layer), these polarization effects can either be dilutive or concentrative. In most applications, the draw solution is applied on the support layer side leading to dilutive ICP and concentrative ECP.

Concentration polarization reduces the difference in osmotic pressure across the active layer and leads inevitably to driving force losses for water permeation. Besides driving force losses, concentrative ECP increases the risk of membrane fouling and scaling. Lower water fluxes as well as turbulent flow conditions can contribute to reducing these risks.

The intensity of dilutive ICP depends on the porosity, tortuosity, and thickness of the support layer (see structural parameter in section 4.4.1) as well as on the diffusivity of the draw solutes and the water flux. Since draw solutes diffuse against the convective water flux, draw solutions of low viscosity and high diffusion coefficients can mitigate dilutive ICP (see draw solution in section 4.2.3). Additionally, highly porous, and thin support layers can lower the extent of driving force losses.

AL-DS membrane orientation can significantly decrease dilutive CP of the draw solution. Since concentrative CP of feed solutes in the support layer is increased, this membrane configuration might only be beneficial in specific applications, where the feed presents low fouling potential).

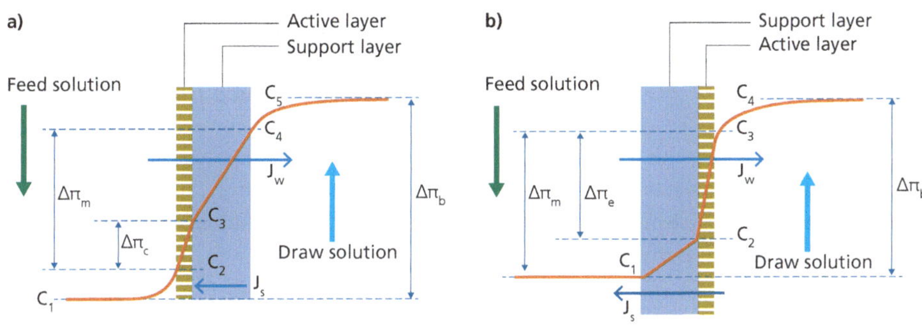

Figure 3 ECP and ICP at a) AL-FS mode, and b) AL-DS mode. Adapted from Wang and Liu (2021).

Pressure limit

Pressure limit is one of many important factors to consider as it affects the choice of flow rate and cross flow velocity sent into each element. This typically already translates into the recommended flow rate range on both feed and draw side. In addition, how much water is transported into the draw solution side is primarily a function of draw solution flow rate and concentration. Even at low draw solution inlet flow, high osmotic pressure difference may result in a large water permeation rate and hence a higher flow rate on the draw side.

System projections are therefore useful to predict the behavior of pressure drop on both feed and draw lines. However, calculating pressure drops in a system can be complicated as this will depend on several factors such as module geometry, array configuration and liquid properties among other factors. Few considerations need thus to be taken when projecting: 1) permissible pressures given by membrane manufacturer should not be exceeded, 2) TMP usually increases during batch concentration (e.g., viscosity increase of feed, feed outflow rate increases due to a lower J_w), or in continuous mode due to fouling, 3) system arrays require special considerations such as accounting for local changes in pressure drops, pressure build-up when more-than-one modules are connected in series, draw solution fed in the system in series or in parallel, etc.

Flow rate limit

Flow rate limit, by extension, is determined by the maximum pressure limit of the module. Flow rate should be selected within manufacturer's recommendation in order not to exceed pressure limit. In addition, users are advised to check for any minimum feed reject flow requirement by manufacturers. In a batch process or semi-batch, feed and bleed process, recovery of feed is time-dependent and not flow-dependent. It is therefore possible to maintain as high cross-flow velocity as possible, while staying within pressure limit, to minimize risks of fouling and scaling. This is especially so when feed streams contain medium to high degree of foulants. In a single-pass continuous process, recovery of the feed is flow-dependent. Designers of the FO process should determine, through projection, whether concentrate flow rate at the last in-series element is below recommendation.

Moreover, draw flow rate in operation should be carefully selected and monitored because it influences transmembrane pressure, permeation flux across FO membrane, and the concentration of draw agent and of possible compounds permeated from the feed side. Usually, for polyamide-based FO membrane, manufacturers may recommend a safe limit of negative TMP, beyond which there poses a risk of delamination of polyamide active layer. Having a high draw flow rate increases the overall permeation. However, an excessively high draw flow rate might raise pressure on the draw solution side and result in a high chance that the negative TMP limit is exceeded.

Flow direction

Flow direction, whether counter-current or co-current, is also a tool available for FO process designers. In co-current operation, feed and draw solutions enter the module through the same end of the module, leading to a constantly reducing driving force along the module length. Counter-current operation enables to maintain a more constant osmotic pressure difference along the lengths of the module (see Figure 4). Additionally, counter-current operation maximizes the average water flux across the FO module or system and the permeate recovery, while minimizing local differences in water flux. This means that the difference between water flux across FO membrane across inlet and outlet of FO system is less for counter-current, as compared to co-current flow direction.

It should however be noted that the selection of the flow mode (i.e., co-current or counter. current) depends on module type. For spiral wound or some plate-and-frame module type, flow path is designed to be in cross flow, where feed and draw solutions are perpendicular to each other. For hollow fiber and tubular membrane type, flow path can be selected to be counter-current or co-current.

Conventionally, a counter-current flow path is the is the preferred option in most applications and experimental setups as it allows maximization of the driving force.

In practice, FO process designers should pay attention to ease of filling up the shell side chamber in counter current mode, assuming that the module is mounted vertically (i.e., feed side flow is upwards and draw side flow is downwards). Modules of larger size which are mounted vertically may however require ingoing streams to enter on the bottom side of the module to remove any trapped air from the module. In such a case, if the draw inlet flow rate is too small, partial filling of shell chamber may occur resulting in underutilized membrane area. In this specific case, operating in co-current operation may be advantageous even though process performance is reduced.

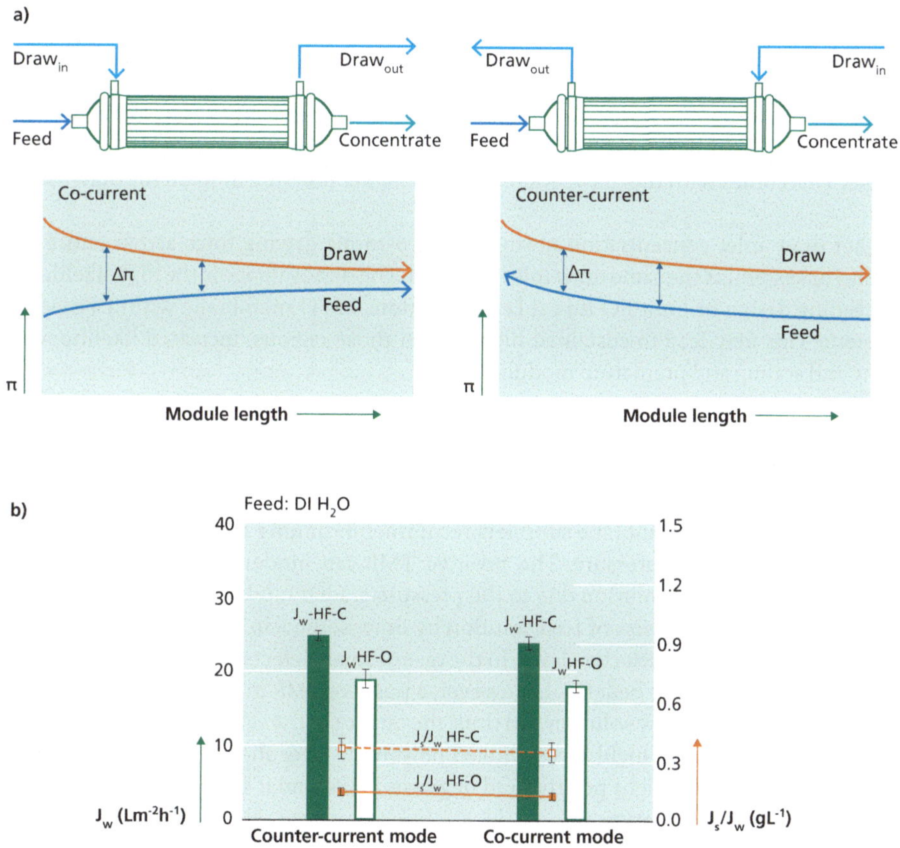

Figure 4 a) Driving force for counter-current and co-current flow direction; b) J_w and J_s/J_w for HF–C (chlorinated membranes) and HF–O (non-modified membranes) in co-current and counter-current when DI water was used as FS. Operating conditions were: Feed flow rate was 100 $L.h^{-1}$, draw flow rate was 25 $L.h^{-1}$, draw concentration was 1 M NaCl and TMP was 0.2 bar. (n = 2). Adapted from: Sanahuja-Embuena *et al.* (2019).

Limiting flux

Lastly, water permeation limit or design flux limit is a major factor affecting the FO design. On the one hand, this is related to feasible feed inlet flow rate for FO module and system. A lower FO feed inlet flow rate limit by manufacturers' recommendation or by system design indicates lower maximum design flux limit. However, this is in fact generally related to fouling potential and reversibility of fouling.

While there is no consensus on what design flux limit for FO membrane should be, there are research reports indicating limiting flux to be between 10-20 LMH and designed flux for reversible fouling to be 5-15 LMH. Here, limiting flux is defined to be the starting flux value at which there is decline of flux over time at constant osmotic driving force difference. Design flux is defined to be the starting flux at which there is minimal flux decline and there is flux restoration upon cleaning if there is any flux decline over time. These flux values vary depending on membrane type and feed quality or foulants present in feed.

Water permeation limits may be controlled by draw inlet flow rate and draw inlet concentration. As mentioned above, a higher draw inlet flow rate means less dilution effect on the draw side, allowing osmotic driving force to be sustained from inlet to outlet of module. This comes with the drawback of having a higher pressure drop on the draw side.

A higher draw inlet concentration means higher osmotic driving force across entire FO modules or system, at the same draw inlet flow rate. The disadvantage is the high likelihood of exceeding design or limit FO flux at certain sections of FO membrane within a module or system. This may lead to sustained high ECP in those regions, increased likelihood of fouling and scaling and premature module failure.

4.3.3 Best practices

Transmembrane Pressure (TMP)

Most forward osmosis membrane suppliers recommend running FO processes under low positive transmembrane pressure. The positive TMP can hinder the transport of draw solutes towards the feed solution due to the pressure gradient, which helps in preventing the immediate contamination of feed solution by draw solutes in the event of membrane breakage or defects on the selective layer. In the case of small defects on the polyamide layer, a positive TMP will also be beneficial. However, a positive TMP may also aid the transport of feed solutes into the draw solution and thus, the quality of the selective layer would need to be checked. A tight and highly cross-linked polyamide layer should not be significantly or drastically affected by slight positive and negative TMPs, and if this happens, it may be a sign of membrane deterioration.

Nevertheless, a negative TMP should be strictly avoided, even for brief periods of time, due to the polyamide layer configuration (where the layer is on the lumen side of the membrane). When a negative TMP is applied, the pressure gradient direction can cause the delamination of the polyamide layer, and consequently, the breakage of the membrane.

During the FO module operation, pressure losses from inlet to outlet for both feed and draw side are expected, regardless of the flow mode selected (i.e., counter-current or co-current), which could provoke negative TMP at the feed outlet or draw inlet locations. It is therefore of paramount importance to maintain a positive TMP at the feed outlet ensuring following the manufacturing guidelines on pressure limits.

In summary, since FO is a virtually pressure-less membrane process, membranes are not designed for high hydraulic pressures on either side of the membrane. Therefore, commonly recommended TMPs are around 0.2 bar. Allowable pressures given by the membrane manufacturers should not be exceeded to ensure safe operation. Here, pressure relief valves in the experimental setup can protect the membrane from maloperation.

Avoiding 'over-recovery'

High recoveries of feed solution can lead to the precipitation and deposition of feed particles on the membrane (fouling and scaling). While membrane fouling is characterized by the deposition of (mainly organic) suspended solids, scaling refers to the precipitation and crystallization due to exceeding salt solubilities. In process configuration consisting of serial connected FO modules, fouling will occur in the first stages while scaling usually occurs in the consequent stages.

Although FO is generally considered a low fouling propensity membrane technology which can handle more difficult-to-treat feed stream, fouling and scaling will ultimately reduce the membrane performance. Indications are a reduced water flux, increased pressure drops on the feed side of the membrane, as well as reduced rejections. Besides an appropriate pre-treatment of the feed solution to remove suspended solids, frequent cleaning-in-place (CIP) can mitigate the deposition of solids on the membrane surface and performance detriment. Scaling should be prevented by estimating the scaling risk of a certain feed composition by using the Scaling Index and avoiding working at water recoveries that could provoke severe compound precipitation.

4.4 DATA ANALYSIS: BASIC FO PROCESS DESIGN

4.4.1 FO Fundamental Equations

In a typical FO process, the equation for water flux flowing from feed side to draw side is given by:

$$J_w = A\left[\left(P_F - P_D\right) = \left(\pi_D^m - \pi_F^m\right)\right] \qquad \text{Eq. 11}$$

Where J_w is water permeation flux, A is the water permeability, P_F and P_D is hydraulic pressure of feed side and draw side respectively, and π is osmotic pressure of draw side and feed side at membrane surface, respectively. In FO operation, hydraulic pressure difference tends to be zero or close to zero.

The salt flux equation is given by:

$$J_S = B\left(C_D^m - C_F^m\right) = B\Delta C_m \qquad \text{Eq. 12}$$

where J_s is sat flux from draw to feed, B is the salt permeability, and C is solute concentration in draw and feed solution at membrane surface, respectively.

The salt transport across the FO membrane is also described by the convection-diffusion model with a diffusive term proportionally related to solute concentration gradient and a convective term related to water permeate flux across the membrane in the opposite direction.

$$J_s = D\frac{dC(x)}{dx} - J_w C(x) \qquad \text{Eq. 13}$$

Where D is the solute diffusion coefficient. The solution of the transport equations above differ depending on the orientation of the membrane.

In active layer facing feed side (AL-FS) mode or FO mode, water permeates from feed side into the support layer on the draw side, leading to dilutive internal and external concentration polarization (i.e., ICP and ECP, respectively). On the feed side, the convective water flux carries solutes from bulk feed solution to membrane surface, at which they are rejected and accumulate, causing concentrative ECP. The solution of the convective-diffusive equation above, for AL-FS mode, become:

$$\Delta C_{m,ALDS} = \frac{C_D^b \exp\left[-\frac{J_w}{k_D}\right] - C_F^b \exp\left[J_w\left(\frac{1}{k_F} + \frac{S}{D_F}\right)\right]}{1 + \frac{B}{J_w}\left\{\exp\left[J_w\left(\frac{1}{k_F} + \frac{S}{D_F}\right)\right] - \exp\left[-\frac{J_w}{k_D}\right]\right\}} \qquad \text{Eq. 14}$$

For active layer facing draw side (AL-DS) mode or PRO mode, water permeates from feed side with solutes that are rejected and accumulate across the support layer, resulting in concentrative ICP and ECP on the feed side. On the draw side, there is dilutive ECP as pure water permeates into the draw side. The solution of the convective-diffusive equation above, for AL-FS mode, becomes:

$$\Delta C_{m,ALDS} = \frac{C_D^b \exp\left[-\frac{J_w}{k_D}\right] - C_F^b \exp\left[J_w\left(\frac{1}{k_F} + \frac{S}{D_F}\right)\right]}{1 + \frac{B}{J_w}\left\{\exp\left[J_w\left(\frac{1}{k_F} + \frac{S}{D_F}\right)\right] - \exp\left[-\frac{J_w}{k_D}\right]\right\}} \qquad \text{Eq. 15}$$

Where k is mass transfer coefficient, and the term '$exp(-J_w/k_D)$' indicates external concentration polarization in general whereas the term $exp[J_w(1/k_F + S/D_f)]$ denotes internal concentration polarization with S being structural parameter of membrane, consisting of porosity and tortuosity term used in modifying solute diffusion coefficient from the bulk solution to the inside support layer.

The mass transfer coefficient k value is dependent on the type of membrane form factor and module. In general, mass transfer coefficient is:

$$k = \frac{Sh \times D}{d_h} \qquad \text{Eq. 16}$$

where Sh is Sherwood number and d_h is hydraulic diameter, both being geometry-dependent.

Combining above stated equations, one is able to calculate the expected water permeation, J_w, and reverse solute flux, J_s, of a FO membrane, given its bulk feed and draw solution characteristics and some basic hydrodynamic information to obtain mass transfer coefficients.

4.4.2 FO Module Mass Balance

To simulate transport inside a membrane module, mass balance equations should be considered. In addition, the effect of volume change due to dissolved solute should also be taken into account. This means the differential term of density and concentration of solute cannot be neglected.

Typically, mass balance equations for pressure, velocity and concentration along module length can be established. For instance, the velocity and concentration differential equation on the feed side can be seen below for a rectangular flat plate channel type.

$$\frac{dv^F}{dx} = \frac{\left[-J_w \times \rho^w + J_s - \frac{d\rho^F}{dc^F} \times J_s\right] \times \frac{1}{H}}{\rho^{F,bulk} - \frac{d\rho^F}{dc^F} \times c^F} \qquad \text{Eq. 17}$$

$$\frac{dc^F}{dx} = \frac{-c^F \times \frac{dv^F}{dx} + \frac{1}{H} \times J_s}{v^F} \qquad \text{Eq. 18}$$

Where ρ^W is density of pure water, v^F is the differential term to account for volume change with solute concentration and H is the height of flat plate flow channel. In other geometries, such as for hollow fiber or tubular types, these terms are referred to inner diameter of hollow fiber (this assumes an inside out FO module with active layer being on the lumen side).

Similarly, the velocity and concentration differential equation on the draw side for a rectangular flat plate channel type can be seen below.

$$\frac{dv^D}{dx} = \frac{\left[J_w \times \rho^w - J_s + \frac{d\rho^D}{dc^D} \times J_s\right] \times \frac{1}{H}}{\rho^D - \frac{d\rho^D}{dc^D} \times c^D} \qquad \text{Eq. 19}$$

$$\frac{dc^D}{dx} = \frac{-c^D \times \frac{dv^D}{dx} + \frac{1}{H} \times J_s}{v^D} \qquad \text{Eq. 20}$$

For a hollow fiber bundle, the *H* hydraulic radius term becomes the following for lumen and shell respectively:

$$H_{lumen} = 4/d_i \qquad \text{Eq. 21}$$

$$H_{shell} = (4 \times n \times d_i)/(n \times d_o^2 - D_i^2) \qquad \text{Eq. 22}$$

where d_i is fiber inner diameter, d_o is fiber outer diameter, D_i is shell housing inner diameter and n is the total number of fibers.

Similarly, the velocity and concentration differential equation on the draw side for a rectangular flat plate channel type can be seen below.

$$\frac{dv^D}{dx} = \frac{\left[J_w \times \rho^w - J_s + \frac{d\rho^D}{dc^D} \times J_s\right] \times \frac{1}{H}}{\rho^D - \frac{d\rho^D}{dc^D} \times c^D} \qquad \text{Eq. 23}$$

$$\frac{dc^D}{dx} = \frac{-c^D \times \frac{dv^D}{dx} + \frac{1}{H} \times J_s}{v^D} \qquad \text{Eq. 24}$$

It should be noted that the sign of the velocity differential equation is reversed in the event of counter current flow.

The pressure drop equation across the module strongly depends on the type of module used. As an example, for the hollow fiber form factor, the analogy of flow through a packed bed with the Ergun equation could be used to model pressure drop across the tube bundle on the shell side.

$$\frac{dP^D}{dx} = \theta \times \alpha \times \left[\frac{150 \cdot (1-\varepsilon)^2 \times \mu^D \times v^D}{\varepsilon^3 \times d_o^2} + \frac{1.75 \times (1-\varepsilon) \times \rho^D \times v^{D^2}}{\varepsilon^2 \times d_o}\right] \qquad \text{Eq. 25}$$

where θ is empirical pressure drop correction factor and α is flow direction (1 for counter current, and -1 for co-current). ε is packing density of hollow fiber bundle, μ is fluid dynamic viscosity, ρ is fluid density, and υ is fluid velocity. *D* denotes draw solution side, which typically flows on the shell side of a hollow fiber module.

Meanwhile, the Hagen-Poiseuille model for pressure drop across cylindrical tube is used for the lumen side pressure drop:

$$\frac{dP^F}{dx} = \frac{32 \times \mu^F \times v^F}{d_i^2} \qquad \text{Eq. 26}$$

where *P* is pressure, *F* denotes the feed solution side, which typically flows on the lumen side of a hollow fiber bundle, μ is fluid dynamic viscosity, υ is fluid velocity and d_i is the fiber inner diameter.

4.4.3 FO Design Considerations

For the batch or feed-and-bleed type of system, feed solution is re-circulated and water extraction happens over time. Sensors may be installed to automate the feed-and-bleed or cycle shutdown operation. In such a system, because the feed side is being concentrated, water flux will start high and decrease over the course of a cycle, assuming that draw inlet flow rate and concentration are constant. Care should be taken to ensure that initial flux is below design flux limit for the given process, and final flux is non-zero so that cycle time is still productive and reverse salt flux into feed batch is minimized.

For a single pass process where the FO feed outlet is expected to reach a desired concentration factor, the number of modules and their array should be designed to achieve recovery outcome, while balancing all design constraints above.

For instance, FO modules may be arranged in parallel to sub-divide flow to be within recommended flow rate. FO modules, and hence membrane area, may be added in series to achieve recovery in single pass, while design flux limit is obeyed. For the same flow rate extraction requirement, added area means lower operating flux, ensuring that it is within design limit. The maximum number of modules in series is dictated by pressure drop across the system, while the maximum number of lines of modules in parallel is dictated by the minimum FO outlet flow rate for each line.

One way to circumvent minimum FO outlet flow rate being below limit is to implement multi-stage design. That means, the flow rate of multiple lines of FO modules of the so-called first stage are combined and redistributed over a smaller number of lines in the second stage. This allows more flow per module when recovery is at the highest point and by design, above module limit by manufacturers' recommendation.

Lastly, it should be noted that process limits should be considered for both flushing or cleaning process as well as FO process. In the former, cross flow velocity on the feed side is highest, and on the later cross flow velocity on the draw side is highest. It should be ensured that design considerations are met for both operation types for successful commissioning of a FO system.

Other design considerations

Beside technical considerations, there are other parameters that FO process designers should pay attention to as it influences the operating cost of such a system. Most directly, increasing the number of FO modules used will increase the cost of membrane replacement and initial capital investment on the system. This will also increase the hold-up volume and volume of flushing water or chemicals required for the cleaning process, even though this tends to take a small fraction of overall operating cost.

If operating flux is still within design flux limit, increasing draw solute concentration results in less membrane required and reduces membrane initial investment. However, this would result in increased reverse salt flux from draw to feed side, increased salt passage into draw regeneration permeate stream and increased energy cost of draw regeneration step.

4.5 APPLICATION EXAMPLES

Textile industry application of FO for lowering water footprint

Textile production is estimated to be responsible for about 20% of global clean water pollution from dyeing and finishing products (Morlet *et al.*, 2017). Given the increasing need for the textile industry to lower the environmental impact it is necessary not only to design appropriate wastewater treatment technologies but also to enable reuse and recycling of water. Here FO can be used for water reclamation using concentrated dyeing salt solutions as draw solutions where the diluted dyeing salt solutions can be used in the dyeing baths directly (see Figure 5).

Following the study of Sheldon *et al.*, (2019) it was evaluated the potential of dye solutions as a novel draw solution by screening, assessing and identifying suitable reactive dyes, e.g., Reactive Black 5 and Basic Blue 41 GRL dyeing solutions were investigated as draw solutions in FO with a dye-to-salt 1:10 mass ratio, see Figure 1. Synthetic seawater (SSW) and two types of textile wastewater (TWW1 and TWW2) were evaluated as feed solutions for water reclamation. Reactive Black 5 and Basic Blue 41 GRL were diluted 10 and 5 times respectively.

With Reactive Black 5 as draw solution and SSW as feed solution a water recovery of 75% was achieved. Using TWW1 and TWW2 as draw solutions, water recovery was around 30%. Using Basic Blue 41 GRL with SSW, TWW1, and TWW2 as feed solutions, water recoveries of 50%, 20% and 20%, respectively, were achieved. The average reverse solute fluxes were between 0.06 and 0.34 g/m^{2}h. Results indicated the potential of FO in the textile industry leading to substantial water savings.

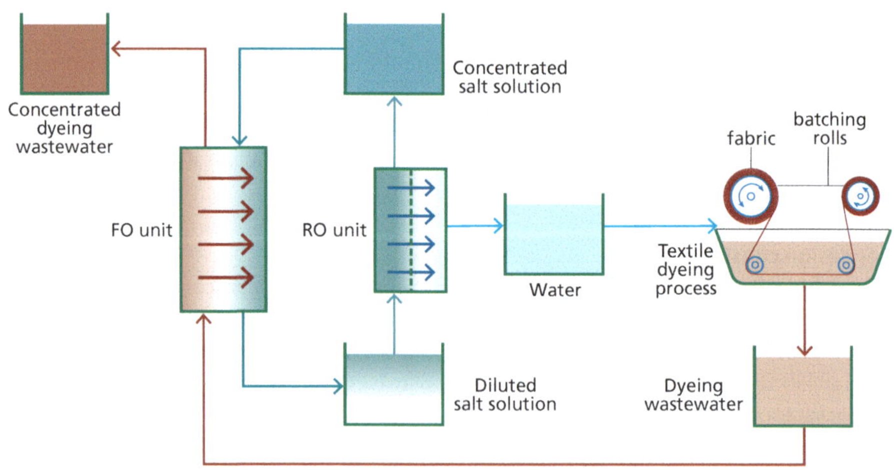

Figure 5 Implementing forward osmosis (FO) into the textile wastewater treatment process can provide high value to an industry segment which is a large consumer of fresh water and one of the biggest polluters. The scheme shows the FO process integrated in a textile wastewater treatment plant using inorganic salt as a draw solution. For using the salt solution as a draw solution there is an integrated reverse osmosis unit for the reuse of the diluted salt.

Concentrating distillery wastewater for subsequent antioxidant retrieval

Alcohol distillation from sugarcane molasses constitutes an important industry in several countries. Molasses-based distillation is a water intensive method with a freshwater consumption in the range of 9-21 L per alcohol and concomitant wastewater production of 7-15 L per L alcohol (Gol, 2014). The resulting wastewater has a high organic load, low pH, and high total dissolved solids. About 2% (w/v) of the wastewater is melanoidins, a product of Maillard reaction obtained from reducing sugars and amino acids during distillation. From a classical wastewater treatment point of view this makes this particular stream problematic as melanoids are not readily biodegradable. However, melaniodins have antioxidant properties which could be a valuable sub-product. The high organic load and the high total dissolved solids makes separation based on classical filtration challenging but due to the inherently low fouling potential FO has attracted attention as a method for up-concentration of this potential antioxidant source.

Singh *et al.* (2018) studied the concentration of distillery wastewater by FO with magnesium chloride hexahydrate ($MgCl_2.6H_2O$) as draw solution. They used a 10% v/v melanoidins model feed solution to optimize the operational parameters. Subsequently they achieved 85-90% melanoidins rejections with as-received distillery wastewater and 3M $MgCl_2.6H_2O$ as draw solution. The water flux was 2.8 L $m^{-2}h^{-1}$ with water recovery over 24 h was around 70% which is significantly higher than reported for RO (35-45%). However, further investigations on membrane fouling and draw solution recovery are required to establish the superiority of FO over RO for the concentration of this type of wastewater.

Concentrating electroplating wastewater

Chromium plating and chromate processes are widespread technologies for electroplating of pristine or nickel-coated plastics as chromium and chromate endow surfaces with special properties such as hardness and corrosion resistance (Korzenowski *et al*, 2018; Sorme *et al.*, 2002). In this process, large quantities of wastewaters, residues, and sludge is generated which can be categorized as problematic waste requiring extensive waste treatment (Sorme *et al.* 2002).

In the study of Bratovcic *et al.* (2022) FO was investigated for concentration of hexavalent chromium (Cr(VI)) in electroplating wastewater from processing plastics to enable the reuse of recovered Cr(VI) in the plating baths, see Figure 6. The feed solution was chromium galvanic wastewater, while the draw solution was an underground brine (close to the factory location) with osmotic pressures of 28 and 226.8 bar, respectively.

Baseline and FO filtrations were performed using Aquaporin Inside(R) membrane hollow fibre FO (AIM™ HFFO) modules with a sequence of baseline, filtration (1.5h) and cleaning (30 min with DI water) steps. During the initial filtration (F1), the water flux decreased on average from an initial value of 28.7 LMH at 46.7 % water recovery to 18.5 LMH. For the second filtration (F2) the water flux decreased from 20.1 LMH at 28.4 % water recovery to 16.8 LMH. The corresponding feed solution (wastewater) volume reduction factors were 1.9 and 1.4 with a concomitant Cr(VI) concentration factor of 1.6 and 1.3 for F1 and F2, respectively. After 1.5 h of filtration, the Cr(VI) rejection was 99.7 % and 95.8 % for F1 and F2, respectively. As the AIM™ HFFO membrane is negatively charged electrostatic repulsion between the membrane surface and the negative ions ($HCrO_4^-$ and $Cr_2O_7^{2-}$) will contribute to the rejection of Cr(VI). The appearance of Cr(VI) in the draw solution indicated a loss of membrane integrity which was ascribed to chemical degradation of the membrane due to oxidation from Cr(VI). Local guidelines for standard chromium discharge from industrial wastewater into the environment is 0.5–1 mg L^{-1}. Since the diluted brine draw solution contained 0.07 gL-1 and 0.65 gL^{-1} of Cr(VI) for F1 and F2 respectively, it cannot be directly discharged into the salt groundwater resource.

In conclusion, brine-driven FO could concentrate chromium galvanic wastewater taking advantage of the high chemical potential gradient provided by the high salinity brine, but the membrane material must be adapted to withstand harsh environments.

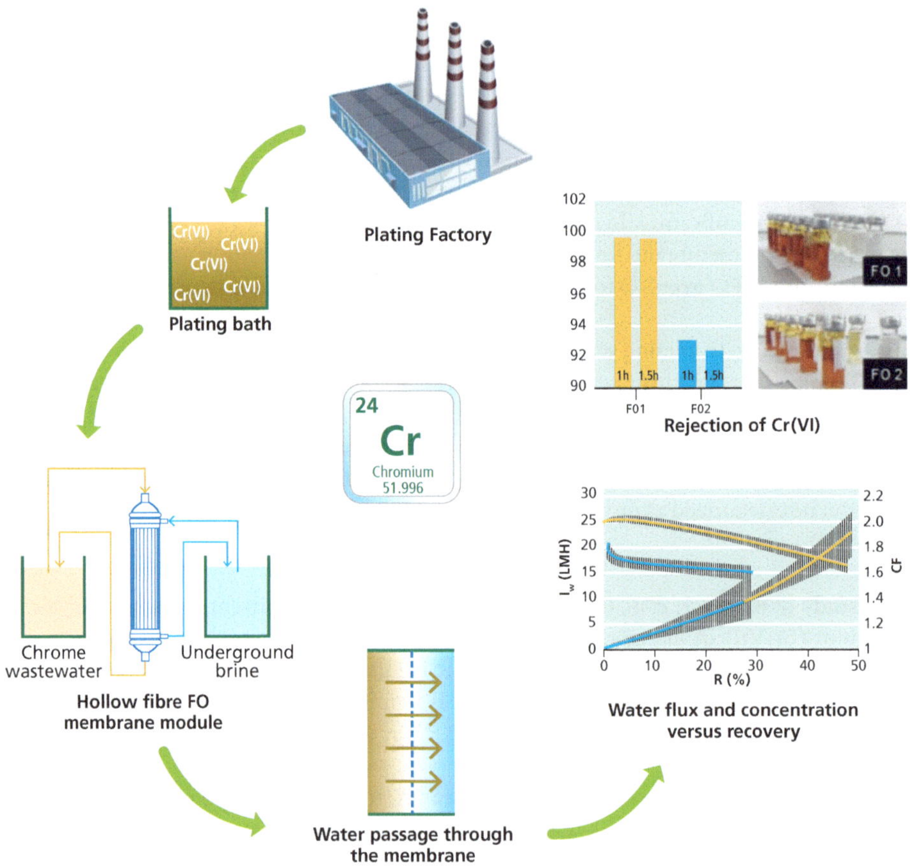

Figure 6 FO tests using Aquaporin Inside membrane hollow fibre FO (AIM™ HFFO) modules for concentration of hexavalent chromium (Cr(VI)) in electroplating wastewater from processing plastics to enable the reuse of recovered Cr(VI) in the plating baths. Chromium galvanic wastewater was used as feed solution while the draw solution was underground brine close to the factory location. The results show that FO can be used in this type of application, but the membrane material must be adapted to withstand harsh environments (Bratovcic *et al*, 2022)

4.6 OUTLOOK

FO is a relatively new technology which presents numerous advantages, especially when a direct FO system can be implemented (i.e., draw solution is available and regeneration is not needed) or when the resulting feed concentrate can bring an added value to the final product. However, FO presents the drawbacks of a developing technology. These are mainly the scarce availability of FO membrane manufacturers, the development of materials which ensure a high water flux, high compound rejection, withstand harsh environments, high selectivity to water and a reduced concentration polarization. The unique system design characteristics required by the FO technology (i.e., draw solution regeneration and membrane configurations) also involve an additional level of system complexity. The availability of non-expensive draw solutions with the desired characteristics and the suitability of these in those applications that require high safety levels, such as in food and pharma industries, are also challenging. However, overall, FO technology can still bring unique advantages in niche applications, although more research in membrane materials and processing are needed to fully understand its capabilities in industry.

4.7 REFERENCES

Baker, R. W. (2004). Membrane Technology and Applications. John Wiley & Sons, Ltd. DOI:10.1002/0470020393

Bratovcic, A., Buksek, H., Helix-Nielsen, C., Petrinic, I., Concentrating hexavalent chromium electroplating wastewater for recovery and reuse by forward osmosis using underground brine as draw solution Chemical Engineering Journal, 431, #133918, 2022.

Gol, 2014. Report by Principal Scientific Advisor to Government of India. Electronic source. http://psa.gov.in/publications-reports/opportunities-green-chemistryinitiatives-molasses-based-distilleries-2014

Im Sung-Ju, Lee H., Jang A. (2021). Effects of co-existence of organic matter and microplastics on the rejection of PFCs by forward osmosis membrane. Environmental Research, 194, 110597

Jørgensen, M. K., Keiding K., Christensen, M.L. (2014). On the reversibility of cake buildup and compression in a membrane bioreactor. Journal of Membrane Science, 455, 152-161

Korzenowski C., Rodrigues M.A.S., Bresciani L., Bernardes A.M., Ferreira J.Z., Purification of spent chromium bath by membrane electrolysis, J. Hazard. Mater. 152 (3) (2008) 960–967.

Morlet, A., Opsomer, R., Herrmann, S., Balmond, L., Gillet, C., and Fuchs, L. (2017). A new textiles economy: Redesigning fashion's future. Ellen MacArthur Foundation.

Mulder, M. (1996). Basic Principles of Membrane Technology. Springer. DOI: 10.1007/978-94-009-1766-8

Peters, Christian D. and Hankins, Nicholas P. (2020). The synergy between osmotically assisted reverse osmosis (OARO) and the use of thermo-responsive draw solutions for energy efficient, zero-liquid discharge desalination, Desalination, Volume 493, 114630

Sanahuja-Embuena, V.; Khensir, G.; Yusuf, M.; Andersen, M.F.; Nguyen, X.T.; Trzaskus, K.; Pinelo, M.; Helix-Nielsen, C. Role of Operating Conditions in a Pilot Scale Investigation of Hollow Fiber Forward Osmosis Membrane Modules. Membranes 2019, 9, 66. https://doi.org/10.3390/membranes9060066

Sheldon *et al.* Water Sci Technol (2019) 80 (6): 1053–1062

Singh *et al.* Water Research 130 (2018) 271-280

Sorme L. and Lagerkvist R., Sources of heavy metals in urban wastewater in Stockholm, Sci. Total Environ. 298 (1-3) (2002) 131–145.

Suwaileh W., Pathak, N., Shon H., Hilal N., Forward osmosis membranes and processes: A comprehensive review of research trends and future outlook, Desalination, Volume 485, 2020, 114455, ISSN 0011-9164, https://doi.org/10.1016/j.desal.2020.114455.

Wang J, Liu X (2021) Forward osmosis technology for water treatment: Recent advances and future perspectives. Journal of Cleaner Production 280: 124354 DOI https://doi.org/10.1016/j.jclepro.2020.124354

Xiao, T., Nghiem, L. D., Song, J., Bao, R., Li, X. & He, T. (2017). Phenol rejection by cellulose triacetate and thin film composite forward osmosis membranes. Separation and Purification Technology, 186 45-54.

doi: 10.2166/9781789062977_0097

Chapter 5

Membrane Distillation

Mohammad Mahdi A. Shirazi, Aalborg University, Denmark

Cejna Anna Quist-Jensen, Aalborg University, Denmark

Aamer Ali, Aalborg University, Denmark

The learning objectives of this chapter are the following:

- Introduction to membrane distillation
- An overview of the membrane materials and setup used in MD
- Main techniques to characterize the membranes for MD
- Present and discuss the main applications of MD
- Provide an overview of the outlook of the process.

5.1 INTRODUCTION

Membrane distillation (MD) is a non-isothermal membrane process (Lawson and Lloyd, 1997a). In MD, a hydrophobic membrane with porous structure separates the feed and permeate channels. The feed channel contains a heated solution with an elevated temperature, while the permeate channel contains a cooling solution with lower temperature (Curcio and Drioli, 2005). This temperature difference can provide a vapor pressure difference between the feed and permeate channels. As the membrane is hydrophobic, the liquid feed does not penetrate the pores (El-Bourawi *et al.*, 2006). However, due to the vapor pressure difference between the hot side and the cold side (i.e., the driving force), the vapor molecules can transfer across the membrane pores (Wang & Chung, 2015). Figure 1 illustrates a general scheme of the MD process.

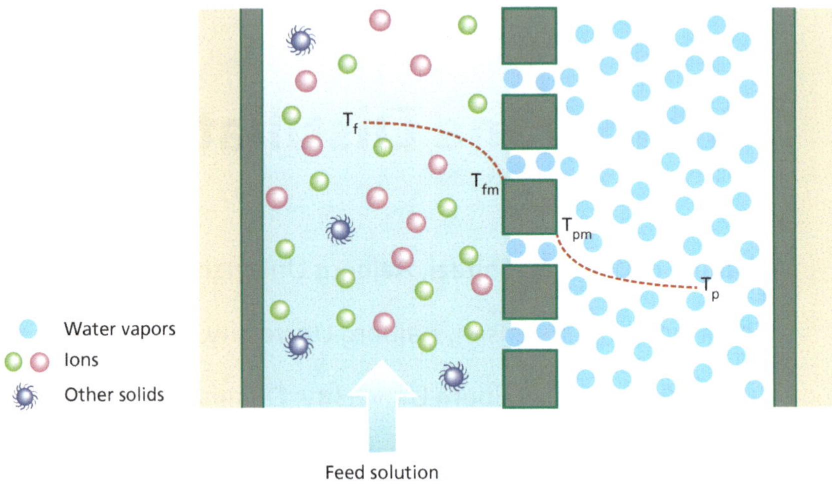

Figure 1 A general scheme of the MD process (T_f: bulk temperature in the feed channel, T_{fm}: temperature on the membrane surface in the feed channel, T_{Pm}: temperature on the membrane surface in the permeate channel, and T_p: bulk temperature in the permeate channel).

MD has numerous attractive advantages for water treatment applications. As only the vapor molecules can pass through the membrane pores a complete solute rejection (i.e., ~100%, theoretically) can be achieved by the MD process. MD uses membranes with pores in the range of 0.1-0.5 μm, which is much larger than the pore size range in the pressure-driven membranes, such as reverse osmosis (RO) and nanofiltration (NF) membranes. Moreover, the operating pressure is much lower in MD compared with RO/NF. These can make MD more cost-effective, less demanding on the membrane properties, and less sensitive to fouling/scaling, as well (Shirazi *et al.*, 2016; Tibi *et al.*, 2020). Furthermore, the low operating pressure allows MD to utilize less expensive materials with better anti-corrosive properties, such as plastics, for module design and fabrication (Hussain *et al.*, 2022). In MD, the process liquid in the feed stream is not necessarily heated up to reach the boiling point and the operating temperature ranges between 40-80 oC. This can reduce the required energy for MD in comparison with the conventional distillation processes. Moreover, working with low temperatures enables MD to utilize low-grade and renewable energy sources, such as solar energy (Ahmed *et al.*, 2020; Drioli *et al.*, 2015b). All these can make MD an attractive alternative for the desalination of seawater and brackish water, however, in lower production rates in comparison with RO. Thus, MD may not still be competitive with RO for freshwater production via seawater desalination for large scales but perform effectively for small-scale desalination. Moreover, MD could find a way as an efficient technology for treating challenging wastewater streams, such as RO brine, textile dyeing wastewater, radioactive contaminated wastewater, etc., where other separation processes cannot perform efficiently (Shirazi & Dumée, 2022). Table 1 highlights the advantages and challenges associated with MD for various applications.

Table 1 Advantages and challenges associated with MD

Advantages	Challenges
Low operating temperature (40-80 °C)	Lack of specific membranes
High non-volatile solute rejection (~100%)	High risk of pore wetting
Very low operating pressure (<0.2 bar)	Heat loss
Less sensitivity to the feed concentration	Temperature polarization
Modular and scalable	Complexity in terms of process optimization
Integrate-able with other separation techniques	Lack of experience in large scale
Competitive production cost with other processes	
Possible to utilize renewable and low-grade energy source	
Less carbon footprint	
Able to treat challenging wastewater streams	

The very first MD experiments were conducted using commercially available microfiltration (MF) membranes, which were made of hydrophobic polymers, such as polytetrafluorethylene (PTFE), polypropylene (PP), and polyvinyl difluoride (PVDF) (Curcio & Drioli, 2005; Shirazi *et al.*, 2013a; Shirazi *et al.*, 2012). However, they suffer from some drawbacks, such as pore wetting, delamination of the active layer, etc., as they were not specifically fabricated for MD (Wang & Chung, 2015). An extensive review of commercial membranes for MD can be found in the literature (Khayet, 2011). The next generation of MD membranes have been fabricated using commercial and synthesized polymers, such as PVDF. The MD membranes can be fabricated using various techniques. The most investigated one is the phase inversion technique (Eykens *et al.*, 2017). However, the membranes by this technique may suffer from some drawbacks, such as limited porosity, and dead-ended pores with twisted structure (i.e., high tortuosity) (Qasim *et al.*, 2021b). Thus, new research directions have been focused on new fabrication techniques, such as electrospinning and 3D printing (Tijing *et al.*, 2014; Tijing *et al.*, 2020). Moreover, extensive research was also carried out on MD modelling and optimization (Ali *et al.*, 2015a; Ali *et al.*, 2016a; Ali *et al.*, 2018; Hitsov *et al.*, 2015; Jantaporn *et al.*, 2017; Olatunji & Camacho, 2018).

The new generation of membrane materials and fabrication techniques, and process development and optimization have enabled MD to extend their applications in new directions for more efficient and greener desalination towards zero-liquid discharge, treating challenging wastewater streams (e.g., wastewater from biological and pharmaceutical processes, metal finishing, and electronic industry), removal of specific gas streams (e.g., H2S) from process water, concentrating fruit juices in the food industry; and recovery of minerals and value-added chemicals from wastewater streams (Tibi *et al.*, 2020; Hussain *et al.*, 2022; Sanaeepur *et al.*, 2022; Julian *et al.*, 2022; Gontarek-Castro *et al.*, 2022; Afsari *et al.*, 2021). Despite of all developments in this separation technique, MD still needs further development for specific membranes with durable properties, high pore wetting resistance, and long-term performance, new module design for higher energy efficiency and lower polarization effect as well as experiencing new and challenging feed samples for treatment.

5.2 MATERIALS, EXPERIMENTAL SET-UP

5.2.1 MD membranes

5.2.1.1 Membrane properties

As MD is a non-isothermal membrane separation and only the water vapor molecules must pass through the pores, the applied membrane should possess some essential requirements (Alklaibi and Lior, 2005).

The most important characteristic of an MD membrane is surface hydrophobicity. The membrane structure can possess a single layer or even multi-layers. However, the layer which is in direct contact with the feed stream must be hydrophobic to repel the liquid and only pass the vapor molecules (Shirazi *et al.*, 2014).

To ensure a sustainable performance without pore wetting, the MD membrane should have high liquid entry pressure (LEP). LEP defines as the minimum required pressure that allows the feed liquid to enter the pores (He *et al.*, 2011). To ensure a proper LEP value, further to high surface hydrophobicity, the membrane pore size should be as small as possible. The typical pore size for MD membranes has been reported in the range of 0.1-0.5 μm (Khayet, 2011). Moreover, the pore size distribution and the maximum pore size should be as narrow as possible and as small as possible, respectively, to provide a high LEP value (McGaughey *et al.*, 2020a).

Porosity is another important parameter for MD membranes. It defines as the void volume fraction which is openly available for transferring the vapor molecules (Gryta, 2007). The membrane porosity is proportional to its permeability. Thus, the more porous the membrane, the higher the permeate flux, regardless of the MD configuration. It is worth noting that the fabrication technique can considerably affect the membrane porosity (Ravi *et al.*, 2020). For example, nanofiber membranes possess higher porosity (≥85%) in comparison with phase-inverted membranes (40-80%) (Tijing *et al.*, 2014).

The pore structure of MD membrane is often assumed to be a straight cylinder for modelling purposes, while it is not true in real life. The deviation of the pore structure from the standard cylindrical shape is defined by the tortuosity factor. Unlike porosity, tortuosity is inversely proportional to the permeability of the MD membrane. Therefore, the lower the tortuosity factor, the higher the permeate flux (Tai *et al.*, 2019; Kim, 2021).

In MD, both mass and heat transfer through the membrane happen simultaneously (Qtaishat *et al.*, 2008). Although a high mass transfer (higher permeate flux) is favorable for MD, high heat transfer through the membrane is considered as heat loss (Phattaranawik *et al.*, 2003a). With reference to the membrane thickness, a thin membrane can have lower mass transfer resistance (higher permeate flux) and higher heat loss, while a thicker membrane can have lower heat loss and higher mass transfer resistance (lower permeate flux). Thus, the membrane thickness should possess an optimum value to compromise the heat and mass transfer in MD. To achieve this, membranes with multi-layer structures have been introduced for MD applications, where a thin hydrophobic membrane provides lower mass transfer resistance and higher permeate flux, while the thicker and less hydrophobic or hydrophilic support layer can reduce the heat loss (Bonyadi & Chung, 2007; Cheng *et al.*, 2018; Shirazi *et al.*, 2020a; Zuo *et al.*, 2017).

Table 2 Guideline for desired properties of MD membranes

Property	Description	Recommendation
Hydrophobicity	The higher the hydrophobicity (i.e., the lower the surface energy), the higher the liquid retention.	As high as possible
Liquid entry pressure (LEP)	The higher the LEP, the more pore wetting resistance.	>250 kPa
Pore size	Larger pore size can provide higher permeate flux; however, it reduces the LEP.	0.1 - 0.45 μm
Porosity	The higher the porosity; the higher the permeate flux, the lower the heat loss, and the lower the mechanical strength.	60-80%
Thickness	The lower the thickness; the higher the permeate flux and heat loss.	30-450 μm
Thermal conductivity	The lower the thermal conductivity, the higher the permeate flux and energy efficiency.	As low as possible
Tortuosity	The higher the tortuosity, the lower the permeate flux.	As low as possible
Chemical resistance	The higher the chemical resistance, the better the membrane integrity.	As high as possible
Mechanical strength	The higher the mechanical strength, the better the membrane integrity.	As high as possible

As mentioned earlier, heat loss through the thermal conduction of the membrane is an important challenge for MD (Phattaranawik *et al.*, 2003b; Susanto, 2011a). Therefore, the thermal conductivity of the membrane material should be as low as possible. Moreover, the higher the porosity, the lower the heat loss through the membrane, as the heat transfer coefficient of the trapped air in the membrane pores is much lower than that of the polymer thermal conductivity (Eykens *et al.*, 2016). Therefore, low thermally conductive polymers can be used for fabricating a highly porous membrane to enhance the MD performance (Shirazi *et al.*, 2020a).

Further to all the above-mentioned characteristics, MD membranes should also possess anti-fouling properties and should be chemically and thermally durable for stable performance in the long-term operation (Chen *et al.*, 2020). Table 2 provides a guideline for the desired properties of MD membranes.

5.2.1.2 Membrane materials

The very first MD experiments were mostly focused on the process (e.g., proving the concept, optimization of operating parameters, enhancing energy efficiency, etc.) (Lawson and Lloyd, 1997b). These experiments were mostly carried out using commercial MF membranes which were made of hydrophobic polymers (Franken *et al.*, 1987a; Fane *et al.*, 1987; Schofield *et al.*, 1987; Kimura *et al.*, 1987; Schofield *et al.*, 1990). However, these membranes have not been specifically designed and fabricated for MD, which is a non-isothermal separation based on the vapor-liquid interface equilibrium. This was found as a serious challenge for optimizing and developing the application of MD. Therefore, a considerable part of research in the next step has been focused on developing novel membranes with enhanced performance in terms of anti-wetting properties, permeate flux, solute rejection, and energy efficiency (Wang & Chung, 2015).

Table 3 Commercial polymers for the fabrication of MD membranes (Wang and Chung, 2015; Qasim *et al.*, 2021a)

Polymer	Chemical structure	Surface energy ($\times 10^{-3}$ Nm^{-1})	Thermal conductivity ($Wm^{-1}K^{-1}$)	Thermal stability	Chemical stability
PTFE	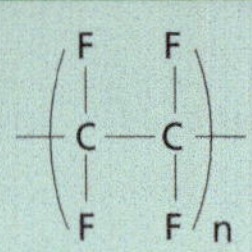	9-20	~0.26	Very good	Very good
PVDF	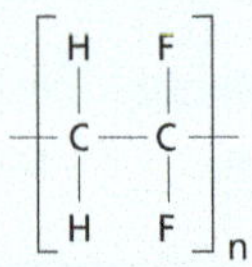	30.3	~0.18	Moderate	Good
PP	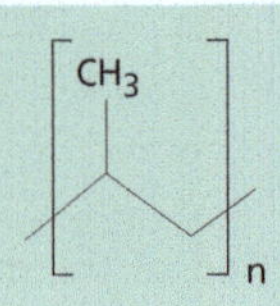	30	~0.14	Moderate	Good
PE	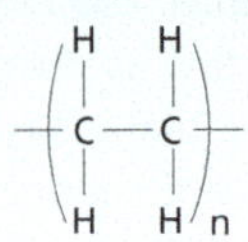	28-33	~0.4	Poor	Good

MD membranes can be fabricated from both polymeric and inorganic materials. However, the considered material must be hydrophobic (i.e., with low surface energy) intrinsically, or by proper modification (Tibi *et al.*, 2020). The common commercial polymers for the fabrication of MD membranes include PTFE, PP, PVDF, and polyethylene (PE) (Qasim *et al.*, 2021b; Wang & Chung, 2015). Table 3 presents the characteristics of some common hydrophobic polymers. As could be observed, PTFE has the lowest surface energy ($9\text{-}20 * 10^{-3}$ $N.m^{-1}$), which can provide a relatively high surface hydrophobicity. Moreover, it is chemically and mechanically stable. However, due to the insolubility of PTFE polymer in the majority of chemical solvents, PTFE membranes should be fabricated using complicated and expensive techniques, such as melt-extrusion (Feng *et al.*, 2018). PVDF has higher surface energy than that of PTFE, which means it is less hydrophobic. However, it has a lower thermal conductivity in comparison with PTFE (Table 3). It is worth noting that PVDF and their derivatives, such as poly(vinylidene fluoride-co-hexafluoropropylene) (PVDF-HFP), are the most investigated polymers for developing MD membranes (Qasim *et al.*, 2021a). This is due to the numerous advantages of these polymers, such as proper solubility in a wide range of solvents and good processability, for membrane fabrication (Kang and Cao, 2014; Ji *et al.*, 2015; Zou and Lee, 2022).

Recently, with developing new techniques for membrane fabrication, such as electrospinning (which can provide nanofibers), new commercial polymers have also been investigated for fabricating the MD membranes, such as polystyrene (PS), high-impact polystyrene (HIPS), styrene-acrylonitrile (SAN), poly (methyl methacrylate) (PMM), acrylonitrile-butadiene-styrene (ABS), styrene-butadiene-styrene (SBS), and polydimethylsiloxane (PDMS) (An *et al.*, 2017; Duong *et al.*, 2018; Niknejad *et al.*, 2021; Sadeghzadeh *et al.*, 2020a; Shirazi *et al.*, 2020b).

Table 4 Comparison of polymeric and inorganic membranes for MD

Item	Polymer membrane	Inorganic membrane
Membrane		
Material	- Widely available	- Available
Fabrication	- Well-developed techniques - Inexpensive	- Developed techniques - Expensive
Mechanical strength	- Flexible - Less durable in long term	- Brittle - Durable in long term
Fouling and cleaning	- Sensible to fouling - Cleaning is challenging	- Sensible to fouling - Flexible in cleaning
Corrosion resistance	- High	- Moderate
Performance		
perating temperature*	- Low to moderate	- Low to high
Permeate flux	- Low (AGMD) to moderate (VMD)	- Moderate (AGMD) to high (VMD)
Solute rejection	- High	- High

* Temperature range: Low (30 °C) to High (90 °C)

Further to polymers, inorganic materials can also be used for the fabrication of MD membranes (Ramlow *et al.*, 2019). The inorganic membranes can be made of a wide range of materials, such as alumina, zirconia, titania, silicon nitride, and metal oxides of iron (Camacho *et al.*, 2013). However, these materials are hydrophilic in nature (e.g., due to the presence of the hydroxyl group). Thus, inorganic membranes should be modified properly for their surface to impart the required hydrophobicity for MD applications (Ferreira *et al.*, 2021). Inorganic membranes, in particular ceramic membranes, are more chemically, mechanically, and thermally stable in comparison with polymeric membranes. Moreover, they can be cleaned several times, even using extreme cleaning techniques, such as chemicals, steam, and high-pressure backwash. However, inorganic membranes are expensive, and brittle, and show higher fouling and scaling tendency in comparison with polymeric membranes (Omar *et al.*, 2022). Table 4 compares the polymeric and inorganic membranes for MD.

5.2.2 Experimental set-up

5.2.2.1 MD configurations

MD has four main configurations, as shown in Figure 2. All these configurations are identical for the feed channel, where the hot solution stream is in direct contact with the hydrophobic surface of the membrane (Adewole *et al.*, 2022).

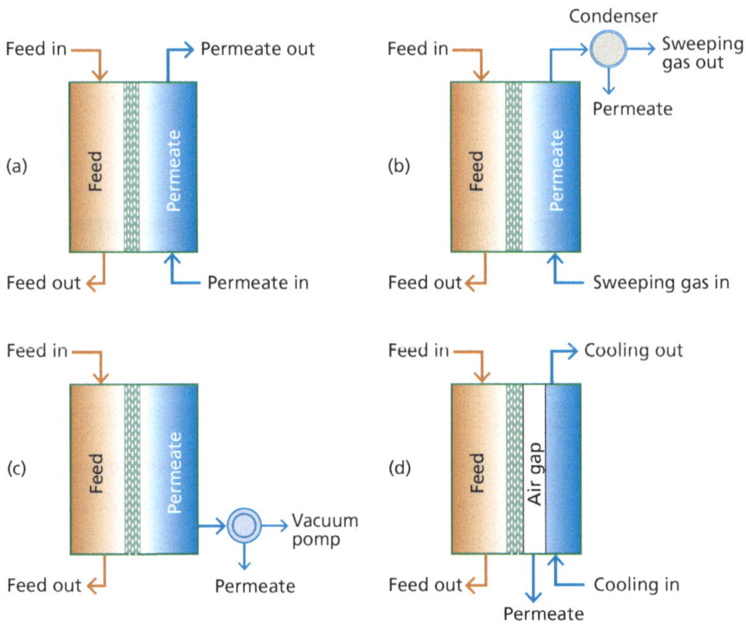

Figure 2 A general scheme of the main MD configurations: (a) direct contact membrane distillation (DCMD), (b) sweeping gas membrane distillation (SGMD), (c) vacuum membrane distillation (VMD), and (d) air-gap membrane distillation (AGMD).

When further to the feed side, the applied membrane is in direct contact with the liquid in the permeate side, which is pure and cold, it is known as the direct contact MD (DCMD) (Figure 2-a). This is the most investigated and simplest MD configuration (Khayet, 2011). DCMD has been used for different applications, such as seawater desalination and wastewater treatment. However, the most important drawback of DCMD is the high heat loss through the membrane thermal conduction (Ashoor *et al.*, 2016; Niknejad *et al.*, 2021).

Table 5 A comparative overview of MD configurations

Configuration	Description	Highlights
DCMD	• Both feed and permeate channels contains liquid streams which are in direct contact with the membrane surfaces.	• Simple modules design. • Internal condensation. • Possibility for internal heat recovery. • Pure/fresh water is required for cooling stream. • The most investigated MD configuration. • Wide range of applications. • High heat loss and low thermal efficiency.
SGMD	• An inert gas stream with low temperature is used for sweeping the vapor molecules in the permeate channel.	• Lower permeate flux compared to VMD. • External condenser is required. • The least investigated MD configurations. • Promising for concentration application, where the permeate can be vented. • The operating condition in the permeate. channel is more effective on flux compared to DCMD. • Good when feed stream contains volatile compounds. • Low heat loss. • Challenging in terms of heat recovery.
VMD	• Vacuum is used to enhance the permeation of vapor molecules in the permeate channel.	• High permeate flux. • Higher risk of pore wetting. • Low heat loss. • Challenging in terms of heat recovery. • External condenser is required. • Good for processing solutions with low. vapor pressure at a desire temperature. • Feasible for large scale.
AGMD	• An air gap is maintained between the membrane and a condensing surface in the permeate channel.	• High energy efficiency. • Internal condensation. • Possibility for internal heat recovery. • Seawater and non-hazardous (pre-treated) wastewater can be used as cooling stream. • Lower permeate flux compared to other configurations. • Wide range of applications. • Feasible for large scale.

The water vapor molecules in the permeate channel can be collected either by imposing a vacuum pressure or by imposing a sweeping gas flow. Under these circumstances, the configurations are known as the vacuum MD (VMD) and the sweeping gas MD (SGMD), respectively (Figures 2-b and 2-c, respectively) (Abu-Zeid *et al.*, 2015; Said *et al.*, 2020). Although SGMD and VMD can provide relatively higher permeate flux in comparison with DCMD along with moderately better energy efficiency, an external condenser is required for both configurations, which can increase the operation cost and energy consumption (Khayet *et al.*, 2003; Huayan *et al.*, 2011). To overcome this challenge, an air gap can be imposed between the membrane and a condensing surface in the permeate channel (Figure 2-d). This MD configuration is known as the air gap MD (AGMD) (Khayet and Cojocaru,

2012). This module design for MD can considerably enhance energy efficiency without needing an external condenser (Kalla *et al.*, 2019). However, the stagnant air gap can impose extra mass transfer resistance for transferring the water vapor molecules from the interface of the membrane support in the permeate channel towards the condensing surface (Shahu and Thombre, 2019a). Table 5 presents the comparison among the main configurations of the MD process.

5.2.3 Process

5.2.3.1 MD system

Figure 3 illustrates a general scheme of the MD system. The system consists of various parts which are introduced briefly.

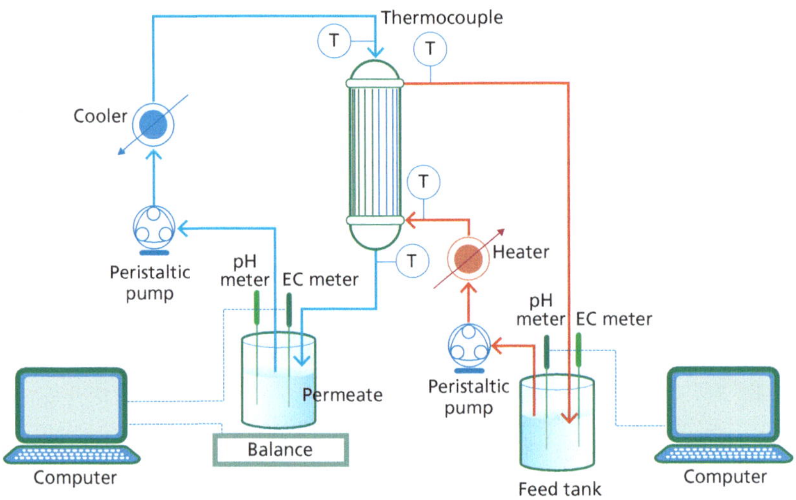

Figure 3 A general scheme of the MD experimental system.

The MD system consists of at least two containers, one to store the feed solution (feed tank) and another one to collect the product (permeate tank). The permeate tank should be placed on a balance to record the variation in the permeate mass and calculate the permeate flux.

The mass transfer and permeate production take place inside the membrane module, which can have different configurations, including plate and frame, tubular, hollow fiber, and spiral wound (Francis *et al.*, 2022). A pump is required to recirculate the feed stream in the close loop of the feed tank-module-feed tank. Depending on the MD configuration, the cooling flow in the permeate channel can be provided by another pump in DCMD and AGMD or by a compressor or blower in SGMD. In the case of VMD, the permeate channel is under vacuum pressure using a vacuum pump.

A heating system is required for heating up the feed stream to a desired temperature. Likewise, a cooling system is required to keep the temperature of the cooling stream low and constant. In the case of the SGMD and VMD configurations, an external condenser is also required to condense and collect the permeate. To monitor and control temperature, at least four thermal sensors should be placed as close as possible to the inlet and outlet points of the MD module. Thermal sensors are shown by 'T' in Figure 3.

The quality of the product in the permeate tank can be monitored using an EC meter and a pH meter. Likewise, the variation of the feed quality can be monitored in the feed tank. Depending on the requirements of the research, other equipment, such as an in-line microscope or an injection, can also be considered.

5.2.3.2 Operating parameters

MD is a non-isothermal separation process, and the feed temperature is a dominant operating parameter in all MD configurations (Curcio and Drioli, 2005). The operating temperature in the feed channel can vary from 40 to 80 °C, depending on the available energy source, the thermal resistance of the membrane, and the system design (Ahmed *et al.*, 2020). Moreover, the temperature of the cooling stream in the permeate channel is also important for DCMD, SGMD, and AGMD configurations. For VMD, however, the vacuum pressure in the permeate channel is an important operating parameter further to feed temperature (Mohammadi *et al.*, 2015; Peydayesh *et al.*, 2015).

Fluid flow rate (i.e., the recirculation rate) in both the feed and permeate channels is another operating parameter. Different values were reported in the literature for the flow rate, depending on the MD module and capacity of the system. One should be considered the inlet pressure at high flow rate. When the fluid flow is provided by a peristaltic pump, the pressure of the fluid flow in the feed and permeate channels is quite low, i.e., very close to the atmospheric pressure. However, if other types of pumps, such as centrifugal or diaphragm pumps, are used, a proper pressure reducer device/tool (e.g., the pressure regulator) should be used before the MD module.

The feed solution in MD can contain various compounds, such as chemicals, alcohols in dilute concentration, different salts (e.g., NaCl, MgCl, $CaCO_3$, Na_2CO_3, Na_2SO_4, etc.), sugars (lactose, sucrose, glucose, etc.), etc. (Shirazi & Kargari, 2019; Ali *et al.*, 2021; Ali *et al.*, 2015, 2018; Park *et al.*, 2020; Peydayesh *et al.*, 2015; Quist-Jensen *et al.*, 2016a; Quist-Jensen *et al.*, 2017; Quist-Jensen *et al.*, 2016; Shirazi *et al.*, 2014). Thus, the feed type and its concentration are other operating parameters, which can affect the MD performance, then they can also be considered.

Depending on the MD configuration, the vacuum pressure, sweeping gas flowrate, and the distance of the air gap in VMD, SGMD, and AGMD configurations, respectively, can also be investigated as other operating parameters. Moreover, membrane properties (i.e., type, material, pore size, etc.) have also been investigated in various research (Abu-Zeid *et al.*, 2015; Chamani *et al.*, 2021; Curcio & Drioli, 2005; Eykens *et al.*, 2017; Hitsov *et al.*, 2015; Shahu & Thombre, 2019a; Shalaby *et al.*, 2022a, 2022b).

5.2.4 MD modules

A typical membrane module is a unit consisting of a membrane mounted in a housing and containing feed inlet, retentate outlet and permeate outlet channels (Yang *et al.*, 2013). As new membrane applications emerge and new module designs are developed, the definition of modules evolves as well. For example, in submerged membrane modules, as the name suggests, the module is directly immersed in the feed solution without an outer casing and has only ports for permeate removal. Other examples are air-gap and permeate-gap MD modules (AGMD and PGMD, respectively), which require additional channels at the inlet and outlet of the cooling flow.

The main purpose of this module is to properly secure the membrane so that it can be used in its designated application. However, a well-designed module must also meet several other requirements. A proper module design should ensure a high packing density of the membrane. Packing density is taken as the size of the surface area of a functional membrane in a given volume. Generally, high packing density is desirable to avoid inefficient use of module housing. However, it should be noted that for hollow fiber membranes, increasing the packing density beyond a critical value can result in stagnant or 'dead' regions of poor heat and mass transfer within the module. important. For plate and frame flat sheet membrane modules, typical packing densities range from 100 to 400 m^2/m^3, whereas hollow fiber membrane modules can have higher packing densities of up to 3000 m^2/m^3 (Peng *et al.*, 2012).

MD systems involve mass transport steps through the feed, membrane, and permeate, with each region having a specific transfer coefficient. To mitigate mass transfer resistance at the boundary layer, appropriate module designs should demonstrate good hydrodynamics, minimize temperature and concentration polarizations, and minimize energy consumption. Modules should maximize heat recovery, be easy and economical to fabricate, and minimize leakage issues. They should also facilitate scale-up and integration into existing processes. The module's performance should be predictable using mathematical models under various operating conditions and feed characteristics. The module must maintain integrity during long-term operation, minimize foulant deposition, and be resistant to heat and chemical degradation.

New variants of flat sheet and hollow fiber membrane modules have been introduced for MD applications. Flat sheet membranes are typically assembled in plate and frame or spiral wound configurations, while hollow fiber modules can be classified into shell and fiber/tube and submerged configurations. These new variants aim to improve process performance by improving heat and mass transport, heat recovery, and membrane area.

The simplest module design for MD experiments is the plate and frame module. In this design, the membrane is placed between two frames and plates. The membranes for this type of module are in flat sheet form. This module can possess different sizes and is useful for lab-scale experiments. However, the membrane area is small, and this module does not have that much chance for industrial applications. Various flow arrangements can also be considered for this module, including the co-current, counter-current, and crossflow. The efficiency of these flow arrangements in terms of the permeate flux in the plate and frame modules is in the order of: crossflow>counter-current>co-current (Shirazi *et al.*, 2014).

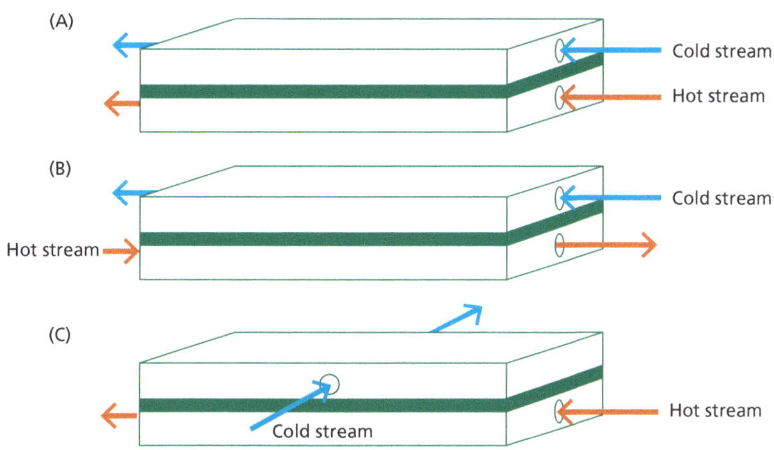

Figure 4 The flow arrangement of hot and cold streams in the plate and frame module; (a) co-current, (b) counter-current, and (c) cross-current flows (Shirazi *et al.*, 2014).

Vacuum multi-effect MD (VMEMD) systems operate under reduced pressure and can achieve higher water recovery rates compared to AGMD. These systems use multiple stages of MD in series, working at lower operating temperatures and pressures. They are compact, high-efficiency systems with solar thermal collectors and solar-photovoltaic sources as heat sources. VMEMD designs enable internal heating and condensation, saving heat energy (Zhao *et al.*, 2013).

The air gap width is a crucial parameter in AGMD module design, determining distillate production rate. It aims to prevent condensing media from contacting the membrane surface, reducing heat loss, and increasing vapor transport distance. The lower limit is determined by thermal efficiency and water bridging. Various membrane configurations, including flat sheet, tubular, hollow fiber, and spiral wound, have been applied in AGMD studies.

The fundamental module design of the AGMD has undergone numerous modifications, including the introduction of spacers in the feed channel and the use of cooling plates on the coolant channel. These modifications have improved the efficiency of heat removal from the coolant and increased system flux, with the flat and channelled plates being effective in increasing system flux. Multi-effect AGMD (ME-AGMD) is another novel approach to improving module performance and industrialization of MD. Pangarkar and Deshmukh developed a new ME-AGMD module for water treatment applications, which performed better in terms of permeate flow and energy utilization (Pangarkar and Deshmukh, 2015). The parallel stage MS-AGMD system generated 2.6 and 3 times more permeate volume than a single-stage system, but its precise energy usage was only 1.5 times that of the single-stage system. Operational modifications of the conventional AGMD module have reduced pressure inside the gap below atmospheric pressure, increasing distillate flux up to 3 times. The generation of vacuum in the gap requires an additional vacuum pump and extra electric energy input. The traditional AGMD module has been modified to an air-cooled AGMD, which has the potential to significantly reduce energy consumption and costs for desalination by minimizing or eliminating plant components associated with coolant flow systems.

The spiral-wound module is a variation of the plate-and-frame configuration, where a membrane envelope with a spacer is wound around a porous tube. It was first used in pressure-driven processes like RO and UF and has since been used in research. The Fraunhofer Institute for Solar Energy Systems has produced single and multi-channel spiral wound membrane modules for MD. Fabrication involves rolling a membrane, condenser foil, and spacer materials around a motorized main spindle, and sealing the channels with resin. Multichannel spiral wound modules are built using multiple pairs of evaporator/condenser flow channels. The main advantage of spiral wound modules is better counter-current circulation of hot and cold streams, allowing for more efficient heat recovery (Winter *et al.*, 2012; Winter *et al.*, 2011).

Traditional flat sheet membranes have been used for AGMD modules, but hollow fiber has gained interest. Two strategies have been proposed for designing hollow fiber AGMD modules: using the inside surface of the module to condense vapor passing through the membrane or condensing the condensing surface inside the shell (Abu-Zeid & ElMasry, 2020; Alpatova *et al.*, 2019; Shahu & Thombre, 2019b). Different variants of the second approach have been proposed, including porous and dense hollow fibers packed inside a module, water gap MD (WGMD), and multiple copper tubes enclosed in the shell of a hollow fiber AGMD module. These approaches have shown promising results in improving heat transfer efficiency and reducing cooling channel replacements.

The second conventional module design for MD is the hollow fiber module, which can have hundreds of hollow fibers in a shell tube. Thus, the hollow fiber module can provide a much larger surface area for MD experiments in comparison with the plate and frame modules. The feed stream and the permeate stream can flow through either the fibers or the shell. In terms of the high salinity brine, for example, it would be better to introduce the feed stream to the shell instead of the fibers, as the membrane cleaning would be easier and more efficient (Quist-Jensen *et al.*, 2017). Same as the plate and frame modules, all MD configurations can be performed using the hollow fiber modules. However, in the case of AGMD, special design considerations may be necessary.

A hollow fiber module is a system consisting of a hollow fiber membrane bundle, cartridge, tube sheets, and side caps. The bundle consists of hollow fiber membranes packed together, with a liquid potting substance forming the tube sheet. The tube sheet acts as a fluid-tight barrier, separating streams flowing through the lumen and shell sides of the module. The packing density of the hollow fiber module is crucial for its productivity, as it directly affects the module's productivity(Mat *et al.*, 2014). The packing density can be arranged in various configurations, such as parallel, crisscross, or other precise geometric arrangements.

In DCMD, hollow fiber modules are typically in shell and tube heat exchanger configuration, with feed flow on one side and permeate on the other. The feed compartment is based on the feed solution properties. Axial flow can be divided into co-current and counter-current flows, with the counter-current flow arrangement being the most widely used configuration for MD applications. Cross flow is often used in axial flow designs to reduce stagnant regions and concentration polarization effects.

Traditional parallel hollow fiber modules are susceptible to high temperature and concentration polarizations, especially at low fluid flow rates. The non-uniformity of fiber packing is a challenge due to the production of parallel fiber bundles. This results in sluggish or dead zones, reduced separation efficiency, and channelling or bypassing in poorly packed zones. To improve uniformity, fibers can be weaved into different structural geometries, such as helical, wavy, or twisted shapes (Ali *et al.*, 2015b; Shahu & Thombre, 2021). This results in more uniform shell flow and less concentration polarization due to fluid mixing. Studies have shown that using these geometries can increase flux enhancement in membrane applications. Yang *et al.*, (2012) compared five types of hollow fiber module designs, revealing that the space-knitted fiber design showed the best performance, with over 90% increase in permeate flux. This configuration improved fluid dynamics and even flow distribution, increased vapor permeability, and reduced thermal polarization with lower energy loss.

Submerged hollow fiber MD modules are increasingly popular in membrane bioreactors (MBRs) due to their simplicity and ease of fabrication (Francis *et al.*, 2015; Meng *et al.*, 2015; Gryta, 2020). These modules eliminate the need for feed stream circulation, reducing electric energy consumption. They have been proposed for VMD and DCMD configurations, allowing for recirculation of either feed or permeate stream while submerged in the other stream. The MDBR system, combining MD with a thermophilic MBR, produces high-quality permeate with a flux two orders higher than competitors. Submerged MD hollow fiber modules have also been proposed for desalination of Red Sea water, using PTFE-based hollow fibers immersed in clean water and hot feed introduced inside.

However, some inherent issues have been identified, such as high temperature and concentration polarizations, fouling, and scaling at the membrane surface. Strategies such as mixing feed solution with a magnetic stirrer, transmembrane vibrations, and low-power ultrasound have been proposed to improve the efficiency of these modules.

Module design in membrane-based membranes (MD) has been influenced by overall contact length, which refers to the membrane length over which hot feed streams stay in contact without intermittent heating. High contact lengths can be achieved by increasing membrane length, connecting multiple modules in series, or increasing membrane length in each envelope of flat sheet membrane. Recovery of latent heat of condensation from permeate can reduce the thermal energy consumption of MD, but in DCMD, a sufficient length is required to keep the feed and permeate in contact. Recent studies have shown that increasing flow channel length can reduce gain-to-output ratios, specific thermal energy consumption, and channel depth (Abu-Zeid *et al.*, 2021; Ali *et al.*, 2016b). (Tsai *et al.*, 2023) proposed multipass hollow fiber membrane modules, which studied the effect of operational modes, number of passes, length, and operating temperature on the performance of the multipass modules. The results may open new windows for future MD modules for better and more energy-efficient performance, particularly for desalination and brine management purposes.

5.3 METHODS

5.3.1 Process measurements and calculations

5.3.1.1 Permeate flux

The permeate flux in MD can be measured as follow:

$$J = \frac{\Delta m}{A \times t} \quad \text{Eq. 1}$$

where Δm, A, and t are the collected permeate mass (kg), membrane surface (m^2), and the interval time (sec), respectively (Drioli *et al.*, 2013).

5.3.1.2 Solute rejection

The solute rejection in the permeate stream can be measured as follow:

$$Rejection(\%) = (1 - \frac{C_p}{C_f}) \times 100 \quad \text{Eq. 2}$$

where C_p and C_f are the solute concentrations in the permeate and feed sides (mg/L), respectively. C_p can be calculated based on the following equation with reference to the dilution effect:

$$C_p = \frac{C_1 m_1 - C_0 m_0}{m_1 - m_0} \quad \text{Eq. 3}$$

where m_0 and m_1 are the initial and final masses of the cold stream. C_0 and C_1 are also the initial and final salt concentrations of the cold stream, respectively (Lu *et al.*, 2016).

5.3.1.3 Logarithmic temperature difference

The logarithmic temperature difference along the MD module is defined as follow (Quist-Jensen *et al.*, 2018):

$$dT_{\ln} = \frac{(T_{fi} - T_{p0}) - (T_{f0} - T_{pi})}{\ln \frac{(T_{fi} - T_{p0})}{(T_{f0} - T_{pi})}} \quad \text{Eq. 4}$$

where T_{fi} and T_{fo} are the inlet and outlet temperatures of the feed stream, and T_{pi} and T_{po} are the inlet and outlet temperatures of the permeate stream, in the membrane module, respectively.

5.3.2 Membrane characterization

5.3.2.1 Physical and morphology properties

(a) Scanning electron microscopy

Membrane surface can affect the MD performance in terms of flux, fouling, etc. Thus, surface morphology should be investigated, specifically for fabricated membranes (Talukder *et al.*, 2022; Wang *et al.*, 2023). Scanning electron microscopy (SEM) is the most investigated method for morphology observation of various materials, including membranes (Khulbe and Matsuura, 2017). SEM has numerous advantages for determining the membrane morphology, such as simplicity and ease of operation. Moreover, as SEM is a non-destructive technique, the samples are not damaged during the analysis and can then be used many times for further imaging (Naresh-Kumar *et al.*, 2012). Figure 5 shows the example of SEM images of a commercial PTFE and a fabricated membrane with nanofiber structure for MD.

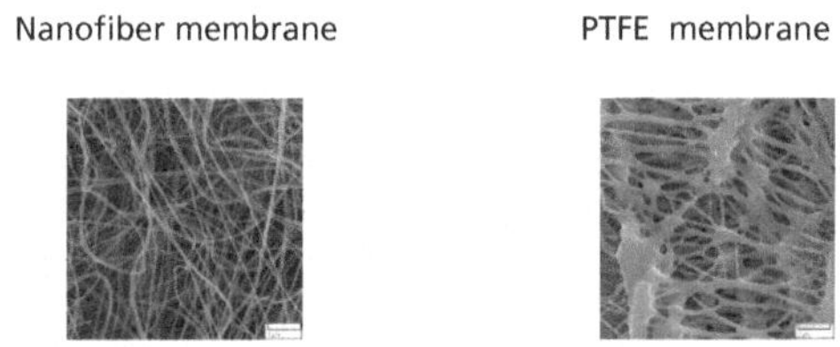

Figure 5 SEM images of a fabricated membrane with nanofiber structure (left) and a commercial PTFE membrane (right) for MD (Shirazi *et al.*, 2020a).

Before conducting an SEM test, sample preparation is required. To prepare the sample, a small piece of membrane is cut, and placed on a stub. As the sample is in small size, tweezers are usually used along with double-sided glue tape for fixing the sample on top of the holder. To prevent charge accumulation, the sample should be sputter coated with a thin layer of a highly conductive metal, such as gold or platinum (Sharma & Bhardwaj, 2019; Vladár & Hodoroaba, 2020). Moreover, if cross-sectional images are required, the sample should carefully be freeze-dried in liquid nitrogen, and then should properly be cut using a razor blade (Zhu *et al.*, 2012; Conners and Banerjee, 2020; Vladár and Hodoroaba, 2020). It is worth noting that the sample should not be touched to avoid adding contamination and footprint.

If more informative images are required, for example for morphology observation and detection of nanoparticles on the membrane surface, field emission SEM (FESEM) with the platinum coating for the sample are recommended, as it can provide images with higher resolution (Lewczuk and Szyry ska, 2021; Kirk, 2017). The SEM or FESEM images can be used for morphological observation, determination of pore size and its distribution, thickness measurement (from the cross-sectional images), investigating the homogeneity, and presence of particles or fouling layer on the membrane surface.

Also, SEM utilizes imaging software to measure the dimensions of, e.g., the size of particles, on the surface at various magnification ranges. Moreover, various external software, such as ImageJ which is an open-source software for image processing, can be used for measuring the pore size, pore size distribution, and porosity (Guillen *et al.*, 2010; Shirazi *et al.*, 2013; Ziel *et al.*, 2008).

(b) Transmission electron microscopy

If clear view of the internal structure of the membrane sample is required, the transmission electron microscopy (TEM) should be applied for imaging the membrane samples. TEM is an alternative to SEM for investigating the structural morphology and crystalline (Sharma *et al.*, 2018; Shyam Kumar *et al.*, 2017). For example, it can be used when nanoparticles are incorporated into the membrane structure. Thus, the internal morphology and distribution of nanoparticles can be investigated using TEM (Qin *et al.*, 2015; Dadari *et al.*, 2022; He *et al.*, 2023). Despite of SEM, which is more practical for surface observation, TEM can provide accurate information about the structure and the body of the membrane sample (Mousa *et al.*, 2022; Talukder *et al.*, 2022; Wang *et al.*, 2023; Wiktor *et al.*, 2017).

(c) Pore size and its distribution

Pore size is an important parameter for MD membranes, as only vapor molecules should pass through the pores (Eykens *et al.*, 2016). For example, although membranes with large pore size are considered to provide higher permeate flux, they also suffer from high pore wetting risk (McGaughey *et al.*, 2020b). Membrane pore size can be measured using various techniques (Nakao, 1994; Tanis-Kanbur *et al.*, 2021; Tung *et al.*, 2014).

As mentioned earlier, the pore size and its distribution can be measured using SEM images of the membrane surface (Ahmed *et al.*, 2015; Sadeghzadeh *et al.*, 2020b; Shirazi *et al.*, 2013a). However, it should be noted that the obtained results are based on surface observation.
The membrane pore size can also be determined using the filtration test. To do this, fine particles with a known particle size distribution can be filtered using the membrane sample. The permeate sample should be tested for the particle content and their sizes. By comparing particle size in the permeate with the original value of the particle size in the feed sample, the pore size range can then be determined. The results of the particle filtration test can then be compared with the obtained results based on the image processing of SEM images for pore size measurement (Gopal *et al.*, 2006; Sadeghzadeh *et al.*, 2020).

More accurate data for pore size and pore size distribution of MD membranes can be provided using the capillary flow porometry technique (Jena and Gupta, 2005a). In this technique, a small piece of membrane sample should be placed in a holder and get wet using a proper wetting liquid of known surface tension, such as Topor or Galwick (Jena and Gupta, 2010; Kolb *et al.*, 2018). Afterward, the different flows of inert gas should be used to displace the liquid inside the pores on the membrane structure. Using this technique, pore size and the pore size distribution can be obtained (AlMarzooqi *et al.*, 2016; Jena and Gupta, 2005b; Li *et al.*, 2006).

(d) Wetting properties

(i) Water contact angle

In MD, the membrane pores must not get wet with the feed solution, and only vapor molecules should be passed through the pores. Therefore, the wetting property of the membrane surface is crucial for MD applications (Chamani *et al.*, 2021). This can be determined using the surface contact angle. According to the standard definition, the membrane surface is hydrophobic if the water contact angle on the sample surface is greater than 90o, while the contact angle lower than this represents the hydrophilicity of the membrane surface (Ismail *et al.*, 2022; Rezaei *et al.*, 2018). Figure 6 illustrates this concept. Figure 7 also shows a typical system for contact angle measurements.

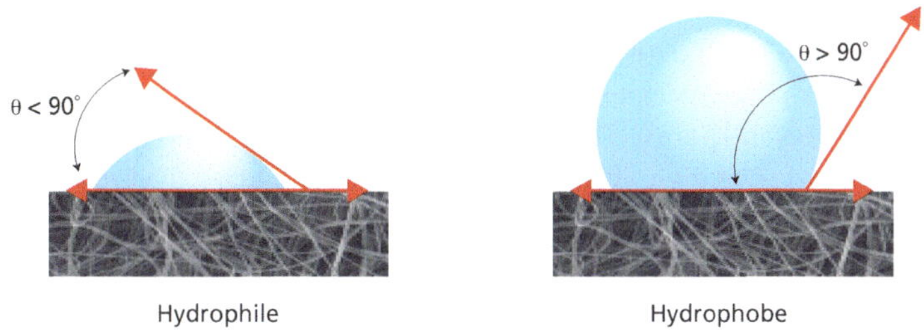

Figure 6 A general scheme of the surface hydrophobicity.

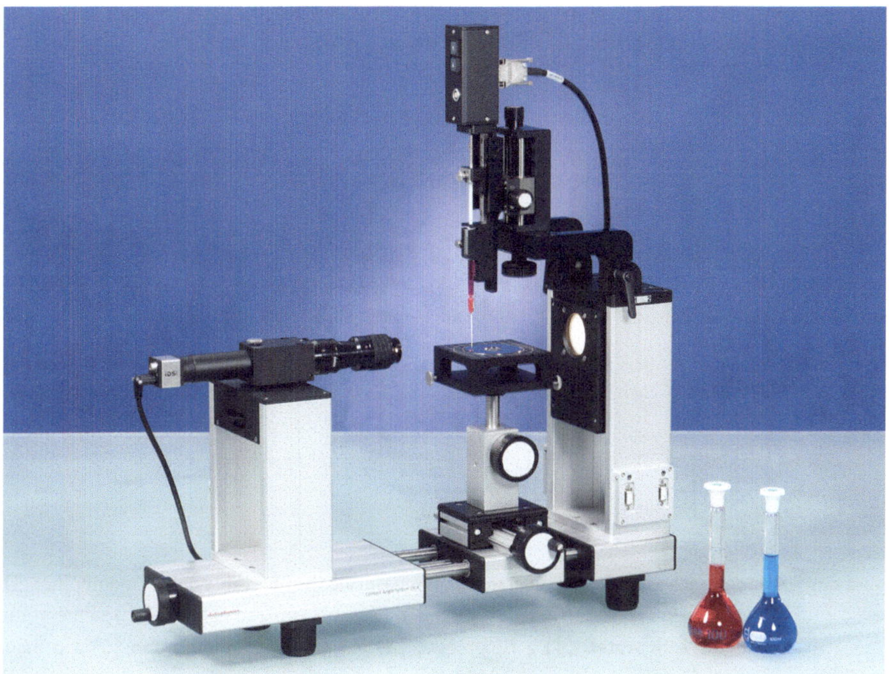

Figure 7 A typical contact angle measurement system (www.kruss-scientific.com).

To measure the contact angle, a 5-μL liquid droplet (usually deionized water) is placed on the membrane surface and a high-resolution speed camera takes the image of the shape of the droplet. The sessile drop technique can then be used for calculating the water contact angle and surface energy of the membrane sample (Franken *et al.*, 1987b; Lu *et al.*, 2019). To have high accuracy in the reported results, it is recommended to conduct the contact angle test at least for five different points on the membrane surface, and then report the average value.

(ii) Liquid entry pressure

Liquid entry pressure (LEP) determines the minimum required pressure to penetrate the feed solution to the pores. When the pressure of the feed stream is greater than LEP, the liquid enters the pores and pore wetting happens (Eykens *et al.*, 2016). Thus, the membranes for MD should possess as high LEP value as possible (Sadeghzadeh *et al.*, 2020b). LEP is proportional to some parameters, including the surface hydrophobicity, surface tension of the feed solution, pore structure, and the pore size. LEP can therefore be calculated as follow:

$$LEP = \frac{-2B_{\gamma l}\cos\theta}{r_{\max}} \qquad \text{Eq. 5}$$

where θ is the surface contact angle (can be measured using the water contact angle test), Υ_l is the surface tension of the feed solution, B stands for the pore geometry factor, and r_{rmax} is the maximum pore size (Silva *et al.*, 2021; Rácz *et al.*, 2014).

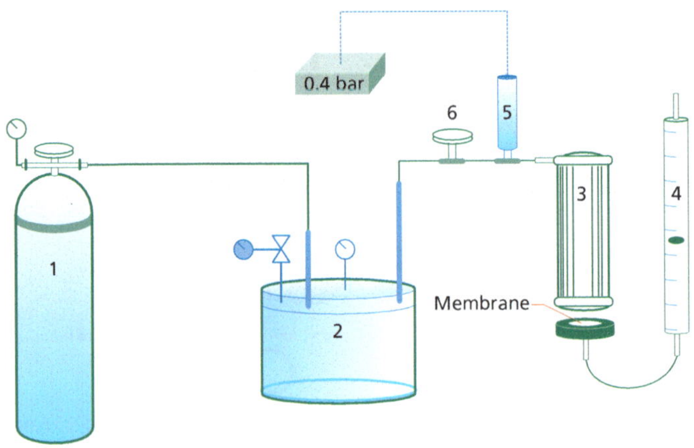

Figure 8 A general scheme of the typical setup for measuring the LEP value of MD membranes. The system consists of (1) an inert gas cylinder, (2) a pressurized container, (3) a membrane cell filled with water, (4) a flowmeter, (5) a digital manometer, and (6) a pressure regulator (Essalhi and Khayet, 2013).

LEP can also be measured experimentally for the membrane samples. Figure 8 illustrates the general scheme of an experimental LEP measurement set-up. To do this, a dry membrane sample should be placed in a plate and frame module (for flat sheet membrane samples), and distilled water should be exposed on the hydrophobic surface of the membrane. The pressure of the module should then be increased stepwise (10 kPa per minute would be recommended followed by a few seconds time laps) using an inert gas (e.g., nitrogen). As soon as the first water droplet is observed on the other side of the membrane sample, the corresponding pressure represents the LEP value (Khayet and Matsuura, 2001; Rácz *et al.*, 2014). It is also recommended to evaluate the LEP using the feed solution further to the distilled water, as the membrane is in direct contact with the feed solution rather than the distilled water in MD experiments (Silva *et al.*, 2021).

(e) Porosity

The membrane porosity can be measured using the gravimetric method. In this technique, a small piece of membrane should be cut and then the dry weight should be recorded. The sample should then be immersed in a proper wetting liquid (e.g., isopropanol alcohol) to get completely wet and re-weighted again (Khayet and Matsuura, 2001). The porosity (ε) of the membrane sample can then be calculated as follow:

$$\varepsilon = \left[1 - \frac{(\frac{w_1 - w_2}{\rho_m})}{(\frac{w_1 - w_2}{\rho_m}) + (\frac{w_2}{\rho_l})} \right] \qquad \text{Eq. 6}$$

where w_1 and w_2 are the weights of the dry and wet samples, respectively. Moreover, ρ_m and ρ_l are the density of the membrane sample and the density of the wetting liquid, respectively (Alkhudhiri *et al.*, 2012). It is worth noting that the weight measurement of the wetted membrane sample should be carried out carefully.

As it was mentioned earlier, the membrane porosity can also be measured using the image processing of SEM images. However, it should be noted that this will be the surface porosity (Sadeghzadeh *et al.*, 2020).

(f) Thickness

Membrane thickness can directly be measured using a precise micro calliper. It is recommended to measure the thickness at least for 10 points and then report the average value, to be sure to minimize the compression effect (Zhang *et al.*, 2017).

Further to this, the membrane thickness can also be measured using the optical microscope along with a proper scale bar (Vicente *et al.*, 2013). More accurate thickness data, however, can be provided by SEM through the cross-sectional imaging (Attia *et al.*, 2018a; Attia *et al.*, 2018b).

(g) Surface roughness and topography

Both the surface roughness and topography can affect the performance of MD membranes. These parameters can be determined using the atomic force microscopy (Khayet *et al.*, 2004). To evaluate the membrane surface using atomic force microscopy (AFM), a sample with defined dimensions should be cut, placed, and then fixed on the top of the holder using the double-sided glue tape (Wyart *et al.*, 2008).

AFM uses a nanometric prob to move along the membrane surface and collect the topographical data using a laser diode and a detector. Imaging can be performed via three different modes, i.e., contact mode, semi-contact mode, and non-contact mode. The generated data should be analysed using a collector system, and then topography images (with Angstrom resolution) can be made. It is worth quoting that the non-contact mode can provide 3D topographic images with higher resolution (Johnson and Hilal, 2015; Hilal *et al.*, 2004). When performing the AFM analysis for a membrane sample, the type of probe (e.g., silicon nitride), scanning environment (e.g., in air at ambient conditions), the specifications of the cantilever and its tip (i.e., length, width, resonance frequency, slope, etc.), the scanning speed (e.g., 5 μm/s at 1 Hz), the applied force (e.g., 0.15 nN), the scan size, and the resolution (250 points per line) are important parameters (Shirazi *et al.*, 2013a). Figure 9 presents the 3D AFM images of three commercial MF membranes for MD applications based on the non-contact mode imaging.

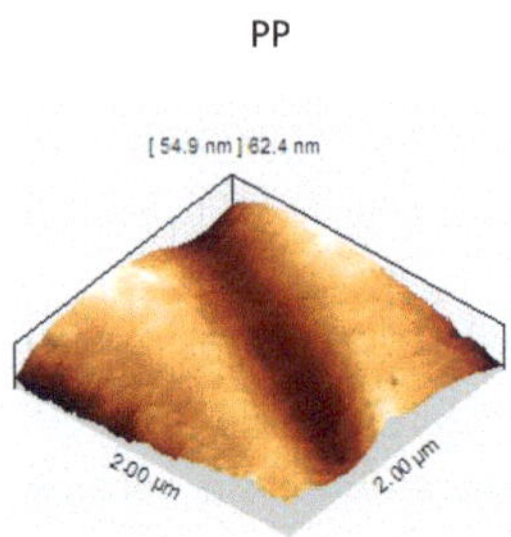

Figure 9 3D AFM image of commercial MF membrane for MD applications (Shirazi *et al.*, 2013b).

Table 6 Typical roughness parameters for evaluating the membrane surface topography (Shirazi *et al.*, 2013a)

Roughness parameter	Expression
Average roughness (R_a)	$R_a = \frac{1}{n}\sum_{i-1}^{n} Z_i$
Root-mean-square roughness (R_q)	$R_q = \sqrt{\frac{1}{n}\sum_{i-1}^{n} Z_i^2}$
Peak-to-valley height (R_z)	$R_z = Z_{\max} - Z_{\min}$
Surface skewness (R_{sk})	$R_{sk} = \frac{1}{nR_q^3}\sum_{i-1}^{n} Z_i^3$
Surface kurtosis (R_{ku})	$R_{ku} = \frac{1}{nR_q^4}\sum_{i-1}^{n} Z_i^4$
Z_i	The height at point *i*
n	Number of points in the image
Z_{max} and Z_{min}	The highest and the lowest Z values

Table 6 introduces some AFM parameters which are useful for evaluating the membrane topography. For example, the average roughness (R_a), which provides an overall view of the surface roughness, is the most reported topography parameter for MD membranes. In other words, the higher the R_a value, the rougher the membrane surface (Johnson & Hilal, 2015; Shirazi *et al.*, 2013b). The skewness factor (R_{sk}) represents the height distribution symmetry. While the positive R_{sk} parameter shows the domination of peaks on the surface, the negative R_{sk} values are associated with a porous surface. The kurtosis factor (R_{ku}) corresponds to the sharpness of the height distributions (Johnson and Hilal, 2015). R_{ku} values greater than 3 represent a surface with sharper height distribution, while values lower than 3 indicate a flat surface. Further detailed descriptions and applications of these parameters for characterizing MD membranes can be found in the literature (Johnson & Hilal, 2015; Shirazi *et al.*, 2013b; Shirazi *et al.*, 2013a, 2013b).

It is worth noting that the AFM parameters and their results are scale and mode dependent. Therefore, AFM images that have been provided with the same scale and the same mode can be compared together. Moreover, compared to SEM analysis, AFM is a non-vacuum analysis technique, and the membrane sample is not coated. Therefore, the AFM results can be closer to the real features of the membrane in real life (Shirazi *et al.*, 2013b).

Mechanical properties

Although MD membranes do not require very strict mechanical properties compared to membranes in pressure-driven processes, such as RO, a minimum mechanical strength is still required for handling and modulation of MD membranes (Essalhi and Khayet, 2014; Essalhi and Khayet, 2013). Tensile test can be employed to evaluate the mechanical strength of MD membranes. Figure 10 illustrates the general scheme of a typical tensile measurement system.

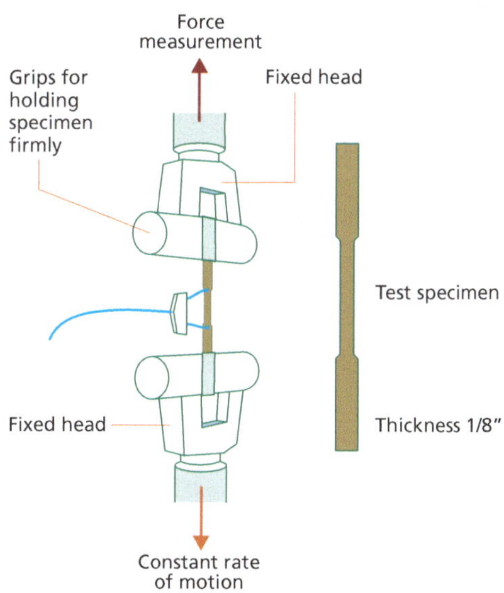

Figure 10 A general scheme of the tensile test (Ismail *et al.*, 2019).

To perform the tensile test, a piece of membrane sample should be cut according to ASTM D882-10 and taped for both ends to secure the grip. The membrane sample should then be placed and fixed in a grip between two jaws. To run the tensile test, some operating parameters are important, such as the load cell (e.g., 10 N), and the cross-head speed of the load cell (e.g., 5 or 10 mm/min). Based on the data recorded in the tensile machine, the stress-strain curves can then be illustrated and investigated (Tijing *et al.*, 2013; Wang *et al.*, 2017).

5.3.2.2 Chemical properties

(a) Elemental composition

Energy-dispersive X-ray spectroscopy (EDS) is a well-known, analytical technique for investigating the elemental composition of MD membranes. It usually performs along with the SEM test. EDS can provide the elemental composition of the membrane surface after modification (e.g., successful adsorption of nanomaterials on the membrane surface) or the composition of the fouling layer on the membrane surface (Kamaz *et al.*, 2019; Shirazi *et al.*, 2020a). It can also be used for investigating the uniform dispersion of nanoparticles on surface or in the membrane matrix. To perform the EDS, when the surface of the membrane sample is bombarded by SEM electron beams, the EDS detector can then detect the emitted X-ray and analyse the elemental composition of the membrane surface (Yang *et al.*, 2018). X-ray photoelectron spectroscopy (XPS) is another practical technique for elemental analysis of MD membranes (Khayet and Matsuura, 2003). XPS can provide useful and detailed information such as the chemical formula and chemical composition of the membrane surface. Thus, XPS can practically be employed to evaluate the efficiency of a proposed surface modification technique (Suk *et al.*, 2006; Wei *et al.*, 2012; Zhao *et al.*, 2017).

(b) Functional groups

Fourier transform infrared (FTIR) spectroscopy is a practical, analytical technique to investigate the functional groups on the membrane surface, either after surface modification or deposition of foulants during the MD test (Korolkov *et al.*, 2018). FTIR can detect different bonding types as well as organic and inorganic functional groups in molecular level. FTIR works based on the absorbance of the wavelength of the IR light of various functional groups in a wide range (400 to 4000 cm^{-1}). Thus, the detector can analyse the absorbed wavelength and identify the target functional group, and its density, as well. When the FTIR graph is plotted, the sharper peaks represent the higher amount of the specific functional group. Likewise, the small peaks represent trace amount of the that group (Belfer *et al.*, 2000; Jhaveri & Murthy, 2016; Rahman *et al.*, 2018). Moreover, same as SEM, FTIR is a non-destructive technique, and thus samples can be used for further measurements.

(c) Chemical structure

When a new membrane is fabricated or modified for investigating the MD performance, it should also be evaluated for the chemical structure. This can be performed using the Raman spectroscopy technique (Intrchom *et al.* 2018; Pouya *et al.*, 2021). Raman technique can also provide informative curves for determining molecular bonds, crystallinity, and the orientation of polymeric chains. Moreover, Raman is a non-destructive analysis technique (Bhadra *et al.*, 2016; Dumée *et al.*, 2011; Huang *et al.*, 2020).

(d) Crystalline structure

Each material can be specified using special X-ray diffraction. This can then be used for investigating the size of atoms, the crystal size, the length of chemical bonds, and the layer spacing (Bunaciu *et al.*, 2015). These have made X-ray diffraction (XRD) a powerful method to determine the structural crystallinity of materials, such as inorganic substances (e.g., metals, ceramics, etc.) (Kahle *et al.*, 2002; Norby, 2006). Thus, this analytical technique is practical for the characterization of inorganic and mixed matrix MD membranes, as polymers usually have a low crystalline structure. Moreover, XRD can be applied for investigating the inorganic scaling on the membrane surface and the produced crystals in the MD crystallization process (Garofalo *et al.*, 2016; Zuo and Chung, 2016; Gryta, 2011; Mokhtar *et al.*, 2014). Using XRD, the change in the chemical bonds and the crystallinity of materials can be detected. When XRD data is plotted, the narrow and sharp peaks indicate ordered material with large particle size, and vice a versa (Petkov, 2008; Flack and Bernardinelli, 2008).

5.3.2.3 Thermal properties

(a) Thermal conductivity

Thermal properties of MD membranes are important for long term performance. This can be highlighted in terms of heat loss through the membrane thermal conduction (Eykens *et al.*, 2016). The thermal conductivity of MD membranes can be calculated as follow:

$$k_m = (1-\varepsilon)k_s + \varepsilon k_g \qquad \text{Eq. 7}$$

where ε is the membrane porosity, and k_s and k_g are the thermal conductivity of polymer and the trapped gas in the pores (e.g., air), respectively. As could be observed, the thermal conduction is proportional to the membrane porosity. Thus, the higher the porosity, the lower the membrane's thermal conductivity. Therefore, membranes with higher porosity, such nanofiber membranes, can perform better in MD experiments due to lower heat loss (Alkhudhiri *et al.*, 2012; Shirazi *et al.*, 2020a). Moreover, the precise examination of the membrane porosity along with accurate conductivity data for the used polymer in membrane fabrication can provide better results for the thermal conduction of MD membranes. However, in terms of mixed matrix MD membranes, other correlations should be investigated (Eykens *et al.*, 2016).

(b) Thermal stability

As the MD membrane works under a range of feed temperatures (40-80 °C), the thermal stability matters in long-term performance (Saffarini *et al.*, 2013; Susanto, 2011b; Gryta, 2005). The thermal stability of MD membranes can be investigated using the differential scanning calorimetry (DSC) method (Ding *et al.*, 2021; Prince *et al.*, 2012). DSC provides very useful structural information of the membrane sample at various temperatures (Khayet *et al.*, 2018; Essalhi *et al.*, 2021; García-Fernández *et al.*, 2015).

5.4 APPLICATIONS AND EXAMPLES

MD is traditionally used for desalination purposes as an alternative to RO or to overcome the limited recoveries of RO and other thermal desalination techniques. Though the most investigated application of MD has been desalination, the low fouling propensity of the process, the ability to handle complex feed solutions, and the fact that separation occurs through temperature-induced phase equilibria at specific locations have led to many other interesting applications of MD being explored (Drioli *et al.*, 2015a). The temperature gradient-based nature of MD also opens new possibilities for use in vapor/gas separation applications where the equilibrium composition contains more volatile components at each temperature. As a result, the scope of the process has expanded beyond traditional desalting applications. MD could also be operated in an osmotic configuration where the mass transport through the membrane pores is driven by the osmotic pressure difference across the membrane. This operational mode is interesting for temperature-sensitive products such as pharmaceutical compounds, juices, dairy products, natural aromatic compounds, and various chemical solutions. MD has also demonstrated the potential to treat the solutions where the final effluent quality must meet strict criteria such as nuclear waste, radioactive water, or water treatment for the semiconductor industry.

In the oil and gas sector, shale gas has been recognized as a game changer due to its abundant availability in different parts of the world. However, the negative environmental impact of shale gas exploration remains a major obstacle to large-scale adaptation. Water produced along with the oil - so-called produced water - is a major contributor to the dangerous environmental impacts of shale gas exploration. The produced water contains a very high proportion of salts, various hydrocarbons, and production chemicals. Handling such complex liquids using state-of-the-art processes is a real challenge. Additionally, the high pressure and temperature of the water generated during the manufacturing process further complicate the immediate handling. MD has been shown to be a potential candidate for treating this water after certain physical processes that remove hydrocarbons from the stream.

Traditional MBR also has the biggest pollution problem. MD as a standalone process or integrated with other processes (such as FO) yielded very interesting results. Similarly, the removal of heavy metals that act as trace contaminants is a challenge for existing RO plants. For example, since boron exists as boric acid, under normal pH conditions it can diffuse through RO membranes, so conventional RO fails to meet the required removal criteria (0.5-2.5 ppm). Current alternative technologies are either expensive or not robust to changing operating conditions. The application of MD successfully removed boron far below the specified limit.

Another area of potential interest for MD is the recovery/removal of phosphorus from agricultural, domestic, and industrial wastewater. The presence of phosphorus in soil is essential for crop growth. Phosphorus influx, on the other hand, causes a condition called eutrophication. Eutrophication is characterized by the excessive growth of algae in the water, reducing oxygen levels and adversely affecting marine life. As phosphorus reserves are limited, MD can be used as a stand-alone process or in combination with other membrane-based processes to not only control the phosphorus content in wastewater, but also to increase phosphorus enrichment. Phosphorus crystals can also be recovered from large streams (Quist-Jensen *et al.*, 2018b; Simoni *et al.*, 2021). The same is true for protein crystallization and crystallization of pharmaceutical compounds by membrane crystallization.

5.5 OUTLOOK

MD has made great progress during the last two decades or so. It is expected that the process will attract further research and commercial interest for sustainable desalination as well as resource recovery from different liquid streams such as desalination and geothermal brines and wastewaters. However, it should be highlighted that further road to progression should not consider MD as the replacement of large-scale reverse osmosis units but rather a complementary process to augment the deficiencies of RO. For instance, MD could be used to concentrate the retentate of RO brine with the ambition to approach zero liquid discharge in seawater and brackish water desalination. Due to its capability to run with solar energy, MD can be a suitable candidate as a standalone desalination process in off-grid water-scarce regions. Due to its capability to concentrate, the process could also be a valuable tool to recover valuable dissolved components from different liquid streams. This is evident from the current interest in the process of the concentration of lithium brines. MD has also demonstrated the potential to produce electricity when operated in pressure retarded mode. This could turn the process into the simultaneous producer of freshwater, electricity, and raw materials (Rahman *et al.*, 2023).

Realization of the full potential of the process, however, is associated with overcoming some key challenges. On the membrane front, the development of membranes with long-lasting hydrophobic and anti-scaling/fouling characteristics is desired. Further research and development in material development and synthesis routes are needed to achieve this goal. As MD is more feasible for the treatment of high-concentrated solutions, the membrane scaling issues are expected to be more significant in MD than the conventional pressure-driven membrane processes such as NF and RO. Therefore, the development of improved techniques to overcome scaling issues is of paramount importance. In this context, the development of anti-scaling membranes as well as appropriate pre-treatment strategies is expected to offer an important contribution. MD is also gaining traction in food processing, and treatment of oily water and organic-rich wastewater where the scaling issues will be significantly higher than the conventionally investigated desalination applications. The membrane and process development, therefore, should also consider the appropriate strategies to tackle fouling issues when tackling such complex feed solutions. The presence of natural organic matter like humic acid, carbohydrates, proteins, lipids, and other low molecular weight species causes organic fouling in MD. Organic matter can adhere to the membrane surface through hydrophobic interactions, chemical affinity, and electrostatic forces. This adherence can cause reduced vapor permeability and can interfere with the hydrophobic character of the membrane. However, currently, very little attention has been devoted to developing strategies to tackle this type of fouling. Due to the low flux and different separation mechanism than the conventional pressure-driven membrane processes, particulate fouling has not received significant attention. However, the solid particles inherently present in the feed or crystals precipitating from the feed can create particulate fouling in MD as reported in the literature (Chimanlal *et al.*, 2022). Therefore, it becomes relevant to develop mitigation strategies for such fouling in MD. The possible strategies include appropriate pre-treatments (e.g., filtration, chemical precipitation), module designs, fouling-resistant membrane configurations and materials and optimized process conditions. Biofouling, or biological fouling, is caused by the accumulation of bacteria and living microorganisms on the surface of a membrane. It leads to the formation of a biofilm, which can significantly reduce membrane performance and productivity. Biofouling is less common in MD compared to other membrane processes, but it still occurs, especially in MD bioreactors. Biofouling in MD inhibits the process through pore wetting and pore blockage mechanisms, allowing particles to penetrate the permeate side and causing distillate contamination. Factors such as feed flow rate, membrane properties, microorganism properties, pH, and feed water source influence the attachment and growth of microorganisms on the membrane surface. To control biofouling in MD, techniques (appropriate pre-treatments, quorum quenching, membrane, and process design) developed for other membrane processes could become of interest.

MD is also becoming increasingly relevant for the treatment of acidic wastewater (e.g., the one originating from the battery recycling process). This will require the development of membranes that are tolerable to exposure to the low-pH solutions.

Future efforts on the process front should focus on minimizing the electric as well as thermal energy consumption of the process. This can be achieved by developing more energy-efficient membranes, process configurations, and module designs. More rational integration of different energy sources (solar, geothermal, and industrial) will also provide an important contribution. In particular, the studies should focus more on the integration of the process with geothermal heat, which is a more stable and broadly available source of energy.

5.6 REFERENCES

Abu-Zeid, M. A.E.R., X. Lu, and S. Zhang. 2021. Influence of Module Length on Water Desalination Using Air Gap Membrane Distillation Process: An Experimental Comparative Study. Arabian Journal for Science and Engineering 46, no. 8 (August 1): 7989–8008.

Abu-Zeid, M.A.E.-R., Y. Zhang, H. Dong, L. Zhang, H.-L. Chen, and L. Hou. 2015. A Comprehensive Review of Vacuum Membrane Distillation Technique. Desalination 356 (January): 1–14.

Abu-Zeid, Mostafa Abd El Rady, and G. ElMasry. 2020. Experimental Evaluation of Two Consecutive Air-Gap Membrane Distillation Modules with Heat Recovery. Water Science and Technology: Water Supply 20, no. 5 (August 1): 1678–1691.

Adewole, J.K., H.M. al Maawali, T. Jafary, A. Firouzi, and H. Oladipo. 2022. A Review of Seawater Desalination with Membrane Distillation: Material Development and Energy Requirements. Water Supply 22, no. 12 (December 1): 8500–8526.

Afsari, M., H.K. Shon, and L.D. Tijing. 2021. Janus Membranes for Membrane Distillation: Recent Advances and Challenges. Advances in Colloid and Interface Science 289 (March): 102362.

Ahmed, F.E., B.S. Lalia, and R. Hashaikeh. 2015. A Review on Electrospinning for Membrane Fabrication: Challenges and Applications. Desalination 356 (January): 15–30.

Ahmed, F.E., B.S. Lalia, R. Hashaikeh, and N. Hilal. 2020. Alternative Heating Techniques in Membrane Distillation: A Review. Desalination 496 (December 15): 114713.

Al-hadidi, A., Kemperman, A., Wessling, M. & van der Meer, W. (2008). The influence of membrane properties on the silt density index. In: EDS (ed.) Membranes in drinking and industrial water. Toulouse: EDS-INSA.

Ali, A., P. Aimar, and E. Drioli. 2015a. Effect of Module Design and Flow Patterns on Performance of Membrane Distillation Process. CHEMICAL ENGINEERING JOURNAL 277: 368–377. http://dx.doi.org/10.1016/j.cej.2015.04.108.

Ali, A., C.A. Quist-Jensen, F. Macedonio, and E. Drioli. 2016. Optimization of Module Length for Continuous Direct Contact Membrane Distillation Process. Chemical Engineering and Processing: Process Intensification 110 (December): 188–200.

Ali, Aamer, C. Quist-Jensen, F. Macedonio, and E. Drioli. 2015. Application of Membrane Crystallization for Minerals' Recovery from Produced Water. Membranes 5, no. 4 (November 24): 772–792.

Ali, Aamer, C.A. Quist-Jensen, E. Drioli, and F. Macedonio. 2018. Evaluation of Integrated Microfiltration and Membrane Distillation/Crystallization Processes for Produced Water Treatment. Desalination 434 (May): 161–168.

Ali, Aamer, C.A. Quist-Jensen, M.K. Jørgensen, A. Siekierka, M.L. Christensen, M. Bryjak, C. Hélix-Nielsen, and E. Drioli. 2021. A Review of Membrane Crystallization, Forward Osmosis and Membrane Capacitive Deionization for Liquid Mining. Resources, Conservation and Recycling 168 (May 1): 105273.

Ali, Aamer, C.A. Quist-Jensen, F. Macedonio, and E. Drioli. 2016. Optimization of Module Length and Membrane Thickness for Membrane Distillation. In 2nd International Workshop on Membrane Distillation and Innovating Membrane Operations in Desalination and Water Reuse, Ravello, Italy. Ravello.

Ali, Aamer, J.-H. Tsai, K.-L. Tung, E. Drioli, and F. Macedonio. 2018. Designing and Optimization of Continuous Direct Contact Membrane Distillation Process. Desalination 426 (January): 97–107.

Alkhudhiri, A., N. Darwish, and N. Hilal. 2012. Membrane Distillation: A Comprehensive Review. Desalination 287 (February): 2–18.

Alklaibi, A.M., and N. Lior. 2005. Membrane-Distillation Desalination: Status and Potential. Desalination 171, no. 2 (January 10): 111–131.

AlMarzooqi, F.A., M.R. Bilad, B. Mansoor, and H.A. Arafat. 2016. A Comparative Study of Image Analysis and Porometry Techniques for Characterization of Porous Membranes. Journal of Materials Science 51, no. 4 (February 27): 2017–2032.

Alpatova, A., A.S. Alsaadi, M. Alharthi, J.G. Lee, and N. Ghaffour. 2019. Co-Axial Hollow Fiber Module for Air Gap Membrane Distillation. Journal of Membrane Science 578 (May 15): 172–182.

An, A.K., J. Guo, E.-J. Lee, S. Jeong, Y. Zhao, Z. Wang, and T. Leiknes. 2017. PDMS/PVDF Hybrid Electrospun Membrane with Superhydrophobic Property and Drop Impact Dynamics for Dyeing Wastewater Treatment Using Membrane Distillation. Journal of Membrane Science 525 (March): 57–67.

Anderson, D. M., Boerlage, S. F. E. & Dixon, M. B. (2017). Harmful Algal Blooms (HABs) and Desalination: A Guide to Impacts, Monitoring and Management. In: UNESCO, I. O. C. o. (ed.) IOC Manuals and Guides No. 78. Paris: UNESCO.

Ashoor, B.B., S. Mansour, A. Giwa, V. Dufour, and S.W. Hasan. 2016. Principles and Applications of Direct Contact Membrane Distillation (DCMD): A Comprehensive Review. Desalination 398 (November): 222–246.

ASTM D4189 - 14 (2014). Standard Test Method for Silt Density Index (SDI) of Water. West Conshohocken, PA: ASTM International.

Attia, H., D.J. Johnson, C.J. Wright, and N. Hilal. 2018a. Comparison between Dual-Layer (Superhydrophobic–Hydrophobic) and Single Superhydrophobic Layer Electrospun Membranes for Heavy Metal Recovery by Air-Gap Membrane Distillation. Desalination 439 (August): 31–45.

———. 2018b. Robust Superhydrophobic Electrospun Membrane Fabricated by Combination of Electrospinning and Electrospraying Techniques for Air Gap Membrane Distillation. Desalination 446 (November): 70–82.

Belfer, S., R. Fainchtain, Y. Purinson, and O. Kedem. 2000. Surface Characterization by FTIR-ATR Spectroscopy of Polyethersulfone Membranes-Unmodified, Modified and Protein Fouled. Journal of Membrane Science 172, no. 1–2 (July): 113–124.

Belfort, G., Davis, R. H. & Zydney, A. L. (1994). The behavior of suspensions and macromolecular solutions in crossflow microfiltration. Journal of Membrane Science, 96, 1-58

Bhadra, M., S. Roy, and S. Mitra. 2016. Desalination across a Graphene Oxide Membrane via Direct Contact Membrane Distillation. Desalination 378 (January): 37–43.

Bonyadi, S., and T.S. Chung. 2007. Flux Enhancement in Membrane Distillation by Fabrication of Dual Layer Hydrophilic–Hydrophobic Hollow Fiber Membranes. Journal of Membrane Science 306, no. 1–2 (December): 134–146.

Bunaciu, A.A., E. gabriela Udri tioiu, and H.Y. Aboul-Enein. 2015. X-Ray Diffraction: Instrumentation and Applications. Critical Reviews in Analytical Chemistry 45, no. 4 (October 2): 289–299.

Camacho, L., L. Dumée, J. Zhang, J. Li, M. Duke, J. Gomez, and S. Gray. 2013. Advances in Membrane Distillation for Water Desalination and Purification Applications. Water 5, no. 1 (January 25): 94–196.

Chamani, H., J. Woloszyn, T. Matsuura, D. Rana, and C.Q. Lan. 2021. Pore Wetting in Membrane Distillation: A Comprehensive Review. Progress in Materials Science 122 (October 1): 100843.

Chen, L., P. Xu, and H. Wang. 2020. Interplay of the Factors Affecting Water Flux and Salt Rejection in Membrane Distillation: A State-of-the-Art Critical Review. Water 12, no. 10 (October 13): 2841.

Cheng, D., J. Zhang, N. Li, D. Ng, S.R. Gray, and Z. Xie. 2018. Antiwettability and Performance Stability of a Composite Hydrophobic/Hydrophilic Dual-Layer Membrane in Wastewater Treatment by Membrane Distillation. Industrial & Engineering Chemistry Research 57, no. 28 (July 18): 9313–9322.

Chimanlal, I., L.N. Nthunya, C. Quist-Jensen, and H. Richards. 2022. Membrane Distillation Crystallization for Water and Mineral Recovery: The Occurrence of Fouling and Its Control during Wastewater Treatment. Frontiers in Chemical Engineering 4 (November 29).

Conners, T.E., and S. Banerjee. 2020. Surface Analysis of Paper. CRC Press.

Curcio, E., and E. Drioli. 2005. Membrane Distillation and Related Operations—A Review. Separation & Purification Reviews 34, no. 1 (January): 35–86.

Dadari, S., M. Rahimi, and S. Zinadini. 2022. Novel Antibacterial and Antifouling PES Nanofiltration Membrane Incorporated with Green Synthesized Nickel-Bentonite Nanoparticles for Heavy Metal Ions Removal. Chemical Engineering Journal 431 (March): 134116.

Ding, Z., Z. Liu, and C. Xiao. 2021. Excellent Performance of Novel Superhydrophobic Composite Hollow Membrane in the Vacuum Membrane Distillation. Separation and Purification Technology 268 (August): 118603.

Drioli, E., A. Ali, S. Simone, F. Macedonio, S.A. AL-Jlil, F.S. al Shabonah, H.S. Al-Romaih, O. Al-Harbi, A. Figoli, and A. Criscuoli. 2013. Novel PVDF Hollow Fiber Membranes for Vacuum and Direct Contact Membrane Distillation Applications. Separation and Purification Technology 115 (August): 27–38.

Drioli, Enrico, A. Ali, and F. Macedonio. 2015. Membrane Distillation : Recent Developments and Perspectives. Desalination 356: 56–84. http://dx.doi.org/10.1016/j.desal.2014.10.028.

Dumée, L., V. Germain, K. Sears, J. Schütz, N. Finn, M. Duke, S. Cerneaux, D. Cornu, and S. Gray. 2011. Enhanced Durability and Hydrophobicity of Carbon Nanotube Bucky Paper Membranes in Membrane Distillation. Journal of Membrane Science 376, no. 1–2 (July): 241–246.

Duong, H.C., D. Chuai, Y.C. Woo, H.K. Shon, L.D. Nghiem, and V. Sencadas. 2018. A Novel Electrospun, Hydrophobic, and Elastomeric Styrene-Butadiene-Styrene Membrane for Membrane Distillation Applications. Journal of Membrane Science 549 (March): 420–427.

El-Bourawi, M.S., Z. Ding, R. Ma, and M. Khayet. 2006. A Framework for Better Understanding Membrane Distillation Separation Process. Journal of Membrane Science 285, no. 1–2 (November): 4–29.

Essalhi, M., and M. Khayet. 2013. Self-Sustained Webs of Polyvinylidene Fluoride Electrospun Nanofibers at Different Electrospinning Times: 1. Desalination by Direct Contact Membrane Distillation. Journal of Membrane Science 433 (April): 167–179.

———. 2014. Self-Sustained Webs of Polyvinylidene Fluoride Electrospun Nano-Fibers: Effects of Polymer Concentration and Desalination by Direct Contact Membrane Distillation. Journal of Membrane Science 454 (March): 133–143.

Essalhi, Mohamed, M. Khayet, N. Ismail, O. Sundman, and N. Tavajohi. 2021. Improvement of Nanostructured Electrospun Membranes for Desalination by Membrane Distillation Technology. Desalination 510 (August): 115086.

Eykens, L., K. de Sitter, C. Dotremont, L. Pinoy, and B. van der Bruggen. 2017. Membrane Synthesis for Membrane Distillation: A Review. Separation and Purification Technology 182 (July): 36–51.

Eykens, Lies, K. de Sitter, C. Dotremont, L. Pinoy, and B. van der Bruggen. 2016. How To Optimize the Membrane Properties for Membrane Distillation: A Review. Industrial & Engineering Chemistry Research 55, no. 35 (September 7): 9333–9343.

Fane, A.G., R.W. Schofield, and C.J.D. Fell. 1987. The Efficient Use of Energy in Membrane Distillation. Desalination 64 (January): 231–243.

Feng, S., Z. Zhong, Y. Wang, W. Xing, and E. Drioli. 2018. Progress and Perspectives in PTFE Membrane: Preparation, Modification, and Applications. Journal of Membrane Science 549 (March): 332–349.

Ferreira, R.K.M., H. Ramlow, C. Marangoni, and R.A.F. Machado. 2021. A Review on the Manufacturing Techniques of Porous Hydrophobic Ceramic Membranes Applied to Direct Contact Membrane Distillation. Advances in Applied Ceramics 120, no. 5–8 (November 17): 336–357.

Flack, H.D., and G. Bernardinelli. 2008. The Use of X-Ray Crystallography to Determine Absolute Configuration. Chirality 20, no. 5 (May 15): 681–690.

Francis, L., F.E. Ahmed, and N. Hilal. 2022. Advances in Membrane Distillation Module Configurations. Membranes 12, no. 1 (January 12): 81.

Francis, L., N. Ghaffour, A.S. Al-Saadi, and G.L. Amy. 2015. Submerged Membrane Distillation for Seawater Desalination. Desalination and Water Treatment 55, no. 10 (September 4): 2741–2746.

Franken, A.C.M., J.A.M. Nolten, M.H.V. Mulder, D. Bargeman, and C.A. Smolders. 1987. Wetting Criteria for the Applicability of Membrane Distillation. Journal of Membrane Science 33, no. 3 (October): 315–328.

García-Fernández, L., M. Khayet, and M.C. García-Payo. 2015. Membranes Used in Membrane Distillation: Preparation and Characterization. In Pervaporation, Vapour Permeation and Membrane Distillation, 317–359. Elsevier.

Garofalo, A., M.C. Carnevale, L. Donato, E. Drioli, O. Alharbi, S.A. Aljlil, A. Criscuoli, and C. Algieri. 2016. Scale-up of MFI Zeolite Membranes for Desalination by Vacuum Membrane Distillation. Desalination 397 (November): 205–212.

Gontarek-Castro, E., R. Castro-Muñoz, and M. Lieder. 2022. New Insights of Nanomaterials Usage toward Superhydrophobic Membranes for Water Desalination via Membrane Distillation: A Review. Critical Reviews in Environmental Science and Technology 52, no. 12 (June 18): 2104–2149.

Gopal, R., S. Kaur, Z. Ma, C. Chan, S. Ramakrishna, and T. Matsuura. 2006. Electrospun Nanofibrous Filtration Membrane. Journal of Membrane Science 281, no. 1–2 (September 15): 581–586.

Gryta, M. 2005. Long-Term Performance of Membrane Distillation Process. Journal of Membrane Science 265, no. 1–2 (November 15): 153–159.

———. 2007. Influence of Polypropylene Membrane Surface Porosity on the Performance of Membrane Distillation Process. Journal of Membrane Science 287, no. 1 (January 5): 67–78.

Gryta, M. 2011. The Influence of Magnetic Water Treatment on CaCO3 Scale Formation in Membrane Distillation Process. Separation and Purification Technology 80, no. 2 (July): 293–299.

———. 2020. The Application of Submerged Modules for Membrane Distillation. Membranes 10, no. 2 (February 1).

Guillen, G.R., T.P. Farrell, R.B. Kaner, and E.M. v. Hoek. 2010. Pore-Structure, Hydrophilicity, and Particle Filtration Characteristics of Polyaniline–Polysulfone Ultrafiltration Membranes. Journal of Materials Chemistry 20, no. 22: 4621.

He, K., H.J. Hwang, M.W. Woo, and I.S. Moon. 2011. Production of Drinking Water from Saline Water by Direct Contact Membrane Distillation (DCMD). Journal of Industrial and Engineering Chemistry 17, no. 1 (January): 41–48.

He, Y., J. Wang, X. Fu, H. Lin, W. Zhang, X. Wang, and F. Liu. 2023. Electrosprayed Thin Film Nanocomposite Polyamide Nanofiltration with Homogeneous Distribution of Nanoparticles for Enhanced Separation Performance. Desalination 546 (January): 116206.

Hilal, N., H. Al-Zoubi, N.A. Darwish, A.W. Mohamma, and M. Abu Arabi. 2004. A Comprehensive Review of Nanofiltration Membranes:Treatment, Pretreatment, Modelling, and Atomic Force Microscopy. Desalination 170, no. 3 (November): 281–308.

Hitsov, I., T. Maere, K. de Sitter, C. Dotremont, and I. Nopens. 2015. Modelling Approaches in Membrane Distillation: A Critical Review. Separation and Purification Technology 142 (March 4): 48–64.

Huang, J., Y. Hu, Y. Bai, Y. He, and J. Zhu. 2020. Novel Solar Membrane Distillation Enabled by a PDMS/CNT/PVDF Membrane with Localized Heating. Desalination 489 (September): 114529.

Huayan, C., W. Chunrui, J. Yue, W. Xuan, and L. Xiaolong. 2011. Comparison of Three Membrane Distillation Confi Gurations and Seawater Desalination by Vacuum Membrane Distillation. Desalination and Water Treatment 28, no. 1–3 (April 3): 321–327.

Hussain, A., A. Janson, J.M. Matar, and S. Adham. 2022. Membrane Distillation: Recent Technological Developments and Advancements in Membrane Materials. Emergent Materials 5, no. 2 (April 5): 347–367.

Intrchom, W., S. Roy, M. Humoud, and S. Mitra. 2018. Immobilization of Graphene Oxide on the Permeate Side of a Membrane Distillation Membrane to Enhance Flux. Membranes 8, no. 3 (August 15): 63.

Ismail, A.F., K.C. Khulbe, and T. Matsuura. 2019. RO Membrane Characterization. In Reverse Osmosis, 57–90. Elsevier.

Ismail, M.F., M.A. Islam, B. Khorshidi, A. Tehrani-Bagha, and M. Sadrzadeh. 2022. Surface Characterization of Thin-Film Composite Membranes Using Contact Angle Technique: Review of Quantification Strategies and Applications. Advances in Colloid and Interface Science 299 (January): 102524.

Jantaporn, W., A. Ali, and P. Aimar. 2017. Specific Energy Requirement of Direct Contact Membrane Distillation. Chemical Engineering Research and Design 128 (December): 15–26.

Jena, A., and K. Gupta. 2010. Advances in Pore Structure Evaluation by Porometry. Chemical Engineering & Technology 33, no. 8 (June 25): 1241–1250.

Jena, Akshaya, and K. Gupta. 2005. Pore Volume of Nanofiber Nonwovens. International Nonwovens. Journal os-14, no. 2 (June 10): 1558925005os–14.

Jhaveri, J.H., and Z.V.P. Murthy. 2016. A Comprehensive Review on Anti-Fouling Nanocomposite Membranes for Pressure Driven Membrane Separation Processes. Desalination 379 (February): 137–154.

Ji, J., F. Liu, N.A. Hashim, M.R.M. Abed, and K. Li. 2015. Poly(Vinylidene Fluoride) (PVDF) Membranes for Fluid Separation. Reactive and Functional Polymers 86 (January): 134–153.

Johnson, D., and N. Hilal. 2015. Characterisation and Quantification of Membrane Surface Properties Using Atomic Force Microscopy: A Comprehensive Review. Desalination 356 (January): 149–164.

Julian, H., N. Nurgirisia, G. Qiu, Y.-P. Ting, and I.G. Wenten. 2022. Membrane Distillation for Wastewater Treatment: Current Trends, Challenges and Prospects of Dense Membrane Distillation. Journal of Water Process Engineering 46 (April): 102615.

Kahle, M., M. Kleber, and R. Jahn. 2002. Review of XRD-Based Quantitative Analyses of Clay Minerals in Soils: The Suitability of Mineral Intensity Factors. Geoderma 109, no. 3–4 (October): 191–205.

Kalla, S., S. Upadhyaya, and K. Singh. 2019. Principles and Advancements of Air Gap Membrane Distillation. Reviews in Chemical Engineering 35, no. 7 (October 25): 817–859.

Kamaz, M., A. Sengupta, A. Gutierrez, Y.-H. Chiao, and R. Wickramasinghe. 2019. Surface Modification of PVDF Membranes for Treating Produced Waters by Direct Contact Membrane Distillation. International Journal of Environmental Research and Public Health 16, no. 5 (February 26): 685.

Kang, G., and Y. Cao. 2014. Application and Modification of Poly(Vinylidene Fluoride) (PVDF) Membranes – A Review. Journal of Membrane Science 463 (August): 145–165.

Khayet, M., and C. Cojocaru. 2012. Air Gap Membrane Distillation: Desalination, Modeling and Optimization. Desalination 287 (February): 138–145.

Khayet, M., M.C. García-Payo, L. García-Fernández, and J. Contreras-Martínez. 2018. Dual-Layered Electrospun Nanofibrous Membranes for Membrane Distillation. Desalination 426 (January): 174–184.

Khayet, M., M.P. Godino, and J.I. Mengual. 2003. Possibility of Nuclear Desalination through Various Membrane Distillation Configurations: A Comparative Study. International Journal of Nuclear Desalination 1, no. 1: 30.

Khayet, M., K. Khulbe, and T. Matsuura. 2004. Characterization of Membranes for Membrane Distillation by Atomic Force Microscopy and Estimation of Their Water Vapor Transfer Coefficients in Vacuum Membrane Distillation Process. Journal of Membrane Science 238, no. 1–2 (July 15): 199–211.

Khayet, M., and T. Matsuura. 2003. Application of Surface Modifying Macromolecules for the Preparation of Membranes for Membrane Distillation. Desalination 158, no. 1–3 (August): 51–56.

Khayet, Mohamed. 2011. Membranes and Theoretical Modeling of Membrane Distillation: A Review. Advances in Colloid and Interface Science 164, no. 1–2 (May): 56–88.

Khayet, Mohamed, and T. Matsuura. 2001. Preparation and Characterization of Polyvinylidene Fluoride Membranes for Membrane Distillation. Industrial & Engineering Chemistry Research 40, no. 24 (November 1): 5710–5718.

Khulbe, K.C., and T. Matsuura. 2017. Synthetic Membrane Characterisation – a Review: Part I. Membrane Technology 2017, no. 7 (July): 7–12.

Kim, W.-J., O. Campanella, and D.R. Heldman. 2021. Predicting the Performance of Direct Contact Membrane Distillation (DCMD): Mathematical Determination of Appropriate Tortuosity Based on Porosity. Journal of Food Engineering 294 (April): 110400.

Kimura, S., S.-I. Nakao, and S.-I. Shimatani. 1987. Transport Phenomena in Membrane Distillation. Journal of Membrane Science 33, no. 3 (October): 285–298.

Kirk, T.L. 2017. A Review of Scanning Electron Microscopy in Near Field Emission Mode. In , 39–109.

Kolb, H.E., R. Schmitt, A. Dittler, and G. Kasper. 2018. On the Accuracy of Capillary Flow Porometry for Fibrous Filter Media. Separation and Purification Technology 199 (June): 198–205.

Korolkov, I. v., Y.G. Gorin, A.B. Yeszhanov, A.L. Kozlovskiy, and M. v. Zdorovets. 2018. Preparation of PET Track-Etched Membranes for Membrane Distillation by Photo-Induced Graft Polymerization. Materials Chemistry and Physics 205 (February): 55–63.

Lawson, K.W., and D.R. Lloyd. 1997. Membrane Distillation. Journal of Membrane Science 124, no. 1 (February): 1–25.

Lewczuk, B., and N. Szyrynska. 2021. Field-Emission Scanning Electron Microscope as a Tool for Large-Area and Large-Volume Ultrastructural Studies. Animals 11, no. 12 (November 27): 3390.

Li, D., M.W. Frey, and Y.L. Joo. 2006. Characterization of Nanofibrous Membranes with Capillary Flow Porometry. Journal of Membrane Science 286, no. 1–2 (December): 104–114.

Lu, K.J., Y. Chen, and T.-S. Chung. 2019. Design of Omniphobic Interfaces for Membrane Distillation – A Review. Water Research 162 (October): 64–77.

Lu, K.-J., J. Zuo, and T.-S. Chung. 2016. Tri-Bore PVDF Hollow Fibers with a Super-Hydrophobic Coating for Membrane Distillation. Journal of Membrane Science 514 (September): 165–175.

Mat, N.C., Y. Lou, and G.G. Lipscomb. 2014. Hollow Fiber Membrane Modules. Current Opinion in Chemical Engineering 4: 18–24.

McGaughey, A.L., P. Karandikar, M. Gupta, and A.E. Childress. 2020. Hydrophobicity versus Pore Size: Polymer Coatings to Improve Membrane Wetting Resistance for Membrane Distillation. ACS Applied Polymer Materials 2, no. 3 (March 13): 1256–1267.

Meng, S., Y.C. Hsu, Y. Ye, and V. Chen. 2015. Submerged Membrane Distillation for Inland Desalination Applications. Desalination 361 (April 1): 72–80.

Mohammadi, T., P. Kazemi, and M. Peydayesh. 2015. Optimization of Vacuum Membrane Distillation Parameters for Water Desalination Using Box–Behnken Design. Desalination and Water Treatment 56, no. 9 (November 27): 2306–2315.

Mokhtar, N.M., W.J. Lau, A.F. Ismail, and B.C. Ng. 2014. Physicochemical Study of Polyvinylidene Fluoride–Cloisite15A® Composite Membranes for Membrane Distillation Application. RSC Adv. 4, no. 108: 63367–63379.

Mousa, H.M., M. Hamdy, M.A. Yassin, M.M. El-Sayed Seleman, and G.T. Abdel-Jaber. 2022. Characterization of Nanofiber Composite Membrane for High Water Flux and Antibacterial Properties. Colloids and Surfaces A: Physicochemical and Engineering Aspects 651 (October): 129655.

Nakao, S. 1994. Determination of Pore Size and Pore Size Distribution. Journal of Membrane Science 96, no. 1–2 (November): 131–165.

Naresh-Kumar, G., B. Hourahine, P.R. Edwards, A.P. Day, A. Winkelmann, A.J. Wilkinson, P.J. Parbrook, G. England, and C. Trager-Cowan. 2012. Rapid Nondestructive Analysis of Threading Dislocations in Wurtzite Materials Using the Scanning Electron Microscope. Physical Review Letters 108, no. 13 (March 30): 135503.

Niknejad, A. S., S. Bazgir, M. Ardjmand, and M.M.A. Shirazi. 2021. Spent Caustic Wastewater Treatment Using Direct Contact Membrane Distillation with Electroblown Styrene-Acrylonitrile Membrane. International Journal of Environmental Science and Technology 18, no. 8 (August 14): 2283–2294.

Niknejad, Ali Sallakh, S. Bazgir, and A. Kargari. 2021. Desalination by Direct Contact Membrane Distillation Using a Superhydrophobic Nanofibrous Poly (Methyl Methacrylate) Membrane. Desalination 511 (September): 115108.

Niknejad, Ali Sallakh, S. Bazgir, A. Sadeghzadeh, and M.M.A. Shirazi. 2021. Evaluation of a Novel and Highly Hydrophobic Acrylonitrile-Butadiene-Styrene Membrane for Direct Contact Membrane Distillation: Electroblowing/Air-Assisted Electrospraying Techniques. Desalination 500 (March): 114893.

Norby, P. 2006. In-Situ XRD as a Tool to Understanding Zeolite Crystallization. Current Opinion in Colloid & Interface Science 11, no. 2–3 (June): 118–125.

Olatunji, S.O., and L.M. Camacho. 2018. Heat and Mass Transport in Modeling Membrane Distillation Configurations: A Review. Frontiers in Energy Research 6 (December 4).

Omar, N.M.A., M.H.D. Othman, Z.S. Tai, T.A. Kurniawan, T. El-badawy, P.S. Goh, N.H. Othman, M.A. Rahman, J. Jaafar, and A.F. Ismail. 2022. Bottlenecks and Recent Improvement Strategies of Ceramic Membranes in Membrane Distillation Applications: A Review. Journal of the European Ceramic Society 42, no. 13 (October): 5179–5194.

Pangarkar, B.L., and S.K. Deshmukh. 2015. Theoretical and Experimental Analysis of Multi-Effect Air Gap Membrane Distillation Process (ME-AGMD). Journal of Environmental Chemical Engineering 3, no. 3 (September 29): 2127–2135.

Park, S.H., J.H. Kim, S.J. Moon, J.T. Jung, H.H. Wang, A. Ali, C.A. Quist-Jensen, F. Macedonio, E. Drioli, and Y.M. Lee. 2020. Lithium Recovery from Artificial Brine Using Energy-Efficient Membrane Distillation and Nanofiltration. Journal of Membrane Science 598 (March): 117683.

Peng, N., N. Widjojo, P. Sukitpaneenit, M.M. Teoh, G.G. Lipscomb, T.-S. Chung, and J.-Y. Lai. 2012. Evolution of Polymeric Hollow Fibers as Sustainable Technologies: Past, Present, and Future. Progress in Polymer Science 37, no. 10 (October): 1401–1424.

Petkov, V. 2008. Nanostructure by High-Energy X-Ray Diffraction. Materials Today 11, no. 11 (November): 28–38.

Peydayesh, M., P. Kazemi, A. Bandegi, T. Mohammadi, and O. Bakhtiari. 2015. Treatment of Bentazon Herbicide Solutions by Vacuum Membrane Distillation. Journal of Water Process Engineering 8 (December): e17–e22.

Phattaranawik, J., R. Jiraratananon, and A.G. Fane. 2003. Heat Transport and Membrane Distillation Coefficients in Direct Contact Membrane Distillation. Journal of Membrane Science 212, no. 1–2 (February): 177–193.

Pouya, Z.A., M.A. Tofighy, and T. Mohammadi. 2021. Synthesis and Characterization of Polytetrafluoroethylene/Oleic Acid-Functionalized Carbon Nanotubes Composite Membrane for Desalination by Vacuum Membrane Distillation. Desalination 503 (May): 114931.

Prince, J.A., G. Singh, D. Rana, T. Matsuura, V. Anbharasi, and T.S. Shanmugasundaram. 2012. Preparation and Characterization of Highly Hydrophobic Poly(Vinylidene Fluoride) – Clay Nanocomposite Nanofiber Membranes (PVDF–Clay NNMs) for Desalination Using Direct Contact Membrane Distillation. Journal of Membrane Science 397–398 (April): 80–86.

Qasim, M., I.U. Samad, N.A. Darwish, and N. Hilal. 2021. Comprehensive Review of Membrane Design and Synthesis for Membrane Distillation. Desalination 518 (December): 115168.

Qin, L., I.A. Mergos, and H. Verweij. 2015. Obtaining Accurate Cross-Section Images of Supported Polymeric and Inorganic Membrane Structures. Journal of Membrane Science 476 (February): 194–199.

Qtaishat, M., T. Matsuura, B. Kruczek, and M. Khayet. 2008. Heat and Mass Transfer Analysis in Direct Contact Membrane Distillation. Desalination 219, no. 1–3 (January): 272–292.

Quist-Jensen, C. A., J.M. Sorensen, A. Svenstrup, L. Scarpa, T.S. Carlsen, H.C. Jensen, L. Wybrandt, and M.L. Christensen. 2018a. Membrane Crystallization for Phosphorus Recovery and Ammonia Stripping from Reject Water from Sludge Dewatering Process. Desalination 440 (August 15): 156–160.

Quist-Jensen, C.A., F. Macedonio, C. Conidi, A. Cassano, S. Aljlil, O.A. Alharbi, and E. Drioli. 2016. Direct Contact Membrane Distillation for the Concentration of Clarified Orange Juice. Journal of Food Engineering 187 (October): 37–43.

Quist-Jensen, Cejna Anna, A. Ali, S. Mondal, F. Macedonio, and E. Drioli. 2016a. A Study of Membrane Distillation and Crystallization for Lithium Recovery from High-Concentrated Aqueous Solutions. Journal of Membrane Science 505 (May): 167–173.

Quist-Jensen, Cejna Anna, F. Macedonio, D. Horbez, and E. Drioli. 2017. Reclamation of Sodium Sulfate from Industrial Wastewater by Using Membrane Distillation and Membrane Crystallization. Desalination 401 (January): 112–119.

Rácz, G., S. Kerker, Z. Kovács, G. Vatai, M. Ebrahimi, and P. Czermak. 2014. Theoretical and Experimental Approaches of Liquid Entry Pressure Determination in Membrane Distillation Processes. Periodica Polytechnica Chemical Engineering 58, no. 2: 81–91.

Rahman, M.M., S. Al-Sulaimi, and A.M. Farooque. 2018. Characterization of New and Fouled SWRO Membranes by ATR/FTIR Spectroscopy. Applied Water Science 8, no. 6 (October 26): 183.

Rahman, S.N., H. Saleem, and S.J. Zaidi. 2023. Progress in Membranes for Pressure Retarded Osmosis Application. Desalination 549 (March): 116347.

Ramlow, H., R.K.M. Ferreira, C. Marangoni, and R.A.F. Machado. 2019. Ceramic Membranes Applied to Membrane Distillation: A Comprehensive Review. International Journal of Applied Ceramic Technology 16, no. 6 (November 13): 2161–2172.

Ravi, J., M.H.D. Othman, T. Matsuura, M. Ro'il Bilad, T.H. El-badawy, F. Aziz, A.F. Ismail, M.A. Rahman, and J. Jaafar. 2020. Polymeric Membranes for Desalination Using Membrane Distillation: A Review. Desalination 490 (September): 114530.

Rezaei, M., D.M. Warsinger, J.H. Lienhard V, M.C. Duke, T. Matsuura, and W.M. Samhaber. 2018. Wetting Phenomena in Membrane Distillation: Mechanisms, Reversal, and Prevention. Water Research 139 (August): 329–352.

Sadeghi Ghari, H., Z. Shakouri, and M.M.A. Shirazi. 2014. Evaluation of Microstructure of Natural Rubber/Nano-Calcium Carbonate Nanocomposites by Solvent Transport Properties. Plastics, Rubber and Composites 43, no. 6.

Sadeghzadeh, A., S. Bazgir, and M.M.A. Shirazi. 2020. Fabrication and Characterization of a Novel Hydrophobic Polystyrene Membrane Using Electroblowing Technique for Desalination by Direct Contact Membrane Distillation. Separation and Purification Technology 239.

Sadeghzadeh, Amirhossein, S. Bazgir, and M.M.A. Shirazi. 2020. Fabrication and Characterization of a Novel Hydrophobic Polystyrene Membrane Using Electroblowing Technique for Desalination by Direct Contact Membrane Distillation. Separation and Purification Technology 239 (May): 116498.

Saffarini, R.B., B. Mansoor, R. Thomas, and H.A. Arafat. 2013. Effect of Temperature-Dependent Microstructure Evolution on Pore Wetting in PTFE Membranes under Membrane Distillation Conditions. Journal of Membrane Science 429 (February): 282–294.

Said, I.A., T. Chomiak, J. Floyd, and Q. Li. 2020. Sweeping Gas Membrane Distillation (SGMD) for Wastewater Treatment, Concentration, and Desalination: A Comprehensive Review. Chemical Engineering and Processing - Process Intensification 153 (July): 107960.

Salinas Rodríguez, S. G., Amy, G. L., Schippers, J. C. & Kennedy, M. D. (2015). The Modified Fouling Index Ultrafiltration Constant Flux for assessing particulate/colloidal fouling of RO systems. Desalination, 365, 79-91.

Sanaeepur, H., A. Ebadi Amooghin, M.M.A. Shirazi, M. Pishnamazi, and S. Shirazian. 2022. Water Desalination and Ion Removal Using Mixed Matrix Electrospun Nanofibrous Membranes: A Critical Review. Desalination 521 (January): 115350.

Schofield, R.W., A.G. Fane, and C.J.D. Fell. 1987. Heat and Mass Transfer in Membrane Distillation. Journal of Membrane Science 33, no. 3 (October): 299–313.

Schofield, R.W., A.G. Fane, C.J.D. Fell, and R. Macoun. 1990. Factors Affecting Flux in Membrane Distillation. Desalination 77 (March): 279–294.

Shahu, V.T., and S.B. Thombre. 2019a. Air Gap Membrane Distillation: A Review. Journal of Renewable and Sustainable Energy 11, no. 4 (July): 045901.

———. 2021. Experimental Analysis of a Novel Helical Air Gap Membrane Distillation System. Water Science and Technology: Water Supply 21, no. 4 (June 1): 1450–1463.

Shalaby, S.M., A.E. Kabeel, H.F. Abosheiasha, M.K. Elfakharany, E. El-Bialy, A. Shama, and R.D. Vidic. 2022. Membrane Distillation Driven by Solar Energy: A Review. Journal of Cleaner Production 366 (September 15): 132949.

Sharma, S., C.N. Shyam Kumar, J.G. Korvink, and C. Kübel. 2018. Evolution of Glassy Carbon Microstructure: In Situ Transmission Electron Microscopy of the Pyrolysis Process. Scientific Reports 8, no. 1 (November 2): 16282.

Sharma, V., and A. Bhardwaj. 2019. Scanning Electron Microscopy (SEM) in Food Quality Evaluation. In Evaluation Technologies for Food Quality, 743–761. Elsevier.

Shirazi, M. M.A., A. Kargari, S. Bazgir, M. Tabatabaei, M.J.A. Shirazi, M.S. Abdullah, T. Matsuura, and A.F. Ismail. 2013. Characterization of Electrospun Polystyrene Membrane for Treatment of Biodiesel's Water-Washing Effluent Using Atomic Force Microscopy. Desalination 329 (November 15): 1–8.

Shirazi, A., M., and A. Kargari. 2019. Concentrating of Sugar Syrup in Bioethanol Production Using Sweeping Gas Membrane Distillation. Membranes 9, no. 5 (May 1): 59. Shirazi, M.J.A., S.

Bazgir, M.M.A. Shirazi, and S. Ramakrishna. 2013. Coalescing Filtration of Oily Wastewaters: Characterization and Application of Thermal Treated, Electrospun Polystyrene Filters. Desalination and Water Treatment 51, no. 31–33 (September): 5974–5986.

Shirazi, M.M.A., A. Kargari, S. Bazgir, M. Tabatabaei, M.J.A. Shirazi, M.S. Abdullah, T. Matsuura, and A.F. Ismail. 2013. Characterization of Electrospun Polystyrene Membrane for Treatment of Biodiesel's Water-Washing Effluent Using Atomic Force Microscopy. Desalination 329 (November): 1–8.

Shirazi, M.M.A., A. Kargari, and M.J.A. Shirazi. 2012. Direct Contact Membrane Distillation for Seawater Desalination. Desalination and Water Treatment 49, no. 1–3.

Shirazi, M.M.A., A. Kargari, and M. Tabatabaei. 2014. Evaluation of Commercial PTFE Membranes in Desalination by Direct Contact Membrane Distillation. Chemical Engineering and Processing: Process Intensification 76.

Shirazi, Mohammad Mahdi A., D. Bastani, A. Kargari, and M. Tabatabaei. 2013. Characterization of Polymeric Membranes for Membrane Distillation Using Atomic Force Microscopy. Desalination and Water Treatment 51, no. 31–33 (September): 6003–6008.

Shirazi, Mohammad Mahdi A., S. Bazgir, and F. Meshkani. 2020a. A Novel Dual-Layer, Gas-Assisted Electrospun, Nanofibrous SAN4-HIPS Membrane for Industrial Textile Wastewater Treatment by Direct Contact Membrane Distillation (DCMD). Journal of Water Process Engineering 36 (August): 101315.

———. 2020b. A Dual-Layer, Nanofibrous Styrene-Acrylonitrile Membrane with Hydrophobic/Hydrophilic Composite Structure for Treating the Hot Dyeing Effluent by Direct Contact Membrane Distillation. Chemical Engineering Research and Design 164 (December): 125–146.

Shirazi, Mohammad Mahdi A., and L.F. Dumée. 2022. Membrane Distillation for Sustainable Wastewater Treatment. Journal of Water Process Engineering 47 (June 1): 102670.

Shirazi, Mohammad Mahdi A., A. Kargari, A.F. Ismail, and T. Matsuura. 2016. Computational Fluid Dynamic (CFD) Opportunities Applied to the Membrane Distillation Process: State-of-the-Art and Perspectives. Desalination 377 (January 1): 73–90.

Shirazi, Mohammad Mahdi A., A. Kargari, M. Tabatabaei, A.F. Ismail, and T. Matsuura. 2014. Concentration of Glycerol from Dilute Glycerol Wastewater Using Sweeping Gas Membrane Distillation. Chemical Engineering and Processing: Process Intensification 78 (April): 58–66.

Shyam Kumar, C.N., V.S.K. Chakravadhanula, A. Riaz, S. Dehm, D. Wang, X. Mu, B. Flavel, R. Krupke, and C. Kübel. 2017. Understanding the Graphitization and Growth of Free-Standing Nanocrystalline Graphene Using in Situ Transmission Electron Microscopy. Nanoscale 9, no. 35: 12835–12842.

Silva, R. de S., H. Ramlow, B. de C. Santos, H.B. Madalosso, R.A.F. Machado, and C. Marangoni. 2021. Membrane Distillation: Experimental Evaluation of Liquid Entry Pressure in Commercial Membranes with Textile Dye Solutions. Journal of Water Process Engineering 44 (December): 102339.

Simoni, G., B.S. Kirkebæk, C.A. Quist-Jensen, M.L. Christensen, and A. Ali. 2021. A Comparison of Vacuum and Direct Contact Membrane Distillation for Phosphorus and Ammonia Recovery from Wastewater. Journal of Water Process Engineering 44 (December 1): 102350.

Suk, D., T. Matsuura, H. Park, and Y. Lee. 2006. Synthesis of a New Type of Surface Modifying Macromolecules (NSMM) and Characterization and Testing of NSMM Blended Membranes for Membrane Distillation. Journal of Membrane Science 277, no. 1–2 (June 1): 177–185.

Susanto, H. 2011a. Towards Practical Implementations of Membrane Distillation. Chemical Engineering and Processing: Process Intensification 50, no. 2 (February): 139–150.

Tai, Z.S., M.H.A. Aziz, M.H.D. Othman, A.F. Ismail, M.A. Rahman, and J. Jaafar. 2019. An Overview of Membrane Distillation. In Membrane Separation Principles and Applications, 251–281. Elsevier.

Talukder, M.E., Md.N. Pervez, W. Jianming, G.K. Stylios, M.M. Hassan, H. Song, V. Naddeo, and A. Figoli. 2022. Ag Nanoparticles Immobilized Sulfonated Polyethersulfone/Polyethersulfone Electrospun Nanofiber Membrane for the Removal of Heavy Metals. Scientific Reports 12, no. 1 (April 6): 5814.

Tanis-Kanbur, M.B., R.I. Peinador, J.I. Calvo, A. Hernández, and J.W. Chew. 2021. Porosimetric Membrane Characterization Techniques: A Review. Journal of Membrane Science 619 (February): 118750.

Tibi, F., A. Charfi, J. Cho, and J. Kim. 2020. Fabrication of Polymeric Membranes for Membrane Distillation Process and Application for Wastewater Treatment: Critical Review. Process Safety and Environmental Protection 141 (September): 190–201.

Tijing, L.D., J.-S. Choi, S. Lee, S.-H. Kim, and H.K. Shon. 2014. Recent Progress of Membrane Distillation Using Electrospun Nanofibrous Membrane. Journal of Membrane Science 453 (March): 435–462.

Tijing, L.D., W. Choi, Z. Jiang, A. Amarjargal, C.-H. Park, H.R. Pant, I.-T. Im, and C.S. Kim. 2013. Two-Nozzle Electrospinning of (MWNT/PU)/PU Nanofibrous Composite Mat with Improved Mechanical and Thermal Properties. Current Applied Physics 13, no. 7 (September): 1247–1255.

Tijing, L.D., J.R.C. Dizon, I. Ibrahim, A.R.N. Nisay, H.K. Shon, and R.C. Advincula. 2020. 3D Printing for Membrane Separation, Desalination and Water Treatment. Applied Materials Today 18 (March 1): 100486.

Tsai, J.H., C. Quist-Jensen, and A. Ali. 2023. Multipass Hollow Fiber Membrane Modules for Membrane Distillation. Desalination 548 (February 15).

Tung, K.-L., K.-S. Chang, T.-T. Wu, N.-J. Lin, K.-R. Lee, and J.-Y. Lai. 2014. Recent Advances in the Characterization of Membrane Morphology. Current Opinion in Chemical Engineering 4 (May): 121–127.

Vicente, J., Y. Wyart, and P. Moulin. 2013. Characterization (two-dimensional - three-dimensional) of ceramic microfiltration membrane by synchrotron radiation: new and abraded membranes. Journal of Porous Media 16, no. 6: 537–545.

Vladár, A.E., and V.-D. Hodoroaba. 2020. Characterization of Nanoparticles by Scanning Electron Microscopy. In Characterization of Nanoparticles, 7–27. Elsevier.

Wang, K., A.A. Abdalla, M.A. Khaleel, N. Hilal, and M.K. Khraisheh. 2017. Mechanical Properties of Water Desalination and Wastewater Treatment Membranes. Desalination 401 (January): 190–205.

Wang, P., and T.-S. Chung. 2015. Recent Advances in Membrane Distillation Processes: Membrane Development, Configuration Design and Application Exploring. Journal of Membrane Science 474 (January): 39–56.

Wang, X., Y. Liu, K. Fan, P. Cheng, and S. Xia. 2023. Nano-Striped Polyamide Membranes Enabled by Vacuum-Assisted Incorporation of Hierarchical Flower-like MoS2 for Enhanced Nanofiltration Performance. Journal of Membrane Science 668 (February): 121250.

Wei, X., B. Zhao, X.-M. Li, Z. Wang, B.-Q. He, T. He, and B. Jiang. 2012. CF4 Plasma Surface Modification of Asymmetric Hydrophilic Polyethersulfone Membranes for Direct Contact Membrane Distillation. Journal of Membrane Science 407–408 (July): 164–175.

Wiktor, C., M. Meledina, S. Turner, O.I. Lebedev, and R.A. Fischer. 2017. Transmission Electron Microscopy on Metal–Organic Frameworks – a Review. Journal of Materials Chemistry A 5, no. 29: 14969–14989.

Winter, D., J. Koschikowski, and S. Ripperger. 2012. Desalination Using Membrane Distillation: Flux Enhancement by Feed Water Deaeration on Spiral-Wound Modules. Journal of Membrane Science 423–424 (December): 215–224. http://linkinghub.elsevier.com/retrieve/pii/S0376738812006175.

Winter, D., J. Koschikowski, and M. Wieghaus. 2011. Desalination Using Membrane Distillation: Experimental Studies on Full Scale Spiral Wound Modules. Journal of Membrane Science 375, no. 1–2: 104–112. http://dx.doi.org/10.1016/j.memsci.2011.03.030.

Wyart, Y., G. Georges, C. Deumié, C. Amra, and P. Moulin. 2008. Membrane Characterization by Microscopic Methods: Multiscale Structure. Journal of Membrane Science 315, no. 1–2 (May): 82–92.

Xing Yang, Hui Yu, Rong Wang, A.G.F., X. Yang, H. Yu, R. Wang, and A.G. Fane. 2012. Optimization of Microstructured Hollow Fiber Design for Membrane Distillation Applications Using CFD Modeling. Journal of Membrane Science 421–422 (December): 258–270. http://dx.doi.org/10.1016/j.memsci.2012.07.022.

Yang, F., J.E. Efome, D. Rana, T. Matsuura, and C. Lan. 2018. Metal–Organic Frameworks Supported on Nanofiber for Desalination by Direct Contact Membrane Distillation. ACS Applied Materials & Interfaces 10, no. 13 (April 4): 11251–11260.

Yang, X., R. Wang, A.G. Fane, C.Y. Tang, and I.G. Wenten. 2013. Membrane Module Design and Dynamic Shear-Induced Techniques to Enhance Liquid Separation by Hollow Fiber Modules: A Review. Desalination and Water Treatment 51, no. 16–18: 3604–3627. https://doi.org/10.1080/19443994.2012.751146.

Zhang, W., Y. Li, J. Liu, B. Li, and S. Wang. 2017. Fabrication of Hierarchical Poly (Vinylidene Fluoride) Micro/Nano-Composite Membrane with Anti-Fouling Property for Membrane Distillation. Journal of Membrane Science 535 (August): 258–267.

Zhao, D., J. Zuo, K.-J. Lu, and T.-S. Chung. 2017. Fluorographite Modified PVDF Membranes for Seawater Desalination via Direct Contact Membrane Distillation. Desalination 413 (July): 119–126.

Zhao, K., W. Heinzl, M. Wenzel, S. Büttner, F. Bollen, G. Lange, S. Heinzl, and N. Sarda. 2013. Experimental Study of the Memsys Vacuum-Multi-Effect-Membrane-Distillation (V-MEMD) Module. Desalination 323 (August): 150–160. http://linkinghub.elsevier.com/retrieve/pii/S0011916412006509.

Zhu, Y., H. Inada, A. Hartschuh, L. Shi, A. Della Pia, G. Costantini, A.L. Vázquez de Parga, *et al.* 2012. Scanning Electron Microscopy. In Encyclopedia of Nanotechnology, 2273–2280. Dordrecht: Springer Netherlands.

Ziel, R., A. Haus, and A. Tulke. 2008. Quantification of the Pore Size Distribution (Porosity Profiles) in Microfiltration Membranes by SEM, TEM and Computer Image Analysis. Journal of Membrane Science 323, no. 2 (October): 241–246.

Zou, D., and Y.M. Lee. 2022. Design Strategy of Poly(Vinylidene Fluoride) Membranes for Water Treatment. Progress in Polymer Science 128 (May): 101535.

Zuo, J., and T.-S. Chung. 2016. Metal–Organic Framework-Functionalized Alumina Membranes for Vacuum Membrane Distillation. Water 8, no. 12 (December 8): 586.

Zuo, J., T.-S. Chung, G.S. O'Brien, and W. Kosar. 2017. Hydrophobic/Hydrophilic PVDF/Ultem® Dual-Layer Hollow Fiber Membranes with Enhanced Mechanical Properties for Vacuum Membrane Distillation. Journal of Membrane Science 523 (February): 103–110.

Part 2

Particulate fouling

doi: 10.2166/9781789062977_0139

Chapter 6

Silt Density Index

Steven J. Duranceau, University of Central Florida, USA

The learning objectives of this chapter are the following:

- Define the silt density index (SDI) and explain its significance
- Present a method that can be used to characterize the particulate fouling potential of reverse osmosis feedwater
- Understand the theory behind the SDI and discuss the basic equations that are used to calculate fouling indices
- To learn how to perform a SDI test using the appropriate equipment and procedures
- Identify and explain the limitations and deficiencies of the SDI as a measure of particulate fouling in synthetic membrane processes.

6.0 ABSTRACT

The most widely accepted and historically used predictor of the fouling potential of reverse osmosis feedwater is the plugging factor (PF), now commonly known as the Silt Density Index (SDI). The SDI procedure was standardized by ASTM International and is intended to be used as a measure of the fouling capacity of feedwater to reverse osmosis systems. The SDI is an index calculated from a test that measures the rate at which a 0.45-micrometer (μm) filter is plugged when subjected to a constant water pressure of 206.8 kPa (30 psi). The SDI gives the percent drop per minute in the flow rate of the water through the filter, averaged over a specified time-period, typically 15 minutes. Because the SDI is a relatively simple procedure and inexpensive to implement, it has been universally applied since the 1960s to assess the particulate fouling tendency of a feedwater intended for treatment by reverse osmosis (RO) membrane processes. Many facilities in the United States rely on automated SDI systems that are microprocessor controlled and fully automatic to allow operators to regularly monitor the feedwater. Care must be taken when employing the SDI with regards to accuracy and reproducibility, as the index is not based on any filtration mechanism and is not proportional with colloidal or particle concentration, in addition to the fact that there is no correlation factor for temperature nor for variations in membrane resistance. Even though there are legitimate concerns regarding the predictive value of the SDI, the index continues to be successfully used in the planning of RO facilities to guide the selection of pretreatment processes, and is often the basis of membrane guarantees and other plant performance contracts.

6.1 DEVELOPMENT OF THE FOULING INDEX

Fouling is a major obstacle to the widespread use of membrane technology. Membrane fouling has a direct impact on process performance because it decreases productivity (permeate flux) over time, increases the amount of energy required to meet water demand, and accelerates the need for membrane cleaning and replacement. Fouling is simply defined as the accumulation of undesired materials, via deposition or adsorption of soluble and particulate matter, onto the surface of the membrane. Membrane fouling can consist of either non-living (inorganic and organic matter) or living organisms (bacteria) and is the primary cause of permeate decline and loss of overall productivity in reverse osmosis and nanofiltration processes.

Early in the development of semi-permeable membranes for water treatment, the need to estimate membrane fouling potential of the raw water was found to be essential to identify pretreatment requirement to prepare the feedwater prior to processing. This is important because effective pretreatment can lower the number of required cleaning events and extend the life of membrane elements. Attempts to correlate fouling propensity of water with turbidity was only slightly successful.

To help solve these issues, the Du Pont de Nemours & Company (Du Pont) introduced its first reverse osmosis permeators for water desalination in 1969 under the trade name 'Permasep' an outcome of the company's research in polymer chemistry and synthetic fibers (Hagley Library, 2022). By 1997, Du Pont had sold over 1.5 billion gallons of desalting capacity, dominating the seawater desalination market for many decades. Since the first hollow-fine fiber membranes were sold by DuPont it was initially believe that performance was hampered by suspended and colloidal matter in the feedwater. Consequently, DuPont developed the Fouling Index, which was later denoted as the Silt Density Index (Schippers *et al.* 2014). Despite Permasep's success, DuPont decided to discontinue the production of hollow-fiber permeators in 1997, primarily attributed to the rise of the spiral-wound membrane configuration's success in the global desalination market (Hagley Library, 2022).

6.2 SILT AS A COMPONENT OF MEMBRANE FOULING

Assalay and colleages (1998) described silt as a solid granular material that is comprised of suspended rock and mineral particles with a size between sand or clay that can accumulate on the membrane surface. Although the SDI is termed as a silt index, this does not mean that the measurement is for silt considerations alone. The SDI is a parameter used to determine the fouling propensity of a source water intended to be processed using reverse osmosis membranes. Sources of membrane fouling can be divided into four principal categories:

- Particulate (silt, inorganic colloids, oxidized iron and manganese, algae, aluminosilicates)
- Microbiological (bacteria)
- Organic fouling (natural or synthetic compounds, oils, grease)
- Scale (limiting salt chemistries)

Since reverse osmosis synthetic membrane processes were first introduced for the treatment of water supplies, it was found that in most instances plugging of the elements was due to blocking filtration by suspended particulate matter (Comstock, 1982). Fouling by particulates (silt) generally impacts the lead membrane elements of any pressure vessel process configuration unlike scale that concentrates in the flow channels of the tail-end or last membrane elements located in a pressure vessel. Scaling is of greater concern with more concentrated feed solutions, therefore the last modules in the process pressure vessel configurations are most affected because they are exposed to the most concentrated feed water. Microbiological and organic-type fouling can occur anywhere within the membrane configuration depending on feedwater quality, pretreatment methods and process operating conditions. Consequently, the SDI is a measurement that can determine the fouling potential of a feedwater for particulate fouling, and may not be as predictive for microbiological, organic or scale type conditions.

6.3 STANDARDIZATON OF THE SILT DENSITY INDEX

In 1982, ASTM International (West Conshohocken, PA), formerly known as the American Society for Testing and Materials, developed the 'Standard Test Method for Silt Density Index (SDI) of Water' designated as D4189-14 by ASTM International (2014) which was first revised on January 30, 1987. According to ASTM International (2014), the SDI test method can be used to '*indicate the quantity of particulate matter in water and is applicable to relatively low (<1.0 NTU) turbidity waters such as well water, filtered water, or clarified effluent samples.*' The test is not applicable to RO, NF or ultrafiltration (UF) permeate. The test essentially consists of filtering water through a 47-mm diameter cellulose-based filter that possesses 0.45-μm at a constant pressure of 30 psi (210 kPa). The standard ASTM SDI test does not contain any correction for testing parameters such as membrane resistance, membrane area, feed temperature and applied pressure. The SDI increases with increasing temperature since the water viscosity is affected; additionally, an increase in pressure and decrease in membrane resistance will increase the measurements result. The SDI test is not an absolute measurement of the quantity of particulate matter.

The method has essentially remained the same procedure since that time and has been proven to be useful from an operating perspective for membrane plant operators (Ruiz-Garcia *et al.* 2015). According to Harn R/O Systems, Inc. (2022), the SDI test gives a calculated number in the range 0 – 6, where 0 is excellent and 6 denotes a very high fouling potential. Most membrane manufacturers require the feed SDI to be below 3.0 to indicate control of colloidal and particulate fouling. SDI values above 3.0 typically indicate periodic cleaning will be required and values above 5.0 could indicate rapid fouling can indicate the need to provide additional pretreatment to protect the membranes during process operation. Although in practice a high SDI typically indicates that fouling may occur, a low value does not guarantee that fouling will not occur.

6.4 METHODS AND PROCEDURES

Manual SDI testing typically requires the following components and items, portions of which are illustrated in Figure 1 (Hydranautics, 2022) and shown in Figure 2:

- SDI test assembly made of high-quality stainless steel or plastic.
- Filter holder to withstand 50 psi (350 kPa) pressure and hold.
- 47 mm nominal, plain filter papers115 to 180 micrometer thickness rated to 0.45 μm, typically white hydrophilic cellulose triacetate or mixed cellulose nitrate type materials.
- Pressure regulator and gauge able to measure 30 psi.
- Feedwater ball valve, plastic.
- Graduated cylinder, 500 mL capacity
- Stopwatch, graduated in hundredths of a minute.
- Thermometer, liquid-in-glass, suitable for measuring the temperature of the water sample; capable of being read to within ±1 °C.
- Dull tweezers.

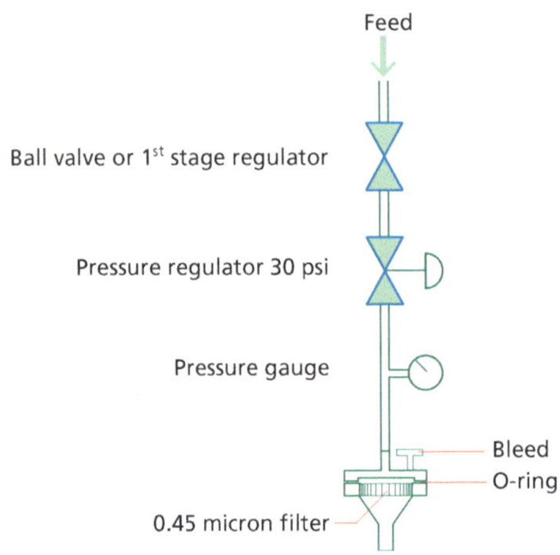

Figure 1 Typical assembly of the apparatus used for SDI measurements
Source: https://membranes.com/wp-content/uploads/Documents/TSB/TSB113.pdf

Several private corporations and original equipment manufacturer's provide in-house procedures, based on ASTM D4189-07, such as the information provided by AquaPhoenix Scientific (2022), Hydranautics (2022) and Harn R/O Systems, Inc. (2022). Note that a booster pump may be required for use in a manual SDI kit if insufficient pressure exists for testing to proceed. To perform an SDI test, the following general procedure can be followed, and is provided in more detail in ASTM International SDI testing method (2014). On-line instruments are also available that automatically and consistently monitor the SDI of RO feedwater that are typically controlled by a microprocessor and rely on a typical 4-20 mA interface. These devices contain side-stream or automatic flush sequences, and often extrapolate the SDI to 75% of the filter plugging condition.

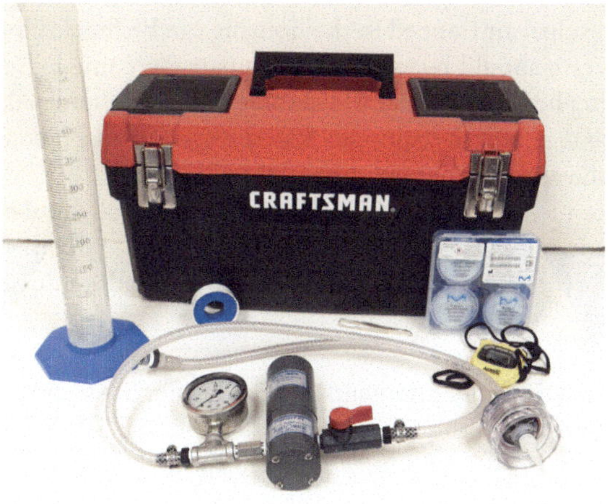

Figure 2 Photo depicting the components of the SDI apparatus: storage box for equipment, 500-mL graduated cylinder, Teflon tape, dull tweezers, SDI filter pads, stopwatch, cell assembly and flexible tubing, pressure regulator and gauge, booster pump.
Source: Courtesy of Harn R/O Systems, Inc., a Division of Komline-Sanderson Corp.

Typically, most reverse osmosis and nanofiltration membrane water treatment facilities make available a well flush valve on the raw water line upstream of the process building. The flush valve allows the raw water to be flushed to waste at process start-up for an operational pre-determined time period to reduce the SDI reading to below 3.0 prior to allowing water to be transferred to the pretreatment system ahead of the membranes. An SDI sample point can easily be installed on the raw water line upstream of the plant inlet valve so that SDI tests can be performed during well flush events. Taking these SDI tests during plant start-up allows operators to determine the length of time needed for individual and collective (or intake) flushing cycles as each facility may have differing source water supply transmission line configurations. It is recommended that an SDI test should be performed at least once a day on the raw water when the process is in operation and the results should be recorded in the operator's daily log.

Step 1: Measure the time required to filter a fixed volume of water through a standard space: 0.45 µm pore size microfiltration membrane at a constant pressure of 30 psi (2.07 bar) per the following procedure. Record this as T_i, or initial T.

a. Connect the test kit less filter paper for pretest flush.
b. Flush the test kit and supply line for 3 to 5 minutes to remove any possible contaminants.
c. Measure the temperature of the water and record the reading.
d. Make sure the O-ring on the filter is in good condition and properly placed. Set the pressure regulator to 30 psig (210 kPa). The set screw on the regulator should be adjusted while there is a small flow. Supply pressure to the regulator should be greater than 40 psig (276 kPa).
e. Open the 47 mm in diameter filter holder and carefully place a 0.45 µm membrane filter into the filter holder using the dull tweezers to avoid damage and touching the filter paper. Screw loosely together.

f. Open the feed valve slightly and adjust the filter housing to overflow, displacing any trapped air. Residual air trapped in the housing can be flushed by opening the small 'bleed' screw; care should be taken else this part can come loose and be easily lost. Tighten the filter housing, open the feed valve completely and make final adjustments to the pressure regulator as required; close the valve and tighten the filter holder the remainder of the way without overtightening.
g. Prepare to take measurements. Open the ball valve and simultaneously, using the stopwatch, begin measuring the time required to fill the 500 mL measuring cylinder. Record the time (t_i). Leave the valve open for continued flow; do not stop the watch or the water flow.

Step 2: Take additional time measurements, normally after 5, 10 and 15 minutes (after silt build up). Measure and record the times to collect additional 500 mL volumes of sample, starting the collection at 5, 10, 15 minutes of total elapsed flow time. This value is recorded as (t_f) with *f* being the time used. Measure the water temperature and check the pressure as each sample is collected. The pressure must remain constant at 30 psig (± 1 psig) and the temperature must remain constant to 1 °C. After completion of the test, the membrane filter may be retained for future reference or additional chemical evaluations of the filtered deposit matter. It is recommended that the date, time, sample location, operator name, SDI value and notes or comments be collected along with the filter pad.

Step 3: Calculate the Plugging Factor (*PF*) after 5, 10 and 15 minutes as determined as shown in Equations 1, 2 and 3, respectively:

$$PF_{5\text{-min}} = (1 - \frac{T_i}{T_5}) \times 100 \quad \text{Eq. 1}$$

$$PF_{10\text{-min}} = (1 - \frac{T_i}{T_{10}}) \times 100 \quad \text{Eq. 2}$$

$$PF_{15\text{-min}} = (1 - \frac{T_i}{T_{15}}) \times 100 \quad \text{Eq. 3}$$

Step 4: The SDI value is then determined at each interval as SDI = *PF/T*. Calculate the Silt Density Index (SDI) as follows using Equation (4):

$$SDI_T = \frac{\%P_{30}}{T} = \frac{\left[1 - \frac{T_i}{T_f}\right] \times 100}{T} \quad \text{Eq. 4}$$

where SDI_T is the Silt Density Index (%/min) at time *T*, t_i is the initial time required to collect 500 mL of sample, t_f is the elapsed filtration time (min) required to collect 500 mL of sample after a test time (typically 15 minutes after the initial measurement), and *%P* is the percent at 30 psi feed pressure.

The ASTM method recommends that if the $\%P_{30}$ exceeds 75% after 5 min then other test methods should be used to analyze for particulate matter. Considering that the $\%P_{30}$ is essentially the percentage of plugging factor, Equation 4 can be rewritten as Equation 5:

$$SDI_T = \frac{\%PF}{T} = \frac{\left[1 - \frac{T_1}{T_2}\right] \times 100}{T} \qquad \text{Eq. 5}$$

SDI measures the percentage of the filtrate flow rate decline per minute and is expressed as a percentage per minute but typically is reported without units. As an example, a SDI of 2.5 would indicate that the SDI filtrate flow was reduced by 2.5 percent per minute during the test. This concept is illustrated graphically in Figure 3 where the filtration flow is presented as a function of time, and V_1 and V_2 are the volumes of the first and second sample:

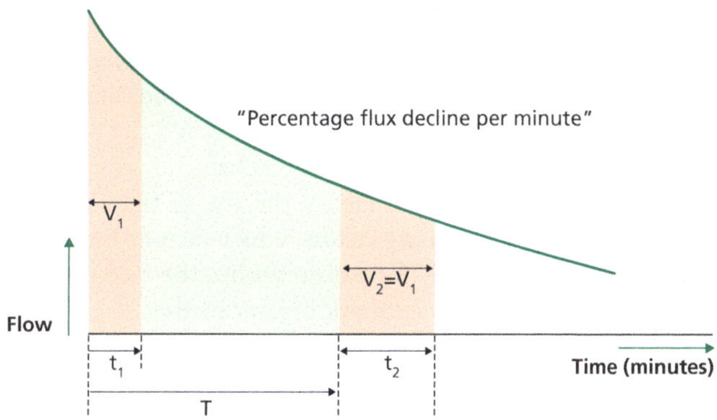

Figure 3 Representation of the filtration flow as a function of time per the SDI test method. (Adapted from Alhadidi and colleagues 2011)

Calculation Examples

1. Calculate the *PF* and SDI for a test where the time measurements indicated a T_i of one minute and T_{15} of 4.0 minutes.
   Solution: The plugging factor is calculated as $PF = 1\text{-}(1/4)\times100 = 75$. On the other hand, the SDI is then calculated as SDI = 75/15 = 5 as a percentage of flux decline per minute. These results indicated that flow had decayed by a factor of four times, indicating that 75% of the 0.45-micron filter has been plugged. As the SDI test is a dead-end filtering scenario, unlike the RO processes that have one input stream and two output streams, the flow across the membrane in the feed-concentrate channel results in concentration polarization that is not evaluated in the SDI; hence the SDI_{15} result should not be used to predict an actual fouling rate for an operating RO process. The SDI value only represents an indication that fouling may occur. For this example, an SDI result of 5 would indicate

that additional pretreatment is most likely required for this source water using RO membrane processes.

2. A municipal utility is considering using a membrane process as an alternative treatment method for its newly developed wellfield. The following information is collected using a SDI testing apparatus on an unfiltered raw ground water sample. Determine the 15-min SDI value.

Table 1 SDI Test Results for Water Utility

Test Time (min)	Water Volume Collected (mL)	Time to Collect 500 mL (sec)
0.0	500	9.0
5.0	500	11
10	500	12
15	500	14

Solution: The SDI for the 15-min test is calculated as SDI = $\{100[1\text{-}(t_i/t_f)]/t\}$ = 100[1-(9sec/14sec)]/15-min} = 2.4 units as a percentage flux decline per minute. Since the SDI is less than 3.0 units, the specific water supply would be considered acceptable as a feedwater for a reverse osmosis treatment process. However, the SDI result would not predict fouling of an RO membrane due to such mechanisms as sparingly salt scaling, the impact of dissolved iron, for example. Hence the use of the SDI results should be supplemented with other predictive fouling factors, which include limiting salt chemistry, contribution of metal oxidation, particle formation within the feed-concentrate channel, and organic deposition due to natural or synthetic organic matter.

6.5 LIMITATIONS OF THE SDI

The procedure outlined in the ASTM International (2014) method should be followed as closely as possible to collect data that has meaning and is reproducible. Test variability (50 - 100%) has been a recognized problem with the SDI method and personnel training in procedural details is a critical factor in obtaining precise and accurate test results. Although the SDI is a popular test that has been empirically correlated with the fouling tendency of desalination processes, its interpretation calls for some expertise on the part of the person carrying out the test. The ASTM method procedure requires several actions to obtain an accurate SDI that include:

- Requirement to flush the equipment prior to use,
- Need to wet the test pad filters prior to use,
- Purging air to avoid air entrapment within the test cell impacting the test pad surface,
- Efforts to avoid touching the membrane filters with hands, and
- Collecting the water temperature before and after each test.

Problems related to the SDI test have been documented by many (Rachman *et al.*, 2013; Alhadidi *et al.*, 2011; Alhadidi *et al.*, 2012; Boerlage 2008, Boerlage 2007; Yiantsios *et al.*,

2005; Schippers and Verdouw, 1980). Issues are many and range from the understanding that the SDI is not based on any one filtration mechanism to the fact that there is no direct correlation between turbidity of a water and it's measured SDI. Also, the SDI has no linear relation with particulate matter and as is not corrected for temperature, pressure and membrane resistance; consequently, values obtained from the method may not be comparable and certainly variable.

Filtration, Flux and Fouling

Although the ASTM International method has proven valuable for determining the fouling potential of seawater supplies for particulate fouling, the SDI may not be represented of conditions that high-pressure RO membranes experience where concentration polarization exists (350 psi to 1,000 psi). It is known that the SDI is not based on any single fouling mechanism hence is not used to predict the rate of fouling in RO systems. This has been illustrated in higher salinity feedwaters where SDI cake filtration is considered the mechanism for particulate fouling as reported by Boerlage (2007).

Because the SDI method operates at 30 psi (210 kPa) in a 'dead-end' mode, cake compression will influence the test results and may not be representative of actual conditions that exist at the active layer of the RO membrane's surface. Furthermore, the SDI makes use of 0.45 μm filters, which are not represented of the pore size in RO and NF spiral-wound membranes that approximate 0.001 μm (Fang, 2013 and Duranceau, 2013; Duranceau, 2021). Another concern in comparing the SDI to actual RO plant operation, is that the flux rate of the test (> 1,600 $L/m^2/hr$ at the start) is far outside the typical values experienced in practice (20 to 25 $L/m^2/hr$) as reported by Schippers and colleagues (1981).

As an example of one of the issues encountered with the testing, Alhadidi and colleagues (2012) found that seawater sources with SDI values less than 3 may still foul the membranes in practice; on the other hand, SDI values greater than 3 have been documented when testing permeate of membrane filtration processing used as seawater pretreatment system feeding a seawater RO process.

Membrane Filters and Holder

There are a variety of chemistries available for use as membrane filters used in collecting SDI measurements, and can include mixed cellulose esters (MCE), mixed cellulose acetate and cellulose nitrates), polyvinylidene fluoride (PVDF), Nylon, polytetrafluoroethylene (PTFE), polyamide (PA) and polyethersulfone (PES). Prior work by Ando and colleagues (2003) compared SDI measurements collected using hydrophilic (MCE, PVDF, PA) that yielded greater values than the counterpart hydrophilic (PTFE) filters.

While there is some debate about which filter pad chemistry is appropriate for various applications, often the membrane element warranty language does not stipulate the material of the SDI filter pad required to maintain compliance with the warranty. Field experience at a variety of municipal facilities has demonstrated that very different SDI results are observed for different filter pad chemistries. Additionally, very different results can be obtained when using the 'same' filter pad chemistries from different manufacturers. Also, SDI results can produce different measurements when using 'sterile' versus 'non-sterile' pads. Furthermore, pH has been shown to impact the SDI measurement, thought to be due to differences in pad chemistry (Mosset *et al.*, 2008).

Nahrstedt and Camargo (2008) studied the effect of filter support on SDI and MFI measurements and found that the filter holder influenced test results, up to 100 percent across measurements. A similar finding was reported by Escobar *et al.*, (2009) as well as Salinas-Rodriguez *et al.*, (2019).

Turbidity, Suspended Solids and the SDI

Water turbidity is an optical characteristic that serves as a measure of the relative clarity of a liquid and is commonly used as an indicator for the general condition of the drinking water. Turbidity is regulated under the U.S. Environmental Protection Agency's secondary drinking water standard for aesthetic reasons, and it is used as an operational control measure as it is an easy field water-quality parameter to measure. Turbidity in water is caused by suspended matter such as clay, silt, and organic matter and by microscopic organisms that interfere with the passage of light through the water (American Public Health Association, 2017). Turbidimeters are calibrated with a formazine standard solution, which was used in experimentation using SDI measurements. Schippers and Verdouw (1980) determined that the absence of a temperature correction for measurement method resulted in higher SDI values at higher temperatures. As has earlier been noted, there exists a non-linear relationship between colloidal particle concentrations and measured SDI values, such that the water being tested appears to be less fouling as the test filter becomes progressively plugged. This concept has been illustrated by Schippers and Verdouw (1980) and is shown in Figure 4.

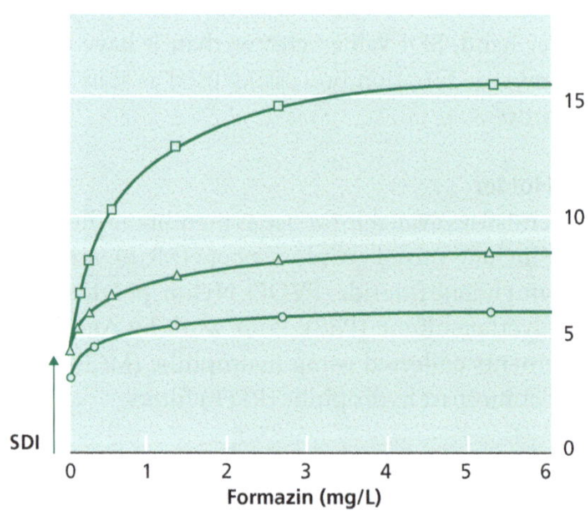

Figure 4 SDI as a function of formazin concentration by test duration. (Schipper and Verdouw 1980)

Surface water often contains high levels of fine suspended solids and therefore can often exhibit higher turbidity values as well as high SDI that can vary depending on rainfall and runoff impacts and may require more extensive pretreatment to achieve acceptable values of SDI and turbidity in feed water from brackish rivers, lakes, bays as well as the ocean. Brackish and briny well water supplies may contain particulates and would be susceptible to metal oxidation and colloid formation if anaerobic groundwater was exposed to an aerobic condition.

Oftentimes water with very low turbidity, less than 0.5 Nephelometric turbidity unit (NTU), can still cause unacceptable membrane fouling, due to the presence of many particles having very small diameter (typically under 3 microns in diameter) which do not efficiently reflect light to show up as higher turbidity. Nonetheless, these small particles can quickly foul spiral wound RO membranes. Sources of silt can include organic colloids, iron corrosion products, precipitated iron hydroxide, algae, and fine particulate matter.

6.6 ALTERNATIVES TO THE SDI

Fouling indices can be categorized in two groups: those operating at constant pressure (the SDI method) and those operating at constant flux. Because of many long-term concerns about the predictive value of the SDI, many have attempted to find either alternatives or find methods that can overcome variations attributed to its use. However, the few recommended alternatives to SDI have proven equally problematic in application as the SDI.

One of the more successful modifications to the SDI approach is the modified fouling index (MFI) proposed by Schippers and co-workers (Schippers and Verdouw 1980). The premise behind this alternative approach is to carry out dead-end filtration tests of relatively short duration, using a permeable membrane rated at 0.45 µm as is the case with the SDI, to establish conditions of cake filtration and determine a quantity indicative of fouling. The MFI-0.45 test uses the same equipment as the SDI test and takes into account that as flow commences through the test pad, initially pore blocking occurs, followed by cake or gel filtration and finally, cake or gel blocking and compression. The modified fouling index (MFI) was suggested to measure the rate of cake formation on the membrane surface that is believed to provide a better prediction of RO fouling phenomena. The development of the MFI in many cases has demonstrated to be a more effective alternative to the SDI because of three primary reasons:

1. A linear relationship with particle concentration exists;
2. The index is corrected for temperature;
3. It is based on the cake filtration mechanism.

However, the MFI poses its own limitations as like the SDI, the alternative test method also operates at constant pressure producing high initial flux values. To overcome this issue, Boerlage *et al.* (2004) developed the concept of a constant flux MFI instead of constant pressure filtration, which was further developed by Salinas-Rodriguez (2011) using polyethersulfone UF membranes. The value of the MFI was confirmed when ASTM International published in 2015 the 'Standard Test Method for Modified Fouling Index (MFI-0.45) of Water' (ASTM International 2015).

A group of researchers developed the concept of a mathematical model that could quantify the influence of pressure, temperature and membrane resistance of the test and normalize SDI measurements (Alhadidi *et al.* 2013). An alternative filtration index was proposed and referred to as the volume-based SDI measurement (SDI-v) that compared the test's initial flow rate to the flow rate after filtering a standard volume of feed water using a 0.45-μm microfiltration membrane. It was found that the SDI-v was independent of the membrane resistance, which could overcome some of the variation typically present in the existing method.

As confirmed in the research by Sioutopoulos and Karabelas (2012), SDI indices operating at constant pressure are affected by the high flux rates at the start of the filtration test and by the compressibility of the formed cake resulting from the very high initial fluxes. The authors point out that there is no method available to predict the rate of fouling in RO/NF plant operation, which is characterized by axial variability of key parameters (local TMP, cross-flow velocity and flux). It was determined that under constant flux filtration over a broad range of fluxes, thin fouling layers, of practical interest in relation to RO membrane operations, exhibited a linear increase with time of the pressure drop across the cake, as demonstrated in Figure 5.

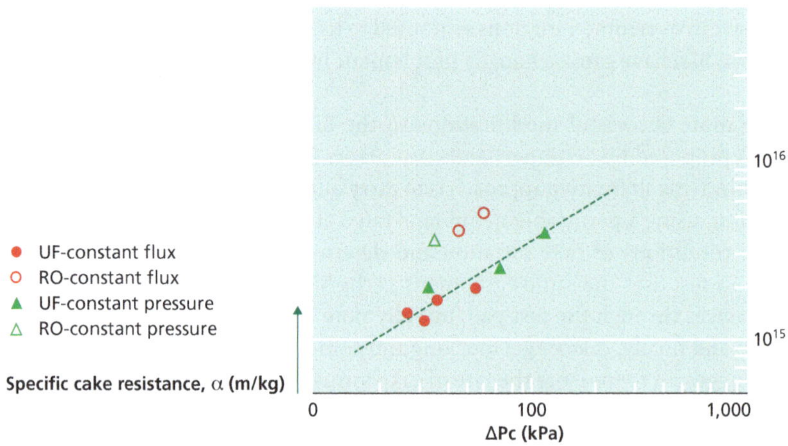

Figure 5 Specific cake resistance versus pressure drop across the cakeunder constant flux and constant pressure conditions. Results shown are for an experimental test water containing 2,000 mg/L total dissolved solids and a mixture of 10 mg/L of 75% humic acid (HA) and 25% sodium alginate (SA). Source: Sioutopoulos and Karabelas (2012).

6.7 SUMMARY

For more than half a decade the SDI has been accepted as a valuable test parameter and applied in the application of RO processes worldwide. The test was developed as an empirical means to measure the fouling propensity of the feed water to RO and NF membrane processes. The SDI represented the fouling potential of synthetic RO and NF membranes due to finely suspended particles present in the feed water.

Despite the success of the SDI, there are disadvantages to the test that limit the accuracy and reproducibility of the method. Limitations included the fact that no correction factor for temperature exists for the method, the test is not based on any single filtration model, there are impacts caused by variations in membrane resistance, and there remains no linear correlation between the measurement and the colloidal or suspended particulate content of the water. Efforts continue to develop enhancements to the SDI (Khirani *et al.*, 2006; Jin *et al.*, 2015).

Regardless of the many identified weaknesses of the SDI, the test method remains a simple procedure to perform at minimal cost that does not need highly skilled professionals to conduct. For this reason, the test will continue to be used (or misused) in the future during the planning and operation of RO and NF facilities worldwide.

6.8 REFERENCES

Alhadidi, A., A.J. B Kemperman, B. Blanket, J.C. Schipper, M. Wessling and W.G.J. van der Meer (2011). Silt density index and modified fouling index relation and effect of pressure, temperature and membrane resistance. Desalination. 273: 48-56.

Alhadidi, A., A.J.B. Kemperman, J.C. Schippers, M. Wessling, and W.G.J. van der Meer (2012). SDI: is it a reliable fouling index? Desalination and Water Treatment. 42: 43-48.

Alhadidi, A., B. Blankert, A.J.B. Kemperman, R. Schurer, J.C. Schippers, M. Wessling, and W.G.J. van der Meer (2013). Limitations, improvements and alternatives of the silt density index. Desalination and Water Treatment. 51 (4-6): 1104-1113.

American Public Health Association (2017). Standard Methods for the Examination of Water and Wastewater, 23rd Edition. R.B. Baird, A.D. Eaton, E.W. Rice (editors). Water Environment Federation, American Water Works Association, American Public Health Association, January 1, 2017, 1,796 pages.

Ando, M, S. Ishiara, H. Iwahori and N. Tada (2003). Peculiar or unexpected behaviour of silt density index of pretreated water for desalination. Proceedings of the International Desalting Association (IDA), World Congress on Desalination and Water Reuse, along with the 12th Annual Conference of the Caribbean Water and Wastewater Association from 28 September to 3 October. Paper BAH 03-071, Paradise Islands, Bahamas.

AquaPhoenix (2022). AquaPhoenix Scientific, 860 Gitts Run Road Hanover, PA 17331; collected from Web [Source: https://www.aquaphoenixsci.com/wp-content/uploads/procedures/processed/Silt%20Density%20Index%20(SDI)%20Test%20Procedure_TK8550-Z.pdf].

Assallay, A.; Rogers, C.D.F.; Smalley, I.J.; Jefferson, I.F. (1998). 'Silt: 2–62 μm, 9–4 '. Earth-Science Reviews. 45 (1–2): 61–88.

ASTM International (2014). D4189-07(2014) Standard Test Method for Silt Density Index (SDI) of Water (Revised), ASTM International, 100 Barr Harbor Drive, PO Box C700, West Conshohocken, PA 19428-2959.

ASTM International (2015). D8002-15 (2015) Standard Test Method for Modified Fouling Index (MFI-0.45) of Water, ASTM International, 100 Barr Harbor Drive, PO Box C700, West Conshohocken, PA 19428-2959.

Boerlage, S.F.E., M.D. Kennedy, M.R. Dickson, D.E.Y. El-Hodali, J.C. Schippers, The modified fouling index using ultrafiltration membranes (MFI-UF): characterization, filtration mechanisms and proposed reference membrane, J. Membrane Science. 197 (2002) 1–21.

Boerlage, S. F. E., M. Kennedy, Z. Tarawneh, R.D. Faber, and J.C. Schippers (2004), Development of the MFI-UF in constant flux filtration, Desalination, 161, 103-113.

Boerlage, S. F. E. (2007), 'Understanding the SDI and Modified Fouling Indices (MFI0.45 and MFI-UF)', IDA World Congress-Maspalomas, Gran Canaria, Spain, October 21-26, 2007, IDAWC/MP07-143.

Boerlage, S. F. E. (2008) Understanding the Silt Density Index and Modified Fouling Indices (MFI0.45 and MFI-UF). Desalination and Water Reuse Quarterly, May to June, 12-21.

Comstock, D. (1982). Testing the membrane plugging factor in reverse osmosis. Journal of the American Water Works Association. 74(9): 486-490.

Duranceau, S.J. (2021). Membrane Fouling Indices: Silt Density Index, Modified Fouling Index and the Mini Plugging Factor Index. Proceedings of the Southeast Desalting Association 2021 Spring Symposium: Together Again for Membranes, The Westin Cape Coral Resort at Marina Village, Cape Coral FL (June 6-9, 2021).

Escobar L, W. Sellerberg, D. Sanchez, F. Pastrana, and A. Wachinski (2009). Detailed analysis of the silt density index (SDI) results on desalination and wastewater reuse applications for reverse osmosis technology evaluation. Proceedings of the international forum on marine science and technology and economic development. Asia-Pacific Desalination conference, Qingdao, China, 8-10 July 2009.

Fang, Y., and S.J. Duranceau (2022). Comparison of Non-Homogeneous and Homogeneous Mass Transfer in Reverse Osmosis Membrane Processes. Desalination and Water Treatment. 51(34-36): 6444-6458 (2013).

Hagley Library (2022). DuPont Permasep Products Records 1969-2003, Hagley Library, 298 Buck Road, Wilmington, DE 19807.

Harn R/O Systems, Inc. (2022). Harn R/O Systems, Inc., a Division of Komline-Sanderson Corporation, 310 Center Ct., Venice, FL, 34285. Collected from the Web [https://blog.harnrosystems.com/how-to-complete-a-silt-density-index-sdi-test-and-why-its-important].

Hydranautics (2022). Hydranautics, A Nitto Group Company, 401 Jones Road, Oceanside, CA 92058. Collected from the Web [https://membranes.com/wp-content/uploads/Documents/TSB/TSB113.pdf].

Jin, Y., J. Younggil, L. Hyunkyung, S. Hong (2015). Fouling potential evaluation by cake fouling index: Theoretical development, measurements, and its implications for fouling mechanisms. Journal of Membrane Science. 490: 57-64.

Khirani, S., R. Ben Aim, M.-H Manero (2006). Improving the measurement of the Modified Fouling Index using nanofiltration membranes (NF–MFI). Desalination 191(2006)1–7.

Mosset, A., V. Bonnelye, M. Petry, and M.A. Sanz (2008). The sensitivity of SDI analysis: from RO feed water to raw water. Desalination. 222: 17-23.

Nahrstedt, A. and J. Camargo-Schmale (2008). New insights into silt density index and modified fouling index measurements. Water Science and Technology: Water Supply. 8(4): 401–411.

Rachman R.M., N. Ghaffour, F. Waly, G.L Amy (2013) Assessment of Silt Density Index (SDI) as fouling propensity parameter in Reverse Osmosis (RO) desalination systems. Desalination and Water Treatment 51: 1091-1103.

Ruiz-Garcia, A., E. Ruiz-Saavedra, S.O. Perez-Baez, J.E. Gonzalez-Gonzalez (2015). Evaluation of the first nine years of operating data of a RO brackish water desalination plant in Las Palmas, Canary Islands, Spain, Desalination and Water Treatment. 55: 2555-2561.

Salinas-Rodríguez, S.G. (2011). Particulate and organic matter fouling of SWRO systems: Characterization, modelling and applications. CRC Press: Balkema, Delft.

Schippers, J. C., J.H. Hanemaayer, C.A. Smolders, and A/ Kostense (1981). Predicting flux decline or reverse osmosis membranes. Desalination, 38, 339-348.

Schippers J.C. and J. Verdouw (1980). The modified fouling index, a method of determining the fouling characteristics of water. Desalination 32: 137-148.

Schippers, J.C., S.G. Salinas-Rodriguez, M.D. Kennedy and S. Boerlage (2014). Why MFI is edging SDI as a fouling index, Desalination and Water Reuse. May-June 2014: 28-32.

Sioutopoulos, D.C. and A.J. Darabelas (2012). Correlation of organic fouling resistances in RO and UF membrane filtration under constant flux and pressure. Journal of Membrane Science. 407 (408): 34-46.

Yiantsios, S.G., D. Sioutopoulos, A.J. Karabelas (2005). Colloidal fouling of RO membranes: an overview of key issues and efforts to develop improved prediction techniques. Desalination 183: 257-272.

doi: 10.2166/9781789062977_0155

Chapter 7

Modified Fouling Index (MFI-0.45)

Sergio G. Salinas-Rodriguez, IHE Delft, The Netherlands

Vanida A. Salgado-Ismodes, IHE Delft, The Netherlands

Almotasembellah Abushaban, UM6P, Morocco

The learning objectives of this chapter are the following:

- Define the theoretical principles of MFI-0.45 constant pressure and its prediction model
- Describe the MFI-0.45's testing set-up, testing protocol and calculation procedure
- Illustrate the application of the MFI-0.45 method

7.1 INTRODUCTION

The modified fouling index (MFI-0.45) was developed to overcome the limitations of the silt density index (SDI). The MFI-0.45 has been standardized by ASTM in its method 'Standard Test Method for Modified Fouling Index (MFI-0.45) of Water' (ASTM D8002 - 15, 2015) and is based on filtration of feed water through a 0.45 μm microfiltration membrane filter in dead-end mode at constant pressure (207 kPa), and importantly, it is based on the cake filtration mechanism (Schippers and Verdouw, 1979, Schippers and Verdouw, 1980).

This MFI-0.45 test can be used to assess the fouling potential of reverse osmosis (RO) / nanofiltration (NF) feed water due particulate matter and is applicable to low and high turbidity waters. Similar to the SDI, the ASTM stipulates that this test is not suitable for assessing ultra-pure water or effluents from most RO and ultra-filtration (UF) systems.

Some of the applications of the MFI-0.45 method are the following:

- It can serve as an indication of the quantity of particulate matter.
- It can be used to determine effectiveness of various treatment processes used to remove particulate matter.
- It can be used to assess the clogging potential of water before infiltration in wells.

7.2 THEORY PARTICULATE FOULING

The flow through a porous medium like a membrane, based on Darcy's law, can be described by:

$$Q_w = \frac{dV}{dt} = \Delta P \times K_w \times A \qquad \text{Eq. 1}$$

where:

Q_w = permeate flow (m^3/hr)
V = total filtered volume water (permeate) (L or m^3)
t = time (hour, minute, second)
ΔP = differential pressure (pressure feed - pressure permeate)
K_w = permeability constant of porous media for water (m^3/m^2-s-bar)
A = surface area of the membrane(s) (m^2)

In membrane technology, flux is defined as the ratio of the permeate flow and surface area of the membrane. It is expressed as:

$$J = \frac{Q_w}{A} = \frac{1}{A} \times \frac{dV}{dt} = \Delta P \times K_w \qquad \text{Eq. 2}$$

Frequently, the concept of resistance (R) is used, instead of permeability:

$$K_w = \frac{1}{\eta \times R_T} \qquad \text{Eq. 3}$$

Where: η is the viscosity of the water and R_T is the total resistance [sum of membrane resistance (R_m), pore blocking (R_p) and cake formation (R_c)].

$$R_T = R_m + R_b + R_c \qquad \text{Eq. 4}$$

Replacing Eq. 3 and Eq. 4 in Eq. 2:

$$J = \frac{1}{A} \times \frac{dV}{dt} = \frac{1}{\eta} \times \frac{\Delta P}{R_m + R_b + R_c} \qquad \text{Eq. 5}$$

Permeability of the clean filter media (R_m) is a function of filter properties such as filter thickness (Δx), surface porosity (ε), pore radius (r_p), and tortuosity (τ) and can be defined using Poiseuille's Law:

$$R_m = \frac{\theta \times \Delta x \times \tau}{\varepsilon \times r_p^2} \qquad \text{Eq. 6}$$

Pore blocking resistance of the membrane (R_b) is defined as the restriction of flow during filtration due to the particles deposited inside pores or blocking the pores entry. The cake resistance (R_c) component in (membrane) filtration can be defined following the Ruth equation (Ruth, *et al.*, 1933), using the concept of 'specific cake resistance' per unit weight (α) (Equation 10). Ruth showed that the resistance of the cake formed during constant pressure filtration is proportional to the amount of cake deposited at the filter medium provided the retention of particles and α are constant. Cake resistance is defined as:

$$R_c = I \times \frac{V}{A} \qquad \text{Eq. 7}$$

and the fouling index (I) is:

$$I = \alpha \times C_b \qquad \text{Eq. 8}$$

Where: I is a measure of the fouling characteristics of the water (m^{-2}). The value of I is a function of the nature of the particles and is proportional to their concentration. C_b is the concentration of particles per unit volume of filtrate (e.g., mg/L) and α is the specific cake resistance per mg cake per m^2 membrane ($m^3/mg/m^2$).

The specific cake resistance is constant for incompressible cakes under constant pressure filtration. Carman (Carman, 1937, Carman, 1938) derived Equation 9 for the specific resistance of a cake composed of spherical particles of diameter d_p from the Kozeny equation including a factor for tortuosity of the voids within the cake. According to this relationship a reduction in the porosity of the cake (ε) or a decrease in particle diameter size (d_p) increases the specific resistance of the deposited cake.

$$\alpha_c = \frac{180 \times (1-\varepsilon)}{\rho_p \times d_p^2 \times \varepsilon^3} \qquad \text{Eq. 9}$$

Combining Eq. 7 and Eq. 5 and integrating at constant ΔP from $t = 0$ to $t = t$, assuming time independent permeability and uniform porosity characteristics throughout the depth of the cake (i.e., no compression of the cake), results in the well-known filtration equation:

$$\frac{t}{V} = \frac{\eta \times R_m}{\Delta P \times A} + \frac{\eta \times I}{2 \times \Delta P \times A^2} \times V \qquad \text{Eq. 10}$$

Where R_m is considered constant throughout the filtration period. Equation 10 gives a straight line when *t/V* vs. *V* is plotted which is used to test the formation of cake filtration. Carmen defined the gradient of the line as:

$$slope = \tan\alpha = \frac{dt/dV}{dV} = \frac{\eta \times I}{2 \times \Delta P \times A^2} \qquad \text{Eq. 11}$$

The fouling index (I) can then be determined from the slope of the linear region in the plot of time/volume vs. volume, which corresponds to cake filtration as illustrated in Figure 1. The MFI is calculated considering the **minimum** I value. From this slope, the I value of the water can be calculated from the actual testing conditions (η, A, ΔP).

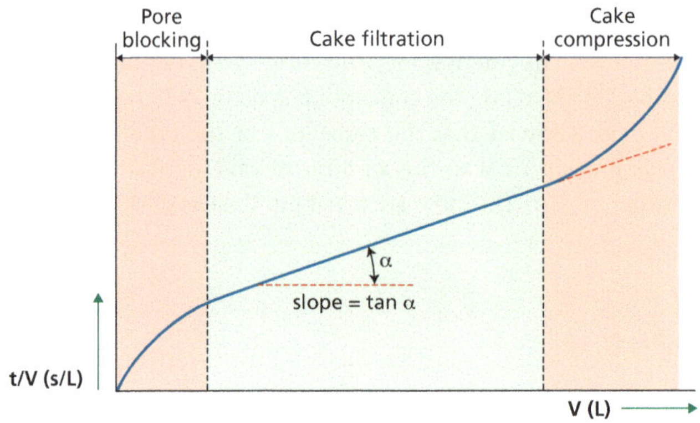

Figure 1 Filtration curve t/V versus V. Adapted from Schippers (1989)

The modified fouling index MFI-0.45 can be obtained after normalizing to reference conditions: pressure (ΔP_o) = 2 bar, membrane area (A_o) = 13.8×10^{-4} m^2, water temperature through water viscosity η at 20 °C.

$$MFI = \frac{\eta_0 \times I}{2 \times \Delta P_0 \times A_0^2} \qquad \text{Eq. 12}$$

Reference conditions for normalization of MFI values

ΔP_o = 2 bar = 200 kPa

A_o = 13.8×10^{-4} m^2 (42 mm effective diameter of a 47 mm diameter filter)

At temperature 20 °C, the viscosity (η_o) is = 0.001 Ns/m^2.

MFI is expressed in units of s/L^2. By doing this the results will be in the same range order of magnitude of SDI in the range 2 to 3 (Schippers, *et al.*, 2014).

Replacing the reference values in the MFI formula helps us to find the conversion factor of MFI into *I*, as follows:

MFI = (0.001 Ns/m^2) × *I* / [2 × 200,000 N/m^2 × (13.8×10^{-4} m^2)^2]

MFI (s/m^6) = 13×10^{-4} × *I* (m^{-2})

MFI (s/L^2) = 13×10^{-8} × *I* (m^{-2})

or

I (m^{-2}) = 7.68×10^{8} × MFI (s/L^2)

An alternative method for calculating MFI is based on the equation:

$$\frac{dt}{dV} = \frac{\eta \times R_m}{\Delta P} + \frac{\eta \times I}{\Delta P \times A^2} \times V \qquad \text{Eq. 13}$$

In this case, the calculated slope is two times higher than in Eq. 11. Thus, this factor needs to be considered in the MFI calculation. This alternative approach has the advantage that possible errors in time and volume at the start of the test will not influence the calculated slope during the course of the test. Nevertheless, very accurate pressure regulator and volume (or flow) measurement devices are needed to obtain accurate MFI values.

7.3 MEASURING MFI-0.45

7.3.1 Filtration set-up and materials

A concept schematic of the filtration apparatus is presented in Figure 2 as it was initially proposed by Schippers and Verdouw (1979). Figure 3 shows the schematic of the filtration set-up as available at the laboratory of IHE Delft.

All parts of the filtration set-up that are in contact with water should be made of high-quality stainless or plastic to prevent contamination by corrosion.

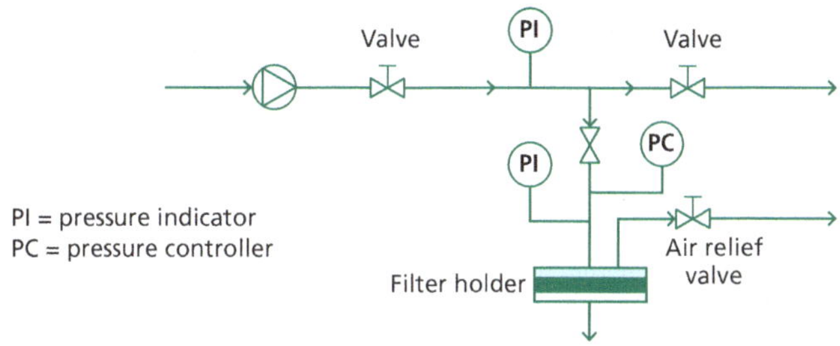

Figure 2 Schematic of the filtration apparatus for measuring MFI-0.45 as initially proposed by Schippers and Verdouw (1979)

The key components of the experimental set-up are: air compressor, pressure vessel, pressure regulator, membrane filter holder, pressure transducer, thermometer, electronic balance, computer and a three-way valve. A three-way valve is used to connect the feed tube with the membrane holder and at the same time with the pressure transducer. The filtration set-up needs to be verified for correct pressure / flow / weight readings, stable constant pressure filtration, no leakages, and avoiding air/bubbles trapped in the system (tubing, feed side of filter holder).

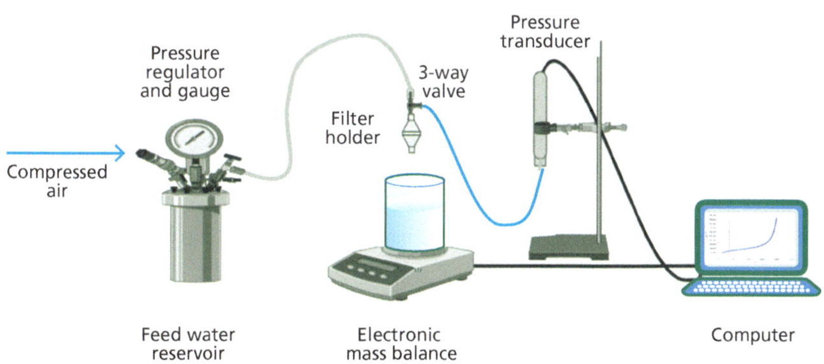

Figure 3 Example of laboratory apparatus for measuring MFI-0.45 at constant pressure with a pressure vessel.

7.3.1.1 Membrane filters

The recommended membrane filters for the MFI-0.45 test are the same as the ones recommended for the SDI test.

The membrane filters need to be white of colour, hydrophilic, with a mean pore size of 0.45 μm, mixed cellulose nitrate and mixed cellulose ester are allowed. The diameter of the filter depends on the size of the membrane filter holder (25 mm or 47 mm).

Table 1 ASTM recommended membrane filter properties for MFI-0.45 and SDI tests

Property	MFI-0.45 ASTM D8002 - 15 (2015)	SDI ASTM D4189 - 14 (2014)
Hydrophobic/ hydrophilic	Hydrophilic	
Material	Mixed cellulose nitrate (50–75 %), mixed cellulose ester* (MCE)	Mixed cellulose nitrate (50–75 %), cellulose acetate
Mean pore size	0.45 μm	
Thickness	115-180 μm	
Bubble point	179-248 kPa	
Pressure	Not mentioned	91.4 - 94.7 kPa
Pure water flow	25-50 s / 500 mL (10-20 mL/s)	
Constant pressure	200 kPa (difference across filter)	207 kPa (gage pressure)

* The standard method mentions acetate but the acronym corresponds to mixed cellulose ester.

Considering that for a range of pressures (91.4-94.7 kPa) the water flow should be around 25-50 seconds per 500 mL, for a nominal filter diameter of 47 mm, the recommended permeability of the filters at 20 °C ranges between 21,911 to 45,405 L/m^2/h/bar; or in terms of membrane resistance (R_m) varies between 7.93×10^9 and 1.64×10^{10} m^{-1}.

7.3.1.2 Filter holder

Filter holders designed to withstand pressures of 3.5 bar are recommended for the test. The filter holder should be equipped with a device for releasing air. The ideal filter support surface (placed at the bottom of the filter holder) is a highly porous one so that it does not influence the effective membrane area during the filtration procedure. Cytiva's Whatman (WHA-10461000 FP 025/1 Polyethersulfone) filter holder for 25 mm filter can be used, although it does not have a fully porous support plate.

It is recommended to measure the head loss across the filter holder and correct for it in the feed side to make sure that the feed pressure of 2 bar is effective in the test.

The filter holder should fit perfectly the size of the membrane filter. Filters are normally available as 25 mm or 47 mm diameter. A smaller size will allow less volume of water needed for the test, easing the sampling, transport and storage of the samples.

7.3.1.3 Feedwater reservoir

The water to be tested needs to be transferred to the feed reservoir which is a pressure vessel like in Figure 4. Commercially there are various volume capacities. The minimum recommended volume for the test is 3.8 L but a larger capacity one is preferred, especially when clean water is tested. The required pressure is achieved by applying compressed air. Nitrogen gas can also be used in case compressed air is not available.

Figure 4 Feedwater reservoir or pressure vessel for laboratory measurement (Sterlitech, 2023)

7.3.1.4 Electronic mass balance

The feed water that passes through the membrane filter needs to be collected in a beaker set on an electronic balance, for instance a Sartorius model Entris with 0.1 g accuracy. The capacity of the scale should be at least the same volume as the feedwater reservoir. The scale needs to connect with a computer via a USB-C or RS232 port to acquire the permeate weight from the balance. Considering the density of water, 1 gram equal to 1 millilitre.

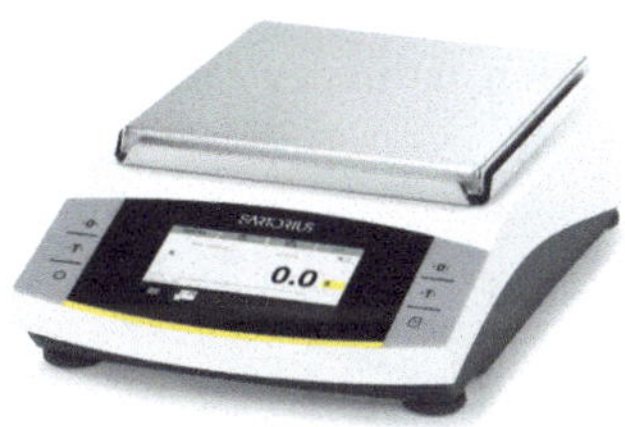

Figure 5 Entris® II Advanced Line Precision Balance 12,200 g|100 mg (Sartorius, 2023)

7.3.1.5 Software for data acquisition

Data sets of filtrate weight collected over time need to be recorded and imported into an MS Excel spread sheet by data acquisition software such as: WinWedge Standard (Taltech, 2023). The sampling frequency can be adjusted according to requirements for calculation (e.g., between 2 and 10 seconds).

The MS Excel spreadsheet for data acquisition can be adapted to include a graph of MFI versus time in order to set the criteria of filtration time; for instance, if the change of slope of the MFI in 5 minutes is less than 5% per minute then the test should be finished.

7.3.1.6 Pressure regulator and gauge

The pressure in the feed water reservoir and just in front of the membrane filter holder needs to be kept at a constant level of 2 bar. This can be set by means of a manually adjustable pressure sustaining valve (Figure 6) which includes a pressure gauge (Figure 6). This valve is manually adjustable by means of a turning wheel (0.02 bar step). The accuracy of adjustment is depending on readings from the gauge.

Figure 6 Pressure regulator (0.02 bar step) (Festo, 2023a) and pressure gauge in the range 0-2.5 bar with 0.05 bar accuracy (Festo, 2023b)

7.3.1.7 Pressure transducer

The pressure transduce is commercially available (PXM409-001BGUSBH, Omega, USA), made of 316L steel typically considered the minimum grade for use in marine environment. The operational pressure range is 0-10 bar with a maximum deviation of 0.08 %.

The pressure transducer has the function to monitor that the pressure remains constant while the test takes place. The location of this device is just in front of the filter holder.

Figure 7 Photo of the Omega pressure transducer with USB output (Omega, 2023)

The signal of the transducer can be read by the computer with help of software such as 'Digital transducer application' provided by the manufacturer of the pressure transducer (Omega, 2023). The pressure transducer is connected to a computer via a USB port.

7.3.1.8 Non-plugging water

For each test a new membrane filter needs to be used. Although the ASTM does not mention cleaning the filters before its use, it is recommended to measure the membrane resistance of the filters and clean them by passing 0.5-1 L non-plugging water (e.g., lab water after 0.2 μm filtration) or preferably ultra-pure water (UPW) through them. UPW is water free of colloidal particles, ions and organic matter.

7.3.2 MFI-0.45 testing procedure

The following procedure is recommended in the standard method (ASTM D8002 - 15, 2015):

1. Set the pressure regulator at 2 bar. The pressure must remain constant during the test (± 1%).
2. Measure temperature of the water.
3. Flush the water to be tested through the apparatus to remove contaminants.
4. Open the membrane filter holder and place a 0.45 µm membrane filter (25 mm or 47 mm in diameter) on the support plate of the holder. It is recommended to handle the membrane filters only with tweezers and avoid touching them with fingers. Use only filters that are packed in the same orientation.
5. Make sure the membrane holder's O-ring is in good condition and properly placed.
6. Close the filter holder.
7. Release air by opening the pressure relief valve and open the small air relief valve on top of the filter holder (in case the filter holder includes an air relief valve or make sure that air is not trapped in the system).
8. Close the relief valve and start recording flow (and preferably pressure as well). Recommended time interval for data acquisition is every 5 seconds.
9. Run the test for 30 minutes to 60 depending on the rate of flow decline. This depends on the volume of the feed water container.
10. After completing the test, the membrane filter may be retained for future reference or analysis.

At the start of the test, the initial flow needs to be controlled to quickly identify if the filter might have been cracked or misplaced in the preparation. The initial flow (or permeability) should be within 10 % of the flow recorded with ultra-pure water (UPW).

It is recommended to report together with the MFI value, the water temperature, the membrane filter material and manufacturer.

7.3.3 MFI-0.45 calculation procedure

The ASTM recommends the following procedure for calculating the MFI-0.45. The volume and time dataset should be plotted as *t/V* versus *V* to identify the linear relation between resistance and the cumulative filtered water volume (see Figure 1, Eq 14), for which the slope (*b*) describes the fouling tendency of a given water (Equation 15).

$$\frac{t}{V} = \frac{1}{Q_{avg}} = \frac{\eta \times R_m}{\Delta P \times A} + \frac{\eta \times I}{2 \times \Delta P \times A^2} \times V \qquad \text{Eq. 14}$$

and

$$b = \frac{\eta \times I}{2 \times \Delta P \times A^2} = \frac{dt / dV}{dV} \qquad \text{Eq. 15}$$

where: t is the filtrations time (s), V is the cumulated permeate volume (L), Q_{avg} is the average flow rate, η is the water viscosity (Ns/m^2), I is the fouling index (m^{-2}), R_m is the membrane resistance (m^{-1}), ΔP is the applied transmembrane pressure (bar or N/m^2), and A is the membrane surface area (m^2).

The slope of the line (b) has been defined as the MFI. This value needs to be normalized to reference conditions of ΔP_0 (200 kPa, 2 bar), η ($\eta_{20°C}$), and A_0 (13.8×10^{-4} m^2 equivalent to 47 mm diameter membrane filter). The term I represents the fouling index for the propensity of particles in water to form a layer with hydraulic resistance.

$$MFI = \frac{\eta_0 \times I}{2\times \Delta P_0 \times A_0^2} \qquad \text{Eq. 16}$$

In conducting the MFI test, the MFI can be determined from the gradient ($\tan\alpha$, b, $(dt/dV)/dV$) of the linear region of **minimum slope** determined in (a plot of) t/V versus V. Then normalizing this slope to standard conditions of temperature (T_{corr}), pressure (P_{corr}) and membrane area (A_{corr}) yields MFI as shown in Equation 17.

$$MFI = \left(\frac{\eta_{20°C}}{\eta_T}\right)\times\left(\frac{\Delta P}{\Delta P_0}\right)\times\frac{d\frac{t}{V}}{dV} = (T_{corr})\times(A_{corr})\times\frac{d\frac{t}{V}}{dV} \qquad \text{Eq. 17}$$

Example – Calculation of MFI-0.45 value

An MFI test was performed on raw seawater at 2 bar, at 20 °C (= 0.001 Ns/m^2) with a filter of 25 mm nominal diameter (21 mm effective diameter). The results of the MFI test are presented in the following figure. Calculate the MFI value.

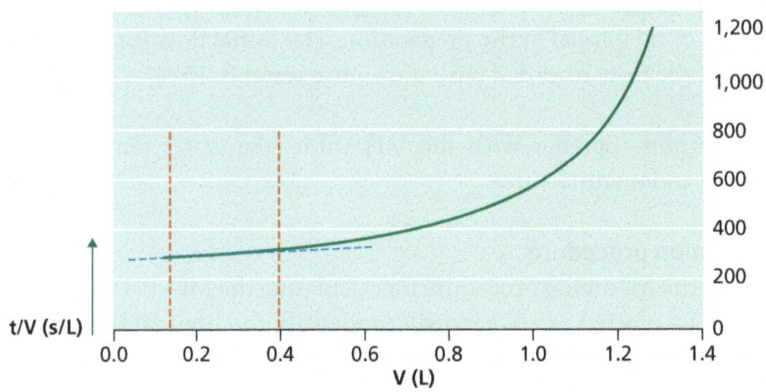

The slope of the linear region (marked by the two red lines) can be calculated.
Slope = 140 s/L^2.
The effective membrane area = $\pi \times (21/2/1000)^2 = 0.0003464$ m^2.
Now we can calculate the MFI-0.45 value.

$$MFI = \left(\frac{\eta_{20°C}}{\eta_T}\right)\times\left(\frac{\Delta P}{\Delta P_0}\right)\times\left(\frac{A}{A_0}\right)^2 \tan\alpha$$

By replacing the reference values and the ones used in the test, we have:
MFI = (0.001 Ns/m^2 / 0.001 Ns/m^2) × (2 bar / 2 bar) × (0.0003464 m^2 / 0.00138 m^2)^2 × 140 s/L^2
MFI = ~11 s/L^2

7.4 MEMBRANE PROPERTIES OF COMMERCIAL MEMBRANES

Salinas Rodriguez *et al.,* (2019) studied three 0.45 μm membrane filters as reported in Table 2. The mixed cellulose nitrate (NC) filter had the highest permeability in comparison to cellulose acetate (CA) and nylon (NN). The measured permeabilities suggest that all the tested filters were according to the recommendations of the ASTM.

Table 2 Filter properties

Code	Filter material	Thickness[1], μm	Bubble point[1], bar	Permeability[2], $L/m^2/h/bar$
CA	Cellulose acetate[3]	106	>2.4	31,020 ± 926 (3%)
NN	Nylon 6.6	144-170	2.2-2.5	22,764 ± 579 (3%)
NC	Mixed cellulose nitrate[3]	150	>2.1	44,438 ± 1,259 (3%)

[1] Information from manufacturer, [2] Measured using filter holder FH4 n=10,
[3] ASTM recommended material.

Top view images of the membrane filters were obtained using a scanning electron microscope (Jeol, JSM-6010 LA). Figure 8 shows similarities in the pore morphology among the three filter materials. NN shows a smoother surface than the other NC and CA. The shape of the pores is not well defined and they are not homogenously distributed over the surface.

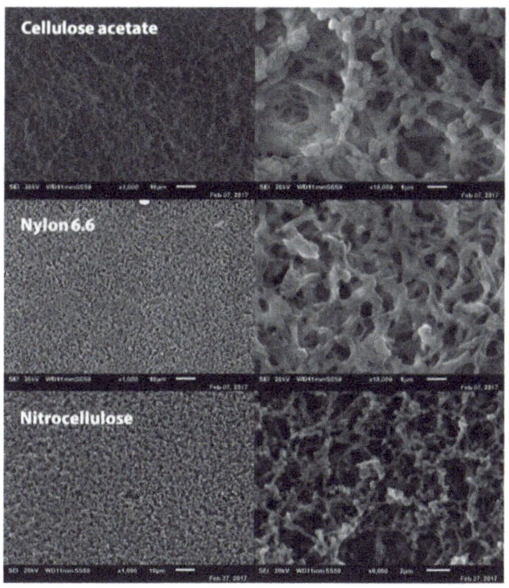

Figure 8 SEM top view images of the three tested filters at different magnifications (1,000x for left images, 10,000x for top right images, and 6,000x for bottom right image). Adopted from Salinas Rodriguez, *et al.* (2019)

7.5 EFFECT OF FILTER MATERIAL ON MFI-0.45

Figure 9 presents the MFI-0.45 values of Delft canal water (DCW) and Formazin solution (NTU= 15) measured with three different membrane materials. Filter properties influencing the measurements are the mean pore size, pore size distribution, surface porosity, thickness, tortuosity, surface charge.

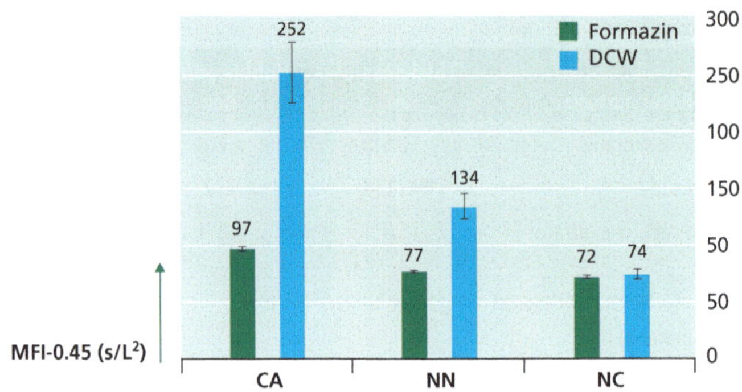

Figure 9 MFI-0.45 values measured with various filter materials (n=10) for Delft canal Water (DCW) and Formazin NTU = 15.

For DCW, $MFI_{0.45}$ value is 252 s/L^2 for the CA, 134 s/L^2 for NN, and 74 s/L^2 for NC. The relative error for CA, NN, and NC is 11 %, 8 %, and 6 %, respectively. NC filters show the lowest variation for MFI-0.45. While for Formazin, the measured MFI-0.45 values are similar for NN and NC (77 s/L^2 for NN and 72 s/L^2 for NC, with low relative errors 0.7 % and 0.8 %, respectively). However, a larger MFI-0.45 value is obtained with the CA filter (97 s/L^2 and 1.3 % relative error). The thinner thickness and clean water permeability of the CA cannot directly explain the higher $MFI_{0.45}$ value compared to the NC and NN filters when testing Delft canal water. It is possible that the surface charge of the filter and the interaction with the particulates and organic matter present in DCW influence the measured fouling potential.

Membrane resistance (R_m) can be used as a general indicator of the membrane properties (e.g., membrane porosity, pore size, and thickness). In the same mentioned study, the NC membranes have the lowest R_m values (9.35×10^9 m^{-1} ± 5.4 %) followed by CA membranes (1.39×10^{10} m^{-1} ± 6.6 %) and NN membranes (1.86×10^{10} m^{-1} ± 4.5 %) with the highest R_m values. No correlation between MFI-0.45 and R_m was observed (Figure 10).

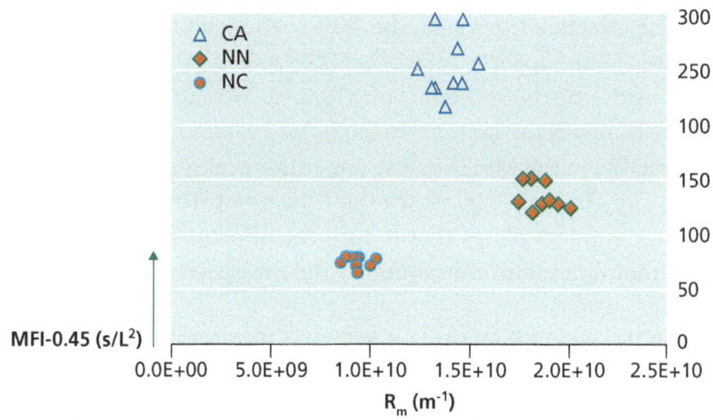

Figure 10 MFI-0.45 as a function on R_m (n=10). Delft canal water.

7.5.1 Effect of membrane support holder

Nahrstedt and Camargo (2008) studied the effect of filter support on SDI and MFI values. They reported that the filter holder had a strong influence on the obtained SDI values. The type of filter holder will determine the effective membrane area during filtration. A difference of more than 100 % was found for the same feedwater depending on the membrane holder used. A similar conclusion was drawn by Escobar *et al.,* (2009) when testing a Millipore holder and a Pall membrane holder.

Salinas *et al.,* (2019) also studied the effect of the filter holder, filter material in SDI and MFI tests for seawater and fresh water samples. In this study they proposed a correction for the effective filtration area by quantifying the actual effective membrane area after filtering a solution of powder activated carbon. This is illustrated in Figure 11 where MFI-0.45 values were measured for Delft Canal Water making use of 4 different filter support holders. Filter holder (FH) 7 is a porous support type, FH 4 (Schleicher & Schuell) and FH2 (Sartorius) and FH1 (Whatman) are concentrically channelled.

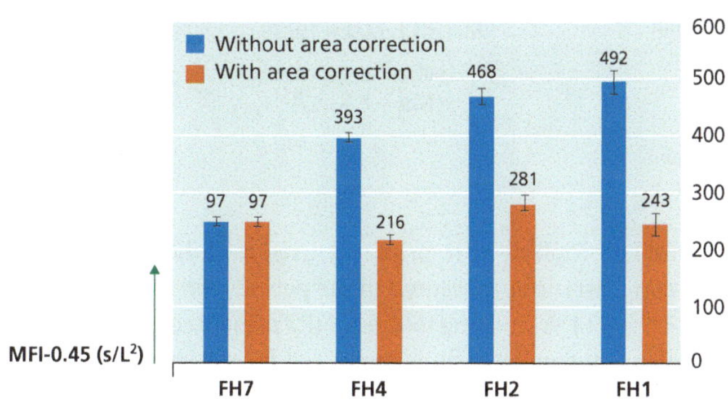

Figure 11 MFI-0.45 values measured with various filter holders (n=10) for Delft canal water with a cellulose acetate filter. FH = filter holder.

The correction factor is the ratio of the actual effective membrane area over the initially available membrane area. This is why a porous support for the membrane filter is recommended for measuring the MFI-0.45 of water.

By correcting for the effective filter area, the MFI-0.45 results obtained with the different filter holders (Figure 11) are closer to each other (247 $s/L^2 \pm 10.8$ %) in comparison with the average without considering the area effect (400 $s/L^2 \pm 27.6$ %). In the MFI formula, the area plays a significant role, hence the large variation in MFI values for filter holders without area correction. Additionally, in the MFI the flow rate influences greatly the fouling potential of a water sample, so any effect that increases the flow rate through the membrane (like the channels in the filter support plates that reduce the effective filter area) will increase the fouling load of the membrane and consequently the measured MFI-0.45 will be higher.

7.6 APPLICATION: WATER QUALITY MONITORING OF NORTH SEA WATER

North Sea water before and after filtration was monitored over almost one year. A summary of the properties of the water is presented in Table 3. The pH and electrical conductivity values remained fairly stable during the period, pH around 8 and EC values of 48-49 mS/cm for the samples. Turbidity measurements ranged between 0.1-1.0 NTU for filtered North Sea water and between 0.9-45 NTU for raw North Sea water. The elevated turbidity values in the raw water can be attributed to high ocean tides and storms in the area.

Table 3 Summary of seawater quality properties in the study period

Parameter	Raw North Sea water	Filtered (glass media, EBCT 30 min) North Sea water
pH	8.0 ± 0.3	
Turbidity, NTU	Min = 0.9; Max = 45 Avg = 10.5 ± 11.0	Min = 0.1; Max = 1.0 Avg = 0.4 ± 0.2
Elec. conductivity, mS/cm	48.1 ± 1.6	
DOC, mg/L	2.1 ± 0.5	
SUVA, L/mg/m	1.8 ± 0.6	
Total algal count, cell/mL	Min = 11, Max = 1,039	-
Chlorophyll-a, µg/L	< 5 (bdl) till end of March. In May ~7.5 µg/L	

bdl = below detection limit

Algal cell counts and chlorophyll-a are primarily used to indicate algal bloom occurrence. The chlorophyll-a concentrations measured in the period were below the detection limit of the method (LOD < 5 µg/L). According to the results presented in Figure 12, the minimum and maximum algal count concentrations measured for raw North Sea water are 11 cell/mL and 1,039 cell/mL, respectively. These values are considered normal due to the low temperature during the period until the end of April when the values rapidly increased, most likely due to a mild algal bloom. An algal cell counts lower than 1,000 cell/mL indicates that there is no algal bloom. Based on the high algal cell numbers at the end of April until mid-May, it can be concluded that a mild algal bloom event took place at the testing location.

From October 2016 to July 2017, the MFI-0.45 values were measured for both raw and filtered North Sea water (Figure 12). SDI_{15}, SDI_{10}, and SDI_5 could not be measured due to clogging of the filter, but the SDI_3 values were reported ranging between 6-26 %/min for filtered and 9-28 %/min for raw North Sea water.

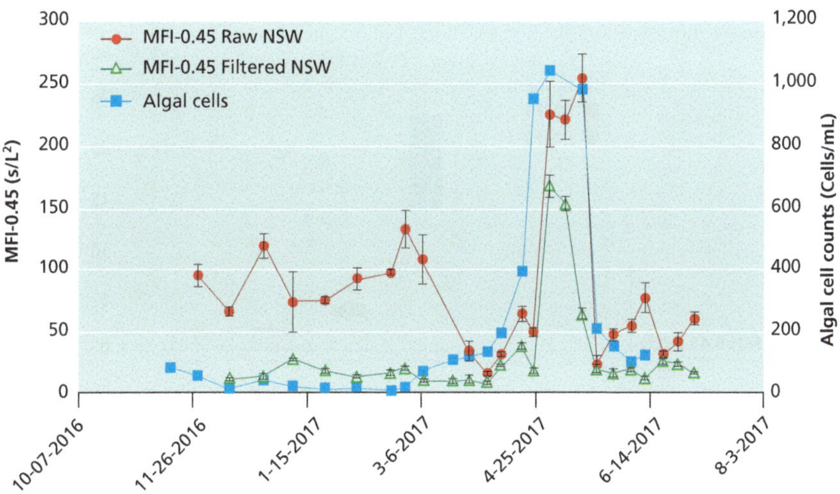

Figure 12 MFI-0.45 values of raw NSW and filtered NSW versus algal cell counts (n=3). CA filter.

MFI-0.45 measured for filtered North Sea water is in the range of 12-170 s/L^2 and for the raw seawater sample between 20-310 s/L^2, Figure 12. Consistently higher MFI-0.45 values were obtained for the raw NSW (up to 8x higher) that has a higher particles content as compared to the filtered NSW.

7.7 MONITORING OF MFI-0.45 IN A FULL-SCALE DESALINATION PLANT

A full-scale seawater desalination plant was monitored with MFI-0.45 over the period of 1 week with daily measurements. The treatment process consists of the following treatment steps (Figure 13): Screens (3 mm) > 2 mg/L $FeCl_3$ inline coagulation > 1st dual media filtration (DMF1, 11-14 m/h) > 2nd dual media filtration (DMF2, 17-19 m/h) > 5 µm cartridge filter (CF) > RO membrane ($R = 43\%$) > Remineralization.

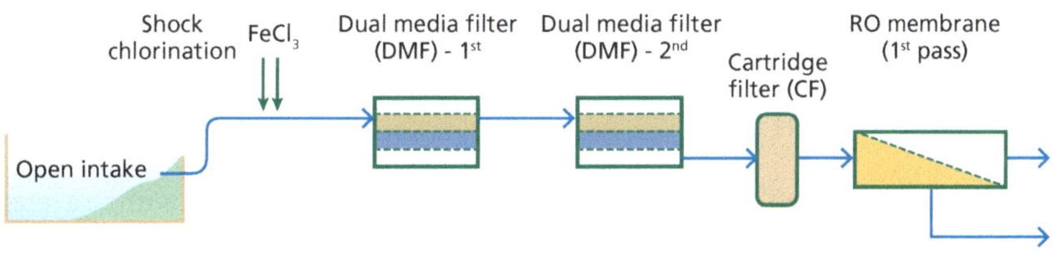

Figure 13 The treatment processes of the 120,000 m^3/d desalination plant.

MFI-0.45 of the influent seawater ranged between 13 and 26.5 s/L^2 with an average of 19.8 s/L^2. About 82% reduction in MFI was noticed through DMF1 where MFI declined from 19.8 s/L^2 to 3.8 s/L^2. A further reduction in MFI was recorded through DMF2 where the reduction percentage increased to about 91%. MFI values after DMF2 ranged from 1.3 to 2.5 s/L^2 while its value after CF ranged from 0.8 to 2.5 s/L^2.

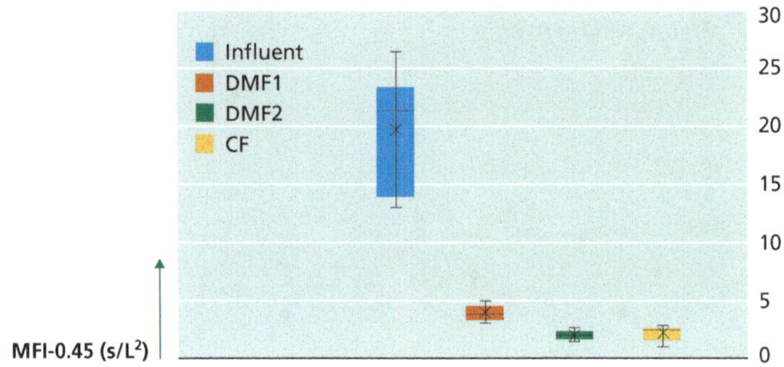

Figure 14 MFI-0.45 through the pre-treatment. Adapted from Abushaban, *et al.* (2021).

MFI after CF (2.1 s/L^2) was slightly higher than the MFI value after DMF2 (1.8 s/L^2). The possible reason for the higher MFI after CF might be the detachment of particles from the old CF.

The particles, larger than 1 μm, through the pre-treatment processes were counted using particles counter device available at the lab of the plant. Particles counter device reports the fractionation of particle sizes between 1 and 20 μm.

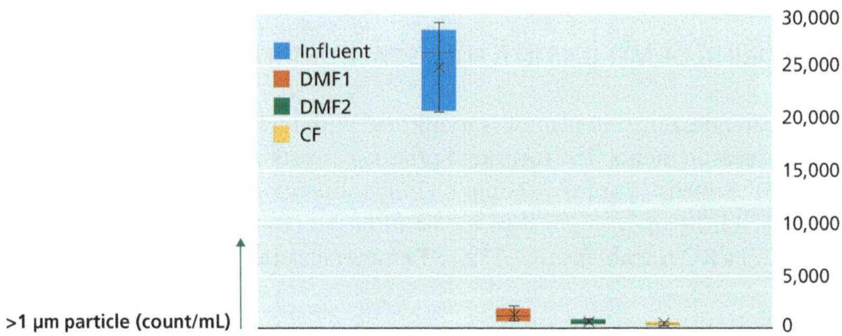

Figure 15 Particle larger than 1 μm counts per mL along the treatment process. Adapted from Abushaban (2019).

The total particles in the influent ranged from 18,777 to 26,561 CNT/mL with an average of 22,700 CNT/mL. The total counts decreased 95.2% through the DMF1 where the particles count declined from 22,700 to 1,080 CNT/mL. This shows that DMF1 has a significant role in the removal of particles which has the highest removal percentage. DMF2 reduced the particles concentration from 1,080 to 470 CNT/mL.

7.8 REFERENCES

Abushaban A (2019) Assessing Bacterial Growth Potential in Seawater Reverse Osmosis Pretreatment: Method Development and Applications CRC Press

Abushaban A, Salinas-Rodriguez SG, Pastorelli D, Schippers JC, Mondal S, Goueli S, Kennedy MD (2021) Assessing Pretreatment Effectiveness for Particulate, Organic and Biological Fouling in a Full-Scale SWRO Desalination Plant. Membranes 11: 167

ASTM D4189 - 14 (2014) Standard Test Method for Silt Density Index (SDI) of Water ASTM International, West Conshohocken, PA.

ASTM D8002 - 15 (2015) Standard Test Method for Modified Fouling Index (MFI-0.45) of Water ASTM International, West Conshohocken, PA.

Carman PC (1937) Fluid flow through granular beds Trans Instn Chem Engrs 15: 32-48

Carman PC (1938) Fundamental principles of industrial filtration (A critical review of present knowledge). Trans Instn Chem Engrs 16: 168-188

Escobar L, Sellerberg W, Sanchez D, Pastrana F, Wachinski A (2009) Detailed analysis of the silt density index (SDI) Results on desalination and wastewater reuse applications for reverse osmosis technology evaluation. Paper presented at the International forum on marine science and technology and economic development - Asia-Pacific Desalination conference, Qingdao, China, 8-10 July 2009 2009

Festo (2023). Precission pressure regulator https://www.festo.com/gb/en/a/162834/. Cited 02 November 2023 2023

Festo (2023). Pressure gauge https://www.festo.com/gb/en/a/162834/. Cited 02 November 2023 2023

Nahrstedt A, Camargo Schmale J (2008) New insights into SDI and MFI measurements. Water Science and Technology: Water Supply 8: 401-412 DOI https://doi.org/10.2166/ws.2008.087

Omega (2023). USB Output Pressure Transducer https://www.omega.nl/pptst/PXM409-USBH.html. Cited 27 October 2023

Ruth BF, Montillon GH, Montonna RE (1933) Studies in Filtration I. Critical Analysis of Filtration Theory. Industrial & Engineering Chemistry 25: 76-82 DOI 10.1021/ie50277a018

Salinas Rodriguez SG, Sithole N, Dhakal N, Olive M, Schippers JC, Kennedy MD (2019) Monitoring particulate fouling of North Sea water with SDI and new ASTM MFI0.45 test. Desalination 454: 10-19 DOI https://doi.org/10.1016/j.desal.2018.12.006

Sartorius (2023). Entris II Precision Balance https://shop.sartorius.com/nl/p/entris-ii-advanced-line-precision-balance-12200-g100-mg-external-adjustment/BCA12201-1S. Cited 02 November 2023 2023

Schippers JC (1989) Vervuiling van hyperfiltratiemembranen en verstopping van infiltratieputten Keuringinstituut voor waterleidingartikelen KIWA N.V., Rijswijk

Schippers JC, Salinas Rodriguez SG, Boerlage SFE, Kennedy MD (2014). Why MFI is edging SDI as a fouling index. Desalination & Water Reuse: 28-32

Schippers JC, Verdouw J (1979). De membraanfiltratie-index als kenmerk voor de filtreerbaarheid van water. 12: 104-109https://edepot.wur.nl/398518

Schippers JC, Verdouw J (1980) The modified fouling index, a method of determining the fouling characteristics of water. Desalination 32: 137-148

Sterlitech (2023). Pressure filtration vessels https://www.sterlitech.com/pressure-filtration-vessel-741060.html. Cited 02 November 2023

Taltech (2023). WinWedge Standard https://www.taltech.com/winwedge/. Cited 02 November 2023 2023

doi: 10.2166/9781789062977_0173

Chapter 8

Modified Fouling Index Ultrafiltration (MFI-UF) Constant Flux

Sergio G. Salinas-Rodriguez, IHE Delft, The Netherlands

The learning objectives of this chapter are the following:

- Define the theoretical principles of MFI-UF constant flux and its prediction model
- Describe the MFI-UF's testing set-up, testing protocol and calculation procedure
- Illustrate the application of the MFI-UF method
- Compare the MFI-UF with other fouling indices

8.1 INTRODUCTION

Membrane fouling due to particulate matter has plagued reverse osmosis (RO) systems since their first use in desalination and remains a persistent issue today. Contrary to the silt density index or modified fouling index (MFI-0.45), the MFI-UF is not, at this date, an ASTM method. Nevertheless, its application has grown in the last 15 years in assessing particulate fouling in full scale and pilot installations such as in seawater and brackish water desalination plants, surface water treatment facilities, and wastewater reuse plants. The application of the MFI-UF is expected to keep growing as a monitoring tool in membrane treatment systems.

Boerlage (2001) proposed the constant flux MFI with a 13 kDa PAN hollow fibre UF membranes and applied it to monitor particulate fouling removal in fresh surface water treatment in a plant of Amsterdam Water Supply and of PWN. Constant flux mode was obtained by manually adjusting the pressure reducing valve of the MFI-UF (constant pressure) equipment to maintain the required applied flux at a constant value.

Since then, the testing set-up was further developed to perform constant flux dead-end filtration, accurate at low flux rates, making use of flat UF membrane filters of various pore sizes. The MFI-UF at constant flux has been applied for many years in its current form as developed by Salinas Rodríguez (2011) and recently further developed by Abunada (2023). The MFI-UF test, as developed by Salinas Rodríguez (2011), has been applied in several research studies assessing particulate fouling in pilot and full scale treatment plants such as presented in Table 1. Tabatabai (2014) applied the MFI-UF to measure algal released organic matter (AOM) removal rates from RO feed water with conventional coagulation (coagulation / flocculation / sedimentation / filtration).

Villacorte (2014) studied the fate of transparent exo-polymer particles (TEP) through the treatment processes of 4 RO plants, where he observed significant correlations between TEP concentration and MFI-UF, suggesting that TEPs likely have a major role in the fouling of UF pre-treatment systems and possibly in seawater RO systems, if not effectively removed by the pre-treatment.

Dhakal (2017) investigated the fouling potential and fouling behaviour of algae and AOM in ultrafiltration membranes. MFI-UF was linearly related to algal cell density and chlorophyll-a concentration, biopolymer concentration, TEP, during the growth phase of the algal species. Abushaban (2019) also applied the MFI-UF to monitor particulate fouling removal along the pre-treatment of full scale seawater RO plants. Zhan *et al.* (2020) applied the MFI-UF as a diagnostic tool of fouling potential of a RO system in ultra-pure water (UPW) production. Abunada (2023) further developed the MFI-UF test by using a 5 kDa membrane and improved the prediction of particulate fouling rate in RO systems.

Some potential applications of the MFI-UF at constant flux are:

- Assessing RO and NF feed water quality with respect to particulate fouling potential.
- Predicting rate of fouling RO and NF membranes due to particles.
- Assessing performance of pre-treatment systems in terms of particulate fouling removal.
- Characterizing MF/UF feed water in predicting development pressure increase during a filtration cycle
- Verifying membrane integrity of MF/UF/RO/NF membrane systems.

Legend Table 1

P = Pilot; FS = Full Scale; UF = ultrafiltration, RO = reverse osmosis; GBF = glass beads filter; FSW = fresh surface water; SW = seawater; TDWW = treated domestic wastewater; On = Onsite; Off = offsite; DMF = dual media filtration; BW = beach well; DAF = dissolved air flotation; Flo = Flotation; SWRO = seawater RO; BWRO = brackish water RO; NF = nanofiltration; MSS = Marine science station; IEX = ion exchange.

Table 1 Examples of pilot and full-scale treatment plants where MFI-UF constant flux was applied

Location	Water	Utility	Type	On-, off-site	Treatment	Year
Toulon, France	SW	Veolia	P	On	UF-RO, DMF-RO	(Salinas Rodriguez and Kennedy, 2009)
Tarragona, Spain	SW	Dow	P	On	UF-RO	(Salinas Rodriguez and Kennedy, 2009, Salinas Rodriguez, *et al.*, 2010, Salinas Rodriguez, *et al.*, 2010)
Kamperland, Netherlands	SW	Evides	P	On	MS – UF – RO	(Althuluth, 2009, Salinas Rodríguez, *et al.*, 2009, Salinas Rodríguez, *et al.*, 2010, Wahyudi, 2010, Salinas Rodríguez, 2011, AlShuaili, 2012)
Emmen, Netherlands	TDWW	Nieuwater	FS	Off	UF-BACF-RO-EDI	(Ekowati, 2011, Ekowati, *et al.*, 2014, Gulrez, 2021)
Baanhoek, Netherlands	FSW	Evides Industry	FS	Off	Coag, UF, RO	(AlShuaili, 2012, Rathnayake, 2013)
Ras Al Khaimah, UAE	SW	Fewa			DAF-UF-RO, DMF-RO	(Dhakal, *et al.*, 2016)
Ras Abu Fontas, Qatar	SW	Kahramaa	FS	On	DAF-DF-UF-SWRO-BWRO	(Dhakal, *et al.*, 2018)
Aqaba, Jordan	SW	MSS Jordan	-	On	Raw water	2018 - 2022
Botlek, Netherlands	FSW	Evides Industry	FS	Off	DAF-IEX-RO	(Dhakal, et al., 2020)
Heemskerk, Netherlands	FSW	PWN	FS	Off	UF-RO	(Bassa, 2021, Abunada, *et al.*, 2023)
Andijk, Netherlands	FSW	PWN	FS	Off,	Coag + Sed + RSF + GAC	(Bassa, 2021, Lindsay, 2022)
Andijk, Netherlands	FSW	PWNT	P	Off, On	UF-RO	(Bassa, 2021, Lindsay, 2022)
Bahia de Palma, Spain	SW	Suez		On	BW-DMF-CF-RO	(Le Bouille, *et al.*, 2021)
Perth, Australia	SW	Veolia	FS	On	Coag-DMF-RO	(Le Bouille, *et al.*, 2023)
Katwijk, Netherlands	FSW	Dunea	P	On	MS, DMF, UF, RO	(Malaichami, 2023, Tahtouh, 2023)
Kamperland, Netherlands	SW	Zeeschelp		Off	GBF, UF	(Mirzaei, 2023)
Baanhoek, Netherlands	FSW	Evides	P	Off, On	Coag, Flo, DMF, GAC, RO and NF	(Ofori, 2023, Qatae, 2023, Yameen, 2023)

8.2 THEORY PARTICULATE FOULING

The flow through a RO membrane can be described by:

$$Q_w = \frac{dV}{dt} = (\Delta P - \Delta \pi) \times K_w \times A \qquad \text{Eq. 1}$$

where:

Q_w	=	permeate flow (m^3/hr)
V	=	total filtered volume water (permeate) (L or m^3)
t	=	time (hour, minute, second)
ΔP	=	differential pressure (pressure feed - pressure permeate)
$\Delta \pi$	=	difference osmotic pressure (osmotic pressure feed – osmotic pressure permeate)
K_w	=	permeability constant for water (m^3/m^2-s-bar)
A	=	surface area of the membrane(s) (m^2)
Q_w/A	=	permeate flow through membrane surface area ($m^3/m^2/h$)
	=	filtration rate ($m^3/m^2/h$), used in rapid sand filtration
	=	flux ($L/m^2/h$) used in membrane filtration
$(\Delta P - \Delta \pi)$	=	net driving pressure (NDP)

In membrane technology, flux is defined as the ratio of the permeate flow and surface area of the membrane. It is expressed as:

$$J = \frac{Q_w}{A} = \frac{1}{A} \times \frac{dV}{dt} \qquad \text{Eq. 2}$$

To simplify the equations, we assume that $\Delta \pi$ is negligible. This assumption is valid for low salinity water only. Then,

$$J = \frac{1}{A} \times \frac{dV}{dt} = \Delta P \times K_w \qquad \text{Eq. 3}$$

Frequently the concept of resistance (R) is used, instead of permeability:

$$K_w = \frac{1}{\eta \times R_T} \qquad \text{Eq. 4}$$

Where: η is the viscosity of the water and R_T is the total resistance [sum of membrane resistance (R_m), pore blocking (R_p) and cake formation (R_c)].

$$R_T = R_m + R_b + R_c \qquad \text{Eq. 5}$$

Replacing Eq. 4 and Eq. 5 in Eq. 3:

$$J = \frac{1}{\eta} \times \frac{P}{R_m + R_b + R_c} \qquad \text{Eq. 6}$$

When we assume that pore blocking does not play a dominant role in RO, then fouling is mainly due to cake formation. As a consequence:

$$J = \frac{1}{\eta} \times \frac{P}{R_m + R_c} \qquad \text{Eq. 7}$$

Permeability of the clean filter media (R_m) is a function of filter properties such as filter thickness (Δx), surface porosity (ε), pore radius (r_p), and tortuosity (τ) and can be defined using Poiseuille's Law:

$$R_m = \frac{\theta \times \Delta x \times \tau}{\varepsilon \times r_p^2} \qquad \text{Eq. 8}$$

The cake resistance (R_c) component in (membrane) filtration can be defined following the Ruth equation (Ruth, *et al.*, 1933), using the concept of 'specific cake resistance' per unit weight (α) (Equation 9). Ruth showed that the resistance of the cake formed during constant pressure filtration is proportional to the amount of cake deposited at the filter medium provided the retention of particles and α are constant. Cake resistance is defined as:

$$R_c = I \times \frac{V}{A} \qquad \text{Eq. 9}$$

and the fouling index (I) is:

$$I = \alpha \times C_b \qquad \text{Eq. 10}$$

Where: I is a measure of the fouling characteristics of the water (m^{-2}). The value of I is a function of the nature of the particles and is proportional to their concentration. C_b is the concentration of particles per unit volume of filtrate (e.g., mg/L) and α is the specific cake resistance per mg cake per m^2 membrane ($m^3/mg/m^2$).

The specific cake resistance is constant for incompressible cakes under constant pressure filtration and can be calculated according to the Carman-Kozeny relationship (Equation 11) (Carman, 1937, Carman, 1938). Carman (Carman, 1937, Carman, 1938) derived Equation 11 for the specific resistance of a cake composed of spherical particles of diameter d_p from the Kozeny equation including a factor for tortuosity of the voids within the cake. According to the Carmen relationship a reduction in the porosity of the cake (ε) or a decrease in particle diameter size (d_p) increases the specific resistance of the deposited cake.

$$\alpha_c = \frac{180 \times (1 - \varepsilon)}{\rho_p \times d_p^2 \times \varepsilon^3} \qquad \text{Eq. 11}$$

As porosity is to the power three, it plays a dominant role. The more compact a cake, the higher the specific cake resistance, and therefore the higher the cake resistance and pressure required to overcome this resistance.

RO plants typically operate at constant capacity and recovery. So, the flux is constant. When membranes foul, the pressure needs to be increased, in order to maintain a constant capacity (and flux) in the system. Rewriting Eq. 7:

$$J = \frac{1}{\eta} \times \frac{\Delta P_t}{R_m + R_c} \qquad \text{Eq. 12}$$

Where: ΔP_t is the pressure at time 't' (which will increase). Rearranging Eq. 2 because flux is constant:

$$\frac{V}{A} = J \times t \qquad \text{Eq. 13}$$

and substituting Eq. 13 in Eq. 9:

$$R_c = I \times \frac{V}{A} = I \times J \times t \qquad \text{Eq. 14}$$

This results in:

$$J = \frac{1}{\eta} \times \frac{\Delta P_t}{R_m + I \times J \times t} \qquad \text{Eq. 15}$$

Rearranging the previous equation, we obtain:

$$P_t = \eta \times R_m \times J + \eta \times I \times J^2 \times t \qquad \text{Eq. 16}$$

Thus, the increase of pressure ΔP_t across the membrane is linearly proportional with time, with the fouling index (I) and with the flux to the power two (J^2). As a consequence, flux has a very dominant effect on the development of ΔP_t.

The modified fouling index with ultrafiltration membranes can be calculated from Eq. 17, where the following reference conditions are considered: pressure (ΔP_o) = 2 bar, membrane area (A_o) = 13.8×10^{-4} m^2, water temperature through water viscosity η at 20 °C. Therefore, conversion of MFI into I results in I (m^{-2}) = 7.6×10^8 × MFI (s/L^2).

$$MFI = \frac{\eta_o \times I}{2 \times \Delta P_o \times A_o^2} \qquad \text{Eq. 17}$$

The fouling index (I) can then be determined from the slope of the linear region in a plot of *pressure vs. time*, which corresponds to cake filtration as illustrated in Figure 1. The MFI is calculated considering the minimum I value.

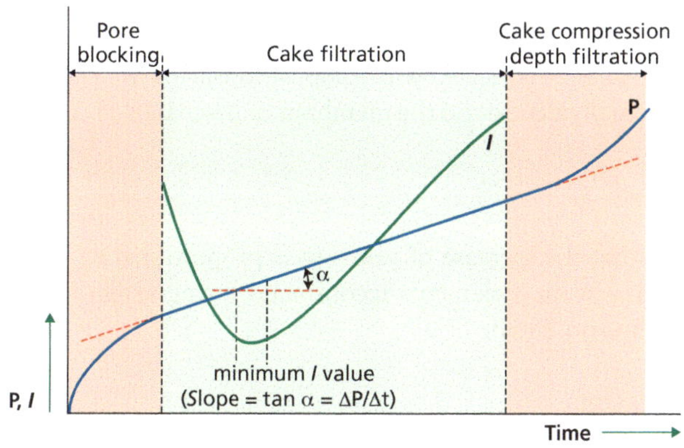

Figure 1 Schematic of the constant flux MFI test illustrating pressure development over time and the identification of the minimum *I* value for MFI calculation. Adapted from (Salinas Rodríguez, 2011)

Substituting the Carman-Kozeny Eq. 11 in Eq. 16 gives Eq. 18. This equation shows that MFI is a function of the dimension and nature of the particles forming a cake on the membrane, and directly dependent on particle concentration in water, as illustrated in Figure 2.

$$MFI = \frac{\eta_o \times 90 \times (1-\varepsilon) \times C_b}{\rho_p \times d_p^2 \times \varepsilon^3 \times \Delta P_o \times A_o^2} \qquad \text{Eq. 18}$$

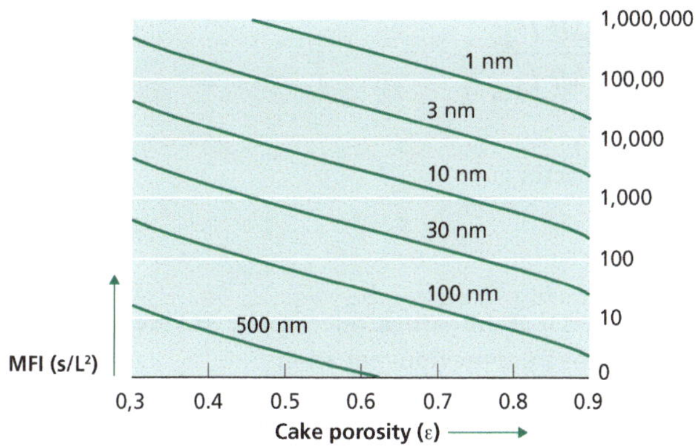

Figure 2 MFI as a function of particle size and cake porosity. Adopted from (Salinas Rodriguez, *et al.*, 2021)

Equation 16 is valid for 'dead-end' filtration. In 'cross-flow' filtration only a part of the particles will deposit on the membrane surface due to the shear-force of the cross-flowing water. Therefore, '*I*' has to be corrected with a deposition factor 'Ω'. This factor is the fraction of particles which actually deposit on the membrane surface ($\Omega \leq 1$). Then, Eq. 16 becomes:

$$\Delta P_t = \eta \times R_m \times J + \eta \times \Omega \times I \times J^2 \times t \qquad \text{Eq. 19}$$

The phenomenon, that the increase of pressure is proportional to $(\text{flux})^2$ explains, why manufactures of spiral wound elements recommend lower design fluxes with feedwater having a higher fouling potential.

8.2.1 Deposition factor

Only a fraction of the RO feedwater is forced to pass through the membrane in cross flow filtration. This fraction of water depends on the recovery at which the RO unit operates. In dead-end filtration all the particles bigger than the membrane's pores will be retained while in the case of cross-flow, only the fraction of water passing through the membrane is affected and the associated fraction of particles may or may accumulate on the membrane surface.

The deposition factor was first proposed by Schippers *et al.* (1980, 1981) in a model to predict flux decline in reverse osmosis systems. It was defined as the fraction of particles deposited, which are present in the water that passes through the RO membrane.
Then, we can obtain the deposition factor equation as function of recovery,

$$\Omega \, ? \, \frac{1}{} + \frac{MFI_{conc}}{MFI_{feed}} \times (1 - \frac{1}{R}) \qquad \text{Eq. 20}$$

Or as function of concentration factor (CF),

$$\Omega \frac{1}{(CF-1)} \times CF - \frac{MFI_{conc}}{MFI_{feed}} \qquad \text{Eq. 21}$$

Where the concentration factor is:

$$CF = \frac{1}{1-R} \qquad \text{Eq. 22}$$

The formula above assumes that the particle rejection by RO membranes is 100 %.
There are possible scenarios from previous equations:

- $\Omega = 0$ means $MFI_{conc} = MFI_{feed} \times CF$ — No particles deposit
- $\Omega = 1$ means $MFI_{conc} = MFI_{feed}$ — All particles deposit
- $\Omega > 1$ means $MFI_{conc} < MFI_{feed}$ — All particles deposited + retention inside pressure vessel (e.g., spacer)
- $\Omega < 0$ means $MFI_{conc} > MFI_{feed} \times CF$ — Particles might be removed/sheared off inside the pressure vessel; earlier deposited particles released; particles formed by bacteria; particle size distribution influence results.

A *positive* deposition factor indicates particles are being accumulated on the membrane surface as they pass through the system while a *negative* factor indicates the number of particles in the concentrate exceeds the concentration in the incoming feedwater (Schippers, 1989, Boerlage, 2001).

8.2.2 The particulate fouling prediction model

The particulate fouling models to predict fouling developed by Schippers are based on the assumption that particulate fouling on the surface of reverse osmosis (or nanofiltration) membranes can be described by the cake filtration mechanism (Belfort and Marx, 1979, Schippers, *et al.*, 1981). The relationship between the MFI measured for a feedwater and the fouling rate predicted for a RO system are presented below. The relationship assumes that scaling, adsorptive blocking and biofouling do not contribute to the fouling observed on the RO membrane.

Generally, RO desalination plants operate at constant flux to meet production requirements. Changes in feedwater temperature are compensated for by adjusting feed pressure. Similarly, fouling resulting in an increase in membrane resistance is compensated for by increasing the feed pressure and hence net driving pressure (NDP). In this case, increase in the NDP can be predicted through equation 19 which already includes the deposition factor Ω. By rearranging this equation, we have:

$$\frac{(NDP_r - NDP_{0r})}{t_r} = J_2 \times \eta \times \Omega \times I_r \qquad \text{Eq. 23}$$

And replacing *I*, we have

$$\frac{\Delta NDP_t}{t_r} = \frac{2 \times MFI \times \Delta P_0 \times A_0^2 \times \Omega \times J_r^2 \times \eta_r}{\eta_{20^\circ C}} \qquad \text{Eq. 24}$$

Where the subscript '*r*' refers to RO system conditions. Based on this equation, a theoretical 'safe MFI' or threshold MFI value for RO feed water can be calculated assuming e.g., an allowable increase in NDP of 15%. To accurately predict the rate of fouling of RO and NF membrane systems, the MFI-UF value should be measured at the same flux rate as the one the membrane system is operating (Salinas Rodríguez, 2011, Salinas Rodríguez *et al.*, 2015, Abunada *et al.*, 2022).

8.3 MEASURING MFI-UF CONSTANT FLUX

8.3.1 Filtration set-up and materials

The MFI-UF constant flux test was proposed initially using a 13 kDa hollow fibre PAN membrane (Boerlage, 2004) and further developed to its current configuration (Salinas Rodríguez, 2011, Salinas Rodríguez *et al.*, 2015, Abunada, 2023) using flat 25 mm diameter PES UF membranes in dead-end filtration (see Figure 3).

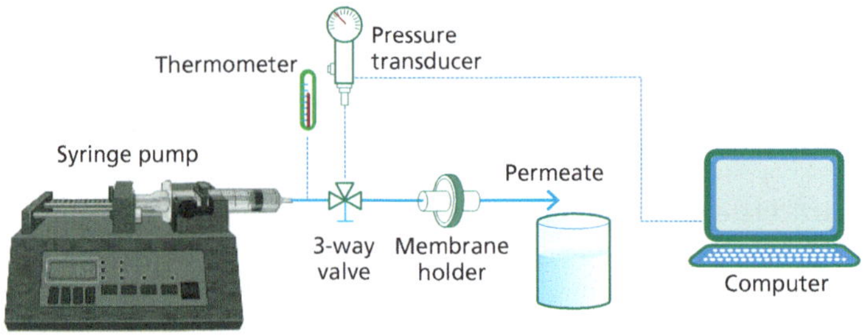

Figure 3 Schematic of the MFI-UF constant flux filtration set-up

The key components of the set-up are: syringe (infusion) pump, membrane holder, pressure transducer, thermometer, computer and a three-way valve. A three-way valve is used to connect the syringe (water sample) with the membrane holder and at the same time with the pressure transducer. The filtration set-up was verified for correct pressure readings, stable constant flux filtration, no leakages, and avoiding air/bubbles trapped in the system (tubing, feed side of filter holder).

The readings from the pressure sensor were verified with a second manometer (pressure gauge) by filtering ultra-pure water (UPW) at various flux rates. A maximum 3 % difference was observed, which is considered acceptable.

The flux rate was verified by monitoring the permeate weight over time with help of an electronic balance. Several flux rates (10 - 400 L/m^2/h) were tested and a maximum difference of 2.8 % was observed between the expected and measured flux with the lower flux rate.

Verification of leakages in the set-up was performed by pressurizing the system (up to 4 bar) without allowing filtration and monitoring the pressure change over time. No leaks were observed at pressures less than 4 bars over time.

The presence of air in the system is not desirable. To verify the effect of air trapped in the system, air was intentionally introduced and filtration was allowed. Erroneous high-pressure values were observed by the effect of air; this could be related to the bubble point of the membranes or related to the compression of air that will produce erratic pressure development.

The parts of the MFI-UF filtration set-up are discussed in the following sections.

8.3.1.1 Membrane filters

Flat circular (25 mm diameter) UF membrane filters are used in the test with molecular weight cut-off (MWCO) in the range 5 – 100 kDa. Two membrane materials are available from Millipore, namely: poly ether sulfone (PES) and regenerated cellulose (RC); nevertheless, the PES membranes have lower membrane resistance than the RC ones and thus are preferred for the test. The specifications of the membrane filters are summarized in Table 2.

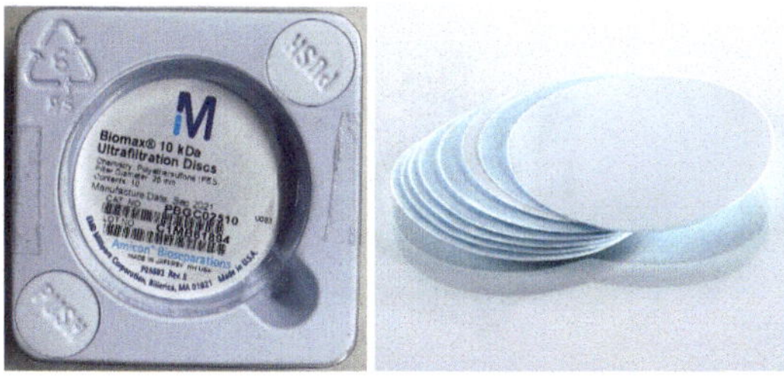

Figure 4 Package of Biomax 10 kDa, 25 mm, PES membrane filters from Millipore. Each package contains 10 membrane filters. Source: Bassa, (2021).

Table 2 Specifications of the UF PES membranes

MWCO, kDa	Code	Mean pore size[1], nm	Surface porosity[2], %	R_m, m^{-1}
5	PBCC02510	8.0	0.6	$6.57\times10^{12} \pm 1.0\times10^{12}$ (15%)[3]
				$2.05\times10^{12} \pm 1.36\times10^{11}$ (7%)[6]
10	PBGC02510	9.2	2.9	$9.65\times10^{11} \pm 1.58\times10^{11}$ (16%)[3]
				$9.44\times10^{11} \pm 1.08\times10^{11}$ (11%)[4]
				$1.12\times10^{12} \pm 1.1\times10^{11}$ (10%)[5]
				$1.02\times10^{12} \pm 4.94\times10^{10}$ (5%)[6]
100	PBHK02510	10.6	10.5	$3.24\times10^{11} \pm 4.17\times10^{10}$ (12.9%)[7]

[1] Abunada, *et al.* (2022) ; [2] (Andyar, 2021, Abunada, *et al.*, 2022); [3] (Bassa, 2021); [4] (Lindsay, 2022); [5] (Qatae, 2023); [6] (Sithole, 2017); [7] (Salinas Rodríguez, *et al.*, 2015)

Salinas Rodríguez *et al.*, (2015) reported that the PES membranes are hydrophilic, have a contact angle of 64° ±4.2°, and are negatively charged above a pH value of 4.5.

Membrane filters are available in packages containing 10 specimens. Each package is numbered with a batch code and production date. For preservation purposes, the manufacturer coats the membranes with glycerine and sodium azide; this coating needs to be removed before the filtration test is performed. A new membrane needs to be used per test. In 2023, the cost per package of membranes is 208 euros (Merck, 2023).

Abunada *et al.*, (2022) using suspensions of polystyrene particles (75 nm) studied the effect of membrane surface porosity on MFI-UF and proposed a preliminary correction factor of 0.5 for the 10 kDa PES UF membrane filters. Nevertheless, properties of real waters may yield different effects than the ones found for inorganic particles.

8.3.1.2 Constant flow pump

An infusion syringe pump was found to be very reliable and accurate for delivering constant flow in the filtration process. The PHD Ultra pump from Harvard Apparatus (see Figure 5) can deliver up to 6 bar of pressure and work in the flow range suitable for the test.

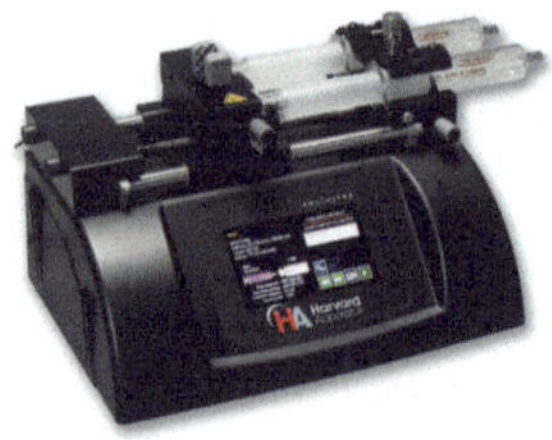

Figure 5 Syringe pump model PHD ULTRA used in the MFI-UF test (Harvard Apparatus, 2023)

The characteristics of the pump are:
- Flow rate range: 1.5 pL/min - 216 mL/min
- Flow rate accuracy: ± 0.25 %
- Force: 75 lb
- Dimensions (H/W/D/) / Weight: 18.42 × 30.48 × 21.59 cm / 4.5 kg
- Battery: None
- Brand: Harvard Apparatus, infusion only PHD Ultra syringe pump
- In 2022 the cost of the pump was about 3,800 euros.

8.3.1.3 Pressure transducer

The chosen pressure transduce is commercially available (PXM409-001BGUSBH, Omega, USA), made of 316L steel typically considered the minimum grade for use in marine environment. The operational pressure range is 0-10 bar with a maximum deviation of 0.08 %.

The pressure transducer has the function to measure and transmit pressure values (up to 1000 readings/second) over time while filtration occurs.

Figure 6 Photo of the Omega pressure transducer with USB output (Omega, 2023)

Some of the properties of this pressure transmitter are:
- Accuracy: 0.08% BSL (linearity, hysteresis and repeatability combined)
- Resolution: Up to 5.5 significant figures
- Power Consumption: 0.35 W typical. USB powered from laptop.
- Wetted Parts: 316L stainless steel
- Weight: 200 g
- In 2023 the cost of the pressure transducer was about 800 euros (Omega, 2023).

The pressure transmitter was verified periodically with a manometer by pushing compressed air.

8.3.1.4 Membrane filter holder

The membrane holder where the membrane filter is placed for the filtration test. It should avoid leakages, and not damage the membrane at all. In this research a holder for 25 mm diameter membranes was used. Several types were tested (Sterlitech - Stainless steel, Whatman GE - Poly propylene, Schleicher & Schuell - Poly propelene) with the Cytiva's Whatman (WHA-10461000 FP 025/1 Polyethersulfone) syringe filter holder for 25 mm filter selected.

Figure 7 Photo of the membrane filter holder. Right is the feed side. O-ring is placed on top of the feed side of the membrane.

This membrane holder was slightly modified by removing the upper inner wall of the membrane holder, so in this was the flow distribution towards the membrane is uniform and only the membrane captures all the particles in the sample water. In the set-up, the filter holder is connected with a three-way valve that connects with the syringe (sample water) and with the pressure transducer.

The membrane holder and syringe should be rinsed with UPW before use. After a lot of use, due to wear and tear, the filter holders need to be replaced as they may give inconsistent results.

8.3.1.5 Syringe

The water sample is placed in a disposable syringe that is attached to the piston pump. The used syringe is a BD Plastipak™ 60 mL. Syringes are cleaned by soaking with lab water before testing. A new syringe can be used every 4 tests and it should be rinsed with UPW before and after its use. The brand and volume of the syringe should be introduced in the settings of the syringe pump.

8.3.1.6 Ultra-pure water

Ultra-pure water (UPW) or lab water was produced from Delft's tap water by a MilliQ instrument (IX 7005) with the following treatment steps: activated carbon filtration, reverse osmosis, electro de-ionization, UV disinfection and 0.22 μm filtration.

UPW or lab water is water with organic matter content less than 5 μg/L and conductivity 0.055 μS/cm (or resistivity, 18.2 MΩ-cm). Besides the UPW being free of natural organic matter and free of ions, it must be free of particles.

In case particles are present in the UPW water, they will be observed by a slight pressure increase during the test to measure the clean water resistance of the membrane filters. If particles are present in the UPW, then pores of the filter might be partly clogged and thus influence the MFI-UF results by decreasing the effective membrane area.

8.3.1.7 Tubing

Tubing should be pressure resistant. The tested operational pressure values were up to 5 bars. Tubing is used to connect the three-way valve with the pressure transducer. Also, it should be resistant to chemicals, aging and abrasion. The tube length is ~50 cm with a diameter of 6 mm. Festo compressed air translucent tubing made of polyurethane or similar product works well. The tubing needs to be filled with water and avoid the presence of air inside of it.

8.3.1.8 Software

The measured signal of the pressure transducer needs to be recorded for further processing. This is done via the software 'Digital transducer application' provided by the manufacturer of the pressure transducer (Omega, 2023). The pressure transducer is connected to a computer via a USB port. Data acquisition is set to every 10 seconds in the MFI-UF test.

8.3.2 Membrane cleaning and conditioning

Membrane filters must be cleaned before performing the MFI-UF test. A surface that is not clean may affect the way that the fouling cake is formed on the membrane. According to the operating instructions provided by the membrane manufacturer, the membranes (PES and RC) are coated with glycerine to prevent the membrane drying out and also sodium azide (NaN_3) to preserve the membrane.

The suggested cleaning consists of flushing the membrane filter with ultra-pure water (UPW) for at least 30 minutes until the pressure stabilizes over time at a high flux rate such as 300 L/m^2/h. This recommendation by Li (2019) was proposed after monitoring the concentration of total organic carbon (TOC) in the permeate water after passing the membrane filters. In Li's study, the higher the flux rate the more TOC was released by the membrane filter; for instance, for a 10 kDa PES membrane, at 300 L/m^2/h about 4 mg-C/L was measured after 30 minutes filtration while at 200 L/m^2/h about 2.5 mg-C/L was measured. Salinas Rodríguez, *et al.* (2015) cleaned the membranes by soaking them 24 h in UPW and flushing 200 mL of UPW which was effective for reducing the DOC to concentrations less than 0.1 mg/L.

The data from the membrane cleaning (pressure and time) should be saved to calculate the membrane resistance (R_m) and MFI-UF of the UPW water. R_m is measured with UPW at any flux rate.

Membrane resistance can be calculated from the following equation:

$$R_m = \frac{P}{\eta \cdot J} \qquad \text{Eq. 25}$$

Where: J is the flux rate (m^3/m^2/s), P is the stable pressure (N/m^2) measured while filtering UPW, and η is the viscosity (Ns/m^2) of the UPW, and R_m is the membrane resistance (m^{-1}). Membrane resistance should be normalized at 20 °C.

The calculated R_m should be within an accepted deviation. Filters with R_m values out of the accepted deviation (e.g., 25 %) should not be used in the test. MFI-UF values more than 150 s/L^2 suggest that the UPW contains particles that might block the pores of the membrane filters.

8.3.3 MFI-UF testing procedure

After the cleaning of the membrane filters and determination of the membrane resistance, the following steps, independently of the membrane MWCO or the flux rate for the test, should be followed:

1. Membrane filter is placed into the membrane holder. The active layer of the membrane is placed facing towards the feed side.
2. Measure the temperature of the water.
3. Place the water sample in the syringe.
4. Connect the syringe to the three-way valve.
5. When connecting the filter holder to the 3-way valve make sure that there are not bubbles trapped inside the feed side of the filter holder. Bubbles can decrease the effective filtration area when in contact with the membrane surface.
6. Filtration flux rate is controlled manually in the pump by defining the flow rate in mL/h. The effective membrane area must be considered when calculating the flux rate.
7. The software for recording the pressure and time values should be started. Both, pump and data logging, must start simultaneously.
8. Criteria to stop the test (either of the following):
   a) When cake filtration is reached (linear trend between pressure and time, or the slope of fouling index and time shows no change in time),
   b) When a minimum fouling index (I) value is observed;
   c) Change in MFI value in last 5 min filtration is less than 5 % per minute.

8.3.3.1 Selection of filtration flux rate

The MFI-UF test can be performed at any flux rate in the range 10-300 $L/m^2/h$. A flux rate of 100 $L/m^2/h$ was proposed as a default value for performing the test. Table 3 presents recommended flux rates for the MFI-UF test depending on the assessment purpose.

The selection of the flux in the MFI test should reflect the treatment unit under consideration. For instance, for profiling a treatment train, a flux rate of 100 $L/m^2/h$ is normally used from source to product water. For purposes of modelling the fouling rate of a membrane system, an MFI-UF flux rate similar to the one of the membrane unit being assessed should be applied or correct for it afterwards. In average, UF systems operate in the range 60-110 $L/m^2/h$, while RO systems operate in the range 15-25 $L/m^2/h$. The measured MFI-UF values depend on the applied flux rate during testing. The higher the flux rate, the higher the measured MFI-UF value. This relationship is linear (Salinas Rodríguez, 2011, Salinas Rodríguez, *et al.*, 2012, Salinas Rodríguez, *et al.*, 2015, Abunada, 2023).

Table 3 Recommended flux rates for MFI-UF depending on purpose

Type of sample	Flux of MFI-UF test, $L/m^2/h$
Prediction of fouling rate of seawater RO	Average of RO system (10-18)
Prediction of fouling rate of brackish water RO	Average of RO system (18-25)
Prediction of fouling rate of RO systems	50, 100, 150, 200 and extrapolate to RO average flux (e.g., 20 lmh).
Prediction of fouling rate of front RO element	25-30
Deposition factor (RO feed and RO concentrate)	Average of RO system or extrapolate from MFI vs. flux relations.
Modelling of UF system, first cycle after backwash	Same flux as UF system
Profiling treatment train	100
Permeate of polymeric and ceramic MF & UF membranes	100
Other samples, e.g., TEP measurement	60 (see Chapter 13)

For the pump, we need to calculate the required flow by considering the effective membrane area (A_{eff}) of the filter, as follows: $Q = J \times A_{eff}$. Considering an effective diameter of 21 mm for the membrane filter, the following flow rates (Table 4) should be used in the test.

Table 4 Fluxes rates, the required flow rates and estimated duration of the MFI-UF test

Flux rate, $L/m^2/h$	Corresponding flow rate, mL/h	Estimated duration of the test, h
200	69.3	0.5
150	52.0	0.75
100	34.6	1
50	17.3	2-3
20	6.9	4-6
10	3.46	6-8

NB. 300 lmh is recommended for the cleaning of the UF membrane filters, while 100 lmh is set for measuring the MFI-UF value.

8.3.4 Calculation procedure

The calculation of MFI-UF is done with the assistance of Microsoft Excel. The data (pressure and time) collected by the software is exported to Excel.

From Eq. 16, the fouling index (I) is calculated by dividing the slope of the pressure vs. time line (linear region) over square flux and water viscosity of the tested sample. As follows:

$$I = \frac{Slope}{J^2 \times \eta} \quad \text{Eq. 26}$$

Where I is the fouling index (m^{-2}), the slope is the minimum slope value from the pressure vs. time (bar/min) corresponding to cake filtration, J is the flux ($L/m^2/h$), η is the dynamic viscosity of the water being tested (Ns/m^2).

To keep MFI-UF values comparable with MFI-0.45, the MFI-UF values are standardized to reference conditions namely: viscosity at temperature of 20 °C (η_o), pressure of 2 bar (ΔP_o) and surface of area of a MFI-0.45 µm micro filter (A_o) as shown in Eq. 27.

$$MFI = \frac{\eta_0 \times I}{2 \times \Delta P_0 \times A_0^2} \quad \text{Eq. 27}$$

Where MFI-UF is the modified fouling index-ultrafiltration (s/L^2), η_o is the dynamic viscosity of water at 20 °C (Ns/m^2), I is the fouling index (m^{-2}), ΔP_o is the reference pressure (bar) = 207 kPa = 2 bar (reference condition from the SDI/MFI-0.45), A_o is the reference membrane area (m^2) = 13.8×10^{-4} m^2.

8.3.4.1 Example of membrane resistance calculation of UPW

In the figure below the pressure values versus time are plotted after converting units from milli bar to Pascal in the vertical axis and from minutes to seconds in the horizontal axis. The pressure profile is horizontal suggesting that the UPW was free of particles. The average pressure can be calculated from the time range between 840 and 2,400 seconds.

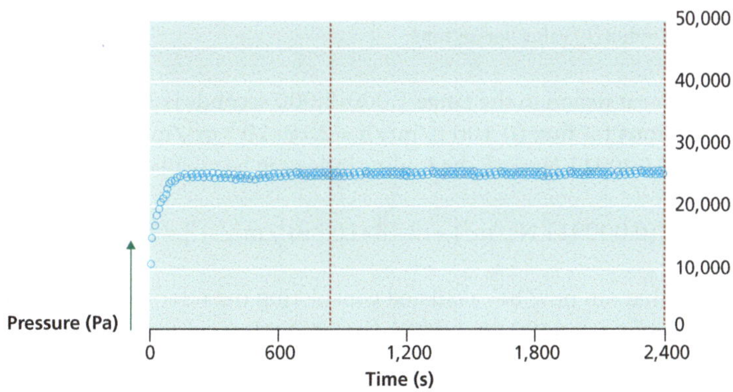

Figure 8 Filtration of UPW through a 10 kDa PES filter at 100 L/m²/h. Temperature 24 °C.

The average pressure is 24,995 N/m^2. The flux (J) is 100 $L/m^2/h$ = 2.78×10^{-5} $m^3/m^2/s$. The viscosity of the water at 24 °C is 0.000911 Ns/m^2.

Replacing the values in the formula, we have:

$R_m = (24{,}975\ N/m^2) / [(0.000911\ Ns/m^2) \times (2.78 \times 10^{-5}\ m^3/m^2/s)]$

Then,

$R_m = 9.87 \times 10^{11}\ (m^{-1})$

The membrane resistance at 20 °C.

The average R_m for the 10 kDa PES filters was reported as 9.65×10^{11} $m^{-1} \pm 1.58 \times 10^{11}$; thus, the calculated R_m falls within the average conditions.

8.3.4.2 Example of MFI-UF calculation

The data (pressure versus time) from the filtration test should be plotted as illustrated in Figure 9.

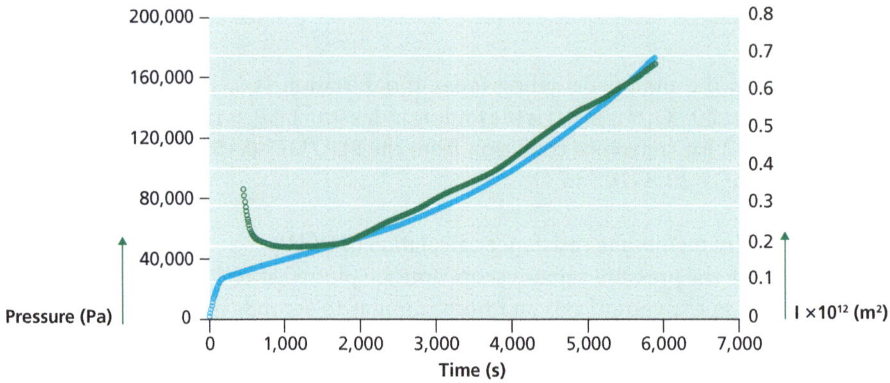

Figure 9 Filtration of Lake water through a 10 kDa PES filter at 100 $L/m^2/h$ and corresponding fouling index (I) value versus time.

The slope of the linear region in the range 1,000-2,000 seconds is 13.58 Pa/s. Considering the testing conditions for flux (J) 100 $L/m^2/h = 2.78\times10^{-5}\ m^3/m^2/s$ and viscosity of the water at 24 °C is 0.000911 Ns/m^2, the fouling index can be calculated:

$I = (13.58\ Pa/s) / [(0.000911\ Ns/m^2) \times (2.78\times10^{-5}\ m^3/m^2/s)^2] = 19.3 \times10^{12}\ m^{-2}$.

The MFI (s/L^2) value can now be calculated considering the reference conditions: $\eta_{20°C}$ = dynamic viscosity of water at 20 °C = 0.001 (Ns/m^2), ΔP_o = reference pressure = 2 bar and reference membrane filter area, $A_o = 13.8 \times10^{-4}\ m^2$.

$$MFI = \frac{\eta_0 \times I}{2 \times \Delta P_0 \times A_0^2}$$

$$MFI = \frac{0.0001\frac{N_s}{m^2}\times(19.3\times10^{12}m^{-2})}{2\times200{,}000\frac{N}{m^2}\times(13.8\times10^{-4}m^2)^2} \times \frac{1}{1{,}000\frac{L}{m^3}\times1{,}000\frac{L}{m^3}}$$

$$MFI = 25{,}370\ s/L^2$$

Reporting decimals is not required due to the accuracy of the test.

The fouling index *I* over time as plotted in Figure 9 can be obtained by using the moving average concept, taking for instance ranges of 3 min for calculating the slope (*I*) values. Another possibility for calculating the I values is to use a polynomial fitting equation to fit the pressure versus time development. By definition, the minimum (*I*) is the first derivative of this equation.

8.3.5 Reproducibility

Replicate measurements of MFI-UF have been reported for various types of water samples. The coefficient of variation ranged from 7% to 12 % as presented in Table 5. From these measurements we can consider a variation of 10 % in MFI-UF values. This was obtained from measurements in raw fresh water samples, samples after treatment and for RO feed water.

Table 5 Average value, standard deviation and coefficient of variation for replicates of MFI-UF of various samples measured with PES membrane filters

Sample	Flux, $L/m^2/h$	MWCO, kDa	n	MFI-UF, s/L^2	Source
A	100	10	5	22,000 ± 2,200 (10 %)	(Bassa, 2021)
B	100	10	6	21,800 ± 2,100 (10 %)	(Bassa, 2021)
C	100	10	5	3,500 ± 400 (11 %)	(Bassa, 2021)
D	100	10	5	2,100 ± 150 (7 %)	(Bassa, 2021)
E	100	10	5	1,410 ± 140 (10%)	(Bassa, 2021)
F	100	10	5	3,536 ± 430 (12%)	(Qatae, 2023)
G	100	10	5	3,900 ± 418 (11%)	(Qatae, 2023)
H	100	10	5	2,481 ± 232 (9%)	(Yameen, 2023)
I	100	10	10	1,655 ± 236 (14%)	(Sithole, 2017
J	50	5	10	2,911 ± 149 (5%)	(Sithole, 2017)
K	100	100	10	3,880 ± 395 (10.3 %)	(Salinas Rodríguez, 2011)
L	CP*	13*	6	2,970 ± 180 (6.1%)	(Boerlage, 2001)

* CP = Constant pressure test at 2 bar with a 13 kDa PAN hollow fibre capillary membrane

8.3.6 Blank and limit of detection

The blank level was measured by pushing lab water (type 2) through the membrane filters. LOD is the concentration or amount corresponding to a measurement level (response, signal) three standard deviation units above the zero analyte (Taverniers, 2004). To measure the LOD the procedure is measuring a minimum of either 10 independent sample blanks (LOD = Average + 3 × StdDev) or 10 independent samples blanks fortified at lowest acceptable concentration (LOD = 3 × StdDev).

Detection/results below this LOD is possible, but has a higher level of uncertainty. By using 3 times (k) the standard deviation and a sample size (n) of at least 10, there is only a 1 % chance that a blank sample will have a higher signal than the LOD. As both k and n decrease, the probability that a blank sample has a higher signal that the LOD increases.

At least ten (10) blanks (lab water) were measured with various membrane MWCOs (e.g., 10, 30, 50 and 100 kDa). The mean of the values of those blanks and the standard deviation was calculated. Results are presented in Table 6.

Table 6 Reported blank and limit of detection values of MFI-UF constant flux

MWCO, kDa	Flux, $L/m^2/h$	Average MFI, s/L^2	Std. Dev.	n	LOD, s/L^2	Source
100	250	13.9	6.8	10	34.3	(Salinas Rodríguez, 2011, Salinas Rodríguez, *et al.*, 2015)
100	250	22.5	6.5	27	42	
50	250	25	12.5	10	62.5	
30	250	35.2	10.5	10	66.7	
10	250	15.3	8.7	30	41.4	
10	100	0.6	1.9	10	6	(Lindsay, 2022)
10	100	108	46	100	246	(Qatae, 2023)
10	100	104.4	11.5	5	138.8	(Yameen, 2023)
100	50	25	34	5	127	(Li, 2019)
100	100	40	15	30	86	
100	200	35	17	10	86	
10	50	70	8	2	94	
10	100	60	12	8	96	

The LOD values ranged between 6 and 246 s/L^2 for the 10 kDa membranes with an average of 105.6 s/L^2. A potential reason for the difference is due to the variation in water quality of the lab water produced over time. Therefore, the LOD was set at 150 s/L^2. Any MFI-UF measurement with value lower than 150 s/L^2 should be noted as below detection limit (bdl).

8.3.7 Sample storage

If possible, it is recommended to measure the MFI-UF values on-site immediately after taking the sample, thus avoiding having to transport and store the samples over time. If that is unavoidable, then glass amber bottles should be applied for sampling and stored at 4 °C for transportation and storage. Measurements should take placed within 24 hours.

Samples should be taken out of the fridge to bring the temperature of the water samples back to the room temperature before MFI-UF testing; nevertheless, this is not strictly necessary as the MFI-UF corrects for water temperature in the calculation process.

Storage effect was measured with water samples from two different locations in the Netherlands (Dordrecht and Andijk) and results are reported in Table 7.

Table 7 MFI-UF values measured at 100 L/m²/h with 10 kDa PES membranes over time

Sample	0 d[4]	1 d	2 d	3 d	4 d	7 d	Avg, s/L²	Std. Dev.	CV, %
[1] A	18,220	27,500	18,300	25,760			22,445	4,884	22
[1] B	16,400	16,860	20,300	20,800			18,590	2,280	12
[1] C	5,470	4,320	3,970	4,470			4,558	643	14
[1] D	3,700	4,110	2,880	3,790			3,620	524	14
[2] Reservoir	17,400	17,850	19,810				18,353	1,281	7
	15,500			11,200	13,700		13,467	2,159	16
	13,100	13,200	13,600		11,500		12,850	926	7
[2] UF feed	4,200	3,200	3,490				3,630	514	14
	4,250			4,300	3,000		3,850	737	19
[2] UF perm	2,150	2,610	1,750				2,170	430	20
	2,295			1,010	1,525		1,610	647	40
[3] Reservoir		21,000		20,000		19,000	20,000	1,000	5
[3] UF perm 100 lmh		1,800		1,900		1,600	1,767	153	9
[3] UF perm 200 lmh		3,000		2,800		2,900	2,900	100	3

[1] (Qatae, 2023); [2] (Lindsay, 2022); [3] (Bassa, 2021); [4] 0 d = same day measurement

The coefficient of variation of the reported results for the 1-7 days of storage ranged between 3 to 40 %, thus, it is recommended to perform the MFI-UF test immediately after taking the sample or within 24 hours.

8.3.8 Concentration of particles

Further evidence that cake filtration occurs during the MFI-UF test can be observed in the results of the MFI-UF as a function of particles concentration in the feedwater. This premise is based on the fouling index, I, being directly related to the concentration of particles C_b (Eq. 10). Thus, I will decrease directly in proportion to an increase in the dilution factor of C_b while the specific cake resistance component (α), characteristic of a feedwater type and independent of concentration, remains constant.

In Figure 10, the results of the MFI-UF with dilutions of Delft canal water at an applied flux of 100 L/m²/h are shown. Linearity was found for the feedwater, with the regression coefficient calculated as 0.989.

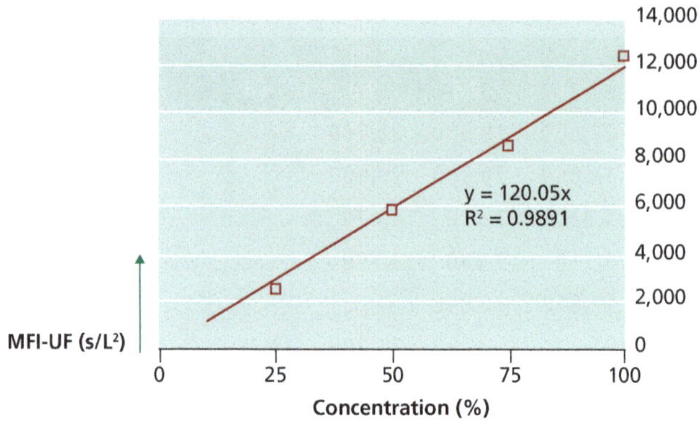

Figure 10. MFI values for dilutions of Delft canal water measured with 100 kDa RC membrane at 100 L/m^2/h. Adopted from (Salinas Rodríguez, 2011)

Abunada, *et al.* (2023) also reported a direct linear correlation between MFI-UF values and the concentration of 150 kDa Dextran, 75 nm polystyrene particles, and Delft canal water.

8.3.9 Membrane material

There are several materials used in ultrafiltration such as: PES, RC, PAN, and PVDF. PVDF is produced mainly for tight MF membranes. PAN and PES are more available for hollow fibre membranes and in various pore sizes. For this research, PES and RC membranes were tested as the range of pore size available was wider.

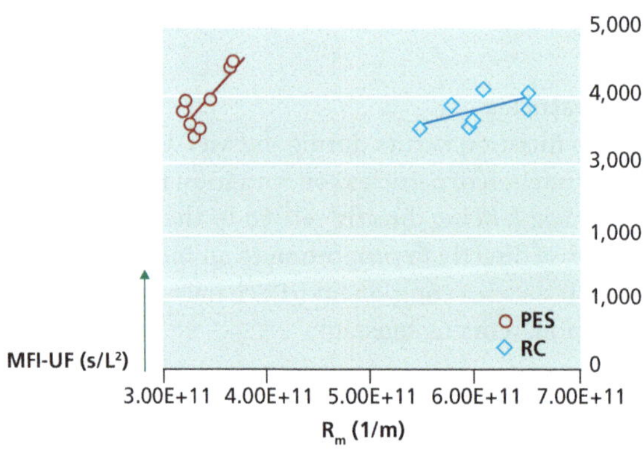

Figure 11 MFI values for Delft canal water measured with 100 kDa PES and 100 kDa RC at 100 L/m^2/h. Adapted from Salinas Rodríguez (2011)

Figure 11 shows the measured MFI values for Delft canal water. For the PES membranes the average was 3,880 s/L^2 ± 395 (10.3 %), and for the RC membranes the average was 3,800 s/L^2 ± 235 (6.3 %). Both membrane materials have an average value close to each other. RC membranes are slightly more uniform than the PES membranes when measuring MFI-UF.

Bassa (2021) when studying MFI-UF values of Ijssel lake water and after ceramic membrane filtration also did not observe a correlation between the 10 kDa PES membrane resistance and MFI-UF. Qatae (2023) confirmed this observation when studying MFI-UF values after dual media filtration.

8.4 VARIABLES AND APPLICATIONS OF THE MFI-UF

8.4.1 Plant profiling and water quality monitoring

The application of various UF membrane filters in a pilot scale desalination plant (Jacobahaven, Netherlands) is illustrated in Figure 12 where 100 kDa, 50 kDa and 10 kDa PES filters were used. The Amiad strainer showed only a small reduction in MFI-UF as expected with a relatively large aperture size of 50 µm. Whereas, the reduction in MFI-UF (and fouling) observed following UF (nominal MWCO of 150 kDa) was much larger i.e., of 94 %, 93 %, and 88 % reduction for 100 kDa, 50 kDa, and 10 kDa MFI-UF test membranes, respectively (Salinas Rodríguez, *et al.*, 2015).

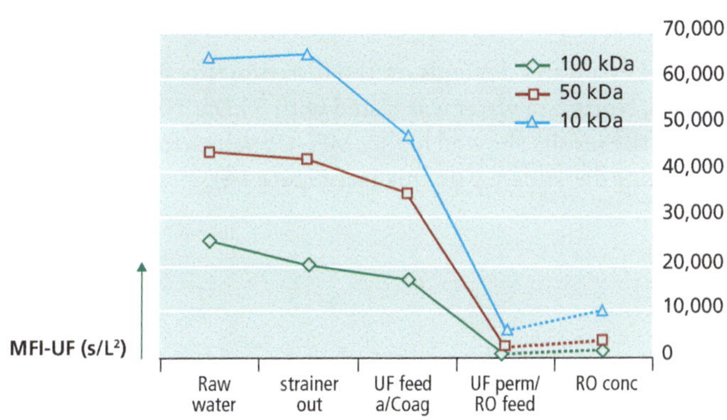

Figure 12 Effect of pre-treatment on MFI-UF at the Jacobahaven seawater pilot plant using PES test membranes of 100, 50 and 10 kDa size (Salinas Rodríguez, *et al.*, 2015)

The percentages in reduction of MFI values after water passing through the ultrafiltration units were 94.3 %, 93.4 % and 87.6 % for 100, 50 and 10 kDa, respectively.

From October 2016 to July 2017, the MFI-UF values were measured for both raw and 2 µm glass-media filtered North Sea water (Figure 13). The measured values with MFI-UF were 75-28 times higher than the ones measured with the MFI-0.45. This illustrates the high fouling potential of particles smaller than 0.45 µm. The high peak at the end of April coincided for the MFI-UF measurements with the high algal cell numbers in the period.

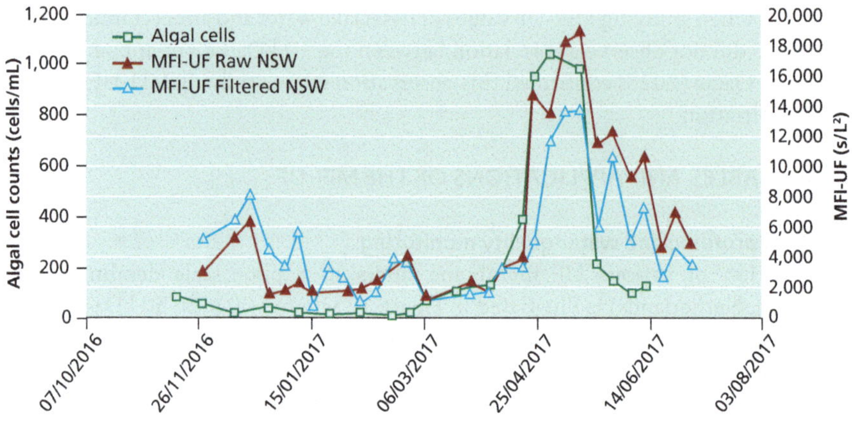

Figure 13 MFI-UF 10 kDa PES measured at 100 lmh for raw NSW and filtered NSW (Sithole, 2017)

8.4.2 Flux rate

The fouling potential of a water depends on the filtration flux during testing. To illustrate this effect in practice, North Sea water was tested with 5 kDa, 10 kDa, 30 kDa and 100 kDa PES membranes and the results showed higher MFI-UF values for the lower MWCO filters, and a remarkably strong dependency on flux (see Figure 14).

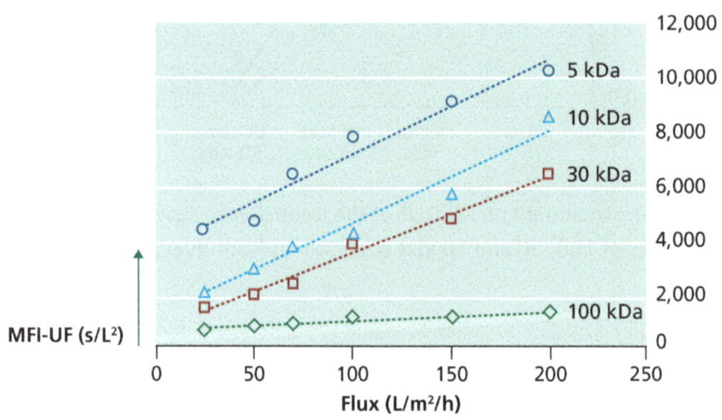

Figure 14 MFI-UF (PES) of North Sea water measured at various flux rates (Salinas Rodríguez, *et al.*, 2019)

When assessing the fouling potential of RO feedwater, it is important to measure the MFI-UF value at the same flux at which the RO is operating, either the average flux for the whole pressure vessel or the flux for the front element. An alternative approach is to find the MFI-UF value at low flux rate by extrapolating from the MFI vs. flux relationship.

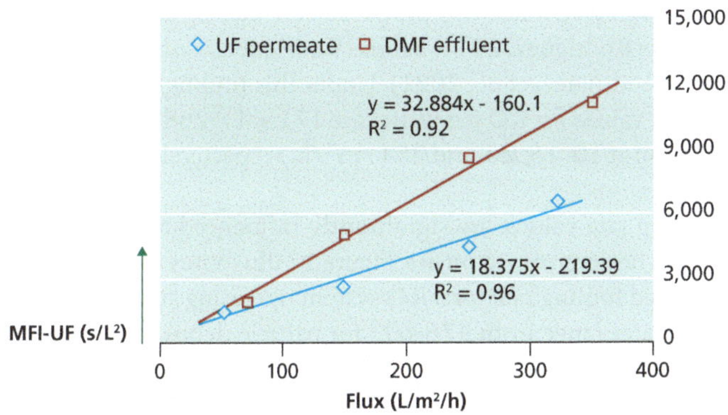

Figure 15 MFI-UF 10 kDa PES values as function of filtration flux of Mediterranean seawater after parallel treatment by dual media filtration and ultrafiltration (Salinas Rodríguez, *et al.*, 2015)

Figure 15 shows the MFI-UF values for Mediterranean seawater (left) and for UF permeate (0.02 µm pore size) and dual media filtration effluent (right) measured at various flux rates from ~50 $L/m^2/h$ up to 350 $L/m^2/h$. Noticeably, the MFI-UF values for both pre-treatments get closer to each other when measured at low flux rates than when measured at high flux rates.

8.4.3 Predicting rate of fouling of seawater RO systems

Figure 16 illustrates the theoretical fouling rate, based on Equation 24, as function of MFI values for various average RO system flux rates for a deposition factor equal to 1 (all particles rejected by the RO membranes will remain on it forming a cake layer).

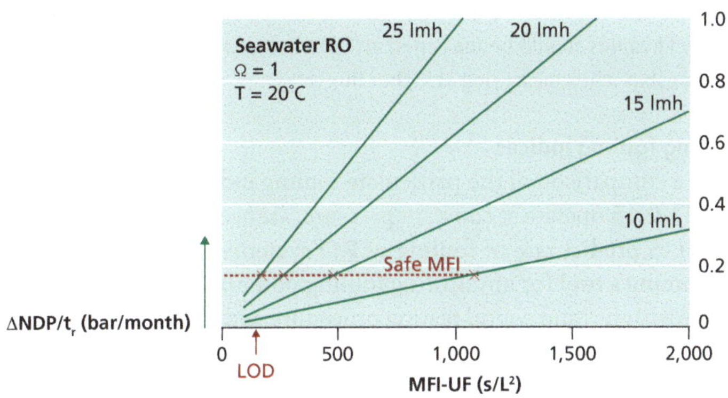

Figure 16 Prediction of fouling rate (bar/month) as function of MFI values for a seawater RO system operating at various flux rates considering Ω = 1 and temperature = 20 °C. Safe MFI = 1 bar NDP increase in 6 months. LOD = 150 s/L^2. NB. MFI values should be measured at 10, 15, 20, or 25 lmh, or can be projected from the flux vs. MFI relationship when measuring at higher flux rates.

A safe MFI value for RO feedwater was proposed considering an increase in NDP of 1 bar in six months (Salinas Rodríguez, 2011, Salinas Rodríguez, *et al.*, 2015, Salinas Rodriguez, *et al.*, 2019, Salinas Rodriguez, *et al.*, 2021). Under this fouling rate (increase of 0.17 bar/month), the safe MFI values for RO feedwater are 171 s/L^2, 268 s/L^2, 476 s/L^2, and 1,071 s/L^2 for RO operation at 10, 15, 20, and 25 L/m^2/h, respectively.

The predicted fouling rate values, are significantly influenced by the particles deposition factor, which need to be determined onsite. Figure 17 illustrates the effect of the deposition factor on the predicted fouling rate for a RO system operating at 15 L/m^2/h. The safe MFI values for RO feedwater range from 476 s/L^2 for particle deposition factor Ω = 1 to 1,904 s/L^2 for Ω = 0.25.

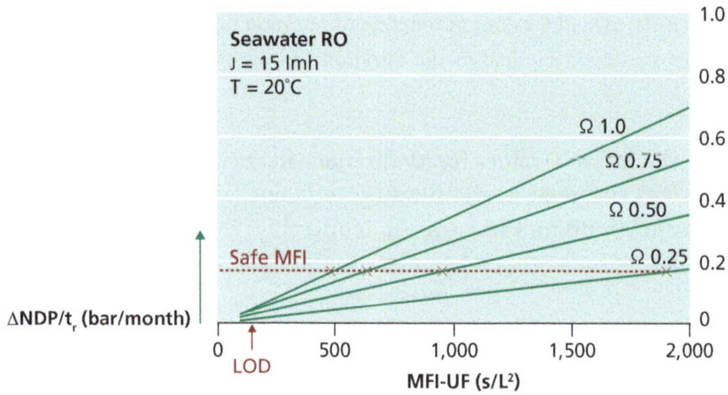

Figure 17 Prediction of fouling rate (bar/month) as function of MFI values for a seawater RO system as function of the particle deposition factor (Ω), considering J = 15 L/m^2/h and temperature = 20 °C. Safe MFI = 1 bar NDP increase in 6 months. LOD = 150 s/L^2. NB. MFI values should be measured at 15 lmh or can be projected from the flux vs. MFI relationship when measuring at higher flux rates.

8.4.4 Comparing fouling indices

Table 8 presents a comparison of the particulate fouling indices that are applied in practice. Both SDI and MFI-0.45 operate at constant pressure, are accepted by the ASTM, but are not sensitive enough to predict rate of fouling of RO systems. MFI-UF constant flux has the potential of becoming a tool for monitoring fouling potential in membrane systems due to its robustness, sensitivity, and sound testing procedure and calculation method.

Table 8 Comparison of particulate fouling indices: SDI, MFI-0.45, MFI-UF

Category	SDI	MFI-0.45	MFI-UF constant flux
Filtration mode	Dead-end	Dead-end	Dead-end
Operation	Constant pressure (2.07 kPa)	Constant pressure (2.00 kPa)	Constant flux (100 L/ m^2/h)
Volume sample	>3.8 L	>3.8 L	60 mL
Pore size	0.45 μm	0.45 μm	10 kDa (others are possible: 5, 50, 100 kDa)
ASTM	D4189-07 (ASTM D4189 - 14, 2014)	D8002-15 (ASTM D8002 - 15, 2015)	-
Filter	Flat, 25 or 47 mm	Flat, 24 or 27 mm	Flat, 25 mm
Fouling mechanism	None	Pore blocking, cake filtration	Cake filtration
Flux rate	>>> 1,000 L/m^2/h	>>> 1,000 L/m^2/h	10 - 350 L/m^2/h
Membrane properties	Hydrophilic, Mixed cellulose nitrate, mixed cellulose acetate, mixed cellulose ester	Cellulose acetate, cellulose nitrate, mixed cellulose ester	Polyethersulfone (PES), regenerated cellulose (RC)
Test recordings	time t_1, time t_2 for a given volume (V).	t/V vs. V (e.g., every 5 s)	P vs. t (I vs. t). (e.g., every 10 s).
Correction/ Normalization	None	Temperature, pressure, (effective) membrane area	Temperature, pressure, membrane area, membrane surface porosity
Formula	$SDI_T = \frac{1-\frac{t_1}{t_2}}{T} \times 100$	$MFI = \frac{\eta_{20}}{\eta} \times \left(\frac{\Delta P}{\Delta P_0}\right)^{1-\omega} \times \left(\frac{A}{A_0}\right)^2 \times \left[\frac{d(t/V)}{dV}\right]$	$MFI = \frac{\eta_{20} \times I}{2 \times \Delta P_0 \times A_0^2}$
Cake compression effect	Not considered	ω (compressibility coefficient included in formula. It requires additional testing)	Controlled by flux rate in the test. Cake filtration identified in calculation.
Prediction model	Theoretically impossible	$t_r = \frac{\Delta NDP_r}{\eta_r \times J_0^2 \times \Omega \times I_r} \; \frac{\Delta J}{2} \times \frac{(2-\Delta J)}{(1-\Delta J)^2}$	$t_r = \frac{(\Delta NDP_r - \Delta NDP_{or})}{J^2 \times \eta \times \Omega \times I_r}$
RO feedwater guideline value	<5 (preferably 3) (DuPont, 2023)	< 4 (preferably 1) (DuPont, 2023)	476 s/L^2 for SWRO at 15 lmh and $\Omega = 1$.
Others	SDI cannot be used to test UF permeate water	MFI-0.45 is not sensitive enough for predicting actual RO fouling. Potential use for predicting clogging of RO feed spacer. It can also be used to predict rate of clogging in infiltration wells.	Safe MFI = 1 bar increase in NDP in 6 months
Duration	~30 min	~30 min	~1 hour (excluding membrane cleaning)

8.5 REFERENCES

Abunada M (2023). Prediction of Particulate Fouling in Reverse Osmosis Systems: MFI-UF Method Development and Application. PhD Dissertation, IHE Delft

Abunada M, Dhakal N, Andyar WZ, Ajok P, Smit H, Ghaffour N, Schippers JC, Kennedy MD (2022) Improving MFI-UF constant flux to more accurately predict particulate fouling in RO systems: Quantifying the effect of membrane surface porosity. Journal of Membrane Science 660: 120854 DOI https://doi.org/10.1016/j.memsci.2022.120854

Abunada M, Dhakal N, Andyar WZ, Li Y, Ajok P, Ghaffour N, Schippers JC, Kennedy MD (2023) Calibrating and Validating the MFI-UF Method to Measure Particulate Fouling in Reverse Osmosis. Membranes 13 DOI 10.3390/membranes13050535

Abunada M, Dhakal N, Gulrez R, Ajok P, Li Y, Abushaban A, Smit H, Moed D, Ghaffour N, Schippers JC, Kennedy MD (2023) Prediction of particulate fouling in full-scale reverse osmosis plants using the modified fouling index – ultrafiltration (MFI-UF) method. Desalination 553: 116478 DOI https://doi.org/10.1016/j.desal.2023.116478

Abushaban A (2019) Assessing Bacterial Growth Potential in Seawater Reverse Osmosis Pretreatment: Method Development and Applications CRC Press

AlShuaili HY (2012). Predicting fouling in seawater reverse osmosis system. MSc, IHE Delft MWI/12.15.

Althuluth M (2009). Further Development of the Modified Fouling Index (MFI-UF) at Constant Flux for SWRO Applications. MSc, UNESCO-IHE MWI 2009-032.

Andyar WZ (2021). Further development of the Modified Fouling Index-Ultrafiltration (MFI-UF) method. MSc, IHIE Delft UWS-WSE/21.21.

ASTM D4189 - 14 (2014) Standard Test Method for Silt Density Index (SDI) of Water ASTM International, West Conshohocken, PA.

ASTM D8002 - 15 (2015) Standard Test Method for Modified Fouling Index (MFI-0.45) of Water ASTM International, West Conshohocken, PA.

Bassa FGM (2021). Evaluating surface water pre-treatment with the MFI-UF for assessing fouling in membrane systems. MSc, IHE Delft UWS-WSE/21.25.

Belfort G, Marx B (1979) Artificial particulate fouling of hyperfiltration membranes II. Analyses and protection from fouling. Desalination 28: 13-30

Boerlage SFE (2001) Scaling and Particulate Fouling in Membrane Filtration Systems Swets&Zeitlinger Publishers, Lisse

Carman PC (1937) Fluid flow through granular beds Trans Instn Chem Engrs 15: 32-48

Carman PC (1938) Fundamental principles of industrial filtration (A critical review of present knowledge). Trans Instn Chem Engrs 16: 168-188

Cytiva-Whatman (2023). Polyethersulfone syringe filter holder for 2.5 cm filter. https://www.sigmaaldrich.com/NL/en/product/aldrich/wha10461000. Accessed 10 November 2023.

Dhakal N (2017) Controlling biofouling in seawater reverse osmosis membrane systems Delft university of technology, CRC Press / Balkema - Taylor & Francis Group, the Netherlands

Dhakal N, Salinas Rodriguez SG, Schippers JC, Kennedy MD (2016) Investigation into Fouling of the Ultrafiltration and Reverse Osmosis Systems: FEWA Desalination Plant, Ras Al Khaimah, UAE. Prepared for Pentair X-Flow. IHE Delft, Delft, pp. 44.

Dhakal N, Schippers JC, Kennedy MD (2018) Investigation into fouling of the Ultrafiltration & reverse osmosis systems: RAF desalination plant, QATAR IHE Delft, Delft, pp. 58.

Dhakal N, Abunada M, Schippers JC, Kennedy MD (2020) Water quality monitoring of a demineralized water plant at Evides. IHE Delft, Netherlands.

DuPont (2023) FILMTEC™ Reverse Osmosis Membranes Technical Manual, Form No. 45-D01504-en, Rev. 16Water solutions, pp. 211.

Ekowati Y (2011). Fouling Prediction in Reverse Osmosis Systems using The Modified Fouling Index - Ultrafiltration (MFI-UF). MSc, IHE Delft MWI 2011/07.

Ekowati Y, Msuya M, Salinas Rodriguez SG, Veenendaal G, Schippers JC, Kennedy MD (2014) Synthetic organic polymer fouling in municipal wastewater reuse reverse osmosis. Journal of Water Reuse and Desalination 4: 125-136 DOI doi:10.2166/wrd.2014.046

Gulrez R (2021). Application of MFI-UF at constant flux in a full-scale RO plant. MSc, IHE Delft UWS-WSE/21.33.

Harvard Apparatus (2023). Standard Infusion Only PHD ULTRA™ Syringe Pumps https://www.harvardapparatus.com/standard-infusion-only-phd-ultra-syringe-pumps.html. Cited 26 October 2023

Le Bouille L, Bassa F, Salinas Rodriguez SG, Philibert M, Kennedy MD (2021) Water quality monitoring at Bahia La Palma Desalination plant IHE Delft, Delft, pp. 20.

Le Bouille L, Salinas Rodriguez SG, Ekowati Y, Philibert M, Kennedy MD (2023) Water quality and fouling monitoring at Perth II Desalination plant IHE Delft, Delft, pp. 22.

Li Y (2019). Further Development of the Modified Fouling Index – Ultrafiltration (MFI-UF) Method: Calibration and Application in Full-Scale UF-RO. MSc, IHE Delft UWS-WSE.19-31.

Lindsay AL (2022). Assessing particulate fouling potential at PWNT Andijk pilot plant via on-site MFI-UF testing. MSc MSc, IHE Delft UWS-WSE.22-17.

Malaichami V (2023). Operational optimization of the UF in terms of hydraulic performance and permeate water quality - A case study for Lake Valkenburg pilot. MSc, IHE Delft UWS-WSE.23-21.

Merck (2023). Ultrafiltration Discs, 10 kDa NMW PBGC02510 https://www.sigmaaldrich.com/NL/en/product/mm/pbgc02510. Cited 27 October 2023

Mirzaei P (2023). Further development of the Seawater Bacterial Growth Potential method MSc, IHE Delft UWS-WSE.23-23.

Ofori M (2023). Predicting Particulate Fouling in Reverse Osmosis systems. MSc, IHE Delft WSD.23-1058145.

Omega (2023). USB Output Pressure Transducer https://www.omega.nl/pptst/PXM409-USBH.html. Cited 27 October 2023

Qatae M (2023). Monitoring particulate fouling potential and organic matter characterization along the pre-treatment line of an NF/RO surface water pilot plant. MSc, IHE Delft UWS/WSE.23.11.

Rathnayake SC (2013). Particulate fouling in reverse osmosis systems : Measuring the deposition factor to predict rate of fouling. MSc, IHE Delft MWI 2013-31.

Ruth BF, Montillon GH, Montonna RE (1933) Studies in Filtration I. Critical Analysis of Filtration Theory. Industrial & Engineering Chemistry 25: 76-82 DOI 10.1021/ie50277a018

Salinas Rodríguez SG (2011) Particulate and organic matter fouling of SWRO systems: Characterization, modelling and applications CRC Press/Balkema, Delft

Salinas Rodríguez SG, Althuluth M, Schurer R, Kennedy MD, Amy GL, Schippers JC (2009) Modified fouling index (MFI-UF) at constant flux for seawater RO applications. Paper presented at the Desalination for the Environment: Clean water and Energy, Baden-Baden, Germany, 17-20 May 2009

Salinas Rodríguez SG, Amy GL, Schippers JC, Kennedy MD (2015) The Modified Fouling Index Ultrafiltration Constant Flux for assessing particulate/colloidal fouling of RO systems. Desalination 365: 79-91 DOI 10.1016/j.desal.2015.02.018

Salinas Rodriguez SG, Kennedy MD (2009) MFI–UF results for DOW Barcelona pilot plant UNESCO-IHE, Delft, pp. 12.

Salinas Rodriguez SG, Kennedy MD (2009) Monitoring of particulate fouling after DMF and UF pre-treatment at Veolia's Toulon pilot plant UNESCO-IHE, Delft.

Salinas Rodríguez SG, Kennedy MD, Amy G, Schippers JC (2010) Flux dependency of particulate fouling by MFI-UF measurements in seawater reverse osmosis systems. Paper presented at the EuroMed 2010: Desalination for Clean Water and Energy, Tel-Aviv, Israel, 3-7 October 2010

Salinas Rodríguez SG, Kennedy MD, Amy GL, Schippers JC (2012) Flux dependency of particulate/colloidal fouling in seawater reverse osmosis systems. Desalination and Water Treatment 42: 155-162 DOI DOI:10.1080/19443994.2012.683104

Salinas Rodriguez SG, Schippers JC, Amy GL, Kim IS, Kennedy MD (2021) Seawater Reverse Osmosis Desalination: Assessment and Pre-treatment of Fouling and Scaling, 1st edn IWA Publishing, London

Salinas Rodriguez SG, Sithole N, Dhakal N, Olive M, Schippers JC, Kennedy MD (2019) Monitoring particulate fouling of North Sea water with SDI and new ASTM MFI0.45 test. Desalination 454: 10-19 DOI https://doi.org/10.1016/j.desal.2018.12.006

Salinas Rodríguez SG, Sithole N, Dhakal N, Olive M, Schippers JC, Kennedy MD (2019) Monitoring particulate fouling of north sea water with SDI, MFI0.45, and MFI-UF. Paper presented at the IDA World Desalination Congress, Dubai, UAE, 20-24 October 2019

Salinas Rodriguez SG, Villacorte LO, Kennedy MD (2010) Water analysis results for Barcelona seawater UF-RO plant. UNESCO-IHE, Delft, pp. 9.

Salinas Rodriguez SG, Villacorte LO, Wahyudi D, Kennedy MD (2010) Water analysis results for DOW Barcelona seawater samples. UNESCO-IHE, Delft, pp. 6.

Schippers JC (1989) Vervuiling van hyperfiltratiemembranen en verstopping van infiltratieputten Keuringinstituut voor waterleidingartikelen KIWA N.V., Rijswijk

Schippers JC, Folmer HC, Kostense A (1980) The effect of pre-treatment of river rhine water on fouling of spiral wound reverse osmosis membranes7th International Symposium on Fresh water from the Sea, pp. 297-306.

Schippers JC, Hanemaayer JH, Smolders CA, Kostense A (1981) Predicting flux decline or reverse osmosis membranes. Desalination 38: 339-348

Sigma-Aldrich (2023). Whatman® FP 025/1 filter holder. https://www.sigmaaldrich.com/NL/en/product/aldrich/wha10461000. Accessed 10 November 2023.

Sithole N (2017). Particulate fouling assessment of North Sea water. MSc, IHE Delft UWS-WSE 2017-33.

Tabatabai SAA (2014) Coagulation and ultrafiltration in seawater reverse osmosis pretreatment CRC Press/Balkema, Delft

Tahtouh L (2023). Monitoring the performance of in-line coagulation with rapid sand filters as a pre-treatment unit for reverse osmosis and for dune's infiltration A case study for lake Valkenburg (Dunea). MSc, IHE Delft UWS-WSE.1085080.

Taverniers I, *et al.* (2004) Trends in quality in the analytical laboratory. II. Analytical method validation and quality assurance. Trends in Analytical Chemistry 23: 535-552

Villacorte LO (2014) Algal Blooms and Membrane Based Desalination Technology CRC Press/Balkema, Leiden

Wahyudi D (2010). Fouling Prediction with the Modified Fouling Index – Ultrafiltration (MFI-UF) at Constant Flux for Seawater Reverse Osmosis (SWRO) Systems. MSc, IHE Delft MWI 2010 – 031.

Yameen A (2023). Prediction of particulate fouling in nanofiltration and reverse osmosis feed by pre-treated surface water using modified fouling index – ultrafiltration (MFI-UF) method. MSc, IHE Delft UWS-WSE.23-18.

Zhan M, Lee H, Jin Y, Hong S (2020) Application of MFI-UF on an ultrapure water production system to monitor the stable performance of RO process. Desalination 491: 114565 DOI https://doi.org/10.1016/j.desal.2020.114565

Part 3

Inorganic fouling and scaling

doi: 10.2166/9781789062977_0207

Chapter 9

Inorganic Fouling: Characterization Tools and Mitigation

Nuria Peña García, Genesys-PWT, Spain

The learning objectives of this chapter are the following:

- Define inorganic fouling in membrane systems
- Description of inorganic fouling components
- Presence and impact of inorganic fouling on membranes
- Tools to avoid inorganic foulant
- Inorganic fouling removal

9.1 INTRODUCTION

There is no doubt about the effect of fouling on the performance of the different membrane processes that can be used in water treatment.

Depending on the type of membrane system used, the meaning of fouling can be different. Microfiltration and ultrafiltration membranes work as barriers to retain suspended matter and that matter should be removed during backwash and cleaning protocols established to remove it. So, in those cases, fouling would correspond to the retained matter which is not properly removed during routine/standard cleaning procedures/cycles.

The rest of water treatment membranes systems were mainly developed to remove salts from water, so any suspended matter reaching their surface will produce fouling. On the other hand, every feed water has a scaling potential which may scale membrane surface if it is not protected by chemical treatment.

Main types of membranes fouling can be classified as shown in Table 1.

Table 1 Membrane fouling types and sub-categories

Organic fouling	Inorganic fouling
Biofilm, organic matter Other organics (oil-greases, chemicals, etc.)	Colloidal - particulate Scaling

The difference between organic and inorganic compounds is mainly referred to the presence of carbon which is as C-H in organics (with some exceptions). Inorganic substances or compounds would include pure elements, salts, many acids and bases, metals and alloys and minerals. There are also a few inorganic compounds containing carbon as oxides, carbides, some carbonates and some cyanides.

Inorganic fouling could be defined as the deposition of inorganic compounds on the membrane surface or inside the membrane pores. Thus, the deposits could be either inorganic compounds with low solubility in water or solutes present in large amounts in water.

When they form supersaturated solutions, eventually precipitate out of the solution and onto the surface of the membrane. Scale formation on membrane surfaces would occur through crystallization/precipitation (Figure 1b) and deposits would be from particulate fouling (Figure 1a). The former mechanism involves the precipitation of ions and their subsequent deposition on the membrane, whereas the latter involves the convective transport of particulates from the bulk of the solution to the membrane surface. In rivers, groundwater, seawater, and municipal wastewater, the main inorganic compounds that contribute to scaling and inorganic fouling are hydroxides, sulfates, carbonates, calcium, magnesium, iron, ortho-phosphates, silicic acids, and silica (Jiang *et.al.*, 2017, Guo W. *et al.*, 2012).

1a Deposit

1b Scaling

Figure 1 Characteristic aspect of a deposit (1a) and scaling (1b) found on membranes surface during autopsies (Images credit by Genesys Membrane Products, S.L.)

Experimental data obtained from the study of reverse osmosis (RO) and ultrafiltration (UF) membranes surface establish that main types of foulant are: biofilm, organics, colloidal matter/particulate, metals, scale.

Figure 2 includes data from autopsies, showing the percentage of membranes showing these foulants as main component and as main reason of membrane failure.

As this figure indicates, approximately 60% of RO membranes show a major presence of inorganic foulant while for UF this percentage is only around 30%. Thus, considering these data, inorganic fouling is a bigger problem on desalination membrane systems than on other membrane filtration process.

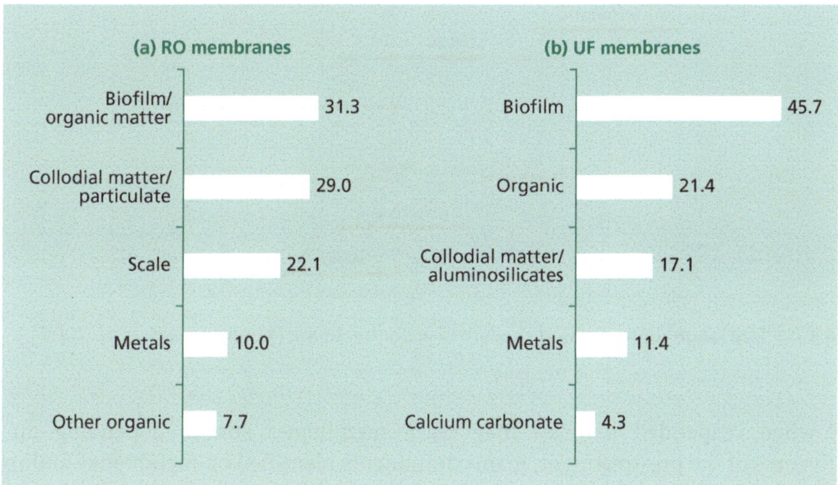

Figure 2 Percentage of membranes showing main presence of different types of foulant on RO membranes (a-left) and UF membranes (b-right). Experimental data obtained from membrane autopsies (1000 data from RO and 100 data from UF).

But even when fouling can be mainly organic on some systems, it is almost impossible to find a pure foulant and they will mostly include an inorganic component (Peña *et al.*, 2013). Composite nature of foulants is the reason that makes them different from one system to other and produces that the impact on membranes and their removal is also different than expected in many cases. At this point, the identification of all the fouling components is essential.

As Figure 1 shows, most common inorganic foulants on membranes are composed of colloidal matter, scaling salts and metals. Same compounds are commonly found as secondary components of biofilm and other organic foulants since they commonly trap these particles coming with the water while they grow or settle on membranes surface.

In this chapter, main characteristics of inorganic foulant components will be covered and a review of the different ways to determine them and how to remove them is included.

9.2 MAIN COMPONENTS OF INORGANIC FOULING

9.2.1 Colloidal matter/particulate

Solids are present in water in three main forms: suspended particles, colloids and dissolved molecules. Figure 3 shows the size range of particles of concern in water treatment (Koohestanian *et al.*, 2008):

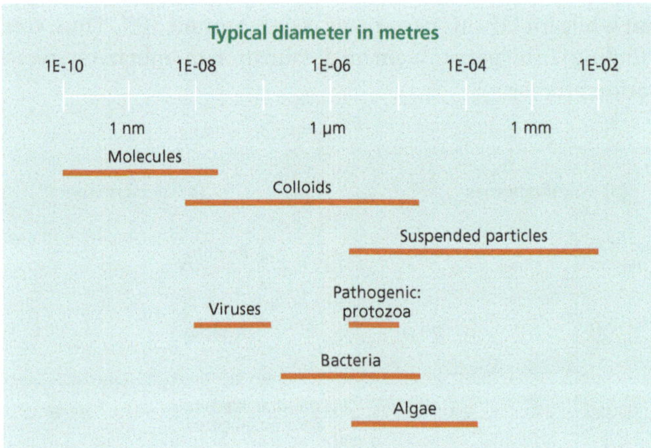

Figure 3 Size ranges of particles of concern in water treatment (Koohestanian, *et.al.* 2008)

Even when suspended particles may reach membranes surface depending on the effectiveness of the pre-treatment, main components identified on membranes fouling are hydrophobic colloids as clay and non-hydrated metal oxides.

Colloids have an assigned size range of 0,001 to 1 µm and constitute a significant component of the particulate matter in natural waters (Koohestanian *et al.*, 2008). In water treatment, colloidal matter is commonly related to higher sizes (<2 µm).

The characteristics of particulate and colloidal material are largely dependent upon its source and thus the unique environmental system under consideration (Miroslaw, 2007). In open ocean systems the majority of particulate matter is biotic in origin and is generated in the upper surface layer of the ocean. This is in contrast to riverine systems that are heavily influenced by erosion of material in the watershed. Large lakes may be qualified as a mix of these two extremes.

Most suspended solids smaller than 0.1 mm found in waters carry negative electrostatic charges. Since the particles have similar negative electrical charges and electrical forces to keep the individual particles separate, the colloids stay in suspension and small particles. The magnitude of the zeta potential (Zp) is usually used to indicate colloidal particle stability. The higher the zeta potential, the greater the repulsion forces between the colloidal particles are and, therefore, the more stable is the colloidal suspension. Coagulation is commonly used to neutralise the negative charges of the suspended solids by using a positively charged element (coagulants based on Fe^{3+} and Al^{3+} for example) and form a gelatinous mass to trap (or bridge) particles thus forming a mass large enough to settle or be trapped in the filter (Kevin *et al.*, Coagulation-Flocculation factsheet).

Pretreatment processes such us coagulation, flocculation and low pressure membrane filtration (microfiltration and ultrafiltration) have been used in front of other membrane processes to remove particles and large colloids, but fouling by small colloidal matter (<2 μm) and fine suspended particles is a main problem (Armstrong *et al.*, 2009).

There are different tools to quantify and determine the amount of matter coming with the water which are very helpful: turbidity, suspended solids, SDI, MFI. SDI and MFI are described in chapters 6 and 7-8, respectively.

Besides, it can be very helpful also to determine the size of the particles composing the suspended matter to determine the best pre-treatment: particle counting.

The principal consequence of membrane fouling by colloidal matter is an increase in hydraulic resistance resulting in a greater energy requirement to operate the process. The formation of highly impermeable deposits on the membrane surface will result in significant problems in maintaining permeate flux with frequent cleaning eventually being required to maintain system operation. The primary effects of fouling by colloidal particles in a membrane system will be seen mainly in the elements in the first positions. However, if this problem remains untreated fouling will gradually affect all membrane elements. The effects will include a reduction in membrane flux (reduction in product flow rate), an increase in salt passage as well as an increase in feed channel pressure drop (ΔP).

Membrane damage through abrasion processes have also been identified during membrane autopsies performed on systems fouled with clay mineral deposits due to the compression of the crystalline structure against the membrane surface by increased operating pressures. In other cases, the increase in ΔP when membrane operates in presence of foulant will produce damage on the areas of contact between the spacer material and membrane surface.

From the different components of colloidal matter/particulate, clays/aluminosilicates, silica (sand particles or colloidal silica) and microorganisms skeletons/inorganic structures are mainly identified in membranes fouling.

Clays

Clay minerals are the most important component of the soil, they are usually ultra-fined particles having less than 2 μm sized particles. Most of the clay minerals are known as hydrous aluminosilicate or hydrous aluminum phyllosilicate (Nascimento *et al.,* 2021).
Mineralogically, clays are divided into 3 principal groups, with more than 30 different minerals within these groups (Armstrong *et al.*, 2009).

Table 2 Main 3 types of clay minerals and their properties (Armstrong, *et al.* 2009)

Clay type	Structure and property	
Kaolinite	The most common clay composed of silicate sheets bonded to aluminium oxide/ hydroxide layers.	Main composition: aluminosilicates
Illite	Structure contains a wide range of cations including Al, Mg, Fe and potassium.	
Montmorillonite/ smectite	Structure includes Ca, Na, Al, Mg and is notable for its ability to take up and lose water.	

As already explained, one of the characteristics of clays as membrane foulant is that they appear mixed with the rest of foulant components, mainly with the organic component. During the operation of the membrane composite foulant compact and age in such a way that it is very difficult to remove from membranes surface and the final impact is an irreversible damage and the consequent lack of retention (Peña *et al.*, 2013). Thus, considering that the small size of the clays/aluminosilicates makes very difficult to achieve a complete removal from water bulk, it is essential to establish preventive cleaning of the systems to avoid them.

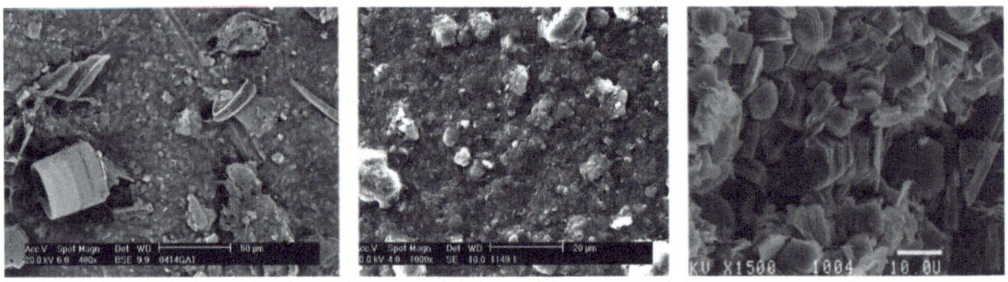

a) Mixture of diatoms and clays **b)** Characteristic mixture of clays and particles **c)** Detail of clay structure

Figure 4 Examples of foulant with presence of clays (Images credit: Genesys Membrane Products S.L.)

Microorganisms with inorganic structures

There are some microorganisms that show a characteristic inorganic skeleton/structure and are commonly found on membranes fouling. Even when the source could be considered organic, they are included in this inorganic classification because what it is commonly identified and detected during the study of membranes is their inorganic structure. Most common microstructures are silica structures from diatoms and calcium carbonate structures from marine algae or marine coralline microstructures.

These microstructures commonly appear embedded in the rest of the fouling as a minor component and they could be removed mixed with the rest of components if the appropriate cleaning procedure is applied. Main risk with these microstructures is that they may produce a severe abrasion on membranes surface and the consequent irreversible damage. Since these structures come in the water bulk, they are mostly found on first position membranes.

It is important to identify the presence of these microstructures during the study of membranes to understand the source of silica and calcium carbonate and to determine if these compounds are as scaling species or as deposits to determine the best way to avoid them. For the removal of these structures it is very important to determine their size and to adapt the pre-treatment to it if they appear in a significant amount.

- **Silica structures-Diatoms** (*Bacillariophyta*) are single cell, siliceous cell wall algae and are the principal component from the phytoplankton division (Smol, *et al.*, 2010). They can exist in all aquatic ecosystems (marine, brackish, fresh waters) including in some moist terrestrial ecosystems (Smol *et al.*, 2010, Wynne *et al.*, 1985). The photographs in Figure 5 show characteristic structures of this kind of algae structures found on membranes. It is common to find not only these full structures, but fragments from them.

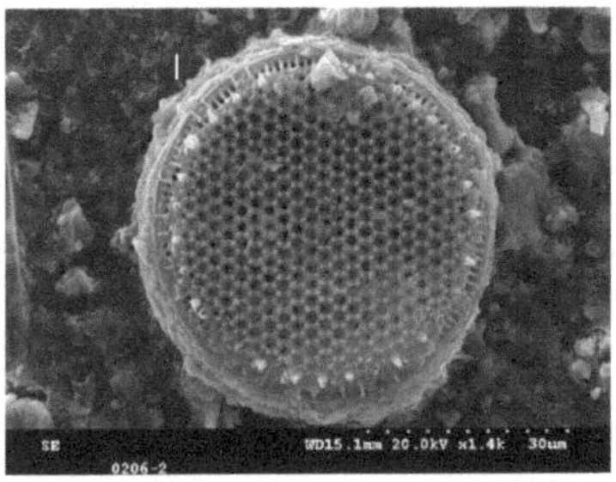

a) Diatom structure

b) Massive mixture of diatoms

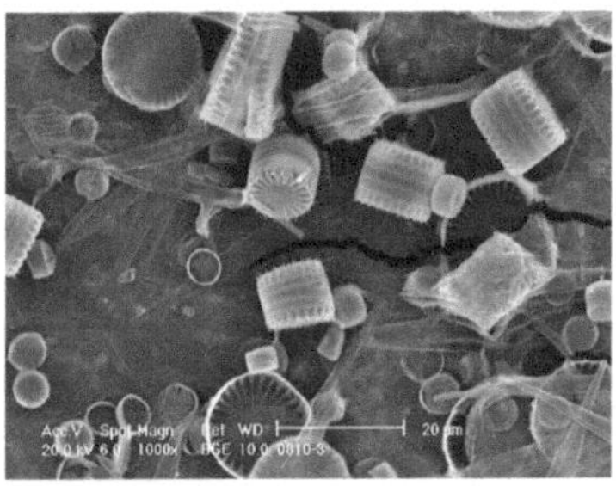

c) Massive presence of diatoms mixed with organic foulant

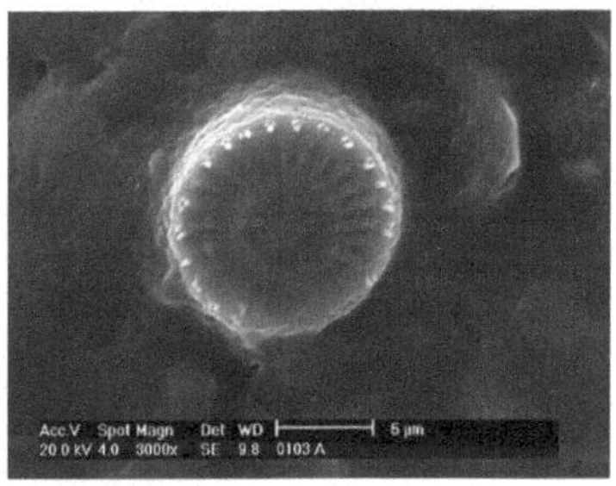

d) Diatom on organic foulant

e) Mixture of diatoms with colloidal matter / clays

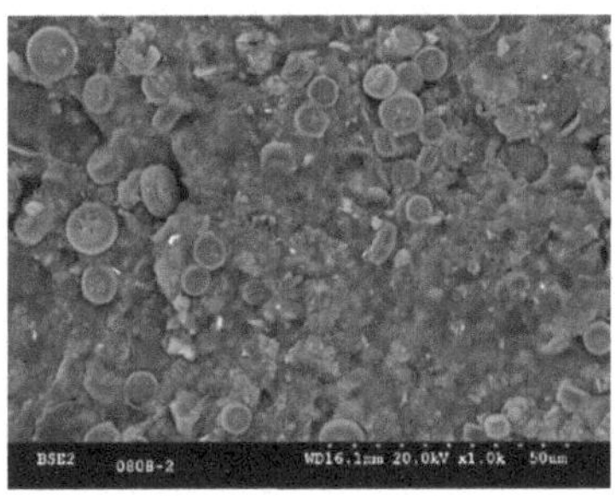

f) Mixture of diatoms with organic foulants and colloidal matter / clays

Figure 5 Examples of different diatom structures found in foulant. (Images credit: Genesys Membrane Products S.L.)

- **Calcareous structures.** Calcium carbonate minerals are the building blocks for the skeletons and shells of many marine organisms. It is very common also to find coccolithophores, which are generally regarded as calcareous scale-bearing marine algae (Figure 6, 2.0–75.0 μm in cell diameter Jordan *et al.*, 2009). In many cases it is easier to distinguish these structures on microfilters and SDI pads.

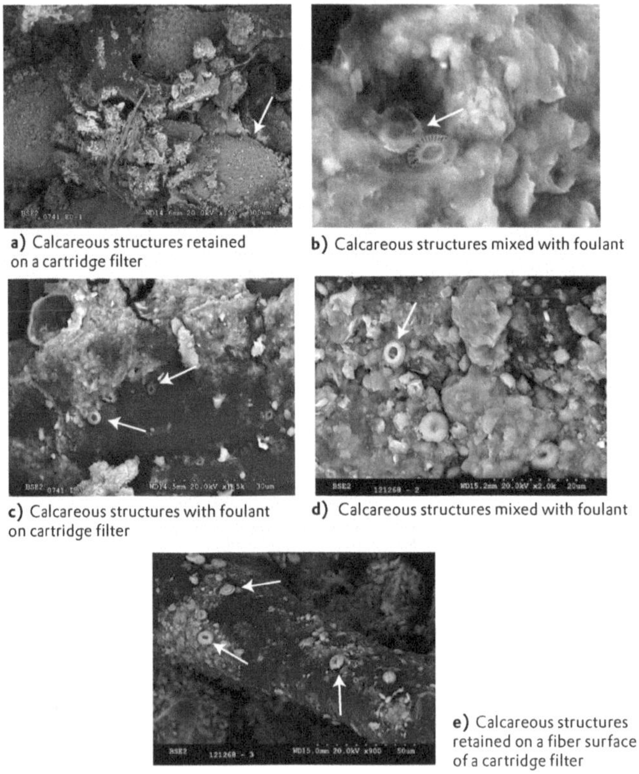

a) Calcareous structures retained on a cartridge filter

b) Calcareous structures mixed with foulant

c) Calcareous structures with foulant on cartridge filter

d) Calcareous structures mixed with foulant

e) Calcareous structures retained on a fiber surface of a cartridge filter

Figure 6 Examples of calcareous structures (Images credit: Genesys Membrane Products S.L.)

9.2.2 Metals

Presence of metals on membranes foulant may have many different sources:

- Metals can be dissolved or as a suspension in the originating body of sea water, well water or surface water.
- Use of coagulants (aluminium and iron salts) during pretreatment
- Corrosion drags/products from pipe materials and equipment
- Complexes with natural organic matter (Schippers, 2021)
- Impurities from chemicals

Effect of metals on membranes is very similar to the rest of inorganic foulant components related to suspended matter: main presence on first position membranes and the effect on membrane performance will mainly depend on the rest of foulant components. On reverse osmosis systems, it is important also to consider the catalytic characteristics of metals and the effect that they many have in oxidation-reduction processes improving the damage of the polyamide.

Considering bibliography and data obtained from autopsies (Peña *et al.*, 2017), main types of metals detected on membranes surface are: iron, manganese and aluminium. Figure 7 shows the percentage of membranes which showed these and other metals as main component of fouling on RO membranes.

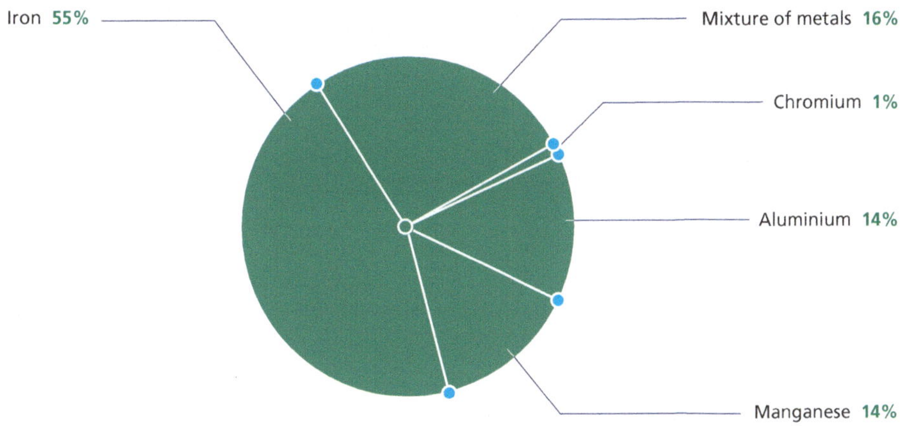

Figure 7 Main metals detected during autopsies (Peña *et al.*, 2017)

However as already explained, besides the main components of foulant, it is essential to consider secondary or minor components. It is actually almost impossible to find a fouling only composed of metals.

On the other hand, when studying different components of fouling on membranes it is very common to detect small presence of metals, especially iron (both as iron oxide particles or as part of aluminosilicates/clays) and also particles from corrosion drags (Fe-Cr-Ni).

Main effect of metallic particles from corrosion will be abrasion of membranes surface (Figure 8).

a) Massive presence of corrosion drags at membrane feed end

b) Presence of corrosion drags on membrane surface feed side

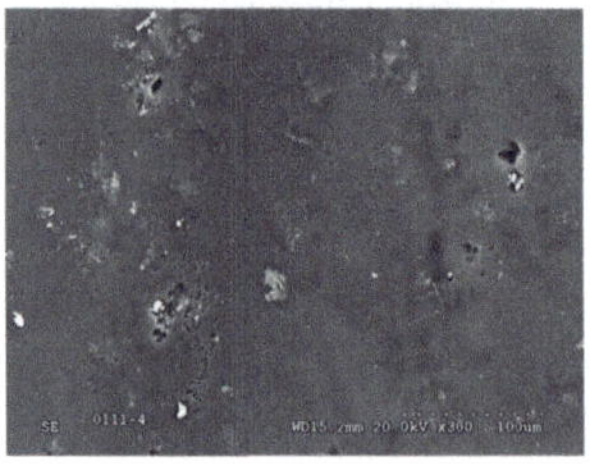

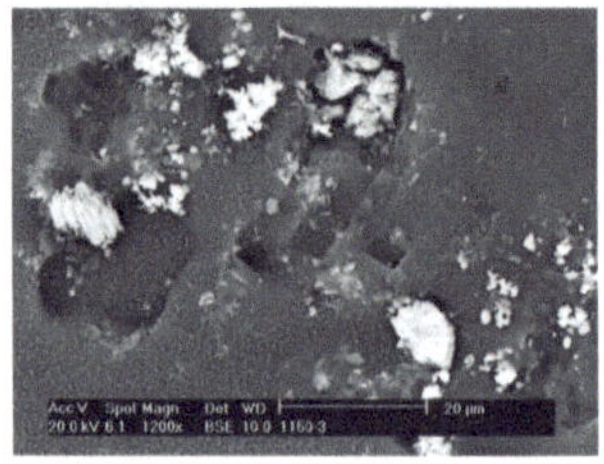

c) Abrasion produced by metallic particles

Figure 8 Examples of membranes with presence of metallic particles from corrosion (Images credit: Genesys Membrane Products S.L.)

A characteristic of main metals found on membranes is that they achieve characteristic color to foulant. The photographs in Figure 9 show some examples and some microphotographs obtained during the study of membranes with significant presence of iron, manganese and aluminium.

a) RO membrane surface with main presence of iron (as iron oxide). It commonly shows a dark orange color.

b) Fibers from a UF module with main presence of iron (as iron oxide). It commonly shows a dark orange color.

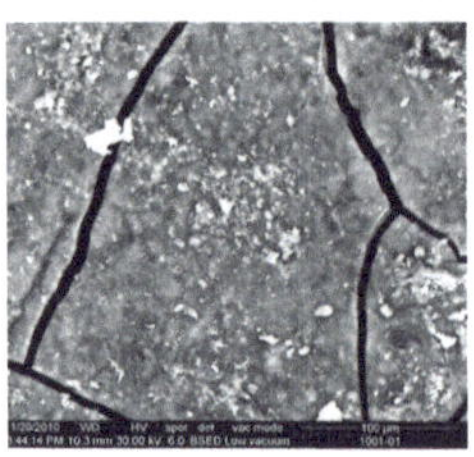

c) Detail of fouling with iron, which corresponds to the brighter component. Iron commonly appears at SEM as very bright and tiny particles mixed with the rest of components.

d) Detail of iron particles - Granular shape.

Figure 9 Membranes and foulant with main presence of iron (Images credit: Genesys Membrane Products S.L.)

Presence of metals in foulant is directly related to a lack of or insufficient pre-treatment. For example, when iron and aluminium are related to the dosing of coagulants, it is essential to optimize dosing and to use the different tools available to control them when water shows a very high variability and makes difficult to get the right concentration.

In the case of iron and manganese, there can be other different sources (Peña *et al.*, 2017) and in those cases it is necessary to use the different technologies available in water treatment for metals removal.

a) Detail of membrane with manganese foulant: it shows a characteristic dark brown color (MnO_2)

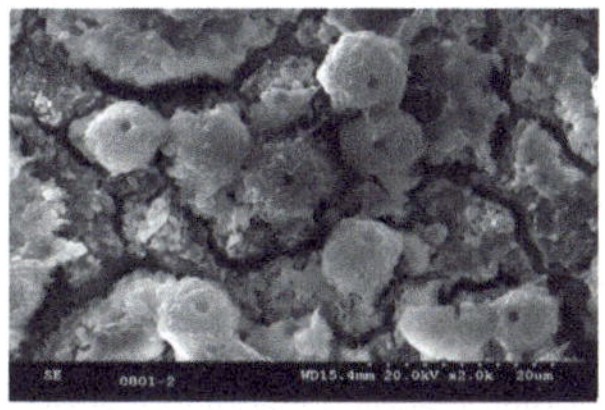

b) Detail of manganese fouling by SEM Manganese commonly shows very characteristic spherical structures mixed with the rest of fouling components.

c) Detail of fouling with main presence of aluminium. It commonly appears mixed with organic matter and shows the color of the mixture due to the different components.

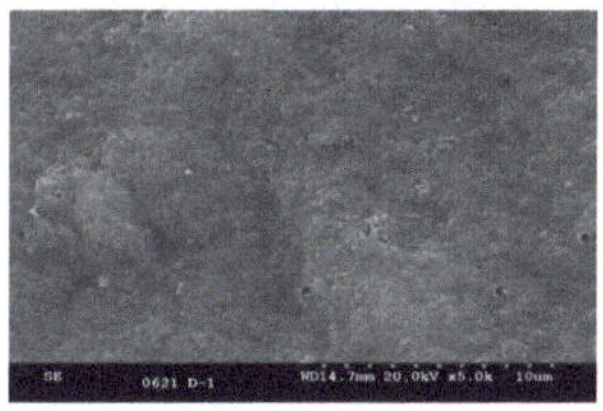

d) Detail of fouling with aluminium by SEM It shows a very unspecific shape since it appears mixed with organic matter. It doesn't show bright intensity as other metals.

Figure 10 Membranes and foulant with main presence of manganese (pictures a and b) and aluminium (pictures c and d) (Images credit: Genesys Membrane Products S.L.)

To get a good control of metals it is necessary to identify the source of the metals and to have data from water analyses from different points of the process treatment.

Following photographs in Figure 11 correspond to SDI pads from different points of pre-treatment, in which a decrease on the initial orange color from iron is distinguished. As it can be observed, the significant presence of iron on the raw water disc is significantly removed after sand filtration but there is not much change after microfiltration.

a) Massive presence of corrosion drags at membrane feed end
b) Presence of corrosion drags on membrane surface - feed side
c) Abrasion produced by metallic particles

Figure 11 SDI pads obtained from different points of a pre-treatment process.

Besides, it is essential to distinguish dissolved metals from oxidized or suspended solids and to consider the specifications of membrane manufacturers about the concentration of metals that can reach the membrane.

It is very important also to consider concentration of metals in water since it is very important to consider them for antiscalant projections. If a lower concentration is considered for antiscalant projection, calculated dosing could be wrong, and it would have scaling consequences.

9.2.3 Scaling

Scaling potential of water is one of the main issues to be considered in water treatment and inevitably needs to be chemically controlled. When there is a failure on this scaling control, the consequence is the precipitation of salts. On reverse osmosis membranes, this scaling is mainly produced on last position membranes where concentrations of sparingly soluble salts are the highest. Besides the issues on membrane performance as drop on production and salt rejection, main risk with scaling is the irreversible damage that may produce by abrasion on membrane surface.

The microphotographs in Figure 12 correspond to main scaling species identified on RO systems and the percentage of membranes in which they were identified as the main component in a study of 500 RO membranes (Peña *et al.*, 2013).

a) Calcium carbonate – 31%

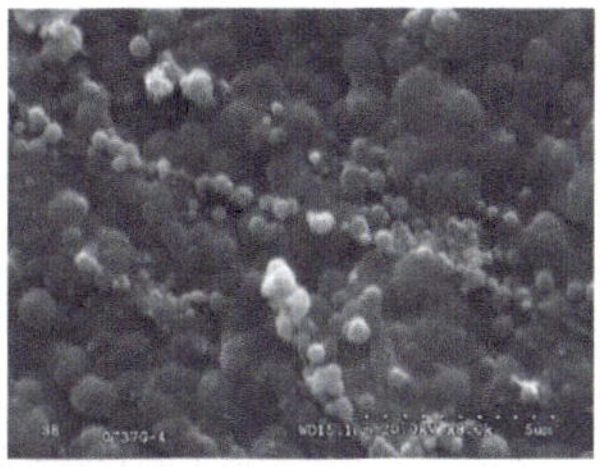

b) Silica – 32%

c) Calcium phosphate – 14%

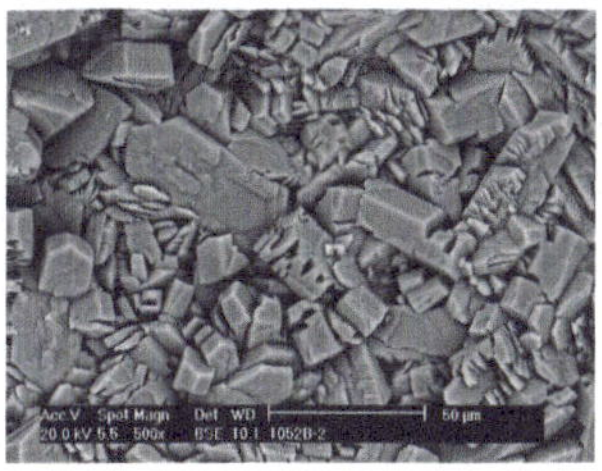

d) Calcium sulphate – 9%

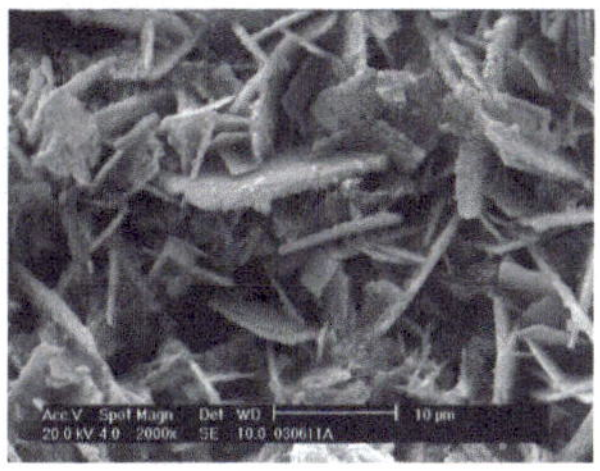

e) Barium sulphate – 4%

Figure 12 Main scaling species identified on RO systems.
(Images credit: Genesys Membrane Products S.L.)

Scaling species are not only identified when water saturation is exceeded. In many cases, precipitation issues are produced due to increases in pH or the use of water with poor quality in hardness during cleaning processes.

The photographs in Figure 13 correspond to UF membranes with calcium carbonate scaling.

a) Presence of calcium carbonate on an UF element fibers surface

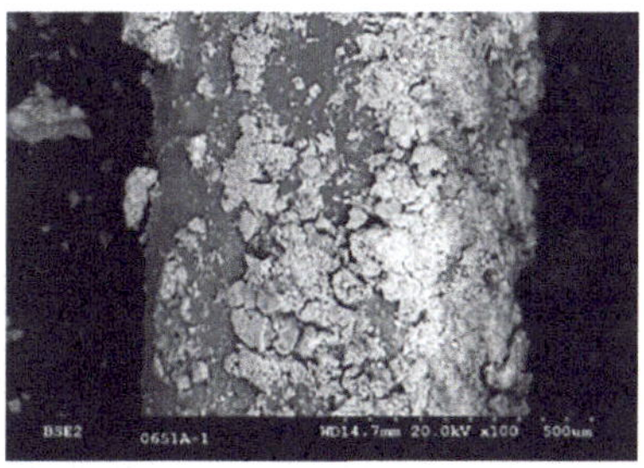

b) Massive presence of calcium carbonate on an UF fiber surface

Figure 13 UF membranes with calcium carbonate scaling.
(Images credit: Genesys Membrane Products S.L.)

As a compliment, microphotographs in Figure 14 correspond to a UF hollow fiber membrane showing scaling on the external surface due to some cleaning issues (blue color), when membrane fouling is composed of aluminosilicates (yellow color from silicon) (Peña *et al.*, 2015):

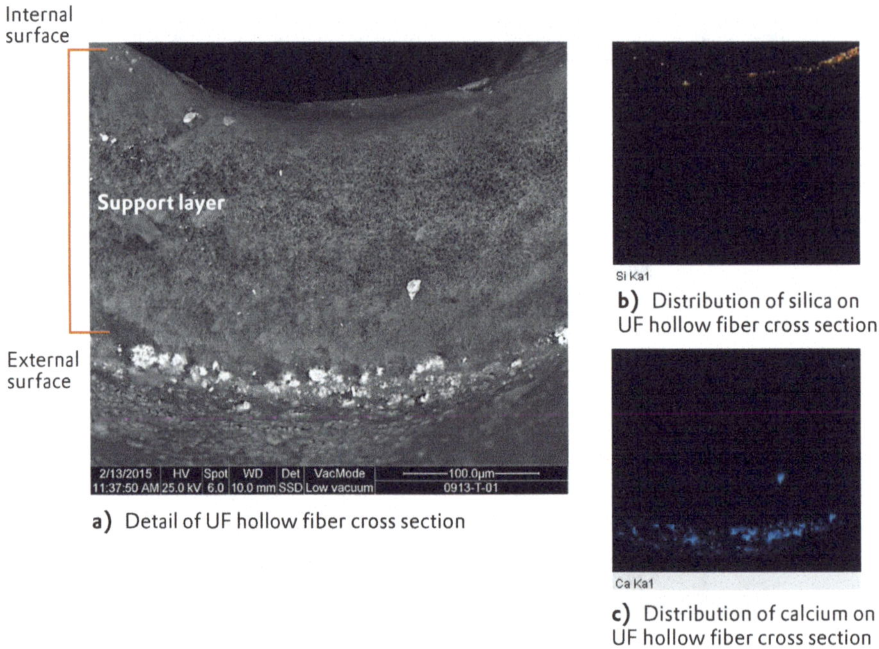

Figure 14 UF hollow fiber membrane showing scaling on the external surface. (Images credit: Genesys Membrane Products S.L.)

It is very important to determine the source of the scaling when it is not due to antiscalant failure since it would be the only way to avoid it. Some tools that may help to identify scaling source are the following:

- On UF systems it is common to find them on the side of the membrane which is directly in contact with the cleaning solution during backwash: out surface for in-out filtration mode and inner surface for out-in fibers. In these cases, scaling would be related to failed cleaning procedures.
- On RO membranes, a scaling due to already precipitated structures would initially be detected on lead positions since it would come from a deposition and not from a precipitation due to an oversaturated solution.
- As previously explained, some species may come from microstructures as diatoms or sea water calcareous structures (see Section 9.2.1).
- To identify the source of the scaling species when they are already formed, the analysis of different components of the pre-treatment such as cartridge filters and SDI pads as well as water samples will certainly help.

9.2.4 OTHER COMPONENTS

Besides the main inorganic elements of foulant already described, and depending on the source of water and membrane applications, other elements could be detected. From them, the following components are very common.

Filtration materials

In many cases, membranes show presence of big particles/grains from filtration materials: silica grains, anthracite, etc.. They concentrate on feed end of first position elements (Figure 15). Unless their presence is massive or they even reach other membranes position, these components are not an issue for membranes performance, but they commonly produce damage by abrasion. Thus, it is very important to check filtration systems and to assure that they do not reach membranes systems.

a) Massive presence of anthracite on membrane feed

b) Massive presence of anthracite on membrane feed

c) Massive presence of filtration material on membrane feed

d) Massive presence of anthracite on membrane feed creates channeling

Figure 15 Detail of different material leakage reaching membranes. (Images credit: Genesys Membrane Products S.L.)

Sea water membranes-Sodium Chloride

Even when it is not an issue as foulant, sea water membranes commonly show sodium chloride as the main inorganic component of foulant and it interferes on the quantification of the organic/inorganic components of foulant (see Chapter 18 – membrane autopsy).

Mainly in biofilm foulant, sodium chloride precipitates during the drying process necessary to carry out some analyses and it is very easily distinguished since it massively appears as dendritic structures when membranes and foulants are studied by SEM-EDX for example.

Besides sodium chloride, common inorganic components of foulant on sea water membranes are magnesium, calcium, phosphorus and sulphur in small percentages.

The microphotographs in Figure 16 show characteristic dendritic structures of sodium chloride.

a) Characteristic dendritic structures of sodium chloride on sea water membranes foulant

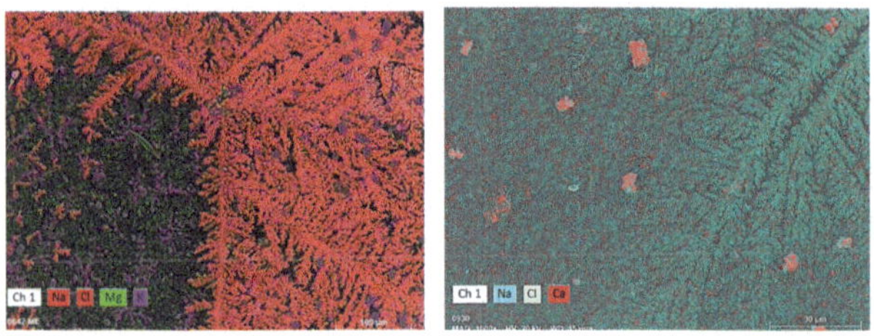

b) Characteristic dendritic structures of sodium chloride with presence of other inorganic components of foulant as magnesium and calcium

Figure 16 Characteristic dendritic structures from sodium chloride sea water on membranes foulant (Images credit: Genesys Membrane Products S.L.)

9.3 METHODS FOR INORGANIC FOULING IDENTIFICATION

It is very important to remember that there is no single analytical tool which can give the complete identification of foulant, mainly due to its composite nature. Then, it is necessary to obtain information from different techniques to get the global information about foulant components.

Tables 4 and 5 include the most common analytical methods used for the identification of inorganic fouling.

To quantify the inorganic content of a foulant, main analytical methods are based on the lower degradation temperature of the organic matter against the inorganic component. Thus, loss of ignition and thermogravimetric analysis (TGA) are the most common methods to quantify the amount or percentage of the organic/inorganic component. These procedures are interesting also as complement for inorganic components identification, since the analysis of the inorganic residue obtained after is a way to concentrate inorganic components and, in some cases, it helps to determine if a specific element is related to the organic or the inorganic component.

Besides, along this chapter it has been mentioned many times how important it is to determine the source of some inorganic components (metals, scaling species, etc). For source identification, it would be necessary to consider not only the analysis of the foulant, but also the analysis of water and other process/pre-treatment components. Table 3 includes some of the samples from water treatment for inorganic water components identification.

Table 3 Analysis of interest from water treatment components for inorganic components identification

Sample		Objective	Analytical technique	Inorganic elements/ components detected
Water	Dissolved components	To determine scaling potential and presence of metals.	Ionic chromatography Spectrophotometry ICP Titration Potentiometry (ISE)	Anions and cations Metals, silica, phosphates, sulphates, silica Metals, elements Calcium, magnesium, chlorides, TA/TAC Fluorides, others.
	Suspended matter	Identification of suspended matter.	SEM-EDX ATR-FTIR Particle counting	By filtering water sample by 0.45 or 0.22 µm, suspended solids can be identified. Determination of particles size.
Filters	Cartridge filter	Identification of suspended matter.	SEM-EDX ATR-FTIR	Identification of inorganic components. SDI pads from raw water, treated and microfiltered water allows to determine components removal and to get a complete characterization of water.
	SDI pads			

Table 4 Analytical techniques for the identification of inorganic components of fouling – Elemental identification

Analytical technique	Observations
Spectrophotometry	- Destructive technique. - To be applied on foulant sample. - It needs digestion of the fouling sample and individual analysis of each component by different methods. - It is necessary to determine in advance which elements are necessary to analyse.
ICP - MS (Inductively coupled plasma Mass Spectrometry) ICP - OES (Inductively coupled plasma Optical Emission Spectroscopy)	- Destructive technique. - To be applied on foulant sample. - It needs digestion of the fouling sample. - Even when this technique allows to make a scan of different elements, in some cases it is necessary to know in advance which are the main since the system must be calibrated with the elements of interest.
XRF (X-ray fluorescence)	- Non-destructive technique. - It can be applied both on the fouling and membrane surface. - It can be applied on the raw but dried sample and to make a scan which gives information about the elements from certain concentration.
SEM-EDX (Scanning Electronic Microscopy – Energy Dispersive X-ray spectroscopy)	- Non-destructive technique. - It can be applied both on the fouling and membrane surface. - It can be applied on the raw but dried sample and to make a scan which gives information about the elements from certain concentration. For some scaling species, considering elements stoichiometric ratio, the % atomic allows stablishing relation between some elements to determine compounds. For example, Ca/S in ratio 1/1 would be as $CaSO_4$ - It allows also studying the distribution of the fouling on the membrane surface and to distinguish different components on the fouling or on different areas. - Abrasion marks can be observed also by this technique.
XPS-ESCA (X-ray Photoelectron Spectroscopy - Electron Spectroscopy for Chemical Analysis)	- Non-destructive technique. - It can be applied both on the fouling and membrane surface. - It can be applied on the raw but dried sample and to make a scan which gives information about the elements from certain concentration. - Energy binding gives information about the oxidation state of the detected elements.

Table 5 Analytical techniques for the identification of inorganic components of fouling – Compounds identification

Analytical technique	Observations
ATR-FTIR (Attenuated Total Reflection Fourier Transform Infrared spectroscopy)	Non-destructive technique. It can be applied both on the fouling and on membrane surface. It can be applied on the raw sample and to make a scan which gives information about functional groups. There are available IR database which allows compounds identification. These databases are commonly provided with IR equipments, although there is a broad offer available on-line, Identification is complicated when there is a mixture of different components.
XRD (X-ray diffraction)	Non-destructive technique. It can be applied both on the fouling and membrane surface. Since the identification is based on the diffraction pattern, it cannot achieve accurate identification if the sample doesn't show crystalline structure. It hardly identifies silica on membranes surface, for example. Difficult identification for mixtures.

9.4 METHODS FOR INORGANIC FOULING REMOVAL

In many systems it is necessary to assume that due to inadequate chemical and physical pre-treatment, some foulant components will continue entering the membrane system and that it is necessary to optimize and to reduce the frequency of required cleaning. To reduce cleaning frequency and minimise membrane damage the operator can achieve optimum deposit removal by cleaning with a technically correct product (Peña *et al.*, 2013).

Table 6 include some basics that need to be considered for inorganic foulant components removal.

For specific cleaning recommendations, Tables 7 to 9 include specifications that can be found at the technical manuals available from some of the main RO and UF membrane manufacturers.

Besides commodity cleaners included in these tables, there are also many different formulated multifunctional cleaners available in the market that can certainly be effective against main inorganic foulants.

Table 6 Basics on cleaning procedures for inorganic foulants removal

Main foulant		General cleaning recommendations
Colloidal matter		Alkaline cleaner applied with temperature Longer contact times will help for the removal Use of optimum flow rates will help.
Metals		Mild acid cleaner applied at room temperature. For aluminium, alkaline cleaner could work also.
Inorganic scaling	Calcium carbonate/ phosphate:	Strong acid cleaner at room temperature.
	Sulphates:	Specific alkaline with surfactant. Temperature Depending on the amount of scaling higher contact time may be needed.
	Silica:	Specific alkaline cleaner with high temperature and long contact time.

Table 7 Cleaning recommendations of some of the leading membrane manufacturers for inorganic colloids

Membrane manufacturer *	RO membranes Recommended cleaner	UF membranes Recommended cleaner (CIP)
DUPONT-DOW FILMTEC	0.1% (W) NaOH and 0.025% (W) Na-DSS, pH 12, 35°C max.	0.2% HCl, 2% citric acid / oxalic acid 0.1% NaOH + 0.2% NaOCl
HYDRANAUTICS	2.0% (w) STPP (sodium tripoliphosphate) ($Na_5P_3O_{10}$) and 0.8% (w) Na-EDTA, pH of 10.0 NaOH, HCl, H_2SO_4, or citric acid 0,1 % NaOH + 0,03% SDS, pH 11,5	NaOH, HCl, H_2SO_4, or citric acid
TORAY	1 - 2 % citric acid adjusted with ammonia (NH_3)	
LG NANO H_2O	NaOH, EDTA / permeate RO NaOH: until 0.1% weight EDTA: until 1.0% weight Citric acid 2.0% weight	
INGE		NaClO H_2O_2 NaOH / acid pH
PENTAIR		NaOCl (active chlorine) 500 ppm max. H_2O_2 1000 ppm max. NaOH pH ≤ 11 Nitric acid pH ≥ 1 Phosphoric acid pH ≥ 1 EDTA pH ≤ 11 Citric acid Enzyme compounds

Table 8 Cleaning recommendations of some of the leading membrane manufacturers for metals

Membrane manufacturer *	RO membranes Recommended cleaner	UF membranes (CIP) Recommended cleaner
DUPONT-DOW FILMTEC	1.0% $Na_2S_2O_4$ (pH:5, 30° C) 2.0% Citric acid 0,5% H_3PO_4 1.0% sulfamic acid	Citric acid, HCl, oxalic acid, sulphuric acid, pH= 2
NITTO-HYDRANAUTICS	2.0% Citric acid 1.0% $Na_2S_2O_4$ (pH: 4-6)	Citric acid or HCl
TORAY	Citric acid 1 – 2 %, adjust with ammonia (NH_3), pH: 2-4	
LG NANO H_2O	2.0% Citric acid, pH: 2-4	
INGE-DUPONT		HCl, H_2SO_4, pH: 1 - 2,5 Citric acid pH: 4
PENTAIR		1% citric acid, 1% oxalic acid, 0,25% ascorbic acid

*Manufacturers technical manuals

Table 9 Cleaning recommendations of some of the leading membrane manufacturers for inorganic salts

Salt species	Membrane manufacturer *	RO membranes Recommended / preferred cleaner
Calcium carbonate / inorganic salts (phosphate, etc)	DUPONT DOW-FILMTEC	0.2 wt% HCl (pH 1 – 2, 35°C) 2.0 wt% citric acid 0.5% H_3PO_4 1.0% $Na_2S_2O_4$
	NITTO-HYDRANAUTICS	2 v/v % citric acid 0,5 v/v % HCl, pH 2,5
	TORAY	Citric acid 1–2 wt%, adjust with ammonia, pH: 2-4
Sulphate scales	DUPONT DOW-FILMTEC	0.1 wt% NaOH 1.0 wt% and Na_4EDTA pH 12, 30°C maximum
	NITTO-HYDRANAUTICS	2 % STPP + 0,8 % Na-DDBS, pH 10 0,5 v/v % HCl, pH 2,5
	TORAY	1 wt% Sodium hexametaphosphate (SHMP), pH 2
Silica	DUPONT DOW-FILMTEC	0.1% (W) NaOH and 0.025% (W) Na-DSS, pH 12, 35°C max. 0.1% (W) NaOH and 1.0% (W) Na_4EDTA, pH 12, 35°C max.
	NITTO-HYDRANAUTICS	0,1 (w/v) % NaOH , pH 11,5

*Manufacturers technical manuals

9.5 REFERENCES

Armstrong M.W., S. Gallego, S.P. Chesters. Cleaning clay from fouled membranes. Desalination and Water Treatment, 10 (2009) 108-114.

Gardner K. H. and Defne S. Apul, Influence of colloids and sediments on water quality, Environmental and Ecological Chemistry – Vol. II – Influence of Colloids and Sediments on Water Quality – Encyclopedia of Life Support Systems (EOLSS).

Guo, W.; Ngo, H.H.; Li, J. A mini-review on membrane fouling. Bioresour. Technol. 2012, 122, 27–34

Jiang, S.; Li, Y.; Ladewig, B.P. (2017). A review of reverse osmosis membrane fouling and control strategies. Sci. Total Environ. 2017, 595, 567–583.

Jonasz M., Georges R. Fournier, The particles size distribution in Light Scattering by Particles in Water, 2007

Jordan R.W. Encyclopedia of Microbiology (Third Edition), 2009, Pages 593-605. https://doi.org/10.1016/B978-012373944-5.00249-2

Koohestanian, M. Hosseini and Z. Abbasian, The Separation Method for Removing of Colloidal Particles from Raw water. American-Eurasian J. Agric. & Environ. Sci., 4 (2): 266-273, 2008.

Kumari N. and Chandra Mohan, Basics of Clay Minerals and their characteristic properties. DOI: 10.5772/intechopen.97672

Nascimento G. M. D. Clay and Clay Minerals. London: IntechOpen; 2021 222 p. Available from: https://www.intechopen.com/books/10949 doi: 10.5772/intechopen.95640.

Peña García N., Javier Rodriguez, Fernando del Vigo, Matt Armstrong, Max Fazel, Stephen Chesters, (2017), Results of a neutral pH cleaner that removes complex fouling and metals from membranes. IDA WWC-Sao Paulo Brazil

Peña García N., Javier Rodriguez, Fernando del Vigo, Stephen Chesters, (2015), UF membranes autopsies: An approach to hollow fiber membranes surface. IDA WWC-San Diego Ca, USA

Peña N., S. Gallego, F. del Vigo & S.P. Chesters. Evaluating impact of fouling on reverse osmosis membranes performance, Desalination and Water Treatment, 51 (2013) 958-961.

Schippers J.C. (2021), Inorganic fouling, Seawater Reverse Osmosis Desalination: Assessment and Pre-treatment of fouling and Scaling (pp.187-206). IWA Publishing. doi: 10.2166/9781780409863_0187

Smol JP, Stoermer EF. The diatoms: applications for the environmental and earth sciences: Cambridge University Press; 2010.

SSWM, Coagulation-Flocculation factsheet, University Course.

Wynne MJ, Bold H. Introduction to the Algae: Structure and Reproduction: Prentice-Hall, Incorporated; 1985.

doi: 10.2166/9781789062977_0229

Chapter 10

Assessing Scaling Potential with Induction Time and a Once-through Laboratory Scale RO System

M. Nasir Mangal, IHE Delft & Berghof Membranes, The Netherlands

Sergio G. Salinas-Rodriguez, IHE Delft, The Netherlands

The learning objectives of this chapter are the following:

- Describe experimental methods for assessing scaling potential
- Present the induction time protocol
- Propose the once-through RO as a system for assessing scaling potential of RO feed water

10.1 INTRODUCTION

Membrane scaling is when one or more sparingly soluble salts (e.g., calcium carbonate, calcium sulphate, silica/metal silicates, barium sulphate, calcium phosphate, etc.) precipitate and form a dense layer on the membrane surface in reverse osmosis (RO) applications. In Figure 1, the scanning electron microscopy (SEM) images of the RO membrane surface without and with scaling are illustrated. Figure 1b is from the RO membrane where calcium carbonate scaling occurred and Figure 1c is the membrane surface scaled with calcium phosphate.

Figure 1 SEM images (500x magnification) of the (a) clean (virgin) RO membrane, (b) RO membrane scaled with calcium carbonate, and (c) RO membrane scaled with calcium phosphate. Adopted from (Mangal, 2023).

Scaling, like other types of membrane fouling, reduces permeate production (due to decreased membrane permeability), raises operational costs (due to higher operating pressure, cleaning costs, etc.), degrades permeate water quality (due to increasing salt passage), and shortens the lifetime of membranes due to frequent membrane cleanings (Kucera, 2010, Mangal, *et al.*, 2021).

Membrane scaling can occur when sparingly soluble salts in RO concentrate become supersaturated, meaning their concentrations exceed their equilibrium (solubility) levels. In RO processes, the increased concentration of sparingly soluble salts in the concentrate is primarily caused by the withdrawal of permeate water from the feedwater. The ratio of permeate water to feedwater is known as recovery which is directly related to membrane scaling. Recovery needs to be as high as possible in RO installations to minimize specific energy consumption. However, at high recovery rates, the concentration of sparingly soluble salts in the concentrate can increase dramatically. For example, for 80% and 90% recovery, the concentration of salts in the concentrate can reach 5 and 10 times their concentration in the feedwater, respectively. If the calcium and phosphate concentrations in the RO feedwater are 200 mg/L and 5 mg/L, respectively, the concentrations in the RO concentrate will be 1000 mg/L and 50 mg/L at 90% recovery, exceeding the calcium phosphate solubility limit and resulting in calcium phosphate scaling. Therefore, in brackish water reverse osmosis (BWRO) processes, scaling is typically the main barrier to operating RO installations at high recoveries.

There are several methods for preventing scaling in RO applications, including acidification of RO feed, lowering RO system recovery, and antiscalant addition. Acidification of RO feedwater was one of the first methods for tackling calcium carbonate scaling in RO processes. However, due to the risks associated with the use of acid, this method is becoming less common. Furthermore, acidification may not be effective for all types of scales; for example, it is very effective in preventing calcium carbonate scaling but not calcium sulphate scaling.

Another method of preventing scaling is to operate RO at low recovery (ratio of permeate water to the feedwater). The recovery of the RO application is reduced in this approach to reduce the supersaturation level of the concentrate water to undersaturated conditions. Low recovery reduces the adverse effect of concentration polarization because there is less solute concentration on the membrane surface, reducing the potential for scale formation. This approach, however, is not very appealing or economical because it results in high specific energy consumption. Furthermore, the large amount of concentrate disposal is a problem. Antiscalant dosing in feedwater is one of the most extensively applied and effective scaling prevention strategies in RO applications (Greenlee, *et al.*, 2010, Kucera, 2010, Antony, *et al.*, 2011, van Engelen and Nolles, 2013, Yu, *et al.*, 2020). Antiscalants are primarily organic compounds containing sulphonate, phosphonate, or carboxylic acid functional groups that hinder the crystallization process, i.e., nucleation and/or growth phase of scaling compounds (Antony, *et al.*, 2011, Boels and Witkamp, 2011, van Engelen and Nolles, 2013). In general, antiscalant prevent scale formation by three mechanisms, namely threshold inhibition, crystal modification and dispersion. Threshold inhibition is when antiscalant molecules adsorb on crystal nuclei and halt their nucleation process, whereas crystal modification and dispersion are the ability of antiscalants to stop the growth and/or agglomeration of crystals and particles. The selection of antiscalants in RO applications depends on the feed water composition as well as other factors such as recovery and discharge regulations.

With the use of antiscalants, the main question which arises is: *How to determine the lowest (optimum) dose of antiscalants to prevent scaling in RO applications*? Operating the RO with the lowest antiscalant dose at which scaling does not occur is highly desirable, since high doses of antiscalant result in additional costs and pose environmental concerns (Boels and Witkamp, 2011). In practice, the antiscalant dose for a given water composition is generally determined using the antiscalant manufacturer's proprietary programs. However, the method used by the manufacturers to calculate the antiscalant dose is unknown and therefore the end-users cannot verify their recommended doses. In general, the suppliers' recommended antiscalant doses are in the range of 2–10 mg/L to prevent scaling in RO processes (Singh, 2005).

Mangal (2023) studied the scaling potential of calcium carbonate and calcium phosphate in brackish water reverse osmosis by assessing the theoretical scaling potential with manual and computer programmes, by measuring the induction time and by developing a new once-through RO system.

The chapter on membrane scaling by Mangal, *et al.* (2021) describes thoroughly the principles of scaling, influencing factors, types of scaling, prediction calculations, scaling indices, use of commercial software, monitoring tools in RO systems, scaling control, and scaling in seawater applications. In this chapter we will focus on the description of the induction time protocol as well as of the once-through lab-scale RO system for scaling studies in RO applications.

10.2 INDUCTION TIME MEASUREMENTS

Induction time is defined as the time elapsed between the emergence of supersaturated conditions and the detection of crystallization. It is composed of three time periods such as relaxation time (t_r), nucleation time (t_n), and growth time (t_g) (Kashchiev, 2000). The time needed to initiate nucleation from time zero to steady state condition is called the relaxation time(Guan, 2009). Nucleation time is defined as the time needed to form a stable nucleus and the period in which detectable crystal are formed from the stable nucleus is known as growth time (Kashchiev, 2000). Induction time depends on the supersaturation level of a water solution, but it is mainly dependent on the precipitation kinetics.

To measure the induction time several methods have been developed, including but not limited to the pH method (Waly, 2011), the conductivity method (Söhnel and Mullin, 1978), turbidity or scattered light method (Prisciandaro, *et al.*, 2001, Abdel-Aal, *et al.*, 2004, Shih, *et al.*, 2006), and the concentration of calcium (Verdoes, *et al.*, 1992). Among these methods, the pH method was reported to be the most accurate one for the induction time measurement of $CaCO_3$ (Waly, 2011). However, for measuring the induction time of (amorphous) calcium phosphate, pH measurements are not that useful especially when high bicarbonate concentrations are present. For calcium phosphate, turbidity measurements could be used, but in the presence of some antiscalants, the formed calcium phosphate particles may not be detected in turbidity measurements (due to their small size) (Mangal, 2023).

10.2.1 Experimental setup

Figure 2 depicts a schematic diagram of the experimental setup which was used by Mangal (2023) to measure the induction time of calcium carbonate. The setup is composed of the following components:

10.2.1.1 Glass reactor

Double wall glass reactor (Applikon, Netherlands) of 3.1 L volume in cylindrical shape (24 cm height and 12 cm internal diameter). It should be air tight with stainless steel lid and a thin rubber.

It has a mixing shaft with two pedals fixed in the stainless-steel lid. The lid plate has 4 circular holes with 13 mm diameter used for the pH probe or other instruments (e.g., turbidity, dissolved oxygen, etc.) when required. Three inlet stainless steel pipes of 0.5 cm diameter.

Air/overflow: There is an air/overflow valve installed at the top of stainless-steel lid. The main function of this valve is for ventilating the system during water circulation and to check that the reactor is totally filled with the calcium chloride and sodium bicarbonate solution by noticing the solution flooding from the overflow pipe.

NaOH addition: This element consists of one vertical pipe with a valve and a rubber on the top of the reactor. The main function of NaOH is to adjust the pH of the targeted solution. Feed line of saline water/demineralised water/acid: It is a 0.5 cm straight pipe inside the reactor to feed the NaHCO3 solution or other saline solution. It connects the reactor with the saline water flask and the demineralised water or acid when cleaning of the reactor is required.

Feed line of $CaCl_2$ /drainage line: This connection (L shaped pipe) allows feeding the reactor with $CaCl_2.2H_2O$ solution from the bottom in order to have homogeneous solution by mixing the solution during feeding. This line also can be used for emptying the reactor from the solution inside after each experiment.

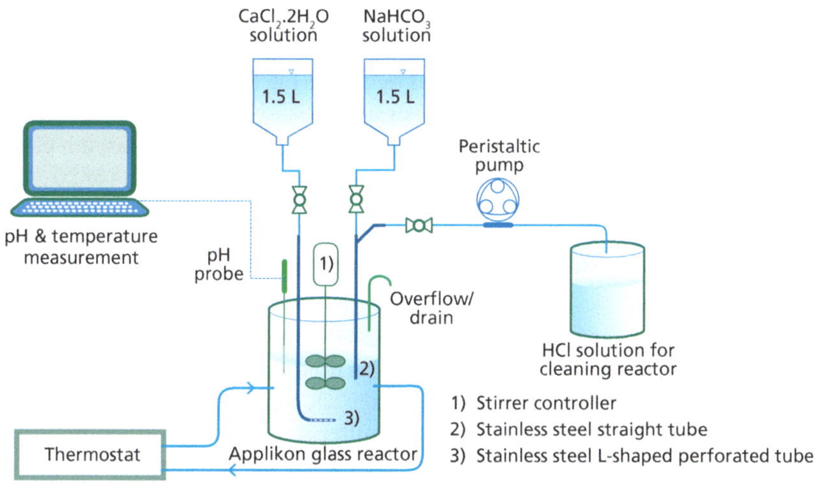

Figure 2 Experimental setup for induction time tests. Adopted from (Mangal, 2023).

10.2.1.2 Stirrer device

It includes Applikon controller and mixing shaft in the reactor. Applikon mixing controller (Applikon type ADI 1032 motor controller) is mainly used for the adjustment of the stirring rate which ranges between 0 and 1250 rpm.

For the induction time experiments the stirring rate is 150 rpm. For cleaning the reactor, the stirring speed is increased up to 1200 rpm.

10.2.1.3 pH meter

The pH meter model is Endress and Hauser Memosens system (0.01 relative accuracy for pH). The pH value and temperature sensor are in one probe. The interval time of pH measurement can be set as low as 1 second.

10.2.1.4 Peristaltic pump

A Master flex peristaltic pumps model 77200 -50 is used to pump the HCl solution (or nitric acid solution) to the reactor for the cleaning purposes.

10.2.1.5 Thermostat

A thermostat is used to keep the temperature at a desired value and constant during an entire experiment. As the Applikon rector is a doubled wall, the thermostat water keeps recirculating between the reactor and the thermostat to maintain a uniform temperature of the solution in the reactor.

10.2.2 Experimental procedure

10.2.2.1 Preparation of artificial brackish water

For the $CaCO_3$ induction time measurements, $CaCl_2.2H_2O$ and $NaHCO_3$ analytical grade salts are mainly used. In addition, NaCl salt is used to achieve a certain TDS level of the water and also to balance the cations concentrations with anions. The following procedure is followed to prepare artificial brackish water:

1. The $CaCl_2.2H_2O$ solution is prepared by dissolving the required amount of the salt for 10 L in 5 L of Milli-Q (2 times concentrated).
2. $NaHCO_3$ and NaCl solution is also prepared by dissolving the required amount of the salts for 10 L in 5 L of Milli-Q (2 times concentrated).
3. To ensure complete dissolving of the salts, the solution is mixed for at least 2 h using a magnetic stirrer at an average speed of 300 rpm and at a room temperature of 20 °C.
4. The two solutions were filtered separately through 0.45 μm Millipore filter using vacuum filtration. This step can be eliminated when Milli-Q water and high-grade chemicals are used for the preparation of $CaCl_2.2H_2O$ and $NaHCO_3$ solutions.

10.2.2.2 Induction time measurement

The following steps are followed to measure induction time of artificial brackish water.

1. Prepare the stock solutions of $CaCl_2.2H_2O$ and $NaHCO_3$ as described in section 10.2.2.1.
2. Calibrate the pH electrodes using 2 standard buffer solutions. The calibration curve can be checked by its slope which has to be between 55 and 59 (Endress Hauser requirement).
3. Install the calibrated pH probe inside the reactor.
4. Check that the reactor is cleaned by filling the reactor with demineralised water and measure its pH value inside the reactor which should be similar to the pH of demineralised water.
5. Calculate the required volume of NaOH or HCL that needs to be added to the solution in order to adjust the initial pH.
6. Switch on the stirring controller and set the stirring speed at 150 rpm.
7. Move 1.55 L of $NaHCO_3$ and NaCl solution to the elevated flask using graduated cylinder and let the solution feeding the reactor by gravity.
8. Switch on the thermostat and adjust the temperature. Keep the solution 10 minutes to stabilize with the new temperature.
9. Add the required volume of NaOH or HCL by using pipette.
10. Close all opening in the reactor except the ventilation valve and calcium chloride feeding line.
11. Switch on the computer and check that it is connected properly with the pH meter.
12. Switch on the pH meter software and set time interval and other settings.
13. Start measuring pH of the solution inside the reactor.
14. Move 1.55 L of $CaCl_2.2H_2O$ solution to the elevated flask using graduated cylinder and let the solution feeding the reactor by gravity.

NB: In case of testing antiscalant (AS), AS dose needs to be mixed with the calcium chloride solution before moving the solution to the reactor.

15. Observe the ventilation valve. The reactor should be filled fully until water reaches to the ventilation valve. Once it reaches the ventilation valve, close the ventilation valve, and close the calcium chloride feeding line.

10.2.3 Calculation of induction time

Figure 3 illustrates an approximate method to determine induction time. As shown, the induction time is the intersection point of the horizontal line of the curve (constant pH) with the slope of the curve when pH starts declining.

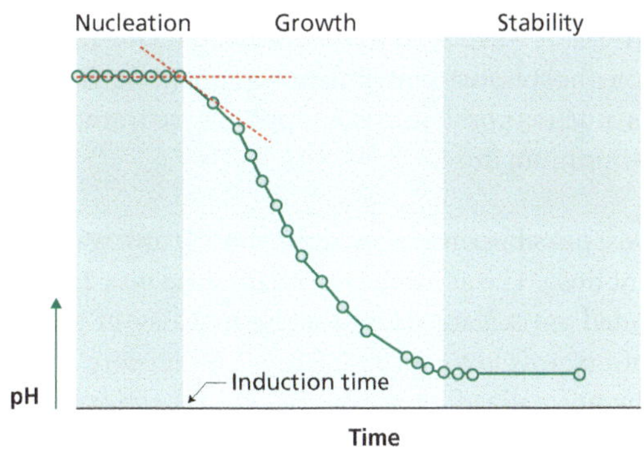

Figure 3 Determination of the induction time.

10.2.4 Cleaning of the reactor

After each experiment, the water inside the reactor has to be drained out from the reactor's stainless-steel pipe (L shaped) using peristaltic pump. Cleaning of the reactor includes all components connected in the reactor such as; pH probe, water connection and fittings, etc. Following is the procedure to clean the reactor:

1. Open the air release/ overflow valve.
2. Connect the drainage pipe with the drainage nozzle and open the valve of the drainage pipe.
3. Once the all water in the reactor is drained, connect the acid line with the acid/ demineralised water feed line and switch on the pump.
4. Fill the reactor with 0.1 M HCl to dissolve any crystals formed during the experiment and increase the stirring speed to 1,200 rpm.
5. Close the acid valve and switch off the pump. Allow acid cleaning for 30 minutes and then empty reactor as mentioned in steps 1-2.
6. Connect the demineralised water line with the acid/demineralised water feed line and open the demineralised water valve in order to flush the reactor.
7. Monitor pH of the demineralised water in the reactor.
8. Drain the water out and repeat step 6 and 7 until the pH of the solution inside the reactor becomes same as the pH of the demineralized water.

10.2.5 Example of application of induction time

This section presents an application of the induction time in investigating the scaling potential of calcium carbonate with and without antiscalant for an RO plant (treating anaerobic groundwater for drinking water production) in the Netherlands and in exploring if (and how much) antiscalant was needed for the plant at a certain recovery.

The detailed information about the RO installation and the composition of the anaerobic groundwater is given elsewhere (Mangal, 2023). The RO plant was operating at 80 % recovery. At this recovery, calcium carbonate (with Langelier saturation index (LSI) of 1.7) was the primary compound which would cause scaling in the RO unit in the absence of antiscalant according to the projection programs of antiscalant suppliers. Therefore, to prevent calcium carbonate scaling in the RO, the supplier's recommended dose of 2.0 mg/L of a phosphonate antiscalant was added to the feedwater. Due to the stringent concentrate discharge regulations, a phosphonate antiscalant was not preferred. It was desirable to lower the antiscalant dose as much as possible as the supplier's recommended dose was considered to be greater than the optimum dose.

To determine the lowest possible antiscalant dose, the RO unit was operated at 80% recovery and varying antiscalant dose. The starting antiscalant dose was 2.0 mg/L (equivalent to the supplier's recommended antiscalant dose) for the first day of the RO operation at 80 % recovery which was afterwards lowered by 0.2 mg/L after every 12 h of operation to a final dose of 0.2 mg/L. The normalized permeability (K_w) for the last element of the last stage was recorded separately. The result is illustrated in Figure 4a. As can be seen, the normalized K_w remained constant when the RO unit was operated for 12 days with an antiscalant dose as low as 0.2 mg/L. Afterwards, in another test, the RO unit was operated at 80 % recovery without antiscalant addition (Figure 4b). As illustrated, the normalized K_w of the last element remained constant for an experimental period of 32 days at 80 % recovery which indicated that there was no need to add antiscalant even when the LSI of the RO concentrate was as high as 1.7. This suggested that calcium carbonate scaling might have been inhibited by some constituents (possibly phosphate and humic substances) present in the feedwater (anaerobic groundwater) that might have functioned as natural antiscalant.

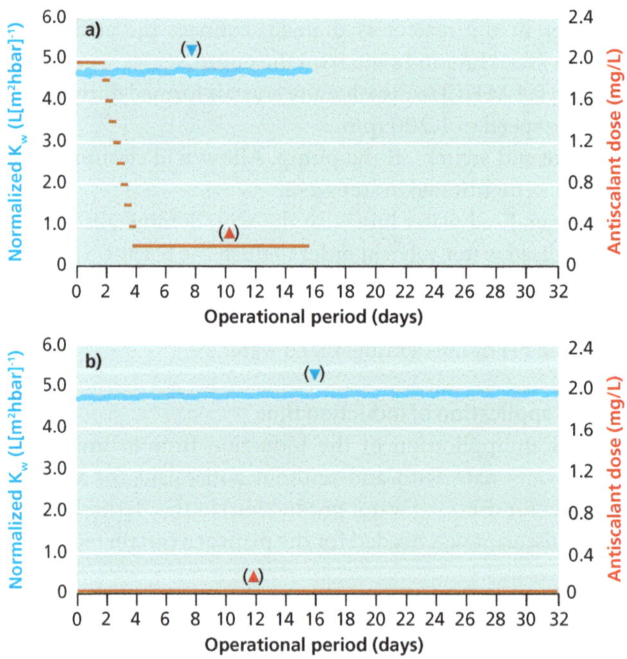

Figure 4 (a) RO operation at 80 % recovery with a phosphonate antiscalant, and (b) RO operation at 80 % recovery without antiscalant addition: (▼) Normalized K_w of the last element of the last stage, (▲) antiscalant dose. Adopted from (Mangal, 2023)

To understand why the RO unit did not scale at high supersaturation levels, induction time measurements were performed with the anaerobic real RO concentrate at 80 % recovery in the absence of antiscalant (Figure 5a). In parallel, induction time measurements were also executed using artificial concentrate solutions. The artificial concentrate solutions were prepared such that the Ca^{2+} and HCO_3^- concentrations were equivalent to the concentrate concentration of real ground water at 80% recovery. In Figure 5a, we show that the measured induction time of the real RO concentrate at 80 % recovery was longer than 168 h (7 days), while for the artificial concentrates corresponding to 80 % recovery, the measured induction time was approximately 1 h. Thus, at the same supersaturation level, the induction time of the real RO concentrate at 80 % recovery was at least 168 times longer than that of the artificial concentrate. When 10 mg/L of antiscalant concentration was added to the artificial RO concentrate, the induction time became longer than 168 h (Figure 5b). This result clearly showed that the formation of calcium carbonate was suppressed in the anaerobic real RO concentrate by some constituents present in the RO feedwater (anaerobic groundwater) which function as natural antiscalants.

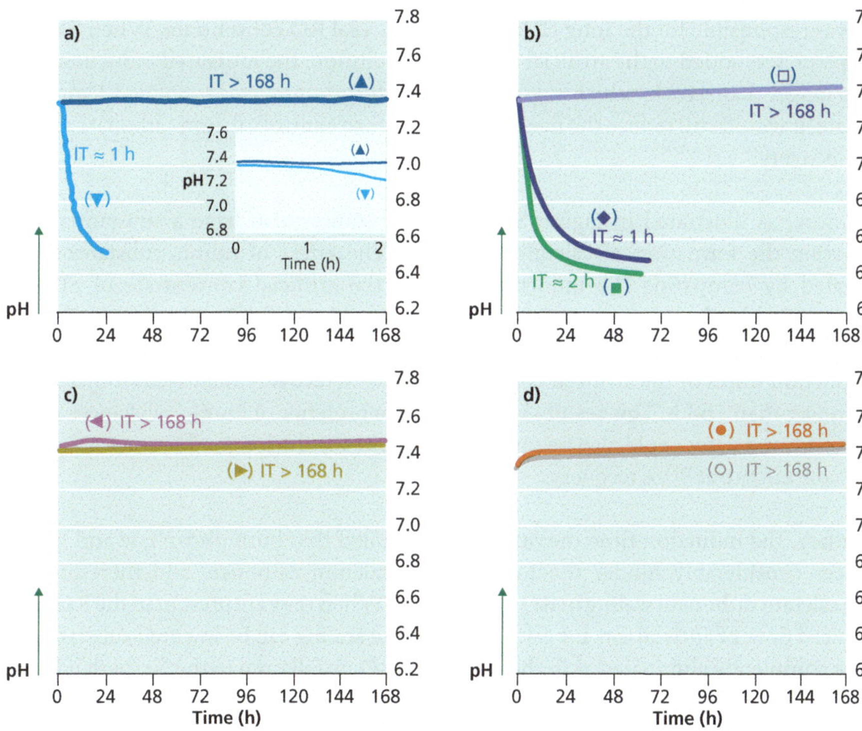

Figure 5 (a) Induction time of the: (▲) real RO concentrate without antiscalant, (▼) Artificial RO concentrate without antiscalant, (b) Induction time of the artificial RO concentrate: (□) with 10 mg/L phosphonate antiscalant, (◆) with 87 mg/L of magnesium ions, (■) with 217 mg/L of sulphate ions, (c) Induction time of the artificial RO concentrate without antiscalant: (◀) with 10 mg/L phosphate ions, (▶) with 5 mg/L phosphate ions, and (d) Induction time of the artificial RO concentrate without antiscalant: (●) with 10 mg/L humic acid, (○) with 10 mg/L fulvic acid. Adopted from (Mangal, 2023)

We consider that the difference between the induction times of the anaerobic RO concentrate and the artificial concentrate is caused by phosphate (ca. 2.0 mg/L) and/or humic substances (ca. 5.0 mg/L) present in the groundwater, as well as by magnesium and sulphate, which are reported in literature to have a positive effect on the suppression of calcium carbonate (Berner, 1975, Bischoff, 1968, Chen *et al.*, 2005). We investigated this hypothesis by varying the composition of the artificial concentrate, evaluating the effect of magnesium and sulphate (Figure 5b), phosphate (Figure 5c), humic substances (Figure 5d).

It was found that magnesium and sulphate did not have a considerable effect on delaying calcium carbonate formation (Figure 5b). The induction time of the artificial RO concentrate at 80% recovery increased to 2 h and 1.2 h with 87 mg/L of magnesium and with 217 mg/L of sulphate, respectively. This showed that neither magnesium nor sulphate was accountable for the long induction time (>168 h) of the real RO concentrate at 80 % recovery. On the other hand, as can be seen from Figure 5c, in the presence of 10 mg/L of phosphate (which was equal to the concentration of phosphate in the real RO concentrate of 80 % recovery), induction time of the artificial concentrate increased from 1 h to a period longer than 168 h which indicated that phosphate was one of the constituents of the feedwater which was responsible for the long IT (>168 h) of the real RO concentrate. When 5 mg/L of phosphate was added to the artificial concentrate solution, the measured induction time was also longer than 168 h, which suggested that if the groundwater contained 1 mg/L of phosphate, it would still reduce the need of antiscalants to control calcium carbonate scaling at 80 % recovery.

Furthermore, as illustrated in Figure 5d, humic substances also have a substantial role in hindering the formation of calcium carbonate. The effect of humic substances was investigated by comparing the induction time of the artificial concentrate of 80% in the absence and in the presence of 10 mg/L humic acid and fulvic acid (obtained from International Humic Substances Society (IHSS)). With 10 mg/L humic acid and fulvic acid, induction times of the artificial concentrate of 80 % recovery increased from 1 h to a period longer than 168 h. This result showed that the presence of humic substances in the anaerobic groundwater could also be one of the reasons for the long induction time of the real RO concentrate at 80 % recovery.

In summary, the induction time measurements revealed that both phosphate and humic substances considerably hinder the formation of calcium carbonate and therefore can prevent calcium carbonate scaling in RO applications. When they are present in the RO feed, the required dose of antiscalants for calcium carbonate scaling can be lowered substantially or can be completely eliminated as in the case of the RO installation in the Netherlands.

10.3 ONCE THROUGH LAB-SCALE RO SYSTEM

10.3.1 Experimental set-up

To evaluate the performance of antiscalants in preventing calcium phosphate scaling in RO systems, a lab-scale RO setup (Figure 6) was used.

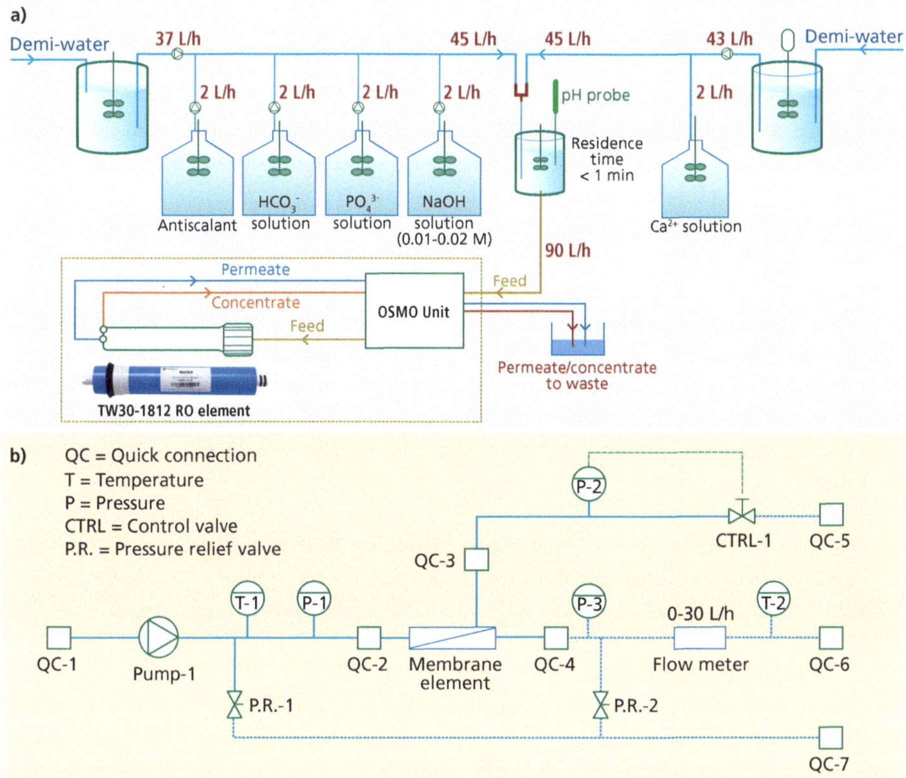

Figure 6 (a) Once-through lab-scale RO setup for calcium phosphate scale inhibition studies, and (b) Piping and instrumentation diagram (P&ID) of the OSMO unit. Adopted from (Mangal, 2023)

The artificial solution was fed at a rate of 90 L/h to a TW30-1812-50 RO element (OsmoPure Water Systems) with the use of an OMSO Inspector unit (Convergence Industry B.V., Netherlands).

The OSMO unit was equipped with a very sensitive flow meter (with high accuracy) which could measure the permeate flow rate of 2 mL/min (0.12 L/h) to 500 mL/min (30 L/h). For each experiment, a new RO element was used.

In all experiments, the initial recovery of the membrane element was in the range of 5–6 % and the permeate flux was between 13–15 L/m^2/h. According to the membrane manufacturer, the minimum ratio of the concentrate flow to the permeate flow should not drop below 5.

In these conditions, the concentrate flow is about 18 times greater than the permeate flow. The cross-flow velocity was in the 10–12 cm/s range. Both permeate and concentrate were directed to the drain.

All experiments should be performed at temperature-controlled conditions of stable conditions like room temperature (20–23 °C).

10.3.2 Experimental protocol

In this setup, as illustrated, antiscalant, HCO_3^-, PO_4^{3-} and NaOH were dosed from stock solutions each at 2 L/h to a stream of demineralized-water (demi-water) with a flow rate of 37 L/h, resulting in a final flow rate of 45 L/h.

The dosage of NaOH was executed from a 0.01–0.02 M stock solution to adjust the pH of the final solution to 7.6. To another stream of demi-water (with a flow rate of 43 L/h), Ca^{2+} was dosed from the stock solution at 2 L/h, also resulting in a final flow rate of 45 L/h. Both streams were then connected to a single pipe resulting in the final flow rate of 90 L/h which had nearly the same composition of the artificial concentrate of 85 % recovery. The final solution (artificial concentrate solution) was introduced to a 4 L reactor in which the artificial concentrate solution was stirred at a rate of 200 rpm with a residence time shorter than 1 min.

The residence time of less than 1 min was achieved by maintaining equal flow rates (90 L/h) of the artificial concentrate solution entering and leaving the reactor and by keeping the volume of the artificial concentrate solution in the reactor to approximately 1.5 L.

10.3.3 Example of application

This section presents an application of the once-through lab-scale RO system to investigate the effectiveness of available calcium phosphate antiscalants against calcium phosphate scaling in the RO installation in the Netherlands. As discussed in the previous section (section 10.2.5), the RO installation could operate at 80% recovery even without antiscalant as calcium carbonate scaling was inhibited by phosphate and humic substances. The drinking water company wanted to increase the RO recovery to 85% recovery (or even higher). However, the permeability of the last stage of the RO unit decreased due to calcium phosphate scaling which was identified in membrane autopsy (Mangal, 2023). Several antiscalants (which were effective against calcium phosphate scaling according to the antiscalant suppliers) were tested. However, none of the antiscalants could prevent calcium phosphate scaling in the RO unit at 85% recovery as the membrane permeability of the last element of the last stage decreased with all antiscalants. As the anaerobic RO concentrate at 85% recovery contained high concentrations of ferrous ion (Fe^{2+} = 55 mg/L), some antiscalant suppliers claimed that the effectiveness of their antiscalants is reduced when iron (II) is present in the RO feed. Therefore, to understand if the available antiscalants can prevent calcium phosphate scaling in the absence of iron (II), once-through lab-scale RO tests (as well as induction time tests) were performed with the artificial concentrate solutions having the same calcium and orthophosphate concentrations that were present in real RO concentrate at 85 % recovery. More precisely, the artificial concentrate of 85 % recovery contained 767 mg/L of Ca^{2+}, 14 mg/L of PO_4^{3-}, and had a pH of 7.6.

Table 1 Induction time measurements (of calcium phosphate) using the artificial concentrate of 85 % recovery without antiscalant addition and with 5.0 mg/L of various antiscalants. Adopted from (Mangal, 2023)

Time	Turbidity (NTU)								
(min)	No AS	AS-A	AS-B	AS-C	AS-D	AS-E	AS-F	AS-G	AS-H
0	0.08	0.08	0.08	0.08	0.08	0.08	0.08	0.08	0.08
1	2.29	1.13	1.01	0.69	0.08	0.08	0.08	0.08	0.08
15	2.94	1.97	2.57	1.62	0.08	0.08	0.12	0.08	0.08
30	3.16	2.36	3.23	2.26	0.08	0.08	0.18	0.08	0.08
45	3.30	2.83	3.82	2.91	0.08	0.08	0.26	0.08	0.08
60	3.38	3.25	4.19	3.60	0.08	0.08	0.39	0.08	0.08

In Table 1, the results of the induction time tests (of calcium phosphate) with and without antiscalant are presented. The actual names of the eight tested antiscalants (dispersant) are replaced with arbitrary names in Table 1. It is worth mentioning that for the induction time of calcium phosphate, turbidity of the artificial solution was measured to detect the formation of calcium phosphate as their formation could not be detected with the pH measurement. As can be seen from Table 1, the turbidity of the artificial RO concentrate of 85% increased without antiscalant as well as with AS–A, AS–B, AS–C, and AS–F which indicated that these antiscalants were not able to prevent the formation of calcium phosphate. Nevertheless, the turbidity of the artificial RO concentrate remained constant with AS–D, AS–E, AS–G, and AS–H which initially gave the impression that the formation of calcium phosphate was inhibited in the presence of those antiscalant. This, however, was not the case. The artificial concentrate solutions with no increase in turbidity were filtered through 100 kDa PES filters. The filters were then examined with SEM (Figure 7). As can be seen, the filter surface was covered with calcium phosphate particles. This revealed that the formation of calcium phosphate in the artificial concentrate of 85 % recovery was not inhibited by AS–D, AS–E, AS–G and AS–H even when no increase in turbidity was observed. The reason for not observing an increase in the turbidity of the artificial concentrate of 85 % recovery could be due to the formation of fewer calcium phosphate particles with very small size by the aforementioned antiscalants.

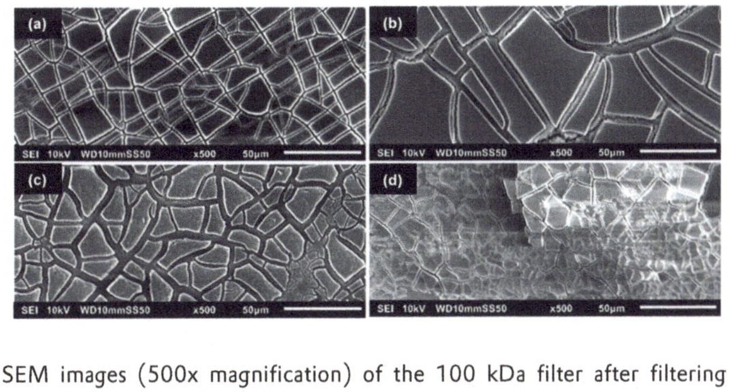

Figure 7 SEM images (500x magnification) of the 100 kDa filter after filtering the artificial concentrate of 85 % recovery in the presence of 33 mg/L of a) AS–D, b) AS–E, c) AS–G, and d) AS–H. Adopted from (Mangal, 2023)

According to some antiscalant suppliers, the induction time results of Table 1 and Figure 7 were not conclusive enough to conclude if those antiscalants are not able to prevent calcium phosphate scaling in RO where the filtration mode is not dead-end but cross-flow. It could be that the adsorbed antiscalants (dispersants) on the formed calcium phosphate particles may diminish their tendency to deposit on the membrane surface in a cross-flow operation. Based on this, it was necessary to test the performance of these antiscalants in a once-through RO system.

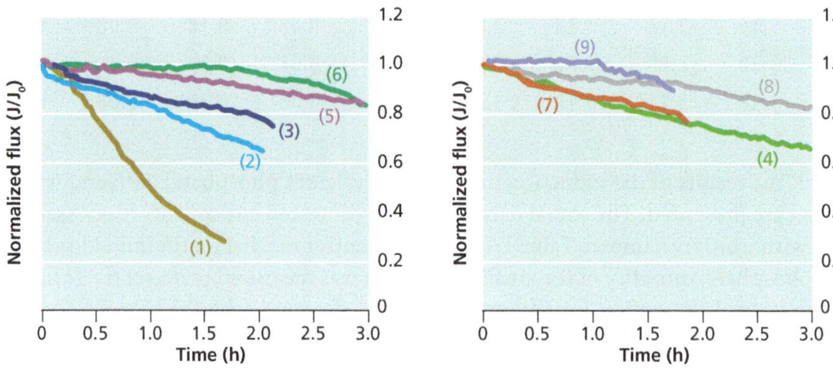

Figure 8 Normalized flux of the membrane element (of the once-through lab-scale RO setup) when fed with the artificial concentrate of 85 % recovery (1) without antiscalant addition, and with 5 mg/L of (2) AS–A, (3) AS–B, (4) AS–C, (5) AS–D, (6) AS–E, (7) AS–F, (8) AS–G, and (9) AS–H. Adopted from (Mangal, 2023)

In Figure 8, the normalized flux of the RO element when fed with artificial concentrate of 85 % recovery without antiscalant addition and with 5 mg/L of various antiscalants is shown. As can be seen, none of the antiscalants could completely prevent the deposition of calcium phosphate particles on the membrane surface since the permeate flux decreased in the presence of each antiscalant. However, it is evident that the rate of flux-decline decreased in the presence of antiscalants. Additionally, one can see that some antiscalants had better performance than others in slowing down the flux decline. For instance, the permeate flux decreased by 25 % with AS-A, approximately 17 % with each AS–B, AS–C and AS–F and about 7 % with each AS–D, AS–G and AS–H in 1.5 h, while no decrease was observed with AS–E in the same duration. However, after 3 h of operation, the permeate flux with AS–E decreased by approximately 15 %. The possible reasons for some antiscalants performing better than others might be due to (i) the formation of fewer particles (which needed longer time to foul the membrane) in the presence of such antiscalants, and/or (ii) less deposition of the formed particles due to the reduction in the deposition propensity of the particles by the antiscalants.

In brief, the results obtained with the once-through lab-scale RO revealed that the available calcium phosphate antiscalants are not effective against calcium phosphate scaling. Furthermore, the once-scale RO system is a useful tool for scaling studies and to identify the performance of antiscalants in preventing a given scaling compound. The once-through lab-scale RO system cannot only be limited to scaling studies, but it can also be implemented to investigate other types of membrane fouling such as organic fouling, biofouling and particulate fouling.

10.4 OUTLOOK AND FINAL COMMENTS

In this chapter, we presented the induction time protocol and the once-through lab-scale RO system for scaling-related research in full-scale RO installations. While there are benefits to both systems, there are limitations as well. The induction time measurements are really useful in investigating if the formation of a scaling specie (e.g., calcium carbonate, calcium phosphate, etc.) can be hindered in the presence of an antiscalant or any other substance (e.g., humic substances, etc.). However, as the induction time measurements are performed in a glass reactor and it doesn't have a membrane or filtration step, it is difficult to correlate its results with the scaling in RO. More precisely, if induction time for a scaling compound is found to be 1 h, it doesn't mean that the induction time (or the onset of crystallization) of that compound in RO installation would be 1 h as well. It could be that scaling in RO occurs much faster than in the glass reactor due to different material, hydrodynamics, etc.

On the other hand, the once-through lab scale RO system is equipped with an RO element and it has similar conditions/properties to an RO element in the full-scale RO system. Therefore, the scaling findings obtained with the once-through lab-scale RO system could be well representative to a full-scale RO plant. Furthermore, as the lab-scale RO system is a once-through system which means that the concentrate and permeate are not recirculated back to the RO feed. This makes the setup more attractive for research, as studies conducted with concentrate/permeate recirculation are often a matter of debate among researchers for not being representative to the conditions of full scale RO systems because of: (i) the residence time in a recycled system is much longer (in the range of hours) than the residence time (< 1 min) of the concentrate in the last stage of full scale RO plants, and (ii) recycling of concentrate back to the feed tank may accelerate the process of scaling as micro(crystals) may be formed. Furthermore, antiscalant manufacturers emphasize that antiscalants may not be as effective in recycled systems as they should be in once-through flow systems (like RO systems) and therefore the performance of antiscalants assessed in recycled systems may not be representative. The drawback of the once-through approach is that it requires an extensive amount of chemicals for longer tests. Nonetheless, by running experiments at greater saturation levels, which means rapid scaling, this problem can be addressed.

It is worth mentioning that the once-through lab-scale RO setup is not limited to laboratory-scale or pilot-scale experiments, as it has the potential for application in full-scale RO systems (as a scale monitor) in identifying the plant's maximum recovery as illustrated in Figure 9. A scale monitor is an additional RO element that is fed with the concentrate of the last stage of a full-scale RO installation, and because the scale monitor provides additional recovery, scaling occurs in the scale-guard prior to the final stage of the full-scale RO installation. The concept of the scale monitor (or scale-guard) is described in more detail elsewhere (van de Lisdonk *et al.*, 2000). Recently, the once-through lab-scale RO unit has been applied in the

RO installation in the Netherlands. The preliminary results (not shared here) are promising. However, the unit needs to be further tested and validated in full-scale RO plants to verify its application as a scale monitor.

This chapter is based on previously published data by Nasir Mangal and co-authors in his Ph.D. dissertation and also from work at IHE Delft. The findings of his dissertation have also been published in peer-reviewed journals (Mangal *et al.,* 2021a; Mangal *et al.*, 2021b; Mangal *et al.*, 2022a; Mangal *et al.*, 2022b).

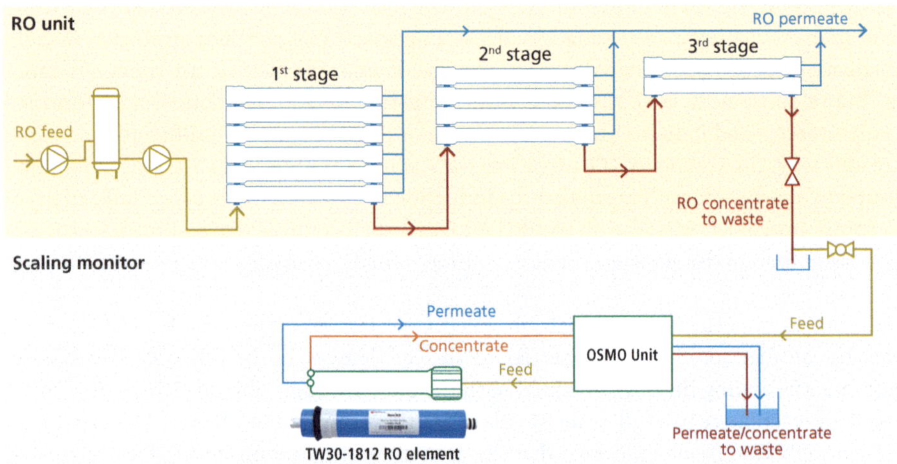

Figure 9 Application of the once-through lab-scale RO setup as a scaling monitor for the identification of maximum recoveries in full-scale RO installations

10.5 REFERENCES

Abdel-Aal EA, Rashad MM, El-Shall H (2004) Crystallization of calcium sulfate dihydrate at different supersaturation ratios and different free sulfate concentrations. Crystal Research and Technology 39: 313-321 DOI 10.1002/crat.200310188

Antony A, Low JH, Gray S, Childress AE, Le-Clech P, Leslie G (2011) Scale formation and control in high pressure membrane water treatment systems: A review. Journal of Membrane Science 383: 1-16 DOI http://dx.doi.org/10.1016/j.memsci.2011.08.054

Boels L, Witkamp G-J (2011) Carboxymethyl Inulin Biopolymers: A Green Alternative for Phosphonate Calcium Carbonate Growth Inhibitors. Crystal Growth & Design 11: 4155-4165 DOI 10.1021/cg2007183

Greenlee LF, Testa F, Lawler DF, Freeman BD, Moulin P (2010) The effect of antiscalant addition on calcium carbonate precipitation for a simplified synthetic brackish water reverse osmosis concentrate. Water research 44: 2957-2969 DOI http://dx.doi.org/10.1016/j.watres.2010.02.024

Guan X (2009) Kinetics Studies of Reactions at Solid-Liquid Interface: Simulation of Calcification VDM Verlag

Kashchiev D (2000) Nucleation Elsevier Science

Kucera J (2010) Reverse Osmosis Membrane Fouling ControlThe Science and Technology of Industrial Water Treatment:247-270.

Mangal MN (2023) Controlling scaling in groundwater reverse osmosis: minimizing antiscalant consumption Veenman, The Netherlands

Mangal MN, Salinas-Rodriguez SG, Blankert B, Yangali-Quintanilla VA, Schippers JC, van der Meer WGJ, Kennedy MD (2021) Role of phosphate and humic substances in controlling calcium carbonate scaling in a groundwater reverse osmosis system. Journal of Environmental Chemical Engineering 9: 105651 DOI https://doi.org/10.1016/j.jece.2021.105651

Mangal MN, Salinas-Rodriguez SG, Dusseldorp J, Blankert B, Yangali-Quintanilla VA, Kemperman AJB, Schippers JC, van der Meer WGJ, Kennedy MD (2022) Foulant Identification and Performance Evaluation of Antiscalants in Increasing the Recovery of a Reverse Osmosis System Treating Anaerobic Groundwater. Membranes 12: 290

Mangal MN, Salinas-Rodriguez SG, Dusseldorp J, Kemperman AJB, Schippers JC, Kennedy MD, van der Meer WGJ (2021) Effectiveness of antiscalants in preventing calcium phosphate scaling in reverse osmosis applications. Journal of Membrane Science 623: 119090 DOI https://doi.org/10.1016/j.memsci.2021.119090

Mangal MN, Yangali-Quintanilla VA, Salinas-Rodriguez SG, Dusseldorp J, Blankert B, Kemperman AJB, Schippers JC, Kennedy MD, van der Meer WGJ (2022) Application of a smart dosing pump algorithm in identifying real-time optimum dose of antiscalant in reverse osmosis systems. Journal of Membrane Science 658: 120717 DOI https://doi.org/10.1016/j.memsci.2022.120717

Mangal N, Salinas Rodriguez SG, Yangali Quintanilla VA, Schippers JC, Kennedy MD (2021) Ch 08 - Scaling. In: Salinas Rodriguez SG, Schippers JC, Amy GL, Kim IS, Kennedy MD (eds) Seawater Reverse Osmosis Desalination: Assessment and Pre-treatment of Fouling and Scaling, 1st edn:207-242.

Prisciandaro M, Lancia A, Musmarra D (2001) Calcium Sulfate Dihydrate Nucleation in the Presence of Calcium and Sodium Chloride Salts. Industrial & Engineering Chemistry Research 40: 2335-2339 DOI 10.1021/ie000391q

Shih W-Y, Gao J, Rahardianto A, Glater J, Cohen Y, Gabelich CJ (2006) Ranking of antiscalant performance for gypsum scale suppression in the presence of residual aluminum. Desalination 196: 280-292 DOI http://dx.doi.org/10.1016/j.desal.2006.04.001

Singh R (2005) Chapter 2 - Water and membrane treatment. In: Singh R (ed) Hybrid Membrane Systems for Water Purification:57-130.

Söhnel O, Mullin JW (1978) A method for the determination of precipitation induction periods. Journal of Crystal Growth 44: 377-382 DOI http://dx.doi.org/10.1016/0022-0248(78)90002-7

van Engelen G, Nolles R (2013) A sustainable antiscalant for RO processes. Desalination and water treatment 51: 921-923 DOI 10.1080/19443994.2012.700787

Verdoes D, Kashchiev D, van Rosmalen GM (1992) Determination of nucleation and growth rates from induction times in seeded and unseeded precipitation of calcium carbonate. Journal of Crystal Growth 118: 401-413 DOI http://dx.doi.org/10.1016/0022-0248(92)90089-2

Waly T (2011) Minimizing the Use of Chemicals to Control Scaling in Sea Water Reverse Osmosis: Improved Prediction of the Scaling Potential of Calcium Carbonate CRC Press/Balkema. Leiden. Netherlands

Waly T, Kennedy MD, Witkamp G-J, Amy G, Schippers JC (2012) The role of inorganic ions in the calcium carbonate scaling of seawater reverse osmosis systems. Desalination 284: 279-287 DOI http://dx.doi.org/10.1016/j.desal.2011.09.012

Waly T, Munoz R, Kennedy MD, Witkamp GJ, Amy G, Schippers JC (2010) Effect of particles on the induction time of calcium carbonate in synthetic SWRO concentrate. Desalination and water treatment 18: 103-111 DOI 10.5004/dwt.2010.1351

Yu W, Song D, Chen W (2020) Antiscalants in RO membrane scaling control. Water research 183: 115985 DOI 10.1016/j.watres.2020.115985

Part 4

Organic fouling

doi: 10.2166/9781789062977_0249

Chapter 11

Practical Considerations of Using LC-OCD for Organic Matter Analysis in Seawater

Barun Lal Karna, Helen Rutlidge, Rita Kay Henderson, Pierre Le-Clech,

University of New South Wales, Australia

The learning objectives of this chapter are the following:

- Understand basic principles of LC-OCD measurement
- Assess the impact of salinity on LC-OCD analysis
- Appreciate the benefits and limitations of LC-OCD measurement for marine samples

11.1 INTRODUCTION

Reverse osmosis (RO) is a well-established treatment process used for the production of drinking water from brackish water and seawater. However, RO systems suffer from fouling of the membrane by dissolved solids, micro-organisms, and dissolved organic matter (Matin *et al.*, 2011; Prihasto *et al.*, 2009). While an integrated membrane system approach involving pre-treatment options has been partially successful in controlling most particulate fouling problems, the more persistent problems of organic and biological fouling persists, influencing the operation and maintenance costs (Prihasto *et al.*, 2009).

Many studies have investigated the characterisation of organic matter (OM) in surface water (Fan *et al.*, 2001; Kennedy *et al.*, 2005; Matilainen *et al.*, 2011; Swietlik *et al.*, 2004); however, the lower concentration of organic matter in seawater initially presented some analytical challenges in performing OM characterisation in seawater (Spyres *et al.*, 2000). In seawater, organics are mainly produced by micro-organisms rather than discharge from industrial effluents, municipal wastewater, runoff rainwater and naturally decaying vegetation (Azam *et al.*, 1983; Cai *et al.*, 2012; Parsons and Strickland, 1961). Both bacteria and algae release neutral and charged polysaccharides, proteins and other organic compounds including lipids, nucleic acids into the seawater environment (Voutchkov, 2008). As OM comprises a heterogeneous mix of molecules at very low concentrations in seawater, high-resolution techniques are necessary to qualify and quantify the marine organic fractions. Furthermore, the high concentrations of Na^+ and Cl^- ions (together ~85%), along with other species of ions (Ca^{2+}, Mg^{2+}, SO_4^{2-}, K^+, HCO_3^-, CO_3^{2-}, Br^- etc.) at low concentration (altogether ~15%) in the seawater matrix adds an additional challenge to their analytical characterisation (Mopper *et al.*, 2007).

Many techniques, such as fluorescence spectroscopy, UV-vis spectroscopy, infrared spectroscopy, nuclear magnetic resonance (NMR), size exclusion chromatography (SEC) have been used to analyse the composition of OM of seawater. The main advantages and disadvantages associated with these techniques are summarised in Table 1. The minimal sample preparation, and range of quantitative composition information obtained from SEC makes it an ideal technique for OM analysis in seawater.

SEC is a widespread method that allows the separation of macromolecular complexes, such as proteins and carbohydrate polymers OM, by exploiting the size and chemical functionality (Mecozzi *et al.*, 2001). Liquid chromatography-organic carbon detection (LC-OCD) is an advanced technique that uses SEC principles and is well suited to measure organic compounds present in a range of aquatic water samples (Huber *et al.*, 2011). This established technique provides wide-ranging information regarding the nature of the organic matter [e.g., size (molecular weight distribution), structure (aromatic or aliphatic), and functionality (charge density i.e., humic and fulvic acids)] with minimal sample preparation (Filloux *et al.*, 2012; Henderson *et al.*, 2011). LC-OCD is a proprietary technique and there is only one supplier of the instrumentation and the associated integration software (DOC-Labor GmbH, 2023).

Table 1 Advantages and disadvantages of analytical techniques of organic carbon for the analysis of marine samples.

Analytical techniques	Advantages	Disadvantages	References
Infrared spectroscopy	- comprehensive structural information. - quick analysis. - accurate wave number.	- requires the isolation of OM from abundant salt from seawater. - the influence of the refractive index of the surrounding environment on the analyses. - only qualitative determination of lipids and hydrocarbons for algae.	(Chon *et al.* 2020; Lee *et al.* 2020; Al-Juboori and Yusaf, 2012)
Nuclear magnetic resonance (NMR)	- simple and fast method for samples with sufficient concentration. - comprehensive structural information. - a non-destructive and non-invasive mean to obtain information regarding the metabolic pathways.	- requires the isolation of OM from abundant salt from seawater. - low sensitivity with nominal detection limits in the millimolar range.	(Al-Juboori and Yusaf 2012; Benner *et al.* 1992; Mopper *et al.* 2007; Jeong *et al.* 2013)
UV-vis spectroscopy	- quick test, - requires minimal sample preparation	- detects aromatic compounds only. - qualitative composition of humic substance.	(Chon *et al.* 2020; Lee *et al.* 2020; Murphy *et al.* 2008; Liu *et al.* 2022
Fluorescence spectroscopy	- high level of sensitivity, - quick test. - requires minimal sample preparation. - characterises marine and terrestrial DOM. - can study the aggregation mechanism of OM, and OM dynamics with the use of flow through cells.	- does not detect aliphatic compounds. - cannot differentiate between overlapping fluorescing components.	(Lee *et al.* 2020; Murphy *et al.* 2008; Liu *et al.* 2022; Liu *et al.* 2023)
Size exclusion chromatography (SEC)	- requires minimal sample preparation. - qualitative and quantitative determination. - separation of organic fractions based on molecular weights. - determination of fatty acid, analysis of carbohydrates and amino acids. - lower temperature during analysis.	- a relative molecular weight technique. - column must be calibrated with polymer standards of known molecular weight, data acquisition and processing is critical.	(Lee *et al.* 2020; Valladares Linares *et al.* 2012; Jeong *et al.* 2022; Tansakul *et al.*, 2011)

The principles behind organic fractionation by LC-OCD are based on three separation processes namely: size exclusion, ion interaction and hydrophobic interaction. Since OM is very heterogeneous, size exclusion is the most dominant mechanism of separation. Chromatography by size exclusion uses the difference in speed of diffusion for smaller and larger molecules. The stationary phase is a packing of porous beads which allows smaller molecules to diffuse into the bead interior. Consequently, large molecules travel faster than small molecules. According to molecular size/weight and polarity, the column separates OM into five fractions: (a) biopolymers (BP), comprising polysaccharides and proteins, (b) humic substances (HS) such as humic and fulvic acids, (c) building blocks (BB) such as hydrolysates of humics, (d) low molecular weight acids (LMWA) such as fraction for all aliphatic low-molar-weight organic acids and (e) low molecular weight neutrals (LMWN) such as alcohols, aldehydes, ketones and amino acids (Huber *et al.*, 2011). A sixth fraction, hydrophobic organic carbon (HOC) is determined as the difference between the total DOC (that passes through a bypass column) and the total amount that elutes from the chromatographic column. These fractions were originally identified and named based on surface water analysis by Huber *et al.* (2011) and have become the conventional labels for the fractions which will be followed in this work, however it is acknowledged that these labels may not be the most appropriate for different water types, including seawater samples. Due to the low concentration of organics in seawater, and the saline matrix that is challenging for traditional single column LC-OCD, the use of dual column LC-OCD has become more common for seawater samples, which is explored further in Section 11.2.2.

In this chapter, the impact of salinity on LC-OCD is explored and some analytical considerations for the use of LC-OCD for saline samples is discussed, including the reproducibility, reliability and sensitivity of LC-OCD in saline environments. Previous applications are also explored to show the potential of LC-OCD for organic characterisation in RO applications.

11.2 LC-OCD ANALYSIS

11.2.1 Instrumentation and chromatogram integration

A LC-OCD instrument uses a phosphate buffer solution that is irradiated in a UV batch reactor to eliminate organic impurities prior to sample addition. A mobile phase (buffer solution at pH of 6.85 requiring 2.5 g KH_2PO_4 + 1.5 g $Na_2HPO_4 \cdot 2H_2O$ to 1 L) is delivered at a flow rate of 1.1 mL/min to an autosampler with 1 mL injection volume for samples with a concentration higher than 1 mgC/L (Huber *et al.*, 2011); otherwise, an injection volume up to a maximum of 4 mL can be used. Original systems used a TSK HW-50S (Toso, Japan) chromatographic column for separation, however due to a change in manufacturing this column is no longer suitable for LC-OCD systems. The first detector after chromatographic separation is nondestructive, fixed wavelength UV-detection (UVD 254 nm), after which the sample passes through an organic carbon detector (OCD) and a dissolved organic nitrogen (DON) detector (Huber and Frimmel, 1991). The OCD uses a Grantzel thin-film reactor in which organics are oxidized to CO_2 by UV before measurement by infrared detection.

Quantification of the fractions is conducted by area integration of chromatograms with ChromCALC software (DOC-LABOR, Germany) (Huber *et al.*, 2011). This software has limitation that raw data integration is done either assuming sample contains humic substances or without humic substances (biopolymers). Evaluation of chromatograms containing HS should be assessed by checking that there is a corresponding humic peak in the UVD chromatogram. The area under the curves is calculated as area units (AU) and then is converted into corresponding concentration in μg/L (ppb) using calibration curves where the calibration of OCD, UVD and OND are undertaken with a mixture of standards comprising potassium hydrogen phthalate and potassium nitrate. More details of the data integration process can be obtained from the literature (Huber *et al.*, 2011).

In addition to the quantitative assessment of concentration by integration, the aromaticity and nominal molecular weight of the HS fraction is calculated, and these values can be plotted in what is known as a HS-diagram (Huber and Frimmel, 1991). This HS-diagram includes a range of water samples, including HS standards of the International Humics Substances Society (IHSS) and natural surface waters, and this is known as the humification pathway, with transitions from aquagenic fulvic acids in the lower left of the diagram to pedogenic fulvic acids to pedogenic humic acids in the upper right of the diagram (Huber *et al.*, 2011). Seawater samples tend to have values in the ranges 0-1.2 L/(mg·m) for aromaticity and 430-610 g/mol nominal molecular weight of the HS fraction.

As an example of the ChromCALC software, the chromatograms of OCD, UVD and OND from a mixture of model compounds are shown in Figure 1 as black, blue and green lines respectively. The model compounds were selected to simulate OM found in seawater and represent the different LC-OCD fractions as per Table 2.

Table 2 Composition of DOM model mixture

Model Compound	LC-OCD Fraction	Concentration (mgC/L)
Bovine serum albumin (BSA)	BP (protein)	0.76
Alginate	BP (polysaccharide)	0.11
Xanthan	BP (polysaccharide)	0.05
Humic acid	HS	0.48
Tryptophan	LMWN	7×10^{-5}
Oxalic acid	LMWA	0.06

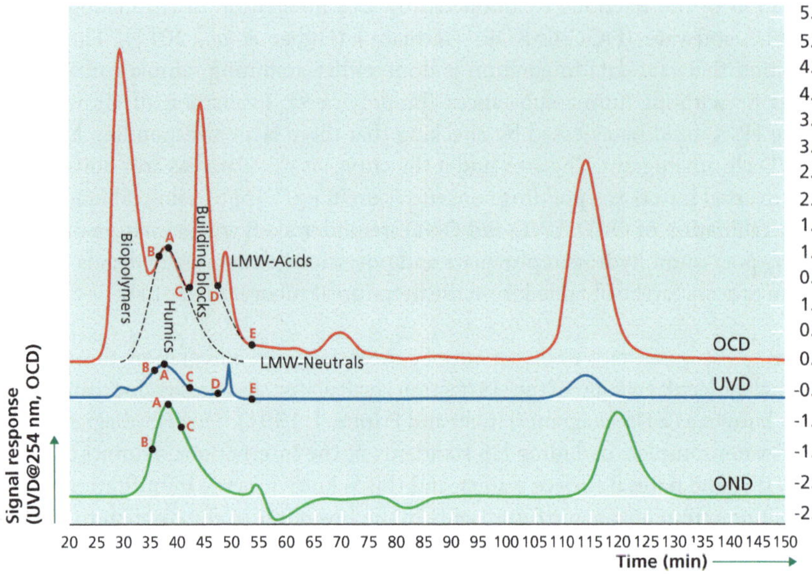

Figure 1 Chromatograms of OCD, UVD and OND for mixture of six model compounds with Milli-Q®, using single column LC-OCD, with each fraction identified. The letters (A-E) represent the various points that are set by the operator as part of the integration, where A is humics peak max., B is humics left slope, C is humics right slope (division between humics and building blocks), D is division between building blocks and acids, and E is division between LWN-acids and LMW-neutrals.

11.2.2 Effect of salinity on organic characterization and calibration

There are known interferences from salts that can impact the LC-OCD calibration and OM characterization in saline matrices (Baghoth *et al.*, 2008; Huber and Frimmel, 1994). The salt content in seawater samples has been shown to reduce the resolution of the individual fractions (Baghoth *et al.*, 2008). To overcome this issue, the use of two chromatographic columns in series (dual LC-OCD) connected in series has been suggested. To better illustrate the impact of the duplication of columns, a comparison between single and dual LC-OCDs for a sample containing model compounds in a NaCl (32 g/L) matrix is shown in Figure 2 (Karna, 2014). The use of dual column leads to longer retention times, and this allows the resolution to be improved with the HS, BB and LMWA peaks clearly separated. As seawater samples often contain low levels of organics, it is therefore recommended to use a dual column LC-OCD to obtain better qualitative and quantitative analysis (Dulaquais *et al.* 2018; Jeong *et al.*, 2016; Karna, 2014).

The LMWN region of a LC-OCD chromatogram can be impacted due to the presence of salts in a sample. It is well understood that chloride causes an interference in nitrate quantification using UV detection (Sah, 1994). This prevents the UVD chromatogram being used to determine nitrate or total nitrogen in salty water. However, there is no salt interference in the BP or HS regions of the OND chromatogram, and the organic nitrogen content can still be accurately determined by LC-OCD (Dulaquais *et al.*, 2018). The LMWN region can also be influenced by salt-induced 'column bleeding', where the salts allow the partial elution of the accumulated hydrophobic material (from previous samples) from the column (Huber and Frimmel, 1994). This can lead to the measured negative concentrations of the HOC fraction that have previously been observed (Karna, 2014).

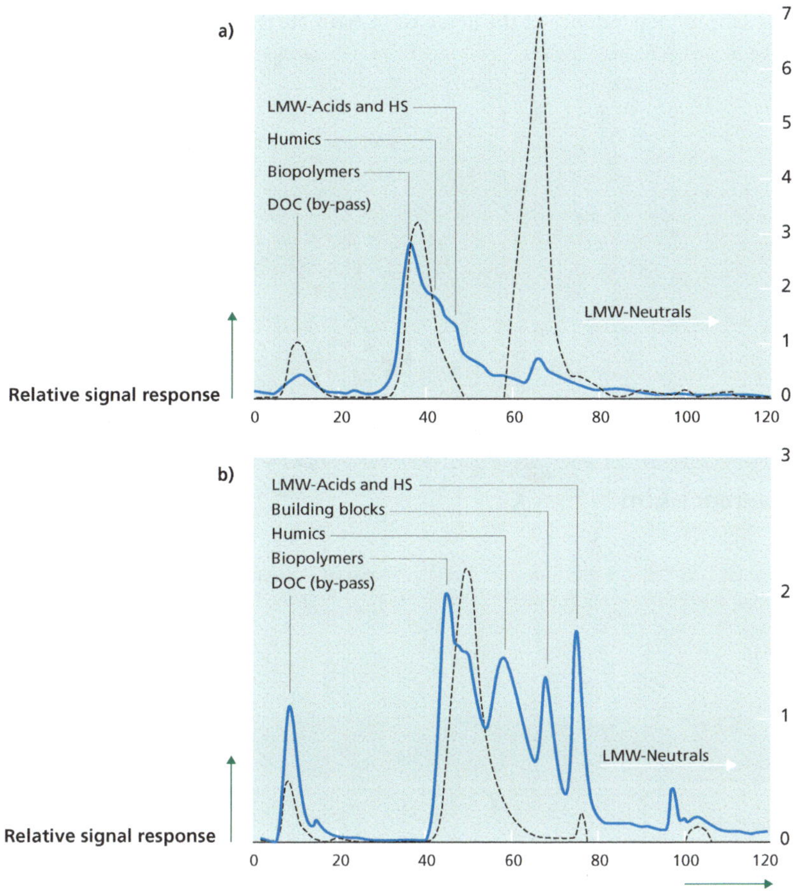

Figure 2 Typical spectrum OCD (blue line) and OND (dotted line) from LC-OCD of model mixture with NaCl (32 g/L) a) in single column and b) in dual column. Figure from (Karna, 2014).

The effect of salinity on the calibration of the individual fractions was investigated using the same model compounds from Section 11.2.1 (Karna, 2014). Mixtures of the model compounds at different concentrations were prepared in Milli-Q® water and NaCl and the calibration in the different matrices was compared, as shown in Figure 3 and Table 3. There were no significant differences observed between the slope of the calibration lines for all model compounds in Milli-Q® water, except for BSA, which was attributed to the poorer oxidation efficiency of the Granztel reactor for the higher molecular weight compounds (Lankes *et al.*, 2009; Li *et al.*, 2019). When the model compounds were mixed in saline solution, there was significant decrease of the gradient observed for humic acid, in addition to BSA. It was suggested that this decrease in the humic acids in the presence of NaCl was due to microflocculation/precipitation of humic substances and subsequent capture on the in-line filter before passing through the LC-OCD column. Dulaquais *et al.* (2018), also found calibrations differences due to salinity and recommended calibrating the LC-OCD system with the same salinity as the samples (if the samples have similar salinity levels) or determining the salinity dependence of the detector to estimate the adjusted calibrations.

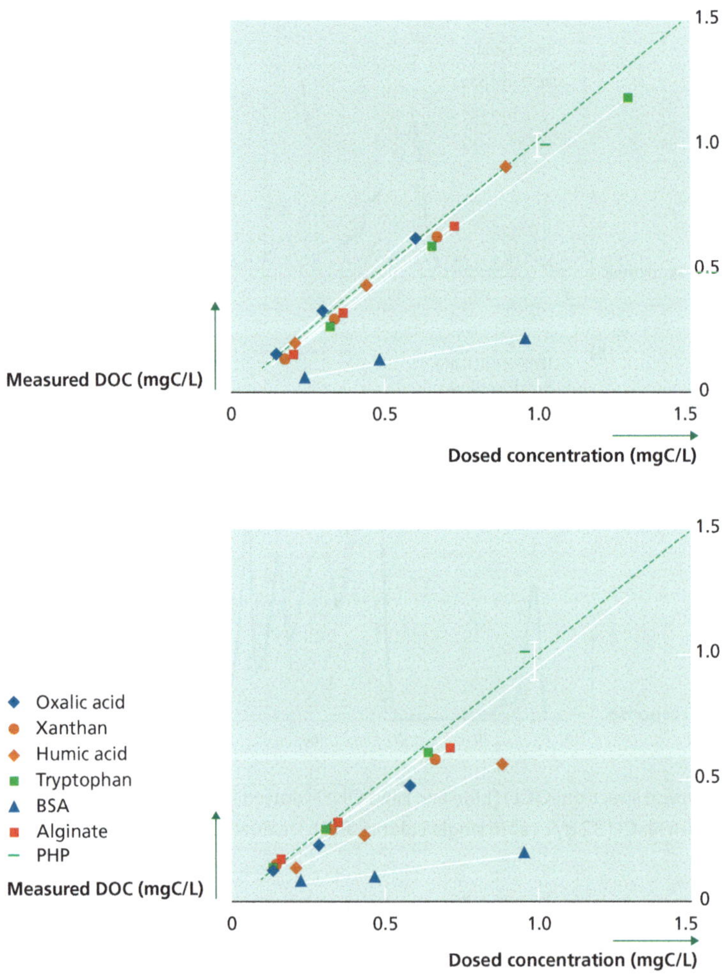

Figure 3 Calibration of individual model compounds measured from chromatographic peak, compared with PHP standard (dotted line in green) in (a) Milli-Q® water (single) and (b) NaCl (32g/L) solution (dual column).

Table 3 Theoretical and measured carbon content of model compounds assessed in freshwater (Milli-Q®) and saline (NaCl) matrix

Model compounds	Chemical formulae	Theoretical Carbon content mg/L	Average carbon content measured from LC-OCD (mg/L)	
			in Milli-Q® (single)	in NaCl (dual)
BSA (BP)	n/a	n/a	0.23	0.18
Alginate (BP)	$(NaC_6H_7O_6)_n$*	0.73	0.67	0.61
Xanthan (BP)	$(C_{35}H_{49}O_{29})_n$*	0.90	0.63	0.56
Humic acid (HS)	n/a	n/a	0.91	0.55
Tryptophan (LMWN)	$H_{12}C_{11}O_2N_2$	1.29	1.19	1.22
Oxalic acid (LMWA)	$H_2C_2O_4$	0.53	0.63	0.46

*assuming n=1

11.2.3 LEVEL OF DETECTION

The instrument detection limit (IDL) and lower level of detection (LLD) of organic compounds have been assessed in two studies one using model compounds (Karna, 2014) and different types of blank samples (Dulaquais *et al.*, 2018), shown in Table 4. In each study, to determine the IDL, ten samples of Milli-Q® water as blank were assessed through LC-OCD and the IDL was calculated as three times the standard deviation of the blank sample. In addition, Dulaquais *et al.* (2018) also assessed the IDL for a seawater blank, through using an UV-irradiated seawater sample. Both studies had a similar IDL for total DOC with Milli-Q® water of 20 and 30 µg/L, and the LLD were similar for most samples, except for alginate and tryptophan, this could be due to lower oxidation efficiencies for higher molecular weight compounds and N-heterocyclic compounds (Li *et al.*, 2019). For the Milli-Q® samples the IDL and LLD for the BP, HS, BB and LMWA were all below 10 µg/L, which indicates that the quantification of these fractions is accurate, even at low concentrations. The higher IDL and LLD for the saline matrix was due to salt interference, which is discussed in Section 11.2.5.

11.2.4 REPRODUCIBILITY OF LC-OCD

The reproducibility of LC-OCD for the low concentration of organics present in seawater has been assessed in a couple of studies, and the results are given in Table 3 (Dulaquais *et al.*, 2018; Karna, 2014). For the Dulaquais *et al.*, (2018) study, the reproducibility was assessed using 10 non-consecutive analyses of a coastal seawater sample; while in the Karna (2014) study, triplicate analysis of model compounds were used. The reproducibility for all fractions was lower than 10%, indicating that LC-OCD can provide reasonably accurate quantification OM. The %RSD values were higher for the model compounds when NaCl was added, indicating the impact of a saline matrix.

Table 4 Instrument detection limit (IDL) and lower level of detection (LLD) for LC-OCD, using Milli-Q® water, and UV-irradiated seawater (Bay of Brest, salinity = 35 g/L). Data taken from (Dulaquais *et al.*, 2018; Karna, 2014). N.B. for the Karna study (Karna, 2014) the LLD was only reported for total DOC instead of the relevant fraction.

Sample	DOC	BP	HS	BB	LMWN	LMWA	Reference
Milli-Q® water	20	2	2	1	10	2	[1]
Irradiated seawater	104	1	1	0	96	6	[1]
Milli-Q® water	30	N/A	N/A	N/A	N/A	N/A	[2]
LLD (μg/L)							
Milli-Q® water	30	5	5	4	19	5	[1]
Irradiated seawater	131	5	3	1	131	18	[1]
BSA (BP)	40	N/A	N/A	N/A	N/A	N/A	[2]
Alginate (BP)	90	N/A	N/A	N/A	N/A	N/A	[2]
Xanthan (BP)	30	N/A	N/A	N/A	N/A	N/A	[2]
Humic acid (HS)	30	N/A	N/A	N/A	N/A	N/A	[2]
Tryptophan (LMWN)	80	N/A	N/A	N/A	N/A	N/A	[2]
Oxalic acid (LMWA)	40	N/A	N/A	N/A	N/A	N/A	[2]

[1] Dulaquais *et al.*, 2018, [2] Karna, 2014

Table 5 Reproducibility for LC-OCD, reported as % relative standard deviation (RSD), using coastal seawater (Bay of Brest, salinity = 35 g/L) and model compounds. BDL indicates that signal was below the detection limit. Data taken from (Dulaquais *et al.*, 2018; Karna, 2014). N.B. for the Karna study (Karna, 2014) the reproducibility was only reported for total DOC instead of relevant fraction.

	Reproducibility (% RSD)						
Sample	DOC	BP	HS	BB	LMWN	LMWA	Study
Sample	DOC	BP	HS	BB	LMWN	LMWA	Study
Coastal seawater	2.9	8.3	2.5	7.5	5.2	BDL	[1]
Alginate (BP)	1.6	0.7	BDL	BDL	BDL	BDL	[2]
Humic acid (HS)	0.9	BDL	0.5	BDL	BDL	BDL	[2]
Alginate (BP) + NaCl	3.3	N/A	N/A	N/A	N/A	N/A	[2]
Humic acid (HS) + NaCl	3.5	N/A	N/A	N/A	N/A	N/A	[2]
Tryptophan (LMWN) + NaCl	12.7	N/A	N/A	N/A	N/A	N/A	[2]
Oxalic acid (LMWA) + NaCl	1.4	N/A	N/A	N/A	N/A	N/A	[2]

[1] Dulaquais *et al.*, 2018, [2] Karna, 2014

11.2.5 CHARACTERISATION OF ORGANIC MIXTURES

The impact of the salt matrices on the organic characterization by LC-OCD has been previously studied, using a mixture of 3 mg/L of six model compounds (as listed in Table 4, and prepared in equal proportion, i.e., 0.5 mg/L by weight) in Milli-Q® and saline matrixes (NaCl, NaCl with added Ca^{2+} and Mg^{2+}, and Red Sea salt) (Karna, 2014). The concentration of NaCl and red sea salt was 32 g/L. There was no significant difference observed between the fraction concentrations prepared in Milli-Q® and NaCl. However, a significant decrease of BP, HS and BB were observed for the red sea salt and NaCl with the added ions in comparison with the NaCl solution alone, while the LMWN and LMWA concentration increased. These differences were attributed to the presence of divalent ions in red sea salt solution, such as Ca^{2+}, which promoted the formation of organic (especially alginate) complexes (Kye *et al.*, 2021) that were potentially retained on the membrane surface (0.45 μm filter) before entering the LC-OCD chromatographic columns. In addition, divalent cations interact specifically with humic carboxyl functional groups and, thus, substantially reducing the humic charge and the electrostatic repulsion between humic macromolecules. Reduced organic interchain repulsion can result in increased organics deposition on the membrane surface (0.45 μm filter) and formation of a densely packed fouling layer before entering the LC-OCD chromatographic columns (Hong and Elimelech, 1997). Therefore, it is important to note that potentially some natural seawater organics will not be measured by LC-OCD as they would be removed prior to entering the LC-OCD column due to fouling of the filter and/or formation of complexes.

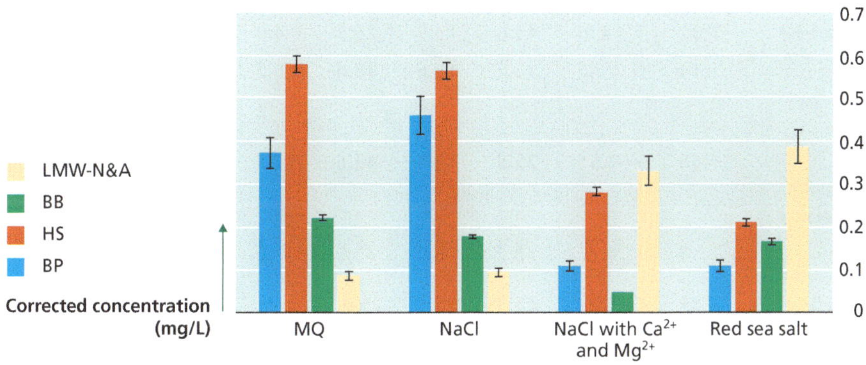

Figure 4 Effect of background matrix on quantification of organic fractions for model mixture in Milli-Q® (single column), in 32 g/L NaCl (dual column), NaCl with added divalent ions (dual column) and red sea salt solution (dual column).

11.2.6 Applications

LC-OCD has been applied to brackish and seawater samples, often with other characterization techniques, with the aim to improve the effectiveness of treatment processes. This has included investigating the fouling behaviour of organic matter in RO and other membrane processes (Fortunato *et al.*, 2020; Jeong *et al.*, 2016; Park *et al.*, 2019; Yin *et al.*, 2019), and the effectiveness of pretreatment techniques in the removal of organic matter (Fortunato *et al.*, 2020; Jeong *et al.*, 2016; Kye *et al.*, 2021; Simon, Rudé *et al.*, 2013; Simon, Penru *et al.*, 2013).

11.2.6.1 OM composition in seawater

In seawater samples, HS and LMWN tend to be the most abundant fractions, with BP and LMWA present in the lowest concentrations (Table 5). Jeong *et al.*, (2016) reported differences in the DOM concentration due to an increase in LMWN in the warm season compared to the cool season. This increase was attributed to increased algal/microbial action in the surface seawater due to higher temperatures which is then photodegraded to LMWN (Jeong *et al.*, 2016).

Due to differing aromaticity and nominal molecular weight values there was grouping of samples based different sea areas at the Korean Peninsula, enabling them to be differentiated on the HS-diagram (Kye *et al.*, 2021).

Table 6 DOM composition from LCOCD for different seawater samples. Low Molecular Weight (LMW) represents the combined LMWN and LMWA fractions.

Sample	DOC (mgC/L)	BP (%)	HS (%)	BB (%)	LMWN (%)	LMWA (%)	LMW (%)	Ref
Rottnest Island, Western Australia	0.98	12.2	52.0	8.2	26.5	1.0	27.5	Rutlidge *et al.*, 2021
Sydney Harbour	1.01	14.9	41.6	13.9	26.7	4.0	30.7	Karna, 2014
NW Mediterranean Sea	1.14	5.6	33.2	17.6	38.4	5.2	43.6	Simon, Rudé *et al.*, 2013
NW Mediterranean Sea (Barcelona)	1.40	7.5	25.9	15.8	45.6	5.2	50.8	Simon, Penru *et al.*, 2013
North Sea	1.35	4.7	38.2	16.0	33.9	7.2	41.1	Salinas, 2011
Singapore	1.18	6.7	42.1	15.5	NR	NR	30.2	Yin *et al.*, 2019
Red Sea	1.33	6.8	10.5	1.5	81.2	NR		Fortunato *et al.*, 2020
Kwinana, Perth, Western Australia	1.50	5.9	34.0	NR	50.3	NR		Jeong *et al.*, 2016
Korean Peninsula	2.02	NR	36.1	NR	32.5	NR		Kye *et al.*, 2021

NR: not reported

11.2.6.2 Fouling behaviour of organic matter

Organic constituents of extracted foulants from RO autopsies have been investigated by LC-OCD, with different fractions being identified. For a specific Australian study, the main organic foulants on the cartridge filters were BP and HS, while the dominant fractions in the incoming raw seawater, HS and LMWN, were the main organic foulants for RO (Jeong *et al.*, 2016). While another study from a desalination plant located on the Red Sea found that LMWN was the main organic foulant for RO, followed by either HS or BP depending on the position in the module (Fortunato *et al.*, 2020). However, Yin *et al.* (2019) found that BP followed by LMWN had a greater impact on RO membrane fouling.

11.2.6.3 Effectiveness of pretreatment methods

LC-OCD has also been used to assess the effectiveness of various pretreatment methods for RO of seawater as per some examples below:

- Ozonation reduced the aromaticity and molecular weight of the OM in seawater samples, with the aromaticity rapidly decreasing and then stabilising after 12 hr (Kye *et al.*, 2021).
- The use of a biofilter in Mediterranean desalination pilot plant resulted in the reduction of the LMWN fraction, followed by BB and BP fractions (Simon *et al.*, 2013).
- The existing pretreatment process (i.e., coagulation, followed by dual media filtration and cartridge filtration) at a Perth Seawater Desalination Plant presented only small removal of OM, with approximately 16% DOM removal, which was predominately the larger BP and HS, as expected (Jeong *et al.*, 2016).

11.3 CONCLUSIONS

LC-OCD is a suitable technique for measuring the organic matter concentration and composition in seawater samples with minimal sample preparation. LC-OCD can reliably measure the low concentrations of organics generally present in seawater, with the concentration of individual fractions being well above the LLD. However, due to the fouling of the filter and the formation of complexes a small portion of natural seawater OM may not be measured by LC-OCD. Due to improved resolution of the chromatogram, it is recommended to use dual chromatographic columns than a single column for saline samples.

11.4 REFERENCES

Al-Juboori, Raed A., and Talal Yusaf. 2012. 'Biofouling in RO system: Mechanisms, monitoring and controlling', Desalination, 302: 1-23.

Azam, F, T Fenchel, JG Field, JS Gray, LA Meyer-Reil, and F Thingstad. 1983. 'The Ecological Role of Water-Column Microbes in the Sea', Marine Ecology Progress Series, 10: 257-63.

Baghoth, S. A., S. K. Maeng, S. G. Salinas Rodríguez, M. Ronteltap, S. Sharma, M. Kennedy, and G. L. Amy. 2008. 'An urban water cycle perspective of natural organic matter (NOM): NOM in drinking water, wastewater effluent, storm water, and seawater', Water Science & Technology, 8: 701-07.

Benner, Ronald, J. Dean Pakulski, Matthew McCarthy, John I. Hedges, and Patrick G. Hatcher. 1992. 'Bulk Chemical Characteristics of Dissolved Organic Matter in the Ocean', Science, 255: 1561-64.

Cai, Yihua, Laodong Guo, Xuri Wang, Allison K. Mojzis, and Donald G. Redalje. 2012. 'The source and distribution of dissolved and particulate organic matter in the Bay of St. Louis, northern Gulf of Mexico', Estuarine, Coastal and Shelf Science, 96: 96-104.

Chon, Kangmin, Namjo Jeong, Hojung Rho, Joo-Youn Nam, Eunjin Jwa, and Jaeweon Cho. 2020. 'Fouling characteristics of dissolved organic matter in fresh water and seawater compartments of reverse electrodialysis under natural water conditions', Desalination, 496: 114478.

DOC-Labor GmbH. 2023. 'Screening of Organics in Natural and Technical Waters', Accessed 9/07/2023. http://doc-labor.de/.

Dulaquais, Gabriel, Johann Breitenstein, Matthieu Waeles, Rémi Marsac, and Ricardo Riso. 2018. 'Measuring dissolved organic matter in estuarine and marine waters: size-exclusion chromatography with various detection methods', Environmental Chemistry, 15: 436-49.

Fan, Linhua, John L. Harris, Felicity A. Roddick, and Nic A. Booker. 2001. 'Influence of the characteristics of natural organic matter on the fouling of microfiltration membranes', Water Research, 35: 4455-63.

Filloux, E., J. Labanowski, and J. P. Croue. 2012. 'Understanding the fouling of UF/MF hollow fibres of biologically treated wastewaters using advanced EfOM characterization and statistical tools', Bioresource Technology, 118: 460-68.

Fortunato, Luca, Abdullah H. Alshahri, Andreia S. F. Farinha, Islam Zakzouk, Sanghyun Jeong, and TorOve Leiknes. 2020. 'Fouling investigation of a full-scale seawater reverse osmosis desalination (SWRO) plant on the Red Sea: Membrane autopsy and pretreatment efficiency', Desalination, 496: 114536.

Henderson, Rita K., Nashida Subhi, Alice Antony, Stuart J. Khan, Kathleen R. Murphy, Greg L. Leslie, Vicki Chen, Richard M. Stuetz, and Pierre Le-Clech. 2011. 'Evaluation of effluent organic matter fouling in ultrafiltration treatment using advanced organic characterisation techniques', Journal of Membrane Science, 382: 50-59.

Hong, Seungkwan, and Menachem Elimelech. 1997. 'Chemical and physical aspects of natural organic matter (NOM) fouling of nanofiltration membranes', Journal of Membrane Science, 132: 159-81.

Huber, S. A., and F. H. Frimmel. 1994. 'Direct Gel Chromatographic Characterization and Quantification of Marine Dissolved Organic Carbon Using High-Sensitivity DOC Detection', Environ Sci Technol, 28: 1194-7.

Huber, Stefan A., Andreas Balz, Michael Abert, and Wouter Pronk. 2011. 'Characterisation of aquatic humic and non-humic matter with size-exclusion chromatography – organic carbon detection – organic nitrogen detection (LC-OCD-OND)', Water Research, 45: 879-85.

Huber, Stefan A., and Fritz H. Frimmel. 1991. 'Flow injection analysis for organic and inorganic carbon in the low-ppb range', Analytical Chemistry, 63: 2122-30.

Jeong, Ganghyeon, Hyeonho Lee, Chang-Min Kim, and Am Jang. 2022. 'Size-dependent transport and fouling formation of organic matters in a pilot-scale PFFO–RO hybrid system for real wastewater treatment', Journal of Cleaner Production, 361: 132233.

Jeong, Sanghyun, Sung-Jo Kim, Chang Min Kim, Saravanamuthu Vigneswaran, Tien Vinh Nguyen, Ho-Kyong Shon, Jaya Kandasamy, and In S. Kim. 2013. 'A detailed organic matter characterization of pretreated seawater using low pressure microfiltration hybrid systems', Journal of Membrane Science, 428: 290-300.

Jeong, Sanghyun, Gayathri Naidu, Robert Vollprecht, TorOve Leiknes, and Saravanamuthu Vigneswaran. 2016. 'In-depth analyses of organic matters in a full-scale seawater desalination plant and an autopsy of reverse osmosis membrane', Separation and Purification Technology, 162: 171-79.

Karna, Barun Lal. 2014. 'Advanced characterisation techniques to assess seawater organic matter removal by dissolved air flotation (DAF)', UNSW Australia.

Kennedy, Maria D., Hyoung K. Chun, Victor A. Quintanilla Yangali, Bas G. J. Heijman, and Jan C. Schippers. 2005. 'Natural organic matter (NOM) fouling of ultrafiltration membranes: fractionation of NOM in surface water and characterisation by LC-OCD', Desalination, 178: 73-83.

Kye, Homin, Kiho Kim, Youmi Jung, Yirga Weldu Abrha, Seong-Nam Nam, Il-hwan Choi, Joon-Wun Kang, and Yeojoon Yoon. 2021. 'Characterization of marine dissolved organic matter and its effect on ozonation', Chemosphere, 277: 130332.

Lankes, Ulrich, Margit B. Müller, Matthias Weber, and Fritz H. Frimmel. 2009. 'Reconsidering the quantitative analysis of organic carbon concentrations in size exclusion chromatography', Water Research, 43: 915-24.

Lee, Yong-Gu, Sangwon Kim, Jaegwan Shin, Hojung Rho, Younggeun Lee, Young Mi Kim, Yongeun Park, Sang-Eun Oh, Jaeweon Cho, and Kangmin Chon. 2020. 'Fouling behavior of marine organic matter in reverse osmosis membranes of a real-scale seawater desalination plant in South Korea', Desalination, 485: 114305.

Li, X., N. R. H. Rao, K. L. Linge, C. A. Joll, S. Khan, and R. K. Henderson. 2019. 'An evaluation of measurement techniques for algal-derived organic nitrogen', Water Research, 165: 114998.

Liu, Jiajian, Li Ling, Qing Hu, Chao Wang, and Chii Shang. 2022. 'Effects of operating conditions on disinfection by-product formation, calculated toxicity, and changes in organic matter structures during seawater chlorination', Water Research, 220: 118631.

Liu, Yang, Xiaofang Liu, Yujian Wen, and Jun Sun. 2023. 'A snapshot on vertical variability of dissolved organic matter in the epilagic zone of the eastern Indian Ocean', Marine Pollution Bulletin, 192: 114985.

Matilainen, Anu, Egil T. Gjessing, Tanja Lahtinen, Leif Hed, Amit Bhatnagar, and Mika Sillanpää. 2011. 'An overview of the methods used in the characterisation of natural organic matter (NOM) in relation to drinking water treatment', Chemosphere, 83: 1431-42.

Matin, Asif, Z. Khan, S. M. J. Zaidi, and M. C. Boyce. 2011. 'Biofouling in reverse osmosis membranes for seawater desalination: Phenomena and prevention', Desalination, 281: 1-16.

Mecozzi, Mauro, Demetria Cardarilli, Eva Pietrantonio, and Marina Amici. 2001. 'Estimation of Similarity in the Qualitative Composition of Humic Substance in Marine Sediments By Means of an Uv-Spectroscopic Library', Chemistry and Ecology, 17: 239-54.

Mopper, Kenneth, Aron Stubbins, Jason D. Ritchie, Heidi M. Bialk, and Patrick G. Hatcher. 2007. 'Advanced Instrumental Approaches for Characterization of Marine Dissolved Organic Matter: Extraction Techniques, Mass Spectrometry, and Nuclear Magnetic Resonance Spectroscopy', Chemical Reviews, 107: 419-42.

Murphy, Kathleen R., Colin A. Stedmon, T. David Waite, and Gregory M. Ruiz. 2008. 'Distinguishing between terrestrial and autochthonous organic matter sources in marine environments using fluorescence spectroscopy', Marine Chemistry, 108: 40-58.

Park, Sanghun, Taewoo Nam, Jeongyeop You, Eun-Sik Kim, Ilhwan Choi, Jongkwan Park, and Kyung Hwa Cho. 2019. 'Evaluating membrane fouling potentials of dissolved organic matter in brackish water', Water Research, 149: 65-73.

Parsons, T. R., and J. D. H. Strickland. 1961. 'On the production of particulate organic carbon by heterotrophic processes in sea water', Deep Sea Research (1953), 8: 211-22.

Prihasto, Noka, Qi-Feng Liu, and Seung-Hyun Kim. 2009. 'Pre-treatment strategies for seawater desalination by reverse osmosis system', Desalination, 249: 308-16.

Rutlidge, Helen, Liza K. McDonough, Phetdala Oudone, Martin S. Andersen, Karina Meredith, Khorshed Chinu, Mark Peterson, and Andy Baker. 2021. 'Characterisation of groundwater dissolved organic matter using LCOCD: Implications for water treatment', Water Research, 188: 116422.

Sah, R. N. 1994. 'Nitrate-nitrogen determination—a critical review', Communications in Soil Science and Plant Analysis, 25: 2841-69.

Salinas Rodríguez SG. 2011. Particulate and organic matter fouling of SWRO systems: Characterization, modelling and applications CRC Press/Balkema, Delft. http://dx.doi.org/10.1201/b11609

Simon, F. Xavier, Ywann Penru, Andrea R. Guastalli, Santiago Esplugas, Joan Llorens, and Sylvie Baig. 2013. 'NOM characterization by LC-OCD in a SWRO desalination line', Desalination and Water Treatment, 51: 1776-80.

Simon, F. Xavier, Elisabet Rudé, Joan Llorens, and Sylvie Baig. 2013. 'Study on the removal of biodegradable NOM from seawater using biofiltration', Desalination, 316: 8-16.

Spyres, Georgina, Malcolm Nimmo, Paul J. Worsfold, Eric P. Achterberg, and Axel E. J. Miller. 2000. 'Determination of dissolved organic carbon in seawater using high temperature catalytic oxidation techniques', TrAC Trends in Analytical Chemistry, 19: 498-506.

Swietlik, J., A. Dubrowska, U. Raczyk-Stanisławiak, and J. Nawrocki. 2004. 'Reactivity of natural organic matter fractions with chlorine dioxide and ozone', Water Research, 38: 547-58.

Tansakul, Chatkaew, Stéphanie Laborie, and Corinne Cabassud. 2011. 'Adsorption combined with ultrafiltration to remove organic matter from seawater', Water Research, 45: 6362-70.

Valladares Linares, Rodrigo, Victor Yangali-Quintanilla, Zhenyu Li, and Gary Amy. 2012. 'NOM and TEP fouling of a forward osmosis (FO) membrane: Foulant identification and cleaning', Journal of Membrane Science, 421-422: 217-24.

Voutchkov, Nikolay. 2008. Pre-treatment Technologies for Membrane Seawater Desalination (Australian Water Association).

Yin, Wenqiang, Xin Li, Stanislaus Raditya Suwarno, Emile R. Cornelissen, and Tzyy Haur Chong. 2019. 'Fouling behavior of isolated dissolved organic fractions from seawater in reverse osmosis (RO) desalination process', Water Research, 159: 385-96.

doi: 10.2166/9781789062977_0265

Chapter 12

Fluorescence Excitation Emission Matrix (EEM) Spectroscopy

Adam C. Hambly & Urban J. Wünsch,

Technical University of Denmark, Denmark

The learning objectives of this chapter are the following:

- Understand the theoretical and historical background of fluorescence EEM spectroscopy
- Outline EEM instrumentation and best practice for method development, with consideration of potential shortcomings and interferences
- Present and discuss research literature for fluorescence EEM spectroscopy as applied to the fouling of membrane-based water treatment systems

12.1 INTRODUCTION

Fluorescence is a form of luminescence, whereby light (energy) is absorbed by a substance at a particular wavelength (excitation), and then emitted at a longer wavelength (lower energy, emission). The difference between the absorption and fluorescence maxima is known as the Stokes shift, and the entire process typically takes place over a very short timeframe. This is one of the characteristics which separates it from phosphorescence, which typically takes place over longer timeframes. These processes are best described in general by a Jablonski diagram (Figure 1), named after the Polish physicist Aleksander Jabłonski (1933).

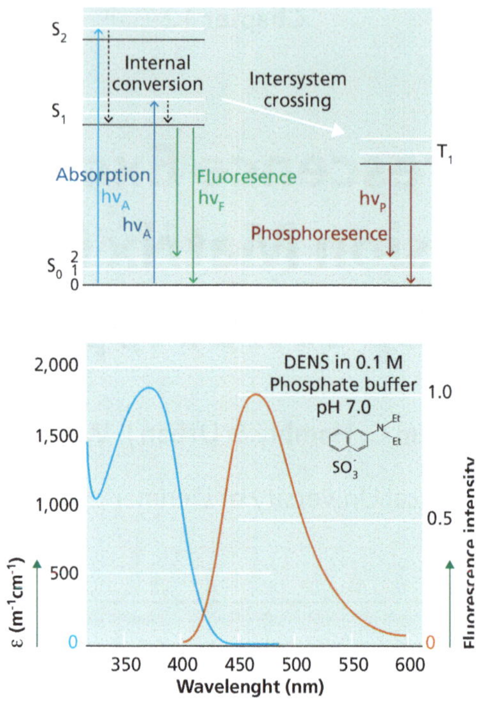

Figure 1 Left Jabłonski diagram; and Right: An example of a compounds absorption (A) and fluorescence emission (F) spectra, where the Stokes shift is the distance between the peaks of the two spectra (Adapted from Lakowicz, 2006b).

Not all substances or compounds are capable of exhibiting fluorescence. However, since it was first referred to in 1852 (Stokes, 1852), fluorescence-based methods have been used to analytically detect and quantify specific compounds with high sensitivity. Fluorescence spectroscopy is often described as being 10 to 1000 times more sensitive than absorbance spectroscopy, due to the nature of the measurement having a 'true' zero. There are many ways which fluorescence can and has been used analytically, from single point to simple 2D emission measurements at a single excitation wavelength, to synchronous fluorescence and excitation-emission matrix (EEM) spectroscopy. Depending on the complexity of the sample, spectra of mixtures such as organic matter, can often be deconstructed into independently varying components using a variety of analytical tools (see section: 11.6 Data Processing). This chapter will focus on fluorescence EEM spectroscopy, and its practical relevance in relation to membrane-based water treatment and fouling in general.

Broadly speaking, the analysis of FDOM (fluorescent dissolved organic matter) with EEM spectroscopy found early use in the natural sciences, the bulk of which was first carried out within oceanography. The methods have also been transferred to freshwater and estuarial research applications, and subsequently have seen applications tied to water treatment within a number of different fields. Fluorescence EEM spectroscopy has now been used to detect and characterise organic content within the aquatic sciences, across a large variety of natural and engineered systems. Oceans (Catalá *et al.*, 2015; Stedmon and Nelson, 2015), seas, lakes (Kothawala *et al.*, 2014; Osburn *et al.*, 2017), and rivers (Baker and Inverarity, 2004), as well as drinking water (Bridgeman *et al.*, 2011), wastewater (Carstea *et al.*, 2016), water recycling (Hambly *et al.*, 2010; Hambly *et al.*, 2015; Henderson *et al.*, 2009; Murphy *et al.*, 2011), aquaculture (Hambly *et al.*, 2015; Spiliotopoulou *et al.*, 2017; Yamin *et al.*, 2017), and desalination (Drozdova *et al.*, 2017; Shutova *et al.*, 2016) are some of the varied aquatic settings in which fluorescence has been applied, from small scale research studies to large scale industrial uses.

Within the context of this chapter, fluorescence EEM spectroscopy has shown particular use in the analysis and understanding of membrane-based systems. As an optimised target compound removal is paramount to the performance of any membrane system, any sort of membrane fouling can therefore limit the systems performance. Feed waters often contain a high level of organic compounds, and as such various forms of organic fouling of the membranes can occur. The performance of a membrane system can be evaluated on the FDOM analysis of different aspects of it, and depending on which aspect it is measuring, the appropriate analytical method will require tweaking. Specific details of fluorescence methods are thus contained in the following sections.

Numerous other aspects of the measurement also need to be addressed before fluorescence data is ready for interpretation. While fluorescence measurements are somewhat simpler when compared to e.g., liquid chromatography coupled to mass spectrometry, analysts must consider many questions before measurements can take place. For example, how should samples be taken and how long can they be stored? How exactly should fluorescence be measured, and which analysis strategy is the best for a given study? The following sections 11.2 to 11.7 provide guidance for the practical aspects behind fluorescence measurements within an aquatic context.

12.2 SAMPLING & STORAGE

Samples taken from different stages of membrane filtration consist of particulate and soluble material in water. A fraction of this material is highly bioavailable to microorganisms, while other fractions resist biodegradation for longer periods (Hu and Ren, 2019; Urgun-Demirtas *et al.*, 2008). When it comes to sampling and subsequent storage prior to measurements, the more bioavailable material requires special attention, as microbes naturally target the most available compounds first and can thus alter the sample quickly (Heinz and Zak, 2018). In general, it is advisable to process samples and perform fluorescence measurements as quickly as possible to avoid storage artefacts. However, the constraints of sampling do not always allow for fast sample processing. In such cases, preserving the sample and knowing about possible storage effects is important. Preservation strategies include filtration, storage in cold and dark conditions, freezing, and chemical poisoning. In contrast, autoclaving introduces changes in FDOM (Andersson *et al.*, 2018).

Good practices for characterizing dissolved materials include 1) removing living organisms as quickly as possible through filtration; 2) storing the sample at temperatures below 10 °C in the dark to minimize biological activity; 3) measuring as quickly as possible but preferably at most within 5 days of sampling, and; 3) maintaining the same procedure for sample processing within a study to keep potential biases constant.

Different filter materials and pore sizes are available for filtration. Amongst the available materials, glass fiber filters are the safest option regarding contamination since they can be ashed (> 4 h at > 450 °C, (Coble *et al.*, 2014)). However, glass fibre filters generally only exist with pore sizes of larger than 0.3 μm. The classic GF/F filter with a pore size of 0.7 μm was used to distinguish particulate from dissolved matter. However, at that pore size, studies have reported bacterial passage rates between 10 and 25 % in marine samples (Morán *et al.*, 1999). It should however be noted that passage of microbes through filters with all common diameters can be observed (Obayashi and Suzuki, 2019; Wang *et al.*, 2007, 2008). If initial cell counts are high and assimilable carbon is abundant, microbial regrowth can quickly change sample character despite the usage of 'sterile' filters (< 0.2 μm). These observations emphasize that storages times should be kept short and effects of microbial passage will depend on the original sample. Lastly, the leaching of filter material into the sample should be investigated for the specific batch of filters used in each study. In the context of wastewater, such leaching is likely negligible but can affect primarily the UVA fluorescence emission range due to the leaching of production-related impurities. Filters should be rinsed with ultrapure water followed by sample prior to obtaining a filtrate for analysis. Filtration should occur slowly to avoid the bursting of cells (Rosenstock and Simon, 1993).

Freezing as a means to slow down biological processes can help to facilitate longer sample storage. While some studies recommend freezing as suitable for samples with low carbon concentrations, significant changes in optical indices, sample absorbance, and fluorescence emission characteristics have been observed (Fellman *et al.*, 2008; Heinz and Zak, 2018; Spencer *et al.*, 2007) concluded that the effects cannot be predicted from the composition of the original sample. Thus, the effects cannot be corrected post-measurement.

12.3 BENCHTOP INSTRUMENTATION

Since fluorescence analyses are increasingly popular, users have a range of choices concerning benchtop instrumentation. However, specifications regarding instrument and software can differ significantly between instruments and affect the measurement experience and resulting data quality.

Commercial instruments on the market today usually feature incandescent or pulsed Xenon lamps. These lamps provide excitation light in the entire ultraviolet-visible spectrum (approx. 220 – 800 nm) and have a relatively continuous emission spectrum. To understand the instrument's limitations, it is important to keep in mind the lamp's output spectrum. For example, incandescent Xe lamps provide little light at wavelengths shorter than 240 nm and resulting emission scans are generally noisier and can be difficult to interpret. Xe flash lamps can provide more light in the ultraviolet range resulting in a wider usable excitation range and a more uniform signal-to-noise relationship across EEMs (Lakowicz, 2006a). However, Xe flash lamps also contain more distinct emission bands, that need to be addressed to obtain spectrally calibrated EEMs. Lastly, incandescent Xe lamps have a lifetime of < 2000 h and thus require more maintenance compared to pulsed light sources.

The optical configuration of spectrofluorometers can differ significantly due to the requirements dictated by the detection system. The classic photomultiplier tube (PMT) is a sensitive photon detector that lacks the ability to distinguish light of different wavelengths but offers superior sensitivity thanks to signal amplification and low noise levels. Spectrofluorometers that utilize PMTs require two monochromators: The first selects a narrow band of light for sample excitation, the second permits a narrow band of fluorescence emission to pass through to the PMT. By scanning through all desired emission wavelengths at all desired excitation wavelengths, an EEM is constructed with typical speeds of 500 nm/min (total time typically between 20-40 min). On the other hand, charge-coupled device detectors (CCDs) allow the simultaneous detection of the entire wavelength range of interest and reduce measurement times considerably.

Due to the necessity to consider non-linearity in fluorescence observations due to the optical density of a sample (discussed below), it is also important to consider the availability of spectrophotometers during the measurement of fluorescence. If potential projects involve field measurements, the use of instruments with a combined absorbance-fluorescence detection system can be advantageous since all required measurements are made within one unit (see Figure 2).

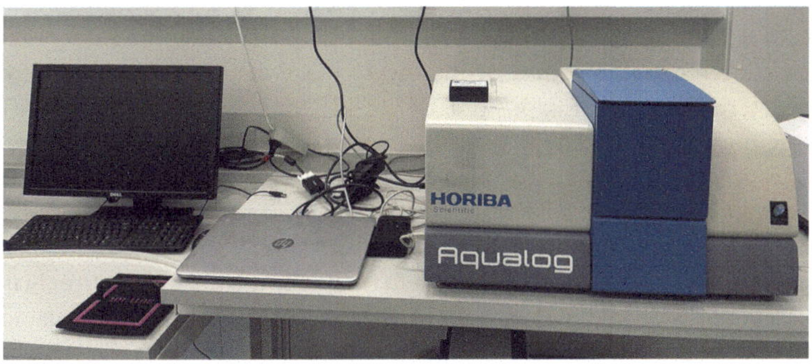

Figure 2 An example of a modern spectrofluorometer instrumentation setup in a research laboratory – in this case an iteration of the Horiba Aqualog with a CCD detector.

As mentioned above, samples generally contain particulate and dissolved material. While both fractions contain fluorescent moieties, different instrument configurations are required to characterize the material. For example, dissolved fluorescent material is quantified after filtration of a water sample through filters in the classic right-angle geometry with a quartz cuvette (Figure 3, left). On the other hand, the measurement of particulate material occurs either directly in the unfiltered, optically thick suspension or by exciting particles directly on a surface. In both cases, the non-transparent nature of the particulate sample necessitates a front-face illumination (Figure 3, right). Front-face measurements of thick suspensions can also occur in cuvettes, but require adapters to either adjust the angle of the incident light relative to the cuvette face or to capture and direct the light at a specific angle toward the cuvette. The remainder of this section will discuss right-angle fluorescence of optically thin solutions since this is by far the most common application of fluorescence.

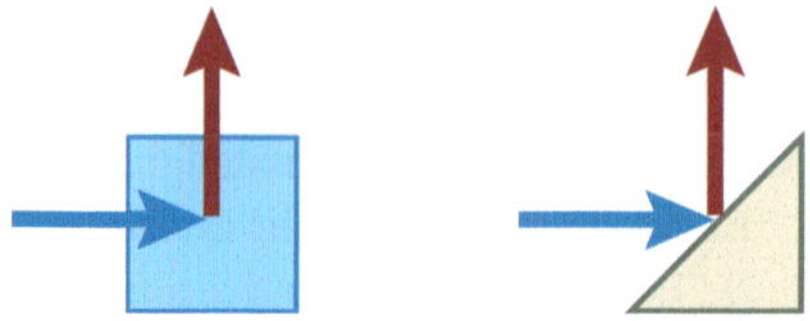

Figure 3 Schematic view (top) of the geometric position of samples and cuvettes.

12.4 QUALITY ASSURANCE

Most of today's spectrofluormeters are spectrally calibrated in full from the factory. However, to generate comparable results, wavelength accuracy and spectral calibration are especially important to monitor over time as the instrument ages. Wavelength accuracy refers to the deviation between true and detected wavelength in nanometres; either the excitation light intercepted by the cuvette or the fluorescence emission captured by the detector. Manufacturers commonly list the specification in the instrument manual and a precision of ± 1 nm is typical. Deviations are monitored by detecting the peak position of Rayleigh scatter at e.g., 467 nm, while the accuracy of the emission detector can be verified by determining the peak position of the 435.8 nm Hg band emitted by common low-pressure mercury vapor lamps (Sansonetti *et al.*, 1996). Monitoring changes in peak positions is especially important after instrument transport.

Spectral calibration refers to the elimination of spectral biases that arise from a biased lamp emission spectrum, and wavelength-dependent monochromator and detector biases. Some of these biases are eliminated with the use of reference detectors, but the remainder of the bias is removed with the use of excitation and emission correction factors that come pre-installed from the factory. It can however be a good idea to verify their appropriateness from time to time. A triangular cuvette of Rhodamine B produces a flat excitation spectrum after the successful elimination of spectral biases (Kopf and Heinze, 1984). The emission calibration is commonly verified with standards available from the National Institute of Standards and Technology (NIST). Different standards cover the ultraviolet-visible emission range and the recorded spectra should fall within the certified values at all times (Gilmore, 2014).

To obtain valid results, the measured sample needs to meet certain criteria to ensure that the instrument is capable of delivering the best results possible. For example, fluorescence counts should not exceed the linear range of the detector. Fortunately, the instrument control software usually warns users when the linearity threshold is exceeded. In such cases, settings can be adjusted (e.g., integration time) or samples diluted.

In cases with high concentrations of chromophores, the sample transmission can be too low to deliver quantitative fluorescence results. Kothawala *et al.*, (2013) defined an absorbance of 0.05 cm^{-1} (89 % transmittance) as the cutoff below which such effects can be safely neglected (Kothawala *et al.*, 2013). On the other hand, an absorbance of 1.5 cm^{-1} (3 % transmittance) was found to be the upper limit after which no quantitative fluorescence results can be obtained even if correction methods are applied (see below). It is thus important that samples meet the second criterion *during* the fluorescence measurements, as no post-measurements for linearity will be possible.

Prior to every study, choosing appropriate measurement settings is important to ensure appropriate fluorescence counts in the relevant ranges of the EEM are accumulated. Regarding range and resolution, the types of observed fluorophores and their properties will govern which parts of the EEM should be captured. For example, if protein-like material, phenolic compounds, and / or oils are present a high resolution in the excitation range below 300 nm is especially important. To distinguish these highly similar fluorophores and quantify their fluorescence, it can also be important to capture emission spectra with a resolution below 3 nm if bandwidth characteristics of monochromators permit this. For instruments with incandescent Xe lamps, it generally makes little sense to capture emission below 240 nm even if sample fluorophores exhibit strong absorbance bands since signal-to-noise ratios deteriorate in the UV region. Capturing the emission up until 800 nm is necessary should the sample contain algae or fluorescent pigments. Moreover, to enable the correction of inner filter effects, it is important to measure the samples absorbance spectrum covering all excitation and emission wavelengths. Otherwise, such corrections can become difficult to implement.

12.5 INTERFERENCES

As an extrinsic property of fluorophores in solution, fluorescence fingerprints are vulnerable to changes due to interferences. Such changes can impact fluorescence yields (per mol of substance) and spectra and are caused by physicochemical properties of the sample. When comparisons between samples are made, it is thus important to consider whether physicochemical properties remain stable or are subject to change. In the following, we will list *some examples* (not all) of the physicochemical parameters known to influence DOM fluorescence.

The sample's temperature can affect the fluorescence intensity observed for a given sample. When solvent temperatures increase, observed fluorescence generally decreases. However, there is no evidence to suggest that spectral characteristics change due to temperature (McKay *et al.*, 2018). This effect is of great importance for *in situ* measurements with sensors since temperature can vary systematically over longer periods of time. However, a compensation is trivial if the sample's temperature is known (Watras *et al.*, 2011). For benchtop instruments, temperature effects are usually not an issue since jacketed cuvette holders and climate-controlled laboratories eliminate the chance for systematic biases.

Changes in pH can lead to spectrally-dependent changes in a sample's fluorescence. Numerous studies have investigated the effect of pH on organic matter fluorescence, and the different results reported in each study hints at sample-dependent, complex effects (e.g. Esteves *et al.*, 1999; Groeneveld *et al.*, 2022; Mobed *et al.*, 1996; Murphy *et al.*, 2018; Spencer *et al.*, 2007). These complex changes make it effectively impossible to compensate the pH-induced interference. It is thus best to avoid sample-to-sample differences in pH to facilitate comparisons between samples.

Beyond temperature and pH, ionic strength, the presence of metal ions, and particle attenuation have also been reported to affect fluorescence measurements. As with all physicochemical properties, it is advisable to obtain reference measurements for samples coming from the system that is subject of a study. This will help to ascertain if issues with certain parameters are to be expected and if so, whether great variation (e.g., pH or ionic strength) might introduce artefacts that complicate interpretation of the fluorescence readings in a given dataset.

12.6 DATA PROCESSING

Fluorescence measurements require several steps of processing before further analyses can take place (Figure 4). While some software offers comprehensive features that contain the most critical steps, we believe it is most advisable to export data from proprietary formats and subsequently use open software environments to process and analyze fluorescence data. This gives the user more control over processing steps, freedom of choice regarding analysis strategies, and maximizes the impact of the conducted research by extracting as much information as possible. Amongst the most common languages for statistical computing, Matlab and R have community-driven software packages (Matlab: drEEM, EEMlab; R: eemR, StaRdom, albatross) that facilitate all steps in Figure 4 and enable a range of multivariate analyses (Micó *et al.*, 2019; Murphy *et al.*, 2013; Pucher *et al.*, 2019). While theoretically possible, an analysis of EEMs in spreadsheet software is not recommended since workflows are not easily automatable.

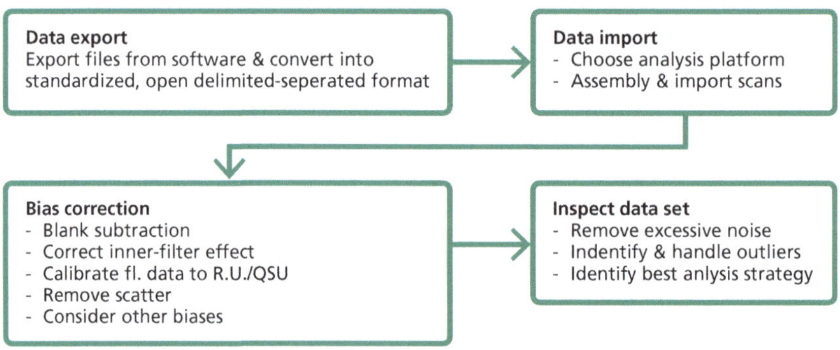

Figure 4 Steps involved in the processing of fluorescence measurements.

After the successful import into the programming environment of choice, the next processing steps concerns the correction or removal of different measurement biases. For example, a blank should be subtracted from each sample fluorescence landscape to remove the impact of background signals and reduce the abundance of scatter. Such blanks should be measured daily and always be made from the sample solvent (e.g. water, buffered water, organic solvent). Blanks can also function as a standard for the calibration of fluorescence signals into Raman Units (Lawaetz and Stedmon, 2009). Next, inner-filter effects (IFEs) are compensated since they disturb the linearity of observed fluorescence due to the partial or complete absorbance of emitted fluorescence by chromophores (Parker and Rees, 1960). Such effects are easily corrected by applying correction factors derived from the sample's absorbance scan – provided the maximum absorbance be below approx. 1.5 cm^{-1} (Kothawala *et al.*, 2013).

The removal of Rayleigh and Raman scatter can be an important step if the subsequent analysis strategy (see below) includes the decomposition of fluorescence EEMs into statistical components according to Beer Lambert's law (Bahram *et al.*, 2006). The open-source software packages mentioned above include functions for scatter removal and thus simplify this task considerably.

12.7 DATA ANALYSIS

Once data is measured and fully processed, the data analysis can occur. Analysis strategies (overview in Table 1) can range from simply comparing fluorescence intensities (known as 'peak picking') and fluorescence indices to multivariate analysis such as parallel factor analysis (PARAFAC).

Table 1 Overview of most common strategies to analyse fluorescence EEMs.

Analysis strategy	Description	References
Peak picking	Extraction of fluorescence intensities from EEMs at defined wavelengths.	(Coble 2007)
Fluorescence regional integration (FRI)	Integration of fluorescence in wavelength regions with predefined interpretation.	(Chen *et al.*, 2003)
Fluorescence indices	Qualitative descriptors of FDOM with insights into humification, aromaticity, freshness and microbial processing.	(Huguet *et al.*, 2009; Maie *et al.*, 2006; Ohno 2002; Parlanti *et al.*, 2000)
Parallel factor analysis (PARAFAC)	Multivariate decomposition of EEMs into components.	(Murphy *et al.*, 2013)

The comparison of fluorescence intensities usually occurs at predefined wavelengths that typically have letters assigned to them (see Figure 5, table 2). For example, peak T, extracted at excitation / emission 275 / 340 is typically ascribed to tryptophan- or protein-like material. Peaks A and C on the other hand are often described as humic-like material. It should be noted that the interpretation of fluorescence peaks should only be informed by comparison with studies of the same sample material and take into account potential issues (e.g. pharmaceuticals fluorescing like amino acids).

Table 2 Position of predefined peaks and fluorescence indices as listed or described in Coble (2007), Maie *et al.* (2006), Huguet *et al.* (2009), Ohno (2002).

Peak / index	$\lambda E_x / \lambda E_m$	Tentative interpretation in natural environments
A	260/400-460	Humic-like, terrestrial
B	275/305	Autochthonous
T	275/340	Autochthonous
M	290-310/370-410	Anthropogenic contaminants
C	320-360/420-460	Humic-like, terrestrial
D	390/509	Humic-like, ubiquitous
Fluorescence index (FI)	370/470 370/520	Distinguishes microbial and terrestrial inputs
Biological index (BIX)	310/380 310/430	Contribution of biological transformations
Humification index (HIX)	$254/\int 435-480$ $254/\int 300-345+254/\int 435-480$	Ratio between protein- and humic-like fluorescence

Fluorescence regional integration (FRI) is a particularly popular technique in engineered systems by which integrals of wavelength regions in the EEM (Figure 5, yellow lines) are tracked across samples. The assignment of these regions is based on model compounds and natural environmental samples (Chen *et al.*, 2003). According to Chen *et al.*, (2003), the five regions as illustrated in Figure 5 are aromatic protein-like material (I + II), fulvic acid-like compounds (III), microbial by-product-like fluorescence (IV), and humic-like material (V). Subsequent to the publication of the FRI approach, multivariate modelling has indicated that underlying fluorescence spectra in regions I, II and IV, as well as III and V overlap spectrally and regional integrals are likely not as specific as the names above suggest (Stedmon *et al.*, 2003).

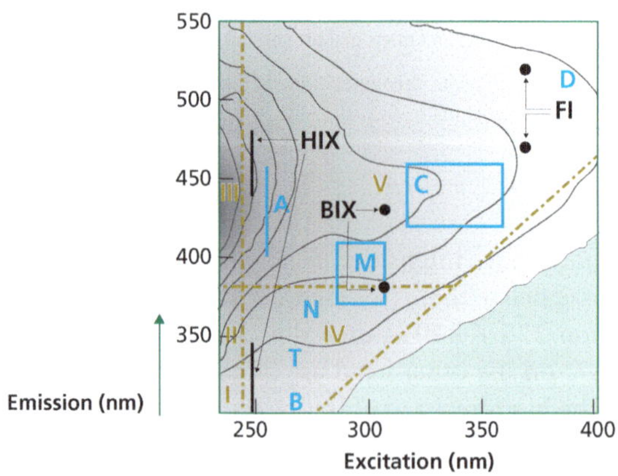

Figure 5 Emission-Excitation Matrix of a sample from a boreal river (Öre river, northern Sweden). Yellow lines and text refer to fluorescence regional integration areas (FRI). Black dots represent fluorescence indices: FI: Fluorescence Index; BIX: Biological Index; HIX: Humification Index). Blue lines and text refer to predefined peak locations.

Several fluorescence-based indices also find frequent application (Figure 5, thick black lines, Table 2). For example, the humification index (HIX) informs about the ratio between protein- and humic-like fluorescence and thus can help to understand qualitative shifts between samples (Ohno, 2002). The biological index (BIX) is a ratio between ultraviolet and visible fluorescence and can indicate the importance of recent biological transformations of material (Huguet *et al.*, 2009). Lastly, the fluorescence index (FI) distinguishes between microbial and terrestrial inputs in aquatic environments (Maie *et al.*, 2006; McKnight *et al.*, 2001). As with peak picking, one should be careful to extrapolate interpretations from studies performed on different sample types to membrane samples. The above-mentioned fluorescence indices were defined in studies of natural aquatic environments but can help to identify qualitative differences between samples.

12.7.1 PARAFAC

Amongst the multivariate analysis techniques, PARAFAC is the most popular technique in the analysis of DOM and this section will provide a short overview over this technique. For tutorials on MCR and PCA, we refer the reader to Bro and Smilde (2014), and de Juan *et al.*, (2014). PARAFAC is a particularly popular model for the decomposition since it naturally follows the analytical principals of fluorescence (Bro, 1997; Murphy *et al.*, 2013; Stedmon and Bro, 2008). Each analyte (termed 'component' in the model) is described as a product of an excitation and emission spectrum multiplied by a concentration factor. The entire EEM is the sum of the fluorescence arising from each of the components. PARAFAC is particularly attractive since it can distinguish spectrally overlapping components and thus allows insights into components that may not be distinguishable in the raw fluorescence data. Also, PARAFAC can isolate systematic signals in noisy measurements and thus help to improve the quality of the results. Moreover, the component spectra can be compared between studies and help to inform the chemical interpretation of the signals.

As a multivariate modelling approach, PARAFAC analyses work best if a number of criteria are met. For example, a minimum number of samples with meaningful compositional variability helps to identify meaningful and robust models. If two or more peaks covary perfectly, the approach may produce questionable models. Similarly, if a dataset consists of too few samples, the algorithm can struggle to identify the underlying components. As discussed above, the fluorescence occurring in each sample can be altered due to interferences. In such cases, it is most likely more fruitful to rely on peak picking and/or the interpretation of fluorescence indices.

12.8 APPLICATION IN MEMBRANE SYSTEMS

As outlined above, there are myriad ways in which fluorescence measurements can be applied to aquatic systems that utilise some form of membrane. Organic matter is a common source of membrane fouling, and it follows that fluorescence measurement of organic matter has gained traction in the analysis and investigation of how organic fouling occurs in these systems. Whether it is microfiltration, ultrafiltration, nanofiltration, or reverse and forward osmosis, using flat sheet, hollow fiber, tubular, or spiral wound constructions, there is nearly always a way in which fluorescence measurements can be, and has been applied.

These methods can assist in evaluating the membrane structure, the process performance in general, and of course the level and character of fouling. However, the construction, matrix, and analytical targets for each of these specific applications will determine how fluorescence EEM spectroscopy can, and cannot, be used in each case.

Section 11.3 has highlighted two physical application differences: (1) front-face fluorescence spectroscopy; and (2) right-angle fluorescence spectroscopy. Both methods can be used to assess and/or predict organic fouling, though in different ways and typically for different applications. Front-face fluorescence EEM spectroscopy currently finds its most common (membrane-related) use for systems such as MBRs (membrane bioreactors), or direct measurement from fouled membrane surfaces. Regardless of whether front-face or right-angle fluorescence is used, the organic character of any fouling will be dependent on both the feed, and the physical and/or chemical characteristics of the membrane in use.

The direct, *in situ* fluorescence analysis of membranes has to date been carried out in different ways in order to understand the main organic components behind the organic fouling of membrane surfaces. For example, Yamamura *et al.,* (2019) used in situ front-face EEM spectroscopy to investigate the organic fouling of PVDF membranes from secondary treated wastewater, in a bench-scale study. They observed increasing intensities with time over three main peaks, and by combining intermittent backwashing with EEM analysis, the authors observed peaks which were predominantly associated with reversible and/ or irreversible fouling. Yu *et al.,* (2019), also utilised front-face fluorescence to detect and quantify model foulants on UF membranes, and concluded that it was a better method than liquid right-angle fluorescence in this particular study due to lower standard deviations observed between repeated measurements (see Figure 6). Pawlowski *et al.,* (2016) used front-face fluorescence to monitor fouling deposition on ion-exchange membranes. In this study, the authors found it to be a useful tool in evaluating reverse electrodialysis processes, and particularly for increasing membrane cleaning efficiencies.

Similar front-face techniques have also been used to investigate activated sludge systems (Huaorng, 2022), to quantify biomass and bioactivity amongst other parameters. From a practical spectroscopy perspective, these can be likened to the sludge and mixed liquor components of MBRs. The fluorescence character of the sludge over time can be linked to the evolution of fouling on the membranes, as the organic matter in particular EPS (extracellular polymeric substances) has been found to be closely related to TMP (Chen *et al.*, 2018). Various iterations of front-face spectroscopy have been used for assessing both the sludge and membrane components of MBR systems *in situ* (Galinha and Crespo, 2022).

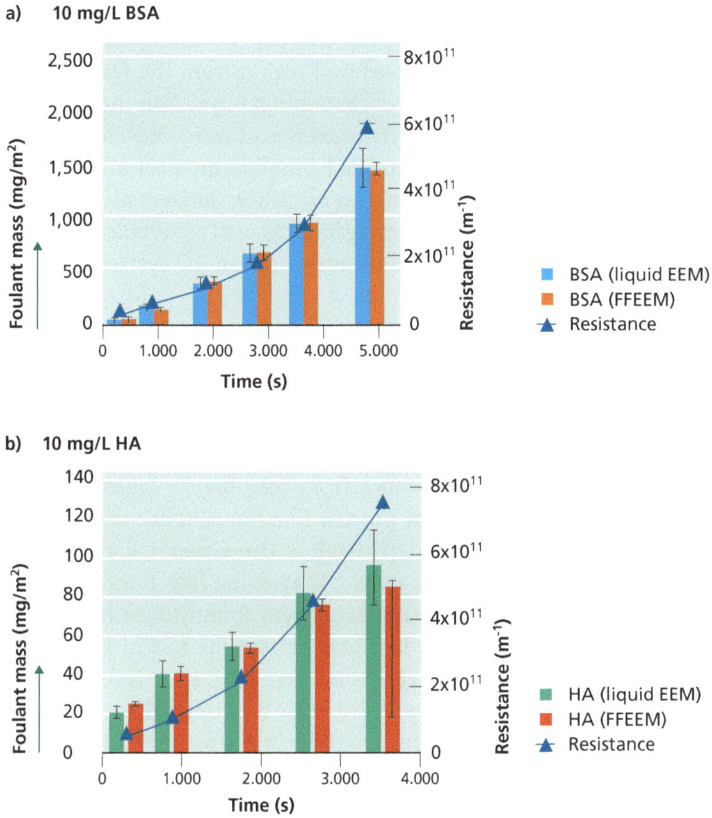

Figure 6 Comparison of fouling by 10 mg/L of (a) Bovine Serum Albumin and (b) Humic Acid measured by liquid right-angle EEM and front-faced fluorescence EEM measurements (adapted from Yu *et al.* (2019)).

Looking beyond front-face fluorescence EEM spectroscopy, the more common method of analysing membrane performance has historically been by a direct measurement of the different liquid streams with right-angle fluorescence. This is due to the higher signal and higher sensitivity that is achieved by this method, though front-face fluorescence spectroscopy has been described and utilised for high absorbance liquid samples for nearly half a century (Eisinger and Flores, 1979). In the case of membrane systems, using the right-angle fluorescence method typically means a comparison of one or more out of the feed, permeate, and concentrate streams. Whether some, or all, of these streams can be analysed with right-angle fluorescence will depend on whether the target matrix adheres to the requirements set out in section 11.4 Quality Assurance. Primarily, the liquid matrix must exhibit low absorbance values, which either are below the threshold where they can be considered negligible, or within the mathematically correctable range. In the latter case, an absorbance measurement must accompany the fluorescence measurement to guarantee accurate values.

For example, Poojamnong (2020) used right-angle fluorescence to compare the feed and permeate EEMs of an MBR treating pulp and paper wastewater. This study found protein-like fluorescence to be the most reduced region from the feed to the permeate, which correlated to the main component of the fouling formed on the UF membranes. A comparison of feed and permeates from RO processes of water recycling plants have also shown that right-angle fluorescence EEM could be used to monitor organics rejection and membrane integrity (Pype *et al.*, 2013; Singh *et al.*, 2009; Singh *et al.*, 2015). Bagastyo *et al.,* (2011) identified humic and fulvic-like organics, as well as soluble microbial products as the main constituents of the concentrate stream within an RO system treating secondary wastewater effluent.

A direct analysis of already fouled membranes can also be performed with right-angle fluorescence EEM spectroscopy. Stripping off individual foulant layers by, for example, backwashing, or acid/base washing, has been used to understand differences in the formation of membrane foulant layers. In 2004, Kimura *et al.* (2004) showed that alkaline cleaning removed primarily protein-like fluorescence from membranes fouled by surface water. Henderson *et al.* (2011) also found that protein-like fluorescence was the predominant fluorescence region that was removed through a three-step UF membrane cleaning procedure. More specifically, in this case a low UV 'tyrosine-like' fluorescence component found within 5 different sources of wastewater, was found to be highly correlated to membrane fouling potential. In a study that investigated the role of DOM in the fouling of membrane bioreactors (MBR) treating wastewater, Tang *et al.* (2010) also found that two protein-like fluorescence components were most correlated to membrane fouling. These and other similar studies highlight how EEM spectroscopy can be used effectively to gain insight into how different foulant layers form, and therefore how to minimise their formation.

Further to the direct observation of various fluorescent membrane fouling components, studies have also investigated different pre-treatment steps to remove these components and hence minimise organic fouling. For example, Wang *et al.* (2017) utilised liquid EEM analysis to compare the performance and effect of 8 different types of powder activated carbon (PAC), as a pre-treatment to UF membrane treatment. They showed that whilst initial fouling was linked to the humic-like fluorescence region, ultimately the ability of PAC to minimise irreversible fouling was linked to how well it absorbed protein-like fluorescence. Aftab *et al.* (2020) applied combinations of different pre-treatment processes to change the FDOM character of landfill leachate, in order to compare how the resulting organics character ultimately affected NF fouling. This study concluded that both fulvic-like and protein-like fluorescence was more linked to irreversible fouling, than was humic-like fluorescence. In slight contrast to this, Xu *et al.* (2022) concluded that both proteins and humics contributed to the initial pore blocking stage, though the study was conducted on synthetic mixtures and model foulant compounds. Through a combination of size exclusion chromatography and liquid fluorescence EEM analysis, Haberkamp *et al.* (2011) also showed that protein-like fluorescence correlated with the extent of hydraulically irreversible fouling of UF membranes by secondary effluents. In this case, the authors also showed that chemical coagulation and biological sand filtration as pre-treatment were both able to significantly reduce membrane fouling. Furthermore, in a study that investigated dissolved air flotation as a potential pre-treatment for membrane desalination, Shutova *et al.* (2016) showed that DAF treatment removed higher proportions of humic-like fluorescence than protein-like fluorescence, yet it still was able to remove between 28% and 58% of protein like fluorescence from real water samples.

Table 3 An overview of publications which have utilised fluorescence within membranerelated treatment studies. N.B. MF (Microfiltration); UF (Ultrafiltration); RED (Reverse electrodialysis); CE (cation exchange); AE (anion exchange); MBR (membrane bioreactor); RO (reverse osmosis); AS (Activated sludge); FF (front-face), RA (right angle), PP (peak picking); PARAFAC (Parallel factor analysis); FRI (fluorescence regional integration); FI (fluorescence indices); PCA (principal components analysis); and PLS (partial least squares)

Reference	Application	Method	Highlights or main findings
Kimura *et al.* (2004)	UF	RA + PP	Polysaccharide-like organic matter, Fe and Mn in surface water responsible for irreversible fouling.
Singh *et al.* (2009)	RO	RA + PP	Humic-like fluorescence most appropriate for distinguishing between stage 1 and stage 2 RO.
Tang *et al.* (2010)	MBR	RA + PP	Protein-like fluorescence correlated positively with membrane fouling.
Henderson *et al.* (2011)	UF	RA + PP/ PARAFAC	Tyrosine-like fluorescence monitoring could be used as an indicator of fouling potential from domestic wastewater.
Bagastyo *et al.* (2011)	RO	RA + PP	Advanced oxidation of RO concentrates more efficient than coagulation & MIEX adsorption.
Haberkamp *et al.* (2011)	UF	RA + PP	Removal of protein-like substances by sand filtration or coagulation resulted in reduced irreversible fouling.
Galinha *et al.* (2011)	MBR	RA/FF + PLS	3 fluorescence components could be used to predict COD concentration in MBR permeate.
Pype *et al.* (2013)	RO	RA + FRI	Fluorescence proposed as surrogate for pathogen removal in RO systems.
Singh *et al.* (2015)	RO	RA + PP	Peak C linked to TMP/fouling.
Shutova *et al.* (2016)	RO	RA + PARAFAC	Humics concentration used to determine optimal coagulant dose.
Pawlowski *et al.* (2016)	RED - CE/AE	FF / RA + PCA	Fluorescence able to monitor fouling development of ion-exchange membrane surfaces.
Vera *et al.* (2017)	UF/RO	RA + PARAFAC	Quantified OM removals through treatment plant, monitoring FDOM composition can optimise treatment conditions due to seasonal variation.
Wang *et al.* (2017)	UF	RA + PP	Humic-like substances contributed to initial membrane fouling, protein-like correlated with irreversible fouling.
Cai *et al.* (2017)	MBR	RA + PARAFAC	Protein-like substances more readily biodegradable than humic-like substances.

Reference	Application	Method	Highlights or main findings
Jacquin *et al.* (2017)	MBR	RA + FRI	Correlations established between LC-OCD-OND and EEM data to quantify protein-like and humic-like substances.
Xiao *et al.* (2018)	MBR	RA + FRI	Identified correlations between characteristic EEM wavelength regions and hydrophobic/hydrophilic DOM components.
Yamamura *et al.* (2019)	MF	FF / RA + PP	Proteinaceous substances responsible for reversible and irreversible fouling, gels mainly contributed to irreversible fouling.
Yu *et al.* (2019)	UF	FF/RA + PARAFAC	FF-EEM method superior to RA-EEM coupled with mass balance for UF foulant determination.
Aftab *et al.* (2020)	NF	RA + PARAFAC	Different pre-treatment led to different quantities and qualities of membrane foulants .
Poojamnong *et al.* (2020)	MBR	RA + FRI	Irreversible foulants mainly comprised of protein-like substances.
Yu *et al.* (2021)	MBR	RA + PP/FI	Combination of protein-like fluorescence and UV280 used to predict fouling MBR potentials.
Xu *et al.* (2022)	UF	RA + PP	Proteins and humics mainly participate in initial pore blocking, polysaccharides mainly participate in later gel/cake layer stage
Yu *et al.* (2022)	AS	FF + PARAFAC	Protein-like substances, NADH, and humic-like substances correlated with MLVSS, intracellular NADH, and humic-like substances in SMP, respectively.
Cifuentes-Cabezas *et al.* (2023)	NF	RA/FF + PP	Fluorescence showed different fouling development between different NF membrane products.

The studies that have been mentioned in this section are but a small proportion of the many research applications of fluorescence EEM spectroscopy within membrane treatment systems to date. While they highlight the method as a clear and practical use for measuring organic fouling on membranes, it is nevertheless imperative to keep in mind that only a fraction of OM is fluorescent. Although the fluorescent fraction is typically considered to be representative of OM as a whole, a better overall picture of OM and OM-based fouling will nearly always be attained when it is applied in combination with other analytical tools.

Within the near future, further advances in optical technology and computer processing power look to be the main impetus' behind further development and wider application of fluorescence sensors in membrane systems. As optical sensors become cheaper and more sensitive, and light sources become more stable with higher output and tighter bandwidths (particularly at lower wavelengths) fluorescence spectroscopy will become even more practical and accessible for both research and industry applications.

12.9 REFERENCES

Aftab B, Cho J, Shin HS, and Hur J (2020) Using EEM-PARAFAC to Probe NF Membrane Fouling Potential of Stabilized Landfill Leachate Pretreated by Various Options. Waste Management. 102: 260–269.

Andersson MGI, Catalán N, Rahman Z, Tranvik LJ, and Lindström ES. (2018) Effects of Sterilization on Dissolved Organic Carbon (DOC) Composition and Bacterial Utilization of DOC from Lakes. Aquatic Microbial Ecology. 82(2): 199–208.

Bagastyo AY, Keller J, Poussade Y, and Batstone DJ (2011) Characterisation and Removal of Recalcitrants in Reverse Osmosis Concentrates from Water Reclamation Plants. Water Research 45(7): 2415–2427.

Bahram M, Bro R, Stedmon CA, and Afkhami A (2006) Handling of Rayleigh and Raman Scatter for PARAFAC Modeling of Fluorescence Data Using Interpolation. Journal of Chemometrics. 20(3–4): 99–105.

Baker A, and Inverarity R (2004) Protein-like Fluorescence Intensity as a Possible Tool for Determining River Water Quality. Hydrological Processes. 18(15): 2927–45.

Bridgeman J, Bieroza M, and Baker A (2011) The Application of Fluorescence Spectroscopy to Organic Matter Characterisation in Drinking Water Treatment. Reviews in Environmental Science and Biotechnology. 10(3): 277–290.

Bro R (1997) PARAFAC. Tutorial and Applications. Chemometrics and Intelligent Laboratory Systems. 38(2): 149–171.

Bro R, and Smilde AK (2014). Principal Component Analysis. Analytical Methods. 6(9): 2812–2831.

Cai W, Liu J, Zhu X, and Liu Y (2017) Fate of Dissolved Organic Matter and Byproducts Generated from On-Line Chemical Cleaning with Sodium Hypochlorite in MBR. Chemical Engineering Journal. 323: 233–242.

Carstea EM, Bridgeman J, Baker A, and Reynolds DM (2016) Fluorescence Spectroscopy for Wastewater Monitoring: A Review. Water Research. 95: 205–219.

Catalá TS, Reche I, Fuentes-Lema A, Romera-Castillo C, Nieto-Cid M, Ortega-Retuerta E, Calvo E, Álvarez M, Marrasé C, Stedmon CA, and Álvarez-Salgado XA (2015) Turnover Time of Fluorescent Dissolved Organic Matter in the Dark Global Ocean. Nature Communications. 6(1): 5986.

Chen W, Qian C, Zhou K-G, and Yu H-Q. (2018) Molecular Spectroscopic Characterization of Membrane Fouling: A Critical Review. Chem. 4(7): 1492–1509.

Chen W, Westerhoff P, Leenheer JA, and Booksh K (2003). Fluorescence Excitation– Emission Matrix Regional Integration to Quantify Spectra for Dissolved Organic Matter. Environmental Science & Technology. 37(24): 5701–5710.

Cifuentes-Cabezas M, Galinha CF, Crespo JG, Vincent-Vela MC, Mendoza-Roca JA, and Álvarez-Blanco S (2023) Nanofiltration of Wastewaters from Olive Oil Production: Study of Operating Conditions and Analysis of Fouling by 2D Fluorescence and FTIR Spectroscopy. Chemical Engineering Journal. 454: 140025.

Coble PG (2007) Marine Optical Biogeochemistry: The Chemistry of Ocean Color. Chemical Reviews. 107(2): 402–418.

Coble P, Lead J, Baker A, Reynolds D, and Spencer R (2014). Aquatic Organic Matter Fluorescence. In P. Coble, J. Lead, A. Baker, D. Reynolds, & R. Spencer (Eds.), Aquatic Organic Matter Fluorescence (Cambridge Environmental Chemistry Series, p. V). Cambridge: Cambridge University Press.

Drozdova AN, Patsaeva SV, and Khundzhua DA (2017) Fluorescence of Dissolved Organic Matter as a Marker for Distribution of Desalinated Waters in the Kara Sea and Bays of Novaya Zemlya Archipelago. Oceanology. 57(1): 41–47.

Eisinger J, and Flores J (1979) Front-Face Fluorometry of Liquid Samples. Analytical Biochemistry. 94(1): 15–21.

Esteves VI, Santos EBH, and Duarte AC (1999) Study of the Effect of PH, Salinity and DOC on Fluorescence of Synthetic Mixtures of Freshwater and Marine Salts. Journal of Environmental Monitoring. 1(3): 251–254.

Fellman JB, D'Amore DV, and Hood E (2008) An Evaluation of Freezing as a Preservation Technique for Analyzing Dissolved Organic C, N and P in Surface Water Samples. Science of The Total Environment. 392(2): 305–312.

Galinha CF, Carvalho G, Portugal CAM, Guglielmi G, Oliveira R, Crespo JG, and Reis MAM (2011) Real-Time Monitoring of Membrane Bioreactors with 2D-Fluorescence Data and Statistically Based Models. Water Science and Technology. 63(7): 1381–1388.

Galinha CF, Carvalho G, Portugal CAM, Guglielmi G, Oliveira R, Reis MAM and Crespo JG (2011) Two-Dimensional Fluorescence as a Fingerprinting Tool for Monitoring Wastewater Treatment Systems. Journal of Chemical Technology & Biotechnology. 86(7): 985–992.

Galinha CF and Crespo JG (2022) Development and Implementation of MBR Monitoring: Use of 2D Fluorescence Spectroscopy. Membranes. 12: 1218.

Gilmore AM (2014) How to Collect National Institute of Standards and Technology (NIST) Traceable Fluorescence Excitation and Emission Spectra. In Fluorescence Spectroscopy and Microscopy: Methods and Protocols, Methods in Molecular Biology, eds. Yves Engelborghs and Antonie J.W.G. Visser. Totowa, NJ: Humana Press, 3–27.

Groeneveld M, Catalán N, Einarsdottir K, Bravo AG, and Kothawala DN (2022) The Influence of PH on Dissolved Organic Matter Fluorescence in Inland Waters. Analytical Methods. 14(13): 1351–1360.

Haberkamp J, Ernst M, Paar H, Palliscgeck D, Amy G, and Jekel M (2011) Impact of Organic Fractions Identified by SEC and Fluorescence EEM on the Hydraulic Reversibility of Ultrafiltration Membrane Fouling by Secondary Effluents. Desalination and Water Treatment. 29(1–3): 73–86.

Hambly A, Henderson RK, Baker A, Stuetz RM, and Khan SJ (2010) Probabilistic Analysis of Fluorescence Signals for Monitoring Dual Reticulation Water Recycling Schemes. Water Science and Technology. 62(9): 2059–2065.

Hambly AC, Arvin E, Pedersen L-F, Pedersen PB, Seredynska-Sobecka B, and Stedmon CA (2015) Characterising Organic Matter in Recirculating Aquaculture Systems with Fluorescence EEM Spectroscopy. Water Research. 83: 112–120.

Hambly AC, Henderson RK, Baker A, Stuetz RM, and Khan SJ (2015) Application of Portable Fluorescence Spectrophotometry for Integrity Testing of Recycled Water Dual Distribution Systems. Applied Spectroscopy. 69(1): 124–129.

Heinz M, and Zak D (2018) Storage Effects on Quantity and Composition of Dissolved Organic Carbon and Nitrogen of Lake Water, Leaf Leachate and Peat Soil Water. Water Research. 130: 98–104.

Henderson RK, Baker A, Murphy KR, Hambly A Stuetz RM and Khan SJ (2009) Fluorescence as a Potential Monitoring Tool for Recycled Water Systems: A Review. Water Research. 43(4): 863–881.

Henderson RK, Subhi N, Antony A, Khan SJ, Murphy KR, Leslie GL, Chen V, Stuetz RM, and Le-Clech P (2011) Evaluation of Effluent Organic Matter Fouling in Ultrafiltration Treatment Using Advanced Organic Characterisation Techniques. Journal of Membrane Science. 382(1): 50–59.

Hu H, and Ren H (2019) Removal of Bioavailable Dissolved Organic Nitrogen in Wastewater by Membrane Bioreactors as Posttreatment: Implications for Eutrophication Control. Bioresource Technology. 271: 496–499.

Huguet A, Vacher L, Relaxans S, Saubusse S, Froidefond JM, and Parlanti E (2009) Properties of Fluorescent Dissolved Organic Matter in the Gironde Estuary. Organic Geochemistry. 40(6): 706–719.

Jablonski A (1933) Efficiency of Anti-Stokes Fluorescence in Dyes. Nature. 131(3319): 839– 840.

Jacquin C, Lesage G, Traber J, Pronk W, and Heran M (2017) Three-Dimensional Excitation and Emission Matrix Fluorescence (3DEEM) for Quick and Pseudo-Quantitative Determination of Protein- and Humic-like Substances in Full-Scale Membrane Bioreactor (MBR). Water Research. 118: 82–92.

de Juan A, Jaumot J, and Tauler R (2014) Multivariate Curve Resolution (MCR). Solving the Mixture Analysis Problem. Analytical Methods. 6(14): 4964–4976.

Kimura K, Hane Y, Watanabe Y, Amy G, and Ohkuma N (2004) Irreversible Membrane Fouling during Ultrafiltration of Surface Water. Water Research. 38(14): 3431–3441.

Kopf U, and Heinze J (1984) 2,7-Bis(Diethylamino)Phenazoxonium Chloride as a Quantum Counter for Emission Measurements between 240 and 700 Nm. Analytical Chemistry. 56(11): 1931–1935.

Kothawala DN, Murphy KR, Stedmon CA, Weyhenmeyer GA, and Tranvik LJ (2013) Inner Filter Correction of Dissolved Organic Matter Fluorescence. Limnology and Oceanography: Methods 11(12): 616–630.

Kothawala DN, Stedmon CA, Müller RA, Weyhenmeyer GA, Köhler SJ, and Tranvik LJ (2014) Controls of Dissolved Organic Matter Quality: Evidence from a Large-Scale Boreal Lake Survey. Global Change Biology. 20(4): 1101–1114.

Lakowicz JR, ed. (2006a) Instrumentation for Fluorescence Spectroscopy. In Principles of Fluorescence Spectroscopy, Boston, MA: Springer US, 27–61.

Lakowicz JR, ed. (2006b) "Introduction to Fluorescence." In Principles of Fluorescence Spectroscopy, Boston, MA: Springer US, 1–26.

Lawaetz AJ, and Stedmon CA (2009) Fluorescence Intensity Calibration Using the Raman Scatter Peak of Water. Applied Spectroscopy. 63(8): 936–940.

Maie N, Parish KJ, Watanabe A, Knicker H, Benner R, Abe T, Kaiser K, and Jaffé R (2006) Chemical Characteristics of Dissolved Organic Nitrogen in an Oligotrophic Subtropical Coastal Ecosystem. Geochimica et Cosmochimica Acta 70(17): 4491–4506.

Mangal MN, Salinas-Rodriguez SG, Dusseldorp J, Blankert B, Yangali-Quintanilla VA, Kemperman AJB, Schippers JC, van der Meer WGJ, and Kennedy MD (2022) Foulant Identification and Performance Evaluation of Antiscalants in Increasing the Recovery of a Reverse Osmosis System Treating Anaerobic Groundwater. Membranes. 12(3): 290.

McKay G, Korak JA, and Rosario-Ortiz FL (2018) Temperature Dependence of Dissolved Organic Matter Fluorescence. Environmental Science & Technology. 52(16): 9022–9032.

McKnight DM, Boyer EW, Westerhoff PK, Doran PT, Kulbe T, and Andersen DT (2001) Spectrofluorometric Characterization of Dissolved Organic Matter for Indication of Precursor Organic Material and Aromaticity. Limnology and Oceanography. 46(1): 38–48.

Micó Tormos P, Garcia-Ballesteros S, Mora Carbonell M, Vincente Candela R, Amat Payá AM, and Arqués Sanz A (2019) EEMlab: A Graphical User-Friendly Interface for Fluorimetry Experiments Based on the DrEEM Toolbox. Chemometrics and Intelligent Laboratory Systems. 188: 6–13.

Mobed JJ, Hemmingsen SL, Autry JL, and McGown LB (1996) Fluorescence Characterization of IHSS Humic Substances: Total Luminescence Spectra with Absorbance Correction. Environmental Science & Technology. 30(10): 3061–3065.

Morán XAG, Gasol JM, Arin L, and Estrada M (1999) A Comparison between Glass Fiber and Membrane Filters for the Estimation of Phytoplankton POC and DOC Production. Marine Ecology Progress Series. 187: 31–41.

Murphy KR, Timko SA, Gonsior M, Powers LC, Wünsch UJ, and Stedmon CA (2018) Photochemistry Illuminates Ubiquitous Organic Matter Fluorescence Spectra. Environmental Science & Technology. 52(19): 11243–11250.

Murphy KR, Hambly A, Singh S, Henderson RK, Baker A, Stuetz RM, and Khan SJ (2011) Organic Matter Fluorescence in Municipal Water Recycling Schemes: Toward a Unified PARAFAC Model. Environmental Science & Technology. 45(7): 2909–2916.

Murphy KR, Stedmon CA, Graeber D, and Bro R (2013) Fluorescence Spectroscopy and Multi-Way Techniques. PARAFAC. Analytical Methods. 5(23): 6557– 6566.

Obayashi Y, and Suzuki S (2019) High Growth Potential of Transiently 0.2-Mm- Filterable Bacteria with Extracellular Protease Activity in Coastal Seawater. Plankton and Benthos Research. 14(4): 276–286.

Ohno T (2002) Fluorescence Inner-Filtering Correction for Determining the Humification Index of Dissolved Organic Matter. Environmental Science & Technology. 36(4): 742–746.

Osburn CL, Anderson NJ, Stedmon CA, Giles ME, Whiteford EJ, McGenity TJ, Dumbrell AJ, and Underwood GJC (2017) Shifts in the Source and Composition of Dissolved Organic Matter in Southwest Greenland Lakes Along a Regional Hydro-Climatic Gradient. Journal of Geophysical Research: Biogeosciences. 122(12): 3431–3445.

Parker CA, and Rees WT (1960) Correction of Fluorescence Spectra and Measurement of Fluorescence Quantum Efficiency. Analyst. 85(1013): 587–600.

Parlanti E, Wörz K, Geoffroy L, and Lamotte M (2000) Dissolved Organic Matter Fluorescence Spectroscopy as a Tool to Estimate Biological Activity in a Coastal Zone Submitted to Anthropogenic Inputs. Organic Geochemistry. 31(12): 1765–1781.

Pawlowski S, Galinha CF, Crespo JG, and Velizarov S (2016) 2D Fluorescence Spectroscopy for Monitoring Ion-Exchange Membrane Based Technologies – Reverse Electrodialysis (RED). Water Research. 88: 184–198.

Pucher M, Wünsch U, Weigelhofer G, Murphy KR, Hein T, and Graeber D (2019) StaRdom: Versatile Software for Analyzing Spectroscopic Data of Dissolved Organic Matter in R. Water. 11(11): 2366.

Poojamnong K, Tungsudjawong K, Khongnakorn W, and Jutaporn P (2020) Characterization of reversible and irreversible foulants in membrane bioreactor (MBR) for eucalyptus pulp and paper mill wastewater treatment using fluorescence regional integration. Journal of Environmental Chemical Engineering, 8(5): 104231

Pype M-L, Patureau D, Wery N, Poussade Y, and Gernjak W (2013) Monitoring Reverse Osmosis Performance: Conductivity versus Fluorescence Excitation–Emission Matrix (EEM). Journal of Membrane Science. 428: 205– 211.

Rosenstock B, and Simon M (1993) Use of Dissolved Combined and Free Amino Acids by Planktonic Bacteria in Lake Constance. Limnology and Oceanography. 38(7): 1521–1531.

Sansonetti CJ, Salit ML, and Reader J (1996) Wavelengths of Spectral Lines in Mercury Pencil Lamps. Applied Optics. 35(1): 74–77.

Shutova Y, Karna BL, Hambly AC, Lau B, Henderson RK, and Le-Clech P (2016) Enhancing Organic Matter Removal in Desalination Pretreatment Systems by Application of Dissolved Air Flotation. Desalination. 383: 12–21.

Singh S, Henderson RK, Baker A, Stuetz RM, and Khan SJ (2009) Distinguishing Stage 1 and 2 Reverse Osmosis Permeates Using Fluorescence Spectroscopy. Water Science and Technology. 60(8): 2017–2023.

Singh S, Henderson RK, Baker A, Stuetz RM, and Khan SJ (2015) Online Fluorescence Monitoring of RO Fouling and Integrity: Analysis of Two Contrasting Recycled Water Schemes. Environmental Science: Water Research & Technology. 1(5): 689–698.

Spencer RGM, Bolton L, and Baker A (2007) Freeze/Thaw and PH Effects on Freshwater Dissolved Organic Matter Fluorescence and Absorbance Properties from a Number of UK Locations. Water Research. 41(13): 2941–2950.

Spiliotopoulou A, Martin R, Pedersen L-F, and Andersen HR (2017) Use of Fluorescence Spectroscopy to Control Ozone Dosage in Recirculating Aquaculture Systems. Water Research. 111: 357–365.

Stedmon CA, and Bro R (2008) Characterizing Dissolved Organic Matter Fluorescence with Parallel Factor Analysis: A Tutorial. Limnology and Oceanography: Methods. 6(11): 572–579.

Stedmon CA, Markager S, and Bro R (2003) Tracing Dissolved Organic Matter in Aquatic Environments Using a New Approach to Fluorescence Spectroscopy. Marine Chemistry. 82(3): 239–254.

Stedmon CA, and Nelson NB (2015). Chapter 10 - The Optical Properties of DOM in the Ocean. In Biogeochemistry of Marine Dissolved Organic Matter (Second Edition), eds. Dennis A. Hansell and Craig A. Carlson. Boston: Academic Press, 481–508.

Stokes GG (1852) XXX. On the Change of Refrangibility of Light. Philosophical Transactions of the Royal Society of London. 142: 463–562.

Tang S, Wang Z, Wu Z, and Zhou Q (2010) Role of Dissolved Organic Matters (DOM) in Membrane Fouling of Membrane Bioreactors for Municipal Wastewater Treatment. Journal of Hazardous Materials. 178(1): 377–384.

Urgun-Demirtas M, Sattayatewa C, and Pagilla KR (2008) Bioavailability of Dissolved Organic Nitrogen in Treated Effluents. Water Environment Research: A Research Publication of the Water Environment Federation. 80(5): 397–406.

Vera M, Cruz S, Boleda MR, Mesa J, Martí-Alonso J, Casas S, Gibert O, and Cortina JL (2017) Fluorescence Spectroscopy and Parallel Factor Analysis as a Dissolved Organic Monitoring Tool to Assess Treatment Performance in Drinking Water Trains. Science of The Total Environment. 584–585: 1212–1220.

Wang H, Ding A, Gan Z, Qu F, Cheng X, Bai L, Guo S, Li G, and Liang H (2017) Fluorescent Natural Organic Matter Responsible for Ultrafiltration Membrane Fouling: Fate, Contributions and Fouling Mechanisms. Chemosphere. 182: 183–193.

Wang Y, Hammes F, Boon N, and Egli T (2007) Quantification of the Filterability of Freshwater Bacteria through 0.45, 0.22, and 0.1 Mm Pore Size Filters and Shape-Dependent Enrichment of Filterable Bacterial Communities. Environmental Science & Technology. 41(20): 7080–7086.

Wang Y, Hammes F, Düggelin M, and Egli T (2008) Influence of Size, Shape, and Flexibility on Bacterial Passage through Micropore Membrane Filters. Environmental Science & Technology. 42(17): 6749–6754.

Watras CJ, Hanson PC, Stacy TL, Morrison KM, Mather J, Hu Y-H, and Milewski P (2011) A Temperature Compensation Method for CDOM Fluorescence Sensors in Freshwater. Limnology and Oceanography: Methods. 9(7): 296–301.

Xiao K, Shen Y, Liang S, Tan J, Wang X, Liang P, and Liang X (2018) Characteristic Regions of the Fluorescence Excitation–Emission Matrix (EEM) To Identify Hydrophobic/Hydrophilic Contents of Organic Matter in Membrane Bioreactors. Environmental Science & Technology. 52(19): 11251–11258.

Xu H, Xu Y, Xiao K, Gao T, Liu Z, Xue W, Wei C, and Huang X (2022) Interplay of Organic Components in Membrane Fouling Evolution: Statistical Evidence from Multiple Spectroscopic Analyses. Journal of Membrane Science 661: 120913.

Yamamura H, Ding Q, and Watanabe Y (2019) Solid-Phase Fluorescence Excitation Emission Matrix for in-Situ Monitoring of Membrane Fouling during Microfiltration Using a Polyvinylidene Fluoride Hollow Fiber Membrane. Water Research. 164: 114928.

Yamin G, Borisover M, Cohen E, and van Rijn J (2017) Accumulation of Humic-like and Proteinaceous Dissolved Organic Matter in Zero-Discharge Aquaculture Systems as Revealed by Fluorescence EEM Spectroscopy. Water Research. 108: 412–421.

Yu H, Li Y, Yang H, Lv Y, Rong H and Qu F (2022) Characterization of Activated Sludge in Wastewater Treatment Processes Using Front-Face Excitation–Emission Matrix (FF-EEM) Fluorescence Spectroscopy. Environmental Science: Water Research & Technology. 8(10): 2265–2276.

Yu H, Wu Z, Zhang X, Qu F, Wang P, and Liang H (2019) Characterization of Fluorescence Foulants on Ultrafiltration Membrane Using Front-Face Excitation-Emission Matrix (FF-EEM) Spectroscopy: Fouling Evolution and Mechanism Analysis. Water Research. 148: 546–555.

Yu J, Xiao K, Qi T, Li Y, Tan J, Wen X, and Huang X (2021) Spectroscopic Sensing of Membrane Fouling Potential in a Long-Term Running Anaerobic Membrane Bioreactor. Chemical Engineering Journal. 426: 130799.

doi: 10.2166/9781789062977_0287

Chapter 13

Transparent Exopolymer Particles

Loreen O. Villacorte, Grundfos, Denmark

Yuli Ekowati, Grundfos, Denmark

Helga Calix Ponce, Denmark

The learning objectives of this chapter are the following:

- Understand the relevance of transparent exopolymer particles (TEP) to membrane filtration processes
- Learn the different methods to measure TEP in fresh and saline water sources
- Describe experimental protocols to quantify TEPs and their precursors
- Discuss application of TEP methods for membrane filtration applications

13.1 INTRODUCTION

Transparent exopolymer particles (TEP) and their precursors has been reported to cause organic or biological fouling in membrane filtration processes such as microfiltration (MF), ultrafiltration (UF), nanofiltration (NF) and reverse osmosis (RO). As the name suggest, TEPs are transparent organic substances, seasonally abundant in marine and fresh surface water environments, particularly during algal blooms. TEPs has been known to exist in marine and lake environments since the early 90's but its link to membrane processes has only been studied since the mid-2000's. Since then, experimental methods have been adopted, modified, developed, and demonstrated to quantify these substances and elucidate their impact to membrane systems.

TEPs largely originate from exudates or detritus of phytoplankton (micro-algae) and bacterioplankton but they can also originate from macro-algae and some species of oysters, mussels, scallops and sea snails. TEPs are generally sticky and highly hydrated gels, comprising mainly hydrophilic, negatively charged, acidic polysaccharides (Mopper *et al.*, 1995). They tend to associate with or absorb proteins, lipids, trace elements and heavy metals from the water (Passow, 2002). This makes them a good platform and hotspot for bacterial growth and likely have an important role in the formation of aquatic biofilms (Alldredge *et al.*, 1993; Passow, 2002; Bar-Zeev *et al.*, 2012a).

Berman and Holenberg (2005) initially proposed the potential role of TEP as a major initiator of biofilm leading to biofouling in reverse osmosis (RO) membranes. Consequently, various studies were conducted to investigate the link between TEP and biofouling in membranes (Figure 1; Bar-Zeev *et al.*, 2009; Villacorte *et al.*, 2009a,b; Villacorte *et al.*, 2017a). Further studies have also demonstrated that TEPs can directly cause organic fouling in MF/UF (Figure 1; Kennedy *et al.*, 2009; Villacorte *et al.*, 2010a,b; 2013; 2015a; Schurer *et al.*, 2012, 2013) and forward osmosis membranes (Valladares Linares *et al.*, 2012).

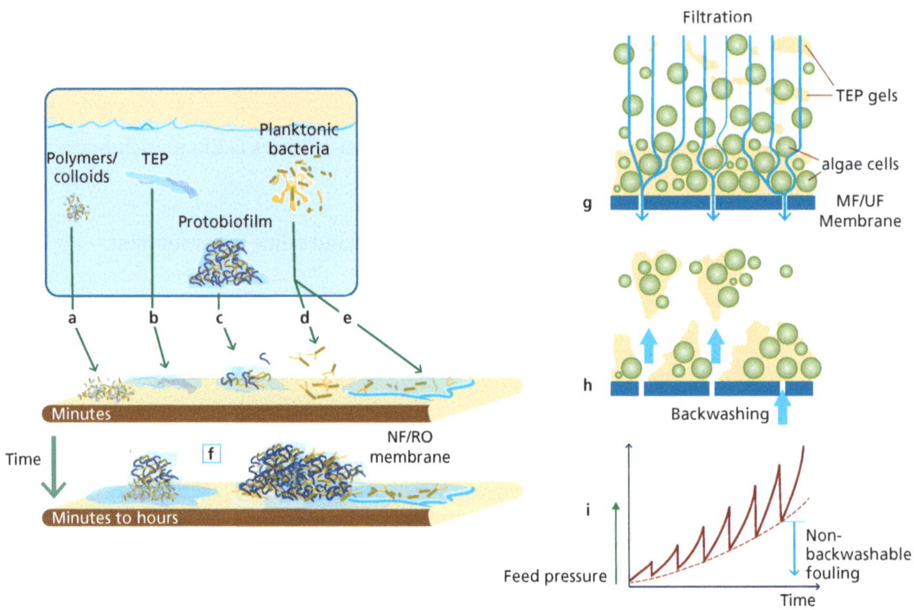

Figure 1 Possible contribution of TEP on biological and organic fouling in membrane systems. In RO/NF, some planktonic bacteria (first colonizers) can attach (d) reversibly on clean surfaces or (e) irreversibly on TEP-conditioned surfaces. When nutrients are available in the water, (f) contiguous coverage of mature biofilm can develop within a short period of time (minutes to hours). In MF/UF, organic fouling can occur during (g) filtration of algal bloom impacted water. TEPs strongly adhere to membrane pores and surfaces causing (h) incomplete removal of cake layer during backwashing and leading to (i) gradual increase in feed pressure or permeability over time. Figures adapted from Bar-Zeev *et al.* (2012a) and Villacorte *et al.* (2021).

Early analytical methods operationally defined TEPs as particles larger than 0.4 μm considering that they were first identified through retention on 0.4 μm pore size membrane filters (Alldredge *et al.*, 1993). They typically contain more than 99% water, which means they can bulk-up to more than 100 times their dried volume (Azetsu-Scott and Passow, 2004; Verdugo *et al.*, 2004). Large TEPs can be directly produced through sloughing of algal cell coatings or through disintegration of large algal colonies (Figure 2). Other TEPs are produced indirectly from colloidal polymers (1-10 kDa) released by phyto-/bacterio-planktons which eventually grow into TEPs (>0.4 μm) through the subsequent process of annealing, gelation, and aggregation (Verdugo *et al.*, 2004; Chin *et al.*, 1998). These sub-micron components (<0.4 μm) which have similar chemical properties as TEPs are collectively known as TEP precursors (Passow, 2000).

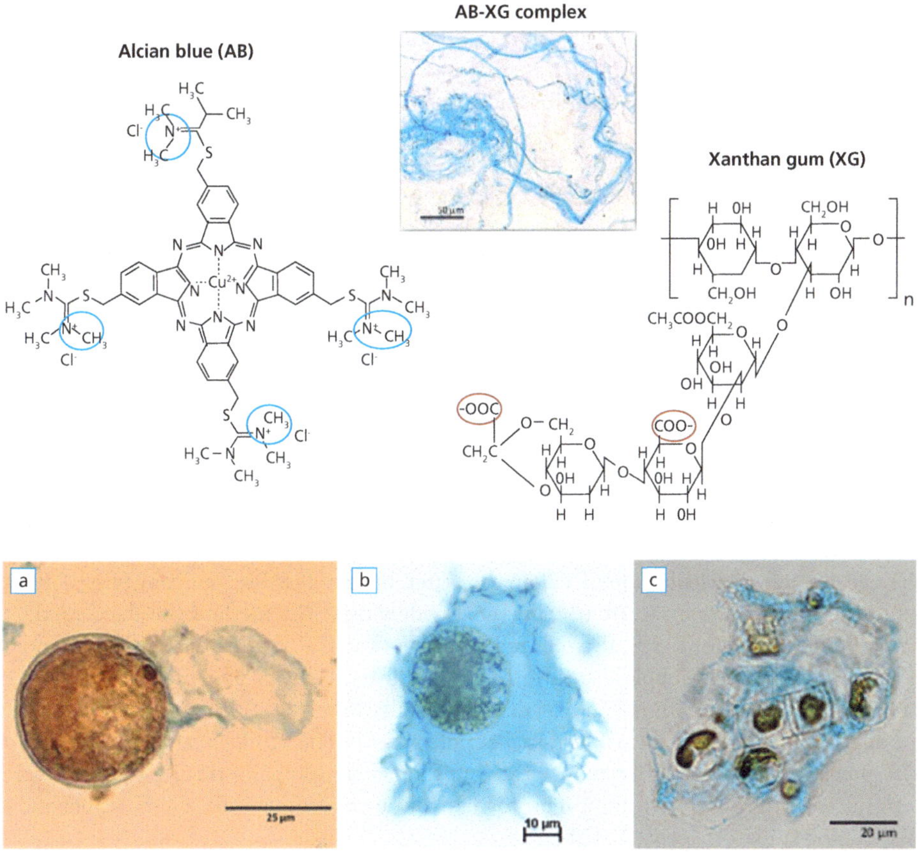

Figure 2 Top figures show the molecular structures of alcian blue (AB) and a standard acidic polysaccharide xanthan gum (XG) and precipitates formed by the reaction of the two compounds at pH 2.5 (adopted from Villacorte *et al.*, 2015b). Bottom figures are optical microscope images of Alcian Blue stained algal cells of (a) *Alexandrium tamarense*, (b) *Lepidodinium chlorophorum* and (c) *Chaetoceros affinis*, where TEPs (stained blue) were released through shedding of cell mucus (b and c) and membrane coatings (a and b). Images a and c adopted from Villacorte *et al.* (2015c) and b from Claquin *et al.* (2008).

13.2 QUANTIFICATION METHODS

Since the discovery of TEPs, various quantification methods have been developed, all of which involved staining with Alcian Blue (AB) dye). The dye is known to be highly selective and forms insoluble complexes with target compounds that cannot be easily reversed by subsequent treatments. AB is also widely available and has been routinely used in medical and biological research. However, despite being one of the most widely used biological stains, the mechanisms involved during reaction of the dye with a specific substrate is still not well understood.

An AB molecule is a tetravalent cation with a copper atom at the center of its core (Figure 2). In aqueous solutions without extra electrolytes, AB can bind with anionic carboxyl, phosphate and half-ester sulphate groups of acidic polysaccharides, resulting in the formation of neutral blue precipitates (Figure 2). It is also known to react with carbohydrate-conjugated proteins such as acidic glycol-proteins and proteoglycans but does not stain nucleic acids and neutral biopolymers.

AB staining can be largely impacted by the type and density of anionic functional groups associated with the material in the sample. Selectivity depends on the pH and ionic strength of the sample solution. In high ionic strength solutions, better interactions between anionic polymers and cationic AB can be expected due to compression of the electrical double layer surrounding the AB molecule. It can also spontaneously aggregate in saline solutions forming AB precipitates not associated with TEP. This is a major drawback of the application of AB for TEP measurements in seawater. To minimize this, AB staining solutions should be pre-filtered and should not be directly applied to solutions with high salinity.

Table 1 shows an overview of the available TEP methods described in literature. The first ever TEP method is a direct quantification through filter retention, AB staining and optical microscopic enumeration (Alldredge *et al.*, 1993). The method can provide information of the size-frequency distribution of TEP in the water, but not feasible for quantifying TEPs < 2 μm and TEP precursors. Currently, the most widely used TEP method was developed by Passow and Alldredge (1995), also referred to this work as $TEP_{0.4\mu m}$. With additional sample preparation steps (e.g., bubble adsorption, laminar shear), it may be possible to measure TEP precursors using this method (Zhou *et al.*, 1998). Villacorte *et al.* (2009a) proposed a slight modification for $TEP_{0.4\mu m}$ by using a smaller pore size filter (0.1 μm) to capture some of the TEP precursors ($TEP_{0.1\mu m}$). To reduce the interference of salinity, a rinsing step was later introduced to the $TEP_{0.4\mu m}$ filtration protocol to dilute or minimize salts on the filter before AB staining (Villacorte *et al.*, 2015b).

Arruda-Fatibello *et al.* (2004) introduced a different approach to measuring TEP by direct staining on water samples. However, it is only applicable for freshwater samples because salts interfere with AB staining. The method by Thornton *et al.* (2007) introduced a similar approach but added a dialysis step for saline samples. Further modification of the method, known as TEP_{10kDa}, was later introduced to address a major practical limitation of the two previous methods by introducing a sample concentration step through 10 kDa membrane filtration (Villacorte *et al.*, 2015b; 2017b). This method can substantially reduce the TEP analysis time for brackish or saline water samples. It also enables the size fractionation of TEPs in the water by filtering through series of membranes with decreasing pore sizes during the concentration step.

More recent developments have shown that TEP can be measured online using an auto-imaging technique (Thuy *et al.*, 2017) or a crossflow filtration unit with integrated spectrophotometer (Sim *et al.*, 2019). Online measurement techniques would be the next logical step towards routine TEP monitoring especially during algal blooms. However, further studies are still needed to verify replicability and reliability of these advanced techniques in the field, particularly regarding the impact of salinity during staining.

Table 1 Overview of quantification methods for TEP and their precursors.

Experimental methods	Description of main steps	TEP analysed	Water type	References
Offline measurement (grab samples)				
Microscopic enumeration	Filtration, AB staining, microscopic counting	TEP>2µm	fresh/saline	Alldredge *et al.* (1993)
$TEP_{0.4\mu m}$	Filtration (0.4µm), AB staining, acid extraction, absorbance measurement	TEP>0.4µm	fresh/saline	Passow & Alldredge (1995); Villacorte *et al.* (2015)
Rapid spectrophotometric	AB staining, centrifugation, absorbance measurement	TEP + precursors	fresh	Arruda-Fatibello *et al.* (2004);
Acidic polysaccharide (APS)	dialysis (1 kDa; if saline water), AB staining, filtration (0.1µm), absorbance measurement	TEP + precursors	fresh/saline	Thornton *et al.* (2007)
$TEP_{0.05\mu m}$	Filtration (0.4 and 0.05µm), AB staining, acid extraction, absorbance measurement	TEP>0.05µm	fresh/saline	Villacorte *et al.*. (2009a)
TEP_{10kDa}	Filtration (10kDa), ultrasonication, AB staining, acid extraction, absorbance measurement	TEP + precursors	fresh/saline	Villacorte *et al.* (2015b)
Inline/Online measurement				
Flow-CAM imaging	AB staining, FlowCAM imaging, image processing and counting	TEP > 5µm	fresh	Thuy *et al.* (2017)
Crossflow TEP monitor	crossflow filtration, AB staining, fiber optic spectrophotometry	TEP + precursors	fresh/saline	Sim *et al.* (2018)

The succeeding sections include detailed descriptions of two relevant methods for measuring TEPs in and fresh and saline water, namely $TEP_{0.4\mu m}$ and TEP_{10kDa}. The $TEP_{0.4\mu m}$ method measures TEPs retained by membrane filters having pores of 0.4 µm and conventionally known as TEP (Passow and Alldredge, 1995). The TEP_{10kDa} method covers transparent exopolymer particles retained by membrane filters with molecular weight cut-off of 10 kDa. Consequently, this method covers both TEP and most (if not all) of their colloidal precursors.

13.2.1 Alcian blue dye preparation

Spectrophotometric TEP methods involve absorbance analyses of AB in either acetic acid solution (pH 2.5) or 80% sulfuric acid solution. Figure 3a shows the spectral scans of AB dissolved in these two matrices. The maximum absorbance of AB in 80% sulphuric acid solution is at 787 nm wavelength while the maximum absorbance of AB in acetic acid solution (pH 2.5) is at 610 nm within the visible light spectrum. The typical absorbance value of AB in sulphuric acid at 787 nm is around twice that of AB of similar concentration dissolved in acetic acid solution at 610 nm (Figure 3b). Regardless of the method used, preparation of AB solution is usually prepared by dissolving AB in acetic acid solution at pH 2.5. The succeeding sections describe the recommended procedure in preparing the AB dye solutions for TEP measurements.

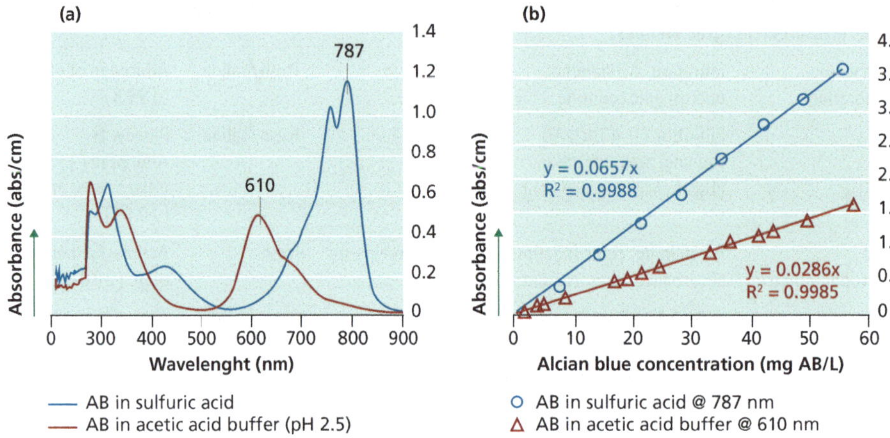

Figure 3 (a) Absorption spectra of Alcian blue 8GX (16 mg AB/L) dissolved in sulphuric acid and acetic acid solutions and (b) peak absorbance values of Alcian blue (AB) solutions at various concentrations (adopted from Villacorte *et al*, 2015b).

Materials and analytical set-up

- Alcian blue 8GX, 0.05 g
- Ultrapure water (UPW), 200 mL
- Magnetic stirrer
- Acetic acid
- pH meter
- Glass beaker
- Filter 0.05 μm (polycarbonate track etched membrane filters 47 mm)
- Clean glass tube (for capturing the filtered solution)
- Vacuum filtration set-up for 47 mm filter
- Vacuum pump
- Demineralised water
- Glass container for the final AB solution

Preparation of AB stock solution (0.025%)

1. Add 200 mL of UPW in a clean glass beaker and stir using a magnetic stirrer at 500 rpm.
2. Add acetic acid to the UPW drop by drop until pH lowers to 2.5.
3. While stirring at 1000 rpm, add 50 mg of AB powder and continue stirring for 18 to 24 hrs at 30 °C. Make sure to cover the top of the beaker with parafilm during this period.
4. Store the stock solution at 4 °C in a closed container. In practice, the stock solution should be used only within 4 weeks after it was prepared.
5. Calculation of the amount of alcian blue powder needed:

$$\%AB = \frac{m(g)}{V(mL)} \times 100\%$$

0.025% AB = x (g) / 200 mL UPW,
then x = 0.05 g AB is added per 200 mL of ultrapure water

Preparation of AB working solution

In each time performing TEP measurement, the amount of AB solution needed should be filtered the same day of the measurement (around 1 mL per measurement plus 10 mL extra).

1. Clean the vacuum filtration set up by filtering 100 mL of UPW at pressure ≤ 0.2 bar. Dispose the filtered water.
2. Place the clean glass tube inside the filter flask to capture the filtered AB stock solution.
3. Mount a 0.05 μm PC filter on the filter holder of the vacuum filtration set-up.
4. Filter the amount of AB stock solution needed at a vacuum pressure ≤ 0.2 bar. Throw away the first few drops of filtered, light blue colored filtrate or until dark blue color filtrate appears.
5. Filter the collected filtrate again through a clean 0.05 μm filter on same filtration set-up.
6. Transfer the twice filtered AB stock solution in a glass container and cover.

13.2.2 $TEP_{0.4\mu m}$ measurement

$TEP_{0.4\mu m}$ was originally developed in the mid-90s by Passow and Alldredge (1995) and later improved by Villacorte *et al.* (2015b) as illustrated in Figure 5 and further described in the succeeding sections.

Materials and analytical set-up

- Alcian blue working solution (see section 13.2.1)
- 47 mm polycarbonate track-etched membrane filters (0.4 μm pore size)
- Clean glass tube (for collecting filtered solution)
- Vacuum filtration set-up for 47 mm filter (see Figure 4)
- Vacuum pump
- Demineralised water (demiwater), for cleaning filtration set-up
- Ultrapure water (UPW), for rinsing membrane filters.
- 50 ml glass beaker
- 80% sulfuric acid solution
- Shaker

- Spectrophotometer
- 1 cm cuvette
- Pipette 5 mL and 1 mL
- Tweezer
- Measuring cylinder

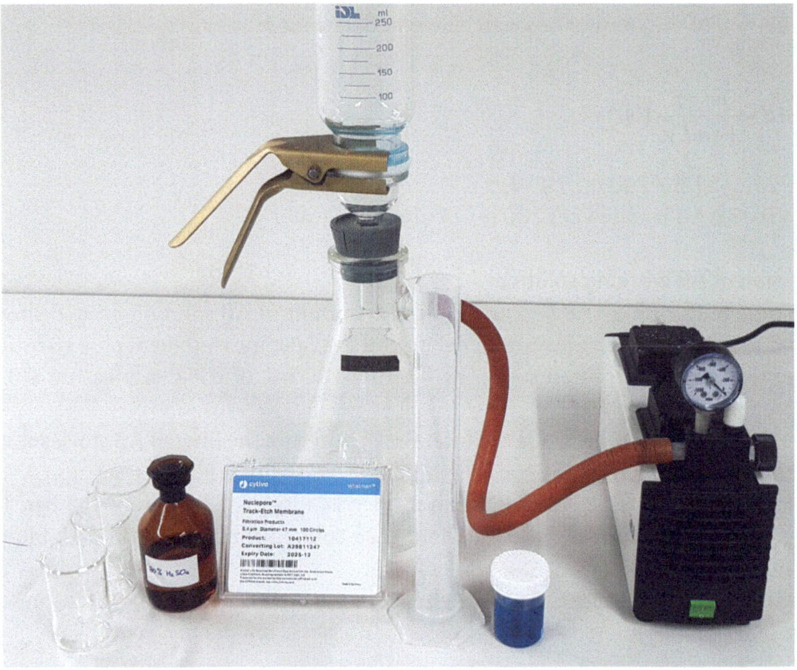

Figure 4 Overview of analytical setup for $TEP_{0.4\mu m}$ measurement.

Sample filtration and staining

1. Collect 0.5-1 L of water samples to be analyzed. A volume (typically >20 mL) of the sample is filtered through a 47 mm diameter PC filter (0.4 μm pore size) by applying a vacuum of 0.2 bar. Note: To remove possible contaminants, rinse PC filters by flushing >200 ml of UPW through it prior to sample filtration.
2. Filter (≤0.2 bar vacuum) 2 mL of UPW through the retained TEP to wash the remaining sample moisture through the filter and replace it with very low salinity water.
3. Pipette 1 mL of the working AB dye solution, apply over the filter, allowed to react with TEP for 10 seconds, and then flush the unreacted dye through by vacuum filtration (<0.2 bar). To remove the remaining unreacted dye, perform a rinsing step by filtering 2 mL of UPW.
4. Using a tweezer, fold the rinsed filter two-fold (while on the filter holder) with the stained TEP in the inner part, and transfer to a 50 mL glass beaker.
5. Add 6 mL of 80% sulfuric acid solution on top on the filter in the beaker, cover beaker with parafilm and mix on an auto-shaker for 2 hours.

6. Transfer part of the acid solution to a 1-cm cuvette and measure the absorbance (A_t) at 787 nm wavelength.
7. Measure the filter blank (A_f) in the same way as Steps 1-6 but filtering TEP-free blank samples (e.g., synthetic water with similar ion concentration as the water sample) instead of actual water samples.
8. Measure sample blank correction (A_s) by filtering water samples in the same way as Steps 1-6 but skipping the AB staining procedure (Step 3).
9. Repeat at least three times Steps 1-8 per sample, per batch of filters and per batch of working AB solution.

To prevent cross-contamination of samples, it is recommended to clean the glass filter holder with sulfuric acid and then with UPW in between tests until no more visible AB remains on the holder. The filtered sample volume in Step 1 can be adjusted (increased or reduced) depending on the initial absorbance results. To get more reliable absorbance results, it is recommended to reduce sample volume if initial absorbance result is ≥0.8 cm^{-1} and increase sample volume if initial absorbance is very close to or lower than the blank absorbance.

Concentration calculation without calibration

The concentration of $TEP_{0.4\mu m}$ in terms of abs/cm/L is calculated as follows:

$$TEP_{0.4\mu m} = \frac{A_t - A_f - A_s}{V_f} \qquad \text{Eq. 1}$$

where (A_t) is the total absorbance of the dye that reacted with TEP and the filter (abs/cm); (A_f) is the absorbance of the dye adsorbed to the filter (abs/cm); (A_s) is the absorbance of unstained sample (abs/cm) and V_f is the volume of sample filtered (L).

Concentration calculation with calibration

$TEP_{0.4\mu m}$ can be further calibrated and expressed in terms of equivalent weight of standard acid polysaccharide - Xanthan gum – as mg X_{eq}/L:

$$TEP_{0.4\mu m} = \frac{A_t - A_f - A_s}{m_{787} V_f} \qquad \text{Eq. 2}$$

where m_{787} is the slope of the calibration line [(abs/cm)/mg X_{eq}] which is determined by calibrating the absorbance of AB corresponding to the mass of the Xanthan gum stained as described in Sections 12.2.4.2 or 12.2.4.3.

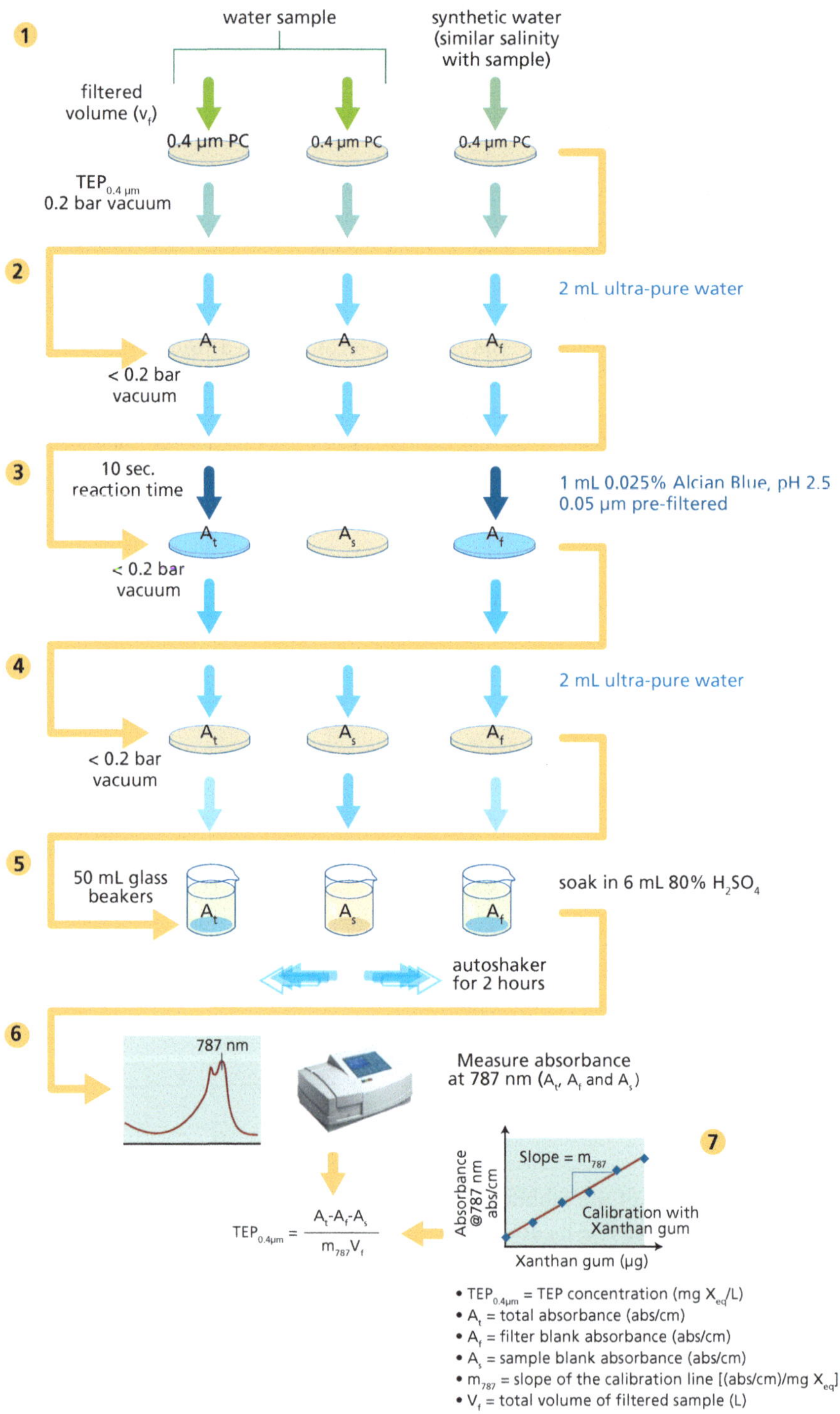

Figure 5 Procedural diagram for measuring $TEP_{0.4\mu m}$ (adapted from Villacorte, 2015).

13.2.3 TEP_{10kDa} measurement

The TEP_{10kDa} method was developed to measure TEP and their precursors (down to 10 kDa). This method was described by Villacorte *et al.* (2015b) and partly based on the method developed by Thornton *et al.* (2007). Figure 7 illustrates the protocol for measuring TEP_{10kDa} and further described in the following sections.

Materials and analytical set-up

- Alcian blue working solution (see Section 13.2.1)
- 10 kDa MWCO flat sheet membrane (25 mm regenerated cellulose)
- 60 mL plastic syringe
- Syringe pump (see Figure 6a)
- Filter holder 25 mm
- Ultrapure water (UPW), for rinsing membrane filters.
- Acetic acid
- Clean plastic cups (for collecting 10 kDa filtered samples)
- Vortex mixer
- Sonicator (see Figure 6b)
- Vacuum filtration set-up for 25 mm filter with stainless steel filter support (see Figure 6c)
- Vacuum pump
- Filter 0.1 μm polycarbonate membrane filters (47 mm diameter)
- clean plastic cups (for capturing 0.1 μm filtered stained sample)
- Demineralised water (demi-water), for cleaning filtration set-up
- Glass beaker
- Spectrophotometer
- Pipette 1 mL, 0.1 mL
- Tweezer
- Measuring cylinder

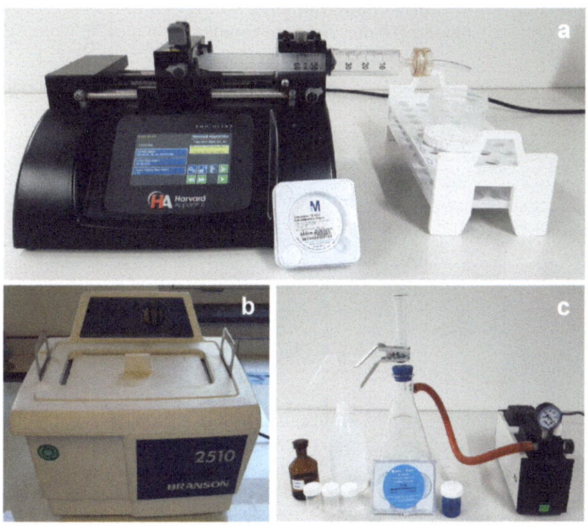

Figure 6 Overview of analytical setup for TEP_{10kDa} measurement: (a) syringe filtration unit, (b) ultrasonication unit and (c) vacuum filtration unit.

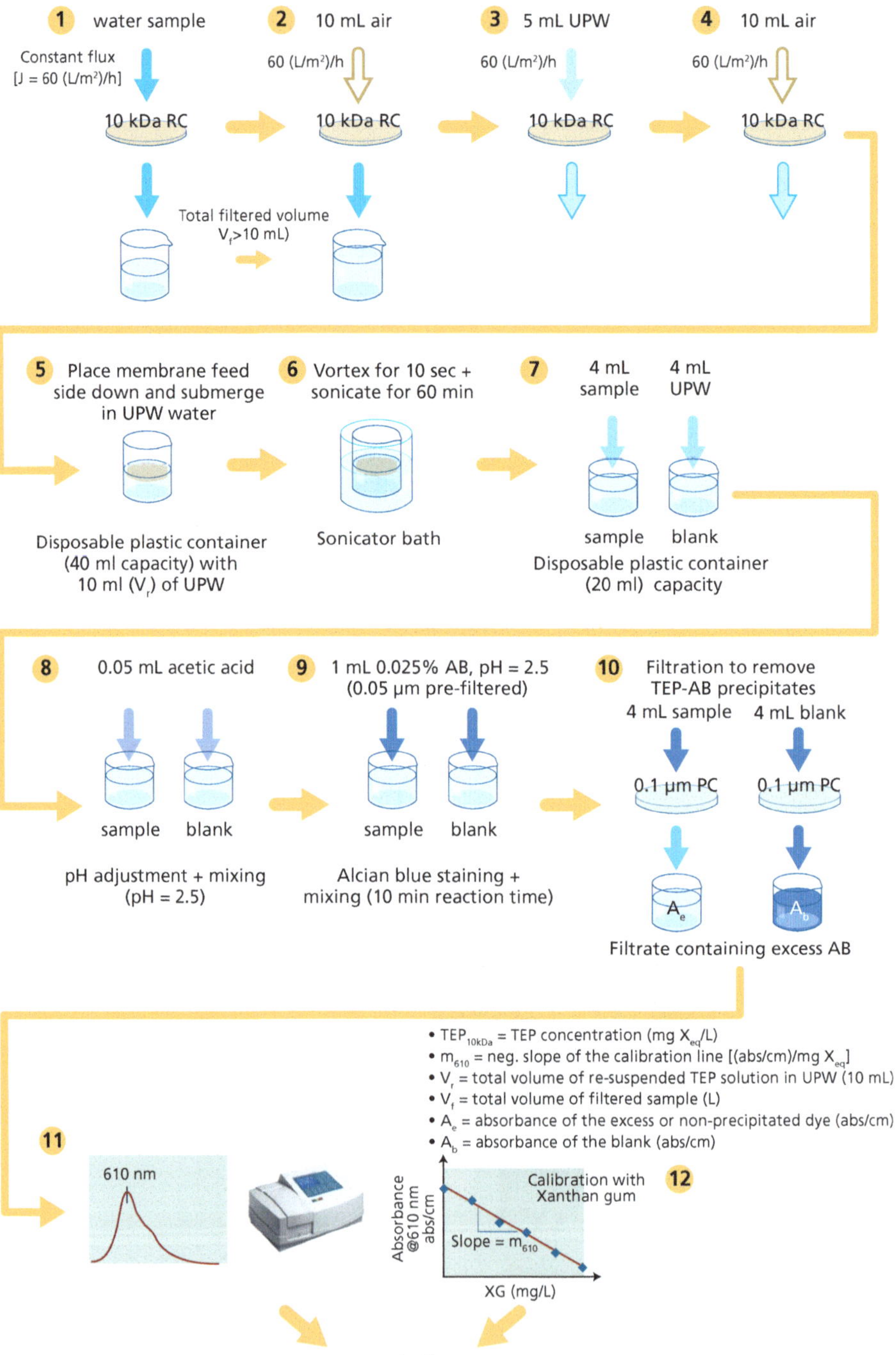

$$TEP_{10\,kDa} = \frac{1}{m_{610}} \frac{V_r}{V_f} (A_e - A_b)$$

Figure 7 Procedural diagram for measuring TEP_{10kDa} (adapted from Villacorte, 2014.)

Sample filtration and TEP extraction

1. Place a clean 10 kDa MWCO regenerated cellulose (RC) membrane (25 mm diameter) to the syringe filter holder and filter 10-100 mL water sample at constant flux (60 L/m^2/h) using a syringe pump while collecting the filtrate in a container. Note: To remove possible contaminants, soak RC membranes for at least 24 hours in UPW and then flushing 5-10 ml UPW prior to sample filtration.
2. After filtering sample, replace the syringe with a new clean syringe containing 5-10 mL of air. Filter air through to the filter (60 L/m^2/h) until all the remaining water in the feed side of the membrane holder has passed through the membrane. Measure total volume of sample in the filtrate container using a measuring cylinder.
3. To rinse out residual saline moisture, filter 5 mL of UPW through the filter holder at 60 L/m^2/h. Filter air again as in step 2 until all the rinse water on the feed side of the membrane holder has passed through the membrane.
4. Carefully remove the membrane from the filter holder and place feed side down, in a clean disposable plastic container filled with 10 mL of UPW.
5. Cover the container, vortex for 10 s and sonicate for 60 min.
6. Transfer 4 mL of the re-suspended TEP solution to a clean 20 mL disposable plastic container. Adjust the sample pH to 2.5 by adding 0.05 mL of acetic acid solution the sample solution.
7. Add 1 mL of the working AB dye solution to the sample, mix vigorously and allowed to react for 10 min.
8. Rinse PC filter by filtering 4 mL of UPW try to remove all the remaining UPW inside of the funnel. Any extra UPW might dilute the TEP-AB filtrate and lower the absorbance reading. Filter 4 mL sample of the TEP-AB solution through a 0.1 μm PC filter by vacuum filtration (0.2 bar). Collect the filtrate in a plastic container (10 mL) as shown in Figure 8.
9. Transfer part of the filtrate to a 1-cm cuvette and measure absorbance (A_e) at 610 nm wavelength using a spectrophotometer.
10. Measure the blank absorbance (A_b) to determine amount of AB stain adsorbed on PC filter by following steps 6-9 but replacing the sample with UPW.

To prevent cross-contamination of samples, it is recommended to clean the syringe filter holder and vaccum filter holder with UPW in between tests until no more visible AB remains on the holders. The filtered sample volume in Step 1 can be adjusted (increased or reduced) depending on initial absorbance results. To get more reliable absorbance results, it is recommended to reduce sample volume if initial absorbance result is ≤0.05 cm^{-1} and increase sample volume if initial absorbance is very close to or higher than the blank absorbance.

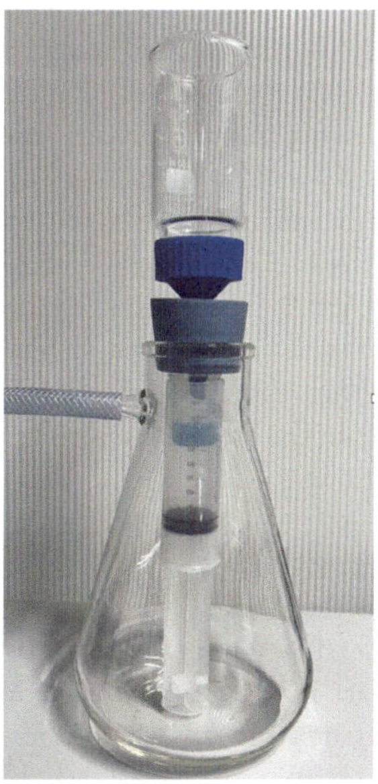

Figure 8 An example of how to assemble the setup to collect residual AB solution directly into a plastic sample container during vacuum filtration.

Concentration calculation (without calibration)

The TEP_{10kDa} concentration in absorbance per cm per liter of filter water (abs/cm/L) is calculated as follows:

$$TEP_{10kDa} = \frac{A_b - A_e}{V_f} \qquad \text{Eq. 3}$$

where A_b is the absorbance of filtered blank (abs/cm), A_e is the absorbance of the excess or un-reacted dye (abs/cm) and V_f is the volume of filtered sample (mL).

Concentration calculation (with calibration)

Ideally, the TEP_{10kDa} concentration can be calibrated and expressed in terms of mg Xanthan equivalent per litre (mg X_{eq}/L):

$$TEP_{10kDa} = \frac{1}{m_{610}} \frac{V_r}{V_f} (A_e - A_b) \qquad \text{Eq. 4}$$

where V_r is the total volume of the re-suspended TEP sample solution (i.e., 10 mL) and m_{610} is the slope of the calibration curve [(abs/cm)(mg X_{eq}/L)] which is determined by calibrating the absorbance of residual AB corresponding to the mass of the Xanthan gum stained on the filter (see section 13.2.4.4).

13.2.4 Method calibration

calibration of absorbance results from TEP analyses (Li *et al.*, 2018). XG is by far the most widely accepted standard among surrogates. However, Thornton *et al.* (2007) argued that alginic acid is a more suitable standard for TEP precursors. For consistency and comparability of results, it is recommended to use XG as the standard for quantifying both TEPs and their precursors.

The first TEP calibration protocol introduced by Passow and Alldredge (1995) involves dry weight measurements to determine the mass of Xanthan gum retained on PC filters. The procedure is tedious and prone to inaccuracies particularly during drying (dust contamination) and weighing (electrostatic force interference) of very low quantities (5-50 µg) of XG. It can be also challenging to prepare a homogeneous and artifact-free solution of XG for the calibration. Various TEP studies have skipped the calibration step entirely, whereby concentrations of TEP are expressed in terms of abs/cm/L (see Section 13.2.3). Without calibration, TEP results cannot be directly compared with results from analyses using different batches of AB staining solution. To overcome the above challenges, improved TEP calibration protocols have been introduced without the drying and weighing steps. These new protocols are described in the succeeding sections.

13.2.4.1 Xanthan gum standard preparation

A standard XG solution is prepared by dissolving XG in UPW solution. For example, to prepare 100 mg/L solution, 50 mg of XG is added to 500 mL of UPW while rapidly stirring with a magnetic stirrer. Rapid stirring is maintained for at least 1 hour until no flocs are visible. Typically, the solution is further homogenized with a tissue grinder. Each volume of 50-100 mL of Xanthan gum solution is homogenized 3 times by fully rotating the pestle 5 times for each batch. A recent study by Bittar *et al.* (2018) found that current commercially available XG is easier to dissolve and forms negligible number of gel-like particles in solution compared with the earlier versions of the XG powder used in the original calibration method (Passow and Alldredge, 1995). Hence, the use of a tissue grinder may no longer be necessary to homogenize the standard XG solution. The following sections describe the different calibration protocols for $TEP_{0.4\mu m}$ and TEP_{10kDa}.

13.2.4.2 $TEP_{0.4\mu m}$ calibration 1

Villacorte *et al.* (2015b) introduced a simpler and replicable standard calibration procedure than the original method for $TEP_{0.4\mu m}$. The calibration steps are as follows:

1. Prepare 4 ml standard solutions containing different concentrations of XG (0, 1, 2, 3, 4 and 5 mg/L) by diluting of XG stock solution (100 mg/L) with UPW.
2. Adjust pH of solutions to pH 2.5 by adding 0.05 mL acetic acid to each solution and then briefly agitated.
3. Take one solution and add 1 mL of pre-filtered AB solution, mix for 10 s and incubate fo 10 min.
4. Filter 4 mL of the reacted solution through a 0.1 μm PC membrane by vacuum filtration (0.2 bar).
5. While in the filter holder, fold the PC membrane with feed side in and carefully transfer to a 50 mL beaker.
6. Add 6 mL of 80% sulfuric acid solution to the beaker, cover it with a parafilm and mix on an auto-shaker for 2 hours.
7. Transfer part of the acid solution to a 1-cm cuvette and measure absorbance at 787 nm using a spectrophotometer.
8. Repeat steps 3-7 for each of the remaining standard solutions.
9. To determine the calibration slope (m_{787}), the mass of Xanthan gum retained on the PC membrane is calculated by multiplying the volume filtered (4 mL) with the concentration of Xanthan in the stained standard solution. The calculated mass is then plotted against the corresponding AB absorbance measured at 787 nm wavelength, whereby the average linear slope is the m_{787}.

13.2.4.3 $TEP_{0.4\mu m}$ calibration 2

Bittar *et al.* (2018) introduced an alternative $TEP_{0.4\mu m}$ calibration procedure as follows:

1. Prepare 75 mg/L of XG stock solution in UPW. To make XG standard solution, pipette 0.125, 0.250, 0.500, 0.750, and 1 mL of XG stock solution, in triplicates, into clean 5-mL polypropylene tubes. Top-up with UPW to bring the final volume of the solutions to 1 mL. The prepared XG standard solutions should contain 9.37, 18.75, 37.50, 56.25, and 75 μg-XG
2. Stain procedural blanks (1 mL ultrapure water in triplicates) and standard solutions by adding 0.5 mL of AB solution (400 mg/L) to the polypropylene tubes, for a final volume of 1.5 mL, and mix by manually agitating the tubes for 1 min.
3. Pour the stained standard solutions directly to the vacuum filtration funnel and filter through 0.22 μm or 0.45 μm polycarbonate filters (25 mm) at <175 mm Hg.
4. Transfer the filters to clean glass vials/beakers with caps/covers.
5. Add 6 mL of extraction solution (80% sulfuric acid) to the vials/beakers and cover immediately. Soak filters in in the solution for 2–20 h while vials are agitated regularly.

6. Measure the absorbance of AB extracted from the filters at 787 nm in a spectrophotometer using a 1-cm cuvette. During absorbance measurement, the spectrophotometer is first blanked with ultrapure water and then the absorbance of acid extraction solution is measured at 787 nm to check for consistency and potential contaminations of each extraction solution batch (0.008–0.150 ± 0.004). The instrument is then further blanked with extraction solution.
7. To determine the calibration slope (m_{787}), the mass of XG retained on the PC membrane is calculated by multiplying the XG stock solution concentration (75 mg/L) with the volume of XG stock solution used for each point in the calibration (0.125, 0.250, 0.500, 0.750, and 1 mL). The calculated mass is then plotted against the corresponding AB absorbance measured at 787 nm wavelength, whereby the average linear slope is the m787.

13.2.4.4 TEP_{10kDa} calibration

For TEP_{10kDa}, Villacorte *et al.* (2015b) developed a new calibration protocol modified from the protocol described by Thornton *et al.* (2007). This new calibration protocol can be performed simultaneously with $TEP_{0.4\mu m}$ calibration as illustrated in Figure 9.

1. Prepare standard solutions (4 mL) containing different concentrations (0, 1, 2, 3, 4 and 5 mg/L) of XG from standard stock solution (Section 12.2.4.1).
2. Adjust sample pH to 2.5 by adding 0.05 mL acetic acid to each solution and then agitated briefly. The solution is then stained by adding 1 mL of pre-filtered AB staining solution, mixed for 10 seconds, and incubated for 10 min.
3. Filter 4 mL of the resulting solution through a 0.1 µm PC membrane by vacuum filtration (0.2 bar). The filtrate is collected, transferred to 1-cm cuvette and absorbance is measured at 610 nm.
4. The concentration of standard XG solution stained with AB is plotted against the measured absorbance (excess dye absorbance) and the average linear line is the calibration slope (m_{610}). Since concentration is inversely proportional to the excess dye absorbance, the calibration slope (m_{610}) is a negative value.

13.2.5 Other considerations

13.2.5.1 Limit of detection

The lower limit of detection (LODmin) of the TEP methods depends on the variability of the blank absorbance. Villacorte (2014) calculated LODmin as follows:

$$TEP_{0.4\mu m} : LOD_{min} = 3\sigma_b\,(1/m_{787})\,(1/V_f)$$

$$TEP_{10kDa} : LOD_{min} = 3\sigma_b\,(1/m_{610})\,(Vr/V_f)$$

where σ_b is the standard deviation of 10 independently measured blank absorbance (abs/cm). The factor 3 corresponds to a significance level of 0.00135, which means that only 0.135% of blank measurements will statistically yield results that fall above the computed detection limit (Harvey, 2000).

The upper limit of detection (LODmax) is the upper threshold of the absorbance which can yield reliable concentration results. For $TEP_{0.4\mu m}$, this limit is determined based on the maximum absorbance at which a linear correlation between absorbance and filtered volume can be observed. For TEP_{10kDa}, this limit is the minimum absorbance at which the excess stain absorbance and the standard concentration has a significant linear correlation. In practice, the recommended absorbance thresholds are 0.8 abs/cm and 0.05 abs/cm for $TEP_{0.4\mu m}$ and TEP_{10kDa}, respectively (Villacorte, 2014)

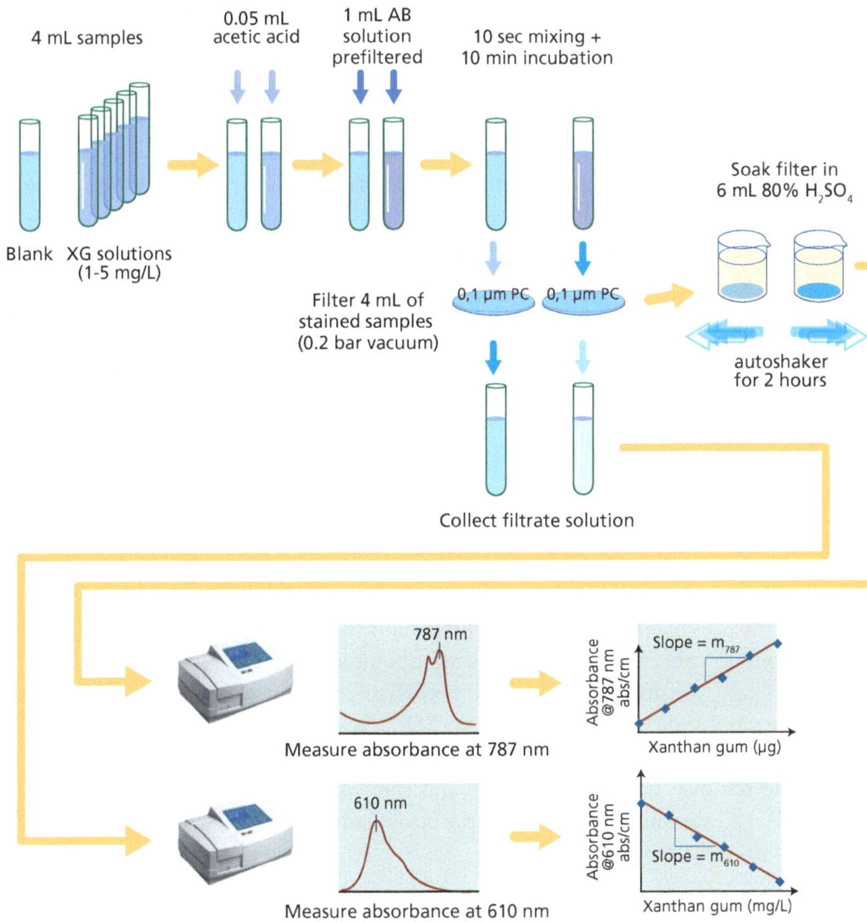

Figure 9 Overview of analytical steps for combined calibration of $TEP_{0.4\mu m}$ and TEP_{10kDa}

Bittar *et al.* (2018) applied a different approached to calculate LOD for $TEP_{0.4\mu m}$. The LODmin, in µg-XG, for each individual curve was calculated according to Corley (2003),

$$TEP_{0.4\mu m} : LOD_{min} = 3 \cdot \frac{\sqrt{Y^1}}{m}$$

where Y^1 is the mean square error of all calibration points and m is the slope of the regression.

The LOD_{max} is defined as the raw absorbance values not exceeding 1, given that the amount of transmitted light detected by the spectrophotometer decreases exponentially (Beer–Lambert Law), corresponding to 10 and 1% for absorbance values between 1 and 2. For $TEP_{0.4\mu m}$, absorbance is typically lower than 1 at ~ 75 µg of AB-stained XG.

13.2.5.2 Impact of storage on TEP concentration

Storing water samples for a prolong period before analysis can lead to significant disparity between the measured and in-situ TEP concentration. TEP loss/gain during storage may be attributed to coagulation, bacterial release, bacterial degradation, adsorption on walls of sample bottles or a combination. Bottle tests conducted by Villacorte (2014) revealed that $TEP_{0.4\mu m}$ concentration may increase over a long period in storage either due to coagulation of TEP precursors or through bacterial TEP release. On the other hand, TEP_{10kDa} concentration can rapidly decrease by up to 45% within the first 3 days of storage resulting from either bacterial degradation or adsorption to walls of sample bottles. Further investigations are necessary to fully understand the mechanisms involved in the TEP loss or increase as well as to develop reliable measures to preserve TEP samples (e.g., sample bottle, freezing, preservative addition). It is therefore important that samples should be analyzed immediately (within 24 hours) after sampling to obtain reliable TEP concentrations. For TEP_{10kDa}, it is recommended to filter and the sample immediately after sampling so the membrane can be stored at 4°C until further analysis.

13.2.6 Application and interpretation

Over the last 15 years, a significant number of experimental studies applied various TEP methods to investigate the impact of TEPs and their precursors in membrane processes. A non-exhaustive overview of these studies is shown in Table 2.

If implemented properly, TEP methods such as $TEP_{0.4um}$ and TEP_{10kDa} can be effective tools in quantifying the presence of TEP and their precursors in the s he pretreatment process or in the feed/product/concentrate streams of a membrane system. $TEP_{0.4\mu m}$ measures mainly TEP while TEP_{10kDa} can measure both TEP and their colloidal precursors. $TEP_{0.4\mu m}$ is a relatively more rapid and cheaper method than TEP_{10kDa} which means it is ideal for routine TEP monitoring in untreated water sources particularly during algal blooms. If the objective is to assess the removal efficiencies of TEP and their precursors through the treatment processes, TEP_{10kDa} measurement is more appropriate method because it covers both TEP and their colloidal precursors.

Villacorte (2014) measured $TEP_{0.4\mu m}$ regularly for 3 years in a seawater desalination plant and illustrated that the occurrence of TEP generally coincide with the seasonal algal bloom based on chlorophyll-a concentration (Figure 10). The study also demonstrated that chlorophyll-a concentration is not a reliable indicator of the abundance of TEP, because some bloom-forming algal species produce more TEP than others (Villacorte *et al.*, 2015c).

Table 2 Overview of membrane related studies applying TEP quantification methods.

Reference	TEP method	Membrane processes	Water type	Scope/highlights of the study
De la Torre *et al.* (2008)	Rapid spectro-photometric	MF/MBR	Wastewater, activated sludge	TEP as potential fouling indicator for MBR systems
Kennedy *et al.* (2009)	$TEP_{0.4\mu m}$	UF	Surface water, treated wastewater	TEP removal in UF membrane with inline coagulation
Villacorte *et al.* (2009b)	$TEP_{0.4\mu m}$ $TEP_{0.05\mu m}$	MF/UF, RO	lake water, seawater	TEP removal in 6 integrated membrane systems including pretreatment
Berman *et al.* (2011)	$TEP_{0.4\mu m}$	UF, RO	Lake water	Investigate role of TEP in aquatic biofilm initiation and membrane fouling
Bar-Zeev *et al.* (2012b)	$TEP_{0.4\mu m}$	RO	Seawater	Removal efficiency of pretreatment for RO
van Nevel *et al.* (2012)	$TEP_{0.4\mu m}$ $TEP_{0.05\mu m}$	UF, RO	Treated wastewater, surface and ground water	Removal efficiencies of UF-RO system and other treatment methods
Valladares Linares *et al.* (2012)	Direct AB staining	FO	Treated wastewater	TEP identified as major foulant in FO
Discart *et al.* (2014)	$TEP_{0.4\mu m}$ $TEP_{0.05\mu m}$	MF	Freshwater	Role of TEP in membrane fouling during algae broth filtration
Villacorte *et al.* (2015c)	$TEP_{0.4\mu m}$ TEP_{10kDa}	MF/UF, RO	Fresh/saline	Characterization of TEP produced by bloom forming algae, from a membrane fouling perspective
Villacorte *et al.* (2015a)	$TEP_{0.4\mu m}$ TEP_{10kDa}	MF/UF	Fresh/saline	MF/UF rejection and fouling potential of algal organic matter from bloom-forming algae
Villacorte *et al.* (2017a)	$TEP_{0.4\mu m}$	RO	Saline	TEP quantification of biofilm extracted from biofouled capillary spiral wound membranes
Li *et al.* (2016)	$TEP_{0.4\mu m}$ $TEP_{0.1\mu m}$	RO	Saline	TEP/TEP precursors monitoring through pretreatment and SWRO processes
Meng and Liu (2017)	$TEP_{0.04\mu m}$ $TEP_{0.05\mu m}$	UF	Model fresh/saline water	TEP-associated UF fouling is more severe with freshwater that with seawater
Zhang *et al.* (2023)	$TEP_{0.4\mu m}$	UF	Freshwater bacterial culture	TEP-induced irreversible fouling in UF can be reduced by the MIEX pretreatment

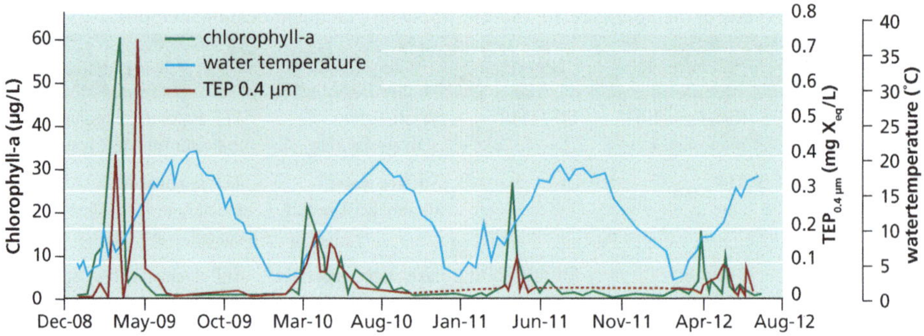

Figure 10 Variations of $TEP_{0.4\mu m}$, chlorophyll-a and water temperature at the intake water of seawater RO plant. Adopted from Villacorte (2014).

Measuring both $TEP_{0.4\mu m}$ and TEP_{10kDa} on the same water samples can offer a better understanding of the TEPs produced by different species of algae in pure culture applications (Villacorte *et al.*, 2015c). As shown in Figure 11, The TEP production and concentrations over time varies substantially for 3 species of bloom-forming fresh and saline water algae.

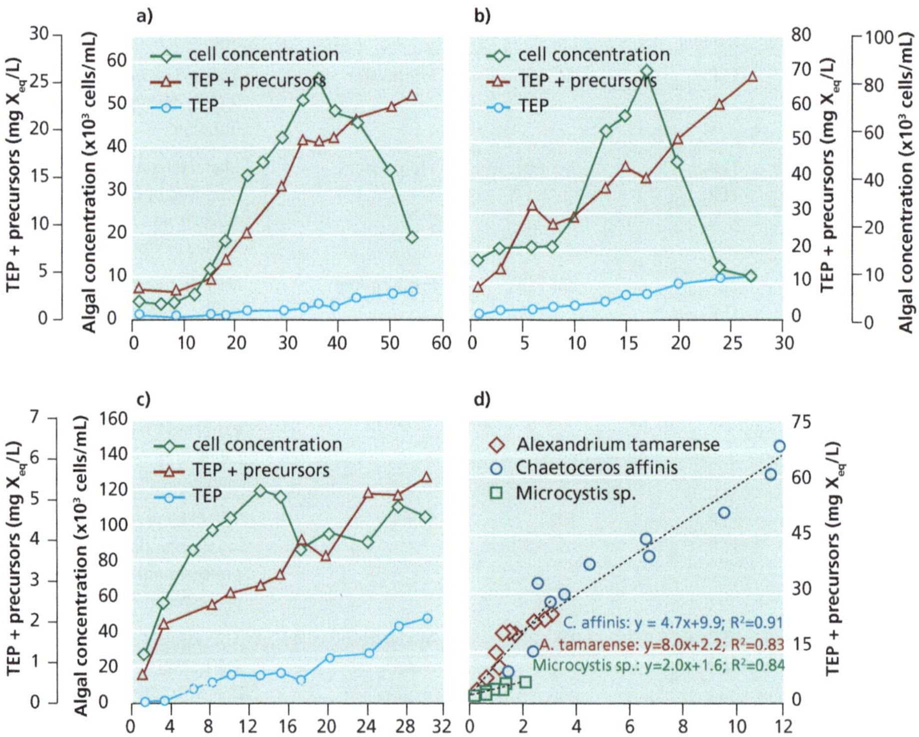

Figure 11 TEP concentration profile over days of incubation for three species of bloom-forming algae: (a) *Alexandrium tamarense*, (b) *Chaetoceros affinis*, (c) *Microcystis sp.*; and (d) correlation between $TEP_{0.4\mu m}$ (TEP) and TEP_{10kDa} (TEP + precursors) concentrations. Adopted from Villacorte *et al.* (2015c).

Monitoring TEP0.4um and TEP10kDa through the pretreatment processes of membrane plant can reveal TEP removal efficiencies of the treatment steps. As shown in Figure 12, TEP can be fully removed by UF, but some TEP precursors can still pass through. A significant correlation was also found for TEP and organic biopolymers measured using the LC-OCD method (see chapter 11).

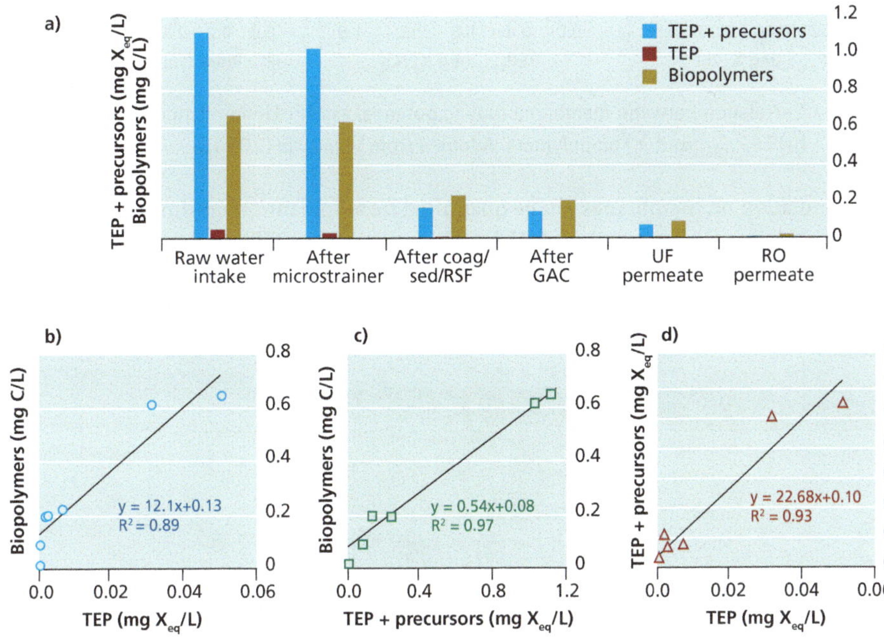

Figure 12 Concentrations of biopolymer (measured by LC-OCD), TEP ($TEP_{0.4\mu m}$) and TEP + precursors (TEP_{10kDa}) in (a) samples collected over the treatment processes of an RO plant plant and (c-d) linear regressions between measured parameters. Note: coag = coagulation + flocculation; sed = sedimentation; RSF = rapid sand filtration; GAC = granular activated carbon; UF = ultrafiltration. Adopted from Villacorte *et al.* (2015b).

Quantifying TEP together with LC-OCD to measure biopolymers and MFI-UF (see chapter 8) to measure the bulk fouling potential, has also been implemented. As shown in Figure 13, a significant correlation was reported between TEP10kDa and MFI-UF, which has better observed correlation than between biopolymers and MFI-UF (Villacorte, 2014). TEP0.4µm showed lower correlation with MFI-UF likely because TEP precursors (< 0.4µm) were not measured in the method.

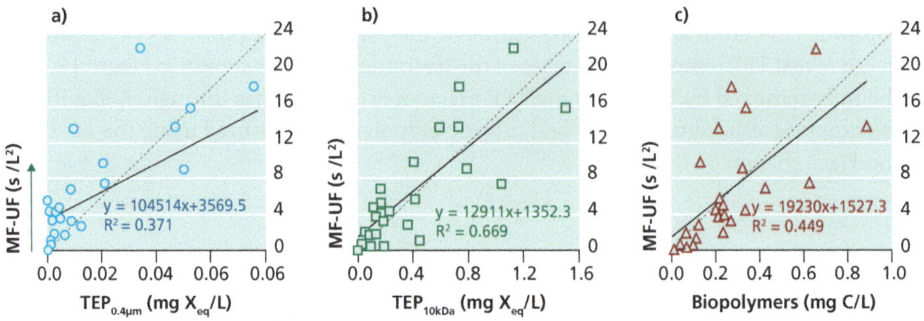

Figure 13 Correlation between membrane fouling potential as MFI-UF in relation to (a) $TEP_{0.4\mu m}$, (b) TEP_{10kDa} and (c) biopolymers. Adopted from Villacorte (2014).

TEP accumulation on membranes can be quantified from foulant or biofilm extracted from a fouled membrane (Villacorte *et al.*, 2017a). Figure 14 shows $TEP_{0.4\mu m}$ concentrations on membrane and spacers in a series of membrane fouling simulator (MFS) experiments. In this application, TEP results can be directly correlated to the hydraulic performance of the membrane (i.e., pressure drop) or bulk organic parameter (i.e., TOC).

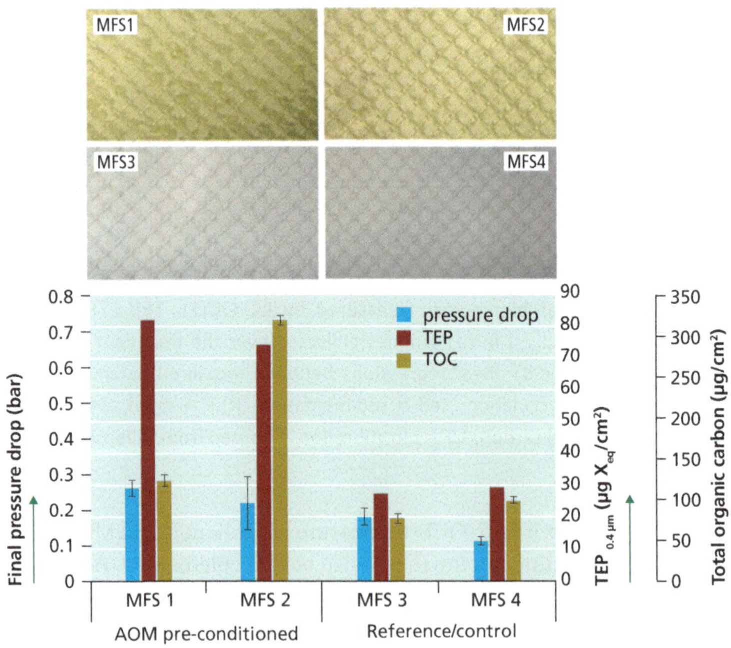

Figure 14 $TEP_{0.4\mu m}$ and TOC concentrations in biofilms on membranes and spacers from membrane fouling simulator (MFS) experiments in relation to the measured feed channel pressure drop. Top images are showing biofilm accumulation on initially clean RO membranes (MFS 3+4) and on algal organic matter (AOM) pre-conditioned RO membrane at day 10 of the experiments. Biofilm samples extracted for TEP/TOC measurements were taken on day 10 for MFS 1 and on day 20 for MFS 2, 3 and 4. Adopted from Villacorte *et al.* (2017a).

13.3 SUMMARY AND OUTLOOK

Various TEP quantification methods were developed over the years, and some are widely applied in membrane filtration studies. The experimental protocols to measure $TEP_{0.4\mu m}$ and TEP_{10kDa} are described in detail in this chapter as a promising tool to semi-quantitatively or qualitatively investigate the impact of TEP to the operational performance of membrane processes. Despite recent improvements, these TEP methods still have some limitations so it should be implemented with proper attention to the protocol used and by someone who is experienced with laboratory analytical techniques. Critical analyses of the results, particularly when comparing results from different sets of samples, different TEP methods and different studies, should include assessing what steps were taken to minimize the impact of the following (Discart *et al.*, 2014; Li *et al.* 2018; Meng *et al.*, 2020):

- variability of staining capacity of AB solutions relative to age or chemical supplier,
- variability of filter characteristics used to retain TEP (e.g., pore size and distribution),
- variability of surrogates or standard solutions used to calibrate the method,
- variability of salinity in the water samples.

Further improvements of the TEP methods should therefore focus on minimizing the impacts of above-mentioned variabilities to the results.

Another practical limitation of current TEP methods is that they are time consuming. $TEP_{0.4\mu m}$ and TEP_{10kDa} analyses can take up 2-4 hours and 3-4 hours per sample, respectively. APS method has shorter analytical time than TEP_{10kDa} for freshwater samples, but it can take up to 24h for saline samples due to the additional dialyses step. If current TEP methods are used as parameter to optimize operation of a membrane processes, the analytical time delay in addition to transport time of samples to the laboratory may prevent the operator to timely mitigate potential fouling issues. New online TEP measurement methods may cut down analytical time to less than 1 hour, but further verifications are needed if such rapid methods have similar/better accuracy than offline methods.

13.4 REFERENCES

Alldredge, A. L., Passow, U., & Logan, B. E. (1993). The abundance and significance of a class of large, transparent organic particles in the ocean. Deep Sea Research Part I: Oceanographic Research Papers, 40(6), 1131-1140.

Arruda-Fatibello, S. H. S., Henriques-Vieira, A. A., & Fatibello-Filho, O. (2004). A rapid spectrophotometric method for the determination of transparent exopolymer particles (TEP) in freshwater. Talanta, 62 (1), 81-85.

Azetsu-Scott, K., & Passow, U. (2004). Ascending marine particles: Significance of transparent exopolymer particles (TEP) in the upper ocean. Limnology and Oceanography, 49(3), 741-748.

Bar-Zeev, E., Belkin, N., Liberman, B., Berman, T., & Berman-Frank, I. (2012b). Rapid sand filtration pretreatment for SWRO: Microbial maturation dynamics and filtration efficiency of organic matter. Desalination, 286, 120-130.

Bar-Zeev, E., Berman-Frank, I., Girshevitz, O., & Berman, T. (2012a). Revised paradigm of aquatic biofilm formation facilitated by microgel transparent exopolymer particles. PNAS, 109(23), 9119-24.

Bar-Zeev, E., Berman-Frank, I., Liberman, B., Rahav, E., Passow, U., & Berman, T. (2009). Transparent exopolymer particles: Potential agents for organic fouling and biofilm formation in desalination and water treatment plants. Desalination and Water Treatment, 3, 136–142.

Berman, T., & Holenberg, M. (2005). Don't fall foul of biofilm through high TEP levels. Filtration and Separation, 42(4), 30-32.

Berman, T., Mizrahi, R., & Dosoretz, C.G. (2011). Transparent exopolymer particles (TEP): A critical factor in aquatic biofilm initiation and fouling on filtration membranes. Desalination, 276, 184-190.

Bittar, T. B., Passow, U., Hamaraty, L., Bidle, K. D., & Harvey, E. L. (2018). An updated method for the calibration of transparent exopolymer particle measurements. Limnology and Oceanography: Methods, 16(10), 621-628.

Chin, W. C., Orellana M. V., & Verdugo P. (1998). Spontaneous assembly of marine dissolved organic matter into polymer gels. Nature, 391, 568–572.

Claquin, P., Probert I., Lefebvre S., & Veron B. (2008). Effects of temperature on photosynthetic parameters and TEP production in eight species of marine microalgae. Aquatic Microbial Ecololy, 51, 1-11.

Corley, J. (2003). Best practices in establishing detection and quantification limits for pesticide residues in foods, p. 59–74. In P. W. Lee [ed.] Handbook of residue analytical methods for agrochemicals, v. 1. Wiley.

de la Torre, T., Lesjean, B., Drews, A., & Kraume, M. (2008). Monitoring of transparent exopolymer particles (TEP) in a membrane bioreactor (MBR) and correlation with other fouling indicators. Water Science and Technology, 58 (10), 1903–1909.

Discart, V., Bilad, M. R., van Nevel, S., Boon, N., Cromphout, J., & Vankelecom, I. F. J. (2014). Role of transparent exopolymer particles on membrane fouling in a full-scale ultrafiltration plant: Feed parameter analysis and membrane autopsy. Bioresource Technology, 173, 67-74.

Harvey, D. (2000) Modern analytical chemistry. McGraw-Hill, USA.

Kennedy, M. D., Muñoz-Tobar, F. P., Amy, G. L., & Schippers, J. C. (2009). Transparent exopolymer particle (TEP) fouling of ultrafiltration membrane systems. Desalination and Water Treatment, 6(1-3) 169-176.

Li, S., Sinha S., Leiknes T., Amy G. L., & Ghaffour N. (2016). Evaluation of potential particulate/colloidal TEP foulants on a pilot scale SWRO desalination study. Desalination, 393, 127–134.

Li, X., Skillman, L., Li, D., & Ela, W. P. (2018). Comparison of Alcian blue and total carbohydrate assays for quantitation of transparent exopolymer particles (TEP) in biofouling studies. Water Research, 133, 60-68.

Meng, S., & Liu, Y. (2017). Transparent exopolymer particles (TEP)-associated membrane fouling at different Na+ concentrations. Water Research, 111, 52-58.

Meng, S., Meng, X., Fan, W., Liang, D., Wang, L., Zhang, W., & Liu, Y. (2020). The role of transparent exopolymer particles (TEP) in membrane fouling: A critical review. Water Research, 181, 115930.

Mopper, K., Zhou, J., Sri Ramana, K., Passow, U., Dam, H. G., & Drapeau, D. T. (1995). The role of surface-active carbohydrates in the flocculation of a diatom bloom in a mesocosm. Deep-Sea Research Part II, 42(1), 47-73.

Passow, U. (2000). Formation of transparent exopolymer particles (TEP) from dissolved precursor material. Marine Ecology Progress Series, 192, 1-11.

Passow, U. (2002). Transparent exopolymer particles (TEP) in aquatic environments. Progress in Oceanography, 55(3), 287-333

Passow, U., & Alldredge, A. L. (1995). A dye-binding assay for the spectrophotometric measurement of transparent exopolymer particles (TEP). Limnology and Oceanography, 40(7), 1326-1335.

Schurer, R., Janssen, A., Villacorte, L. O., & Kennedy, M. D. (2012). Performance of ultrafiltration and coagulation in an UF-RO seawater desalination demonstration plant. Desalination and Water Treatment, 42(1-3), 57-64.

Schurer, R., Tabatabai, A., Villacorte, L. O, Schippers, J. C., & Kennedy, M. D. (2013). Three years operational experience with ultrafiltration as SWRO pretreatment during algal bloom. Desalination and Water Treatment, 51 (4-6), 1034-1042.

Sim, L. N., Suwarno, S. R., Lee, D. Y. S., Cornelissen, E. R., Fane, A. G., & Chong, T. H. (2019). Online monitoring of transparent exopolymer particles (TEP) by a novel membrane-based spectrophotometric method. Chemosphere, 220, 107-115.

Thornton, D. C., Fejes, E. M., DiMarco, S. F., & Clancy, K. M. (2007). Measurement of acid polysaccharides in marine and freshwater samples using alcian blue. Limnology and Oceanography: Methods, 5(2), 73-87.

Thuy, N. T., Huang, C. P., & Lin, J. L. (2017). Visualization and quantification of transparent exopolymer particles (TEP) in freshwater using an auto-imaging approach. Environmental Science and Pollution Research, 24, 17358-17372.

Valladares-Linares, R., Yangali-Quintanilla, V., Li, Z., & Amy, G. L. (2012). NOM and TEP fouling of a forward osmosis (FO) membrane: Foulant identification and cleaning. Journal of Membrane Science, 421–422 (2012) 217–224.

van Nevel, S., Hennebel, T., De Beuf, K., Du Laing, G., Verstraete, W., & Boon, N. (2012). Transparent exopolymer particle removal in different drinking water production centers. Water Research, 46(11), 3603-3611.

Verdugo, P., Alldredge, A. L., Azam, F., Kirchman, D. L., Passow, U., & Santschi, P. H. (2004). The oceanic gel phase: a bridge in the DOM–POM continuum. Marine Chemistry, 92, 67-85.

Villacorte, L. O. (2014). Algal blooms and membrane-based desalination technology. ISBN 978-1-138-02626-1, CRC Press/Balkema, Leiden.

Villacorte, L. O., Boerlage, S. F. E., & Dixon, M. B. (2021). Ch 6 – Algal blooms and RO desalination. In: Salinas-Rodriguez, S. G., Schippers, J. C., Amy, G. L., Kim, I. S., & Kennedy, M. D. (eds.) Seawater Reverse Osmosis Desalination: Assessment & Pre-Treatment of Fouling and Scaling (pp. 145–176). IWA Publishing, London.

Villacorte, L. O., Ekowati, Y., Calix-Ponce, H. N., Kisielius, V., Kleijn, J. M., Vrouwenvelder, J. S., Schippers, J. C., & Kennedy, M. D. (2017a.). Biofouling in capillary and spiral wound membranes facilitated by marine algal bloom. Desalination, 424, 74-84.

Villacorte, L. O., Schippers, J. C. & Kennedy, M. D. (2017b). Appendix 3. Methods for measuring transparent exopolymer particles and their precursors in seawater. In: Anderson, D. M., Boerlage, S. F., & Dixon, M. B. (eds.) Harmful Algal Blooms (HABs) and Desalination: A Guide to Impacts, Monitoring and Management. IOC Manuals and Guides No. 78 (pp. 501–507). Intergovernmental Oceanographic Commission of UNESCO, Paris.

Villacorte, L. O., Ekowati, Y., Winters, H., Amy, G. L., Schippers, J. C., & Kennedy, M. D. (2015a). MF/UF rejection and fouling potential of algal organic matter from bloom-forming marine and freshwater algae. Desalination, 367, 1-10.

Villacorte, L. O., Ekowati, Y., Calix-Ponce, H. N., Schippers, J. C., Amy, G. L., & Kennedy, M. D. (2015b). Improved method for measuring transparent exopolymer particles (TEP) and their precursors in fresh and saline water. Water Research, 70, 300-312.

Villacorte, L. O., Ekowati, Y., Neu, T. R., Kleijn, J. M., Winters, H., Amy, G. L., Schippers, J. C., & Kennedy, M. D. (2015c). Characterisation of algal organic matter produced by bloom-forming marine and freshwater algae. Water Research, 73, 216-230.

Villacorte, L. O., Ekowati, Y., Winters, H., Amy, G. L., Schippers, J. C., & Kennedy, M. D. (2013). Characterisation of transparent exopolymer particles (TEP) produced during algal bloom: a membrane treatment perspective. Desalination and Water Treatment, 51 (4-6), 1021-1033.

Villacorte, L. O., Schurer, R., Kennedy, M. D., Amy, G. L., & Schippers, J. C. (2010a). Removal and deposition of Transparent Exopolymer Particles (TEP) in seawater UF-RO system. IDA Journal, 2, 45-55.

Villacorte, L. O., Schurer, R., Kennedy, M. D., Amy, G. L., & Schippers, J. C. (2010b). The fate of transparent exopolymer particles in integrated membrane systems: a pilot plant study in Zeeland, the Netherlands. Desalination and Water Treatment, 13, 109-119.

Villacorte, L. O., Kennedy, M. D., Amy, G. L., & Schippers, J. C. (2009a). Measuring Transparent Exopolymer Particles (TEP) as indicator of the (bio)fouling potential of RO feed water. Desalination and Water Treatment, 5, 207-212.

Villacorte, L. O., Kennedy, M. D., Amy, G. L., & Schippers, J. C. (2009b). The fate of Transparent Exopolymer Particles (TEP) in integrated membrane systems: Removal through pretreatment processes and deposition on reverse osmosis membranes. Water Research, 43 (20), 5039-5052.

Zhang, Z., Zhang, T., Wang, L., Chen, M., Zhao, B., Li, J., Ma, C., Chu, X., & Zhang P. (2023). Irreversible membrane fouling caused by free TEP: Mitigation performance and mechanism of the integrated MIEX/UF process. Journal of Water Process Engineering, 54, 103919.

Zhou, J., Mopper, K., & Passow, U. (1998). The role of surface-active carbohydrates in the formation of transparent exopolymer particles by bubble adsorption of seawater. Limnology and Oceanography, 43(8), 1860-1871.

Part 5

Biological fouling

doi: 10.2166/9781789062977_0315

Chapter 14

Genomics Tools to Study Membrane-Based Systems

Muhammad Ali, The University of Dublin, Ireland

Lucia Ruiz Haddad, KAUST, Saudi Arabia

Mohamed Fauzi Haroon, Moderna, USA

Pascal E. Saikaly, KAUST, Saudi Arabia

The learning objectives of this chapter are the following:

- Designing an experimental plan for omics-based study for membrane-based systems
- Omics tools to investigate microbial communities of membrane-based systems
- Introduction to state-of-the-art bioinformatics tools and reference database resources
- Application of genome-resolved metatranscriptomics approach to study the activity of microorganisms

14.1 INTRODUCTION

The performance of membrane-based operations is significantly decreased by the accumulation of microbes and chemical foulants on the membrane surface. Thus, biofouling is the main operational challenge for membrane-based processes (Flemming *et al.*, 2011). Various molecular biological techniques are utilized to understand the behaviour of microbes in membrane-based water treatment systems. These techniques enable the analysis of microbes' taxonomy, morphology, physiology, and ecology. The molecular methods can be applied to study membrane-based systems for fast, reliable and cheap identification of relevant microorganisms. The most common molecular methods include polymerase chain reaction (PCR), quantitative polymerase chain reaction (qPCR), fluorescence in situ hybridization (FISH), denaturing gradient gel electrophoresis (DGGE), and terminal restriction fragment length polymorphism (T-RFLP) (Lovley, 2003; Nielsen *et al.*, 2009). However, these techniques fail to address the function of microorganisms and the mechanisms underlying their interactions due to limited throughput as compared to their high-throughput sequencing-based counterparts and cannot resolve abundance and activity for functionally important microbes in low concentrations (Hugenholtz, 2002). Now, the cost of culture-independent next-generation sequencing (NGS) technologies rapidly decreases (Figure 1), significantly improving our understanding and functions in various natural (Alqahtani *et al.*, 2019; Bougouffa *et al.*, 2013; Garcias-Bonet *et al.*, 2018; Speth *et al.*, 2017; Wang *et al.*, 2013) and engineered (Ali *et al.*, 2020b; Ali *et al.*, 2019; Matar *et al.*, 2021; Rehman *et al.*, 2019) ecosystems.

Most environments harbour a diverse range of organisms (genomes). The metagenome is the entire DNA content of an environment, so it includes the collection of individual genomes. Metagenomes allow to 1) Obtain a gene catalogue of the environment; 2) Screen for genes that confer novel functions (Ali *et al.*, 2020a; Garcia Martin *et al.*, 2006); 3) Discover and characterize new bacterial candidate divisions, and 4) Reconstruct microbial genomes via metagenome assembly and binning (Albertsen *et al.*, 2013), offering novel insights into the microbial functions and metabolic pathways involved in complex microbial systems. On the other hand, metatranscriptomics is the entire RNA content of a given environment. High-throughput sequencing can be applied for metagenomics or metatranscriptomics, where all the DNA or expressed genes (mRNA) from a certain community are sequenced to resolve abundance and activity for functionally important microbes in low concentrations. Metatranscriptomics analysis presents unique challenges due to RNA's short half-life and the variability in sequencing coverage caused by the molecule's secondary structure. Unlike metagenomes, metatranscriptomes can shed light on the activity of environmental populations. As a result, metagenomes and metatranscriptomes from the same samples can be analyzed together to identify 'active' microorganisms and their gene expression patterns caused by the metabolic activities of the microbial community (Ali *et al.*, 2020b; Shaw *et al.*, 2020).

Metagenomics, metatranscriptomics, and next-generation sequencing (NGS) revolutionize microbiology and ecology research. However, the most challenging aspect, particularly for newcomers, is the bioinformatics analysis of the massive sequencing data (Desai *et al.*, 2012). Therefore, this chapter covers high-performance bioinformatics computing methods, tools, and pipelines. Besides, the experimental design and sample preparations are critical preliminary steps in determining the dependability, comparability, and (cost-)effectiveness of the sequence data and analytical result (Desai *et al.*, 2012; Goodrich *et al.*, 2014; Knight *et al.*, 2012; Kunin *et al.*, 2008; Thomas *et al.*, 2012). For example, several factors can cause bias in sequence data or render all experimental efforts less significant (Kunin *et al.*, 2008; Scholz *et al.*, 2012). As a result, this chapter will provide a comprehensive technical guide for using metagenomics in the study of microbiology and ecology of membrane-based systems (Figure 2).

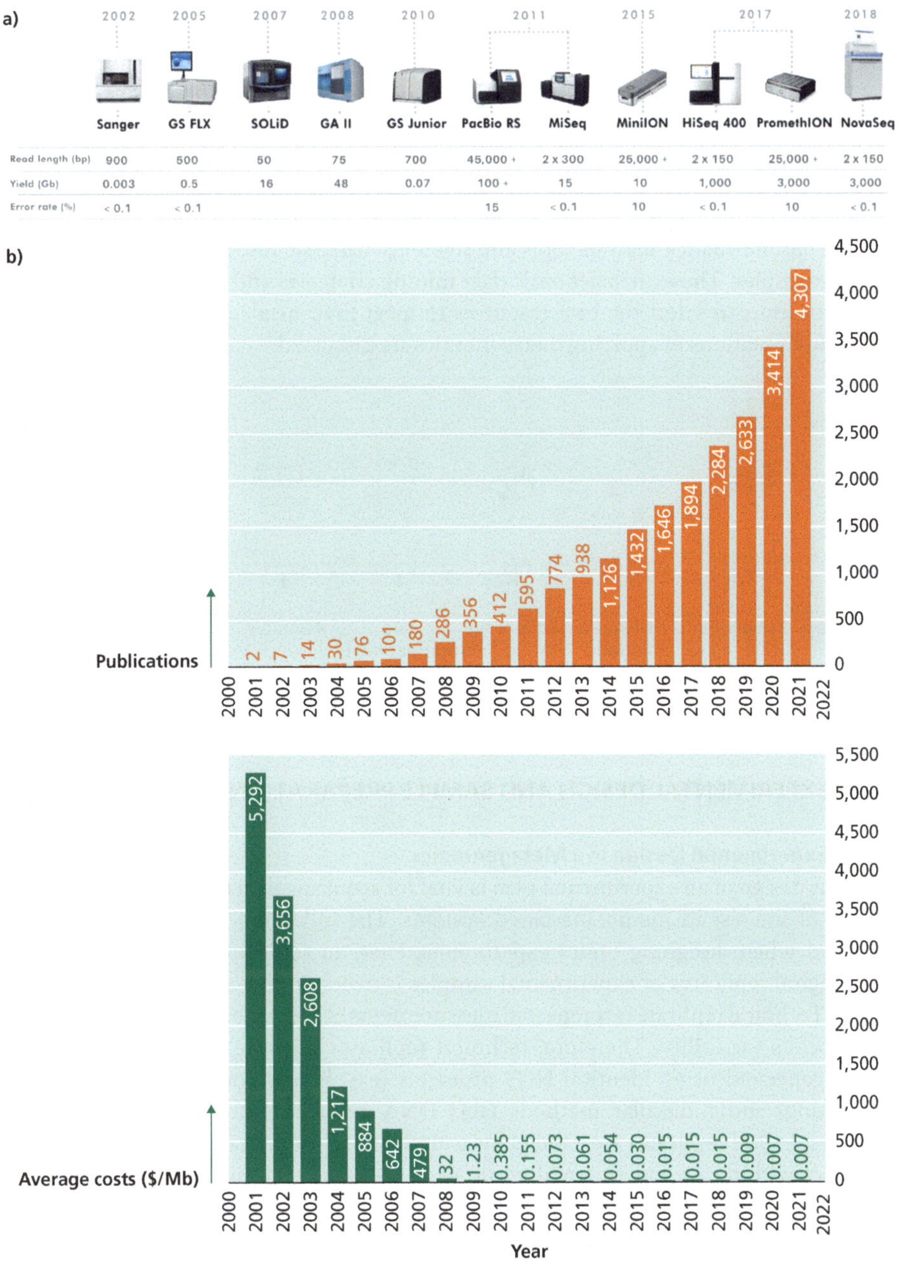

Figure 1 A) Timeline for sequencing technology development. The data is adopted from this link https://figshare.com/articles/dataset/developments_inNGS/100940?file=5470844. B) Cost of per raw megabase of DNA Sequencing. https://www.genome.gov/about-genomics/fact-sheets/DNA-Sequencing-Costs-Data and Number of publications using metagenomics tools. Data obtained from PubMed: metagenom*

This chapter presents the relevant concepts and protocols required for the experimental design, sample preparation, library construction, and NGS sequencing. These steps are crucial for metagenomics analysis of microbial communities in membrane-based bioengineered systems operated for research or industrial aims (e.g., bioenergy production and bioremediation). Additionally, the methods for annotating metagenomic and metatranscriptomics data are compared, and their conditions of applicability are discussed. Likewise, the chapter provides an overview of the most cited, user-friendly, rigorous, and complete bioinformatics analysis tools and reference database resources, along with some analysis examples. These metagenomic data mining strategies and resources will make it simpler for users to select the best resources to meet their needs. Finally, the significant pitfalls and limitations of applying metagenomics are discussed.

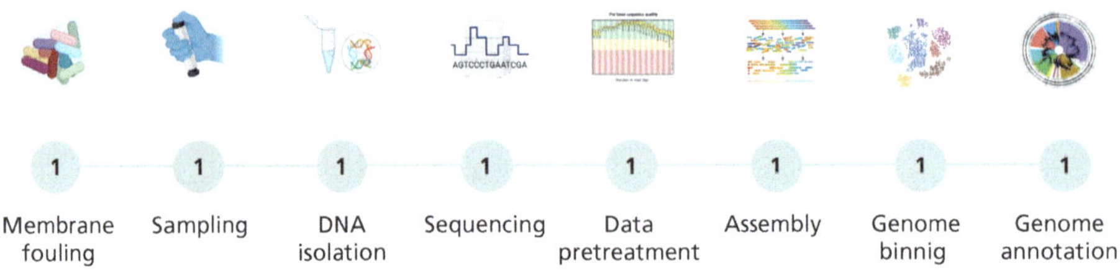

Figure 2 Step-by-step workflow for metagenomics experiment. 'Created with BioRender.com'

14.2 EXPERIMENTAL DESIGN AND SAMPLE PREPARATION

14.2.1 Experimental Design in a Metagenomics

A rigorous design of an experimental plan is vital for acquiring data to answer the scientific question of interest in membrane-based systems. The following key aspects should be considered when designing omics experiments. First, an adequate number of technical and biological replicates of experimental samples is necessary for statistically meaningful analyses. Technical replicates are repeated measurements of the same sample, demonstrating the protocol's variability. Therefore, technical replicates measure the reproducibility of the data generated using identical NGS protocols (e.g., library construction and sample multiplexing) and molecular methods (e.g., DNA extraction and PCR amplification). Besides, biological replicates are parallel measurements of biologically distinct samples (at least three) within one experimental group (e.g., the same treatment or condition). This captures the random biological variation within a group, which may be a subject of study or a noise source. Understanding biological variation allows an accurate assessment of the effectiveness and differences between different experimental groups. Besides replicates, positive and negative controls may be required to ensure no contamination occurs among samples. Many metagenomics-based studies failed to address this issue adequately. Now that the sequencing costs have been significantly reduced, the economic constraints should not be used to compromise experimental rigour.

Moreover, decisions must be made at different experimental stages, including selecting the following: physiochemical parameters, biological parameters, sequencing platform, sample preservation, quality control, bioinformatics tools, and reference databases. Physiochemical parameters include temperature, pH, conductivity, alkalinity, turbidity, dissolved oxygen, metals, nutrients, etc. Biological parameters include the sample size and the scales (temporal or spatial). The sequencing platform will be determined by the desired depth, sequence length, and error rate (Fig. 1A). While sampling preparation includes methods for fixing or storing samples on-site (RNA latter or freezing). Finally, quality control allows for increased amount and purity of DNA or RNA, especially when yield is low or microbial diversity is high. This could be accomplished through amplification, assembly, and binning steps. On the other hand, compiling metadata is an equally important step in the process. Correlating reactor performance with the structure and functions of microbial communities requires synchronous preservation of biological samples and regular monitoring of reactor performance (removal rates, production rate, biodegradation, and accumulation) in different operating conditions.

14.2.2 Sample Collection and Preservation

Sample collection and preservation are the first critical steps in an omics-based study. For sampling from a membrane-based system, the following factors should be considered: biomass type (e.g., an attached vs. suspended biomass), bioreactor configuration (e.g., plug flow vs. mixed), modification (e.g., presence of carriers or baffles), mixing state and operating time and condition. These factors affect biomass distribution and evenness, granule size, substrate gradients, and mass transfer, leading to a heterogeneous microbial composition within a membrane-based system. In certain membrane systems, when extracting DNA material, it may be necessary to obtain a sample by cutting a piece of the membrane, typically measuring at least 1 – 5 cm^2. Multiple samples can be collected separately to reflect microbial heterogeneity, or a composite sample can be taken to represent average microbial profiles. DNA extraction of sludge/slurry samples is preserved at −80 °C (Mason *et al.*, 2012) or fixed in 50% ethanol before storage at −20 °C (Zhang *et al.*, 2012). In contrast, centrifugation and membrane filtration are viable methods for recovering microbial cells in samples containing less biomass such as effluent samples (Thomas *et al.*, 2012). For mRNA-based studies, two preservation methods are frequently used: 1) Flash freezing with liquid nitrogen and 2) Immersion in RNAlater, which is widely used to recover high-quality mRNA (Riesgo *et al.*, 2012).

14.2.3 DNA Extraction

An optimized protocol for DNA/RNA extraction is essential to any analysis of microbial composition and function. The microbes differ enormously in their resistance to different lysing methods (Thomas *et al.*, 2012). Hence, microbes with cell walls that are difficult to lyse will effectively seem less abundant if sub-optimal extraction protocols are used. Thus, the method for DNA/RNA extraction needs to be robust to cope with the challenges.

The protocols and kits used for DNA/RNA extraction should be consistent throughout a study to increase the reliability and comparability of data (Goodrich *et al.*, 2014). These kits should be used following the manufacturer's instructions. Likewise, in 2012 Thomas and co-workers established a guide for isolating high-quality nucleic acid for NGS library preparation and sequencing. Following the extraction of DNA and RNA, their quality must be determined using various quality indicators. Generally, the ratio of absorbance at 260 and 280 nm is used to assess the purity of DNA and RNA. A ratio of 1.8 is commonly accepted as pure DNA, and a value of 2.0 is recognized as pure RNA. (Lucena-Aguilar *et al.*, 2016). At the same time, lower ratios indicate the presence of non-nucleic acid content, such as protein, phenol, or other contaminants that absorb strongly at or near 280 nm. Besides, the absorbance ratio of 260/230 can be used as a secondary measure of nucleic acid purity. The typical values for nucleic acid samples are between 2.0-2.2 (Lucena-Aguilar *et al.*, 2016). If the ratio is lower, it may indicate the presence of contaminants that absorb at 230 nm. These absorbance parameters should be checked with spectrophotometers or qubit.

14.2.4 Library Preparation

Library construction transforms the RNA/DNA into a format compatible with the sequencing platforms. Various physical, enzymatic, and chemical methods can be used for fragmentation and size selection required for library preparation on the different sequencing platforms (see Fig. 1A for desired read length). The library preparation strategy, the sequence length, and the sequencing platforms' length and depth are all essential considerations for sequencing. Moreover, technical details on sequencing library construction have been well documented (Head *et al.*, 2014). However, library preparation generally shares similar principles and considerations regardless of the sequencing platform. For example, two objectives are to maximize library complexity (e.g., a lower ratio of artificial duplicate reads may indicate higher complexity) and to minimize PCR or other amplification-based biases (e.g., less amplification and more sample RNA/DNA). Therefore, a paired-end (PE) library is recommended over a single-end library. A PE improves the performance of metagenome (e.g., scaffolding and chimera detection) (Peng *et al.*, 2012), binning (e.g., tracking multiple-copy genes) (Albertsen *et al.*, 2013), and enables the use of computational tools designed to consider PE relationships (Imelfort *et al.*, 2014).

14.2.5 Sequencing platforms

The expected sequence depth is related to the biodiversity and complexity of microbial samples. Generally, the soil (Roesch *et al.*, 2007) and sediments (Roesch *et al.*, 2007) harbor more diverse microbial species than the bioengineered ecosystems (Dueholm *et al.*, 2022; Wang *et al.*, 2012a; Wu *et al.*, 2019). For biological wastewater treatment systems, higher biodiversity is generally detected in 1) full-scale than lab-scale bioreactors (Matar *et al.*, 2021), 2) biofilm than suspended sludge and 3) activated sludge than anaerobic sludge (Ali *et al.*, 2019; Trego *et al.*, 2020). Recent attempts to assemble large complex soil metagenomes suggest that 80% of the sequencing data could not be assembled due to the low coverage. For instance, even 300 Gb sequencing data are insufficient to deeply cover a localized soil sample (Howe *et al.*, 2014). In contrast, in an enriched microbial system, more than 45% of the metagenomics reads could be effectively assembled (Albertsen *et al.*, 2013).

The following equation can calculate the required sequencing depth for a given genome coverage in metagenomic samples.

$$\text{Sequencing Depth} = (\text{Genome Size} \bullet \text{Coverage} \times 100)/(\text{Relative Abundance})$$

Here,
Genome Size = Bacterial genome length (usually between 2.5- 5 Mbp)
Coverage = Required depth coverage (between 5 to 10×)
Relative Abundance = Expected relative abundance of targeted bacteria species. The relative abundance value might be estimated from previous studies on similar ecosystems.

For example, a sample with a relative abundance of 0.5% for the target bacteria species and a genome size of 3Mbp will require 6 Gb of sequencing depth to achieve 10× coverage. Furthermore, a sequencing depth of 6Gb and a PE read length of 150bp will produce 300bp (150bp × 2) per reading, 20 million of paired reads((6Gb) / (300bp)), and 40 million of single reads((6Gb) / (150bp)). Illumina has recently been the most popular commercial sequencing platform due to its high data throughput and low per-base cost (van Dijk *et al.*, 2014). Notably, the short but high-quality PE sequences (100–150 bps) generated by Illumina's platforms could be used to rectify (> 99.9% accuracy) (Koren *et al.*, 2012) and concatenate the low-quality but exceptionally long reads (e.g., 20 kb) from third-generation sequencing platforms, (van Dijk *et al.*, 2014) such as PacBio RS and nanopore technologies.

14.3 BIOINFORMATICS ANALYSIS

Bioinformatics studies biological data in conjunction with computer science and statistics. This field is constantly evolving and developing new tools to make metagenomics/ metatranscriptomics analysis and genetic engineering simpler and more reliable. The main steps required for bioinformatics analysis are listed below

14.3.1 Data Pre-treatment

Raw sequencing reads must be pre-treated to ensure that only high-quality sequences (clean reads) are used for the downstream analysis. Before pre-treatment, FASTQC can be used to check the data quality (base quality, GC content, ambiguous bases, length distribution, sequence duplication levels, and adapter content). In general, data pre-treatment includes the following steps: adaptors/linkers removal, demultiplexing (assign reads to samples using index reads or barcodes), quality control, dereplication (identifying unique sequences and abundances), and reads overlapping (for PE library sequencing) etc.

14.3.2 Amplicon-based approach

Amplicon-based sequencing is a useful tool for understanding microbial community composition and diversity. It is a cost-effective and computationally simple method (Bodilis *et al.*, 2012). However, an amplicon-based analysis could not always classify microbes to a lower taxonomic level, such as genus or species. Also, generally, it targets hypervariable region(s) of the 16S rRNA gene, which does not provide the functional identity of the microbe (Siegwald *et al.*, 2017). Though, efforts are made to manually curate the existing 16S rRNA gene database (SILVA, (Pruesse *et al.*, 2012)) to provide information about the physiology (function) and ecology of the microorganisms present in bioengineered ecosystems (Albertsen *et al.*, 2015; Dueholm *et al.*, 2022). For amplicon sequencing data processing, there are numerous open-source packages available, such as QIIME (Caporaso *et al.*, 2010), USEARCH (Edgar, 2013), MOTHUR (Schloss *et al.*, 2009), etc.

14.3.3 Metagenomics, read-based approach

Bioinformatics analyses of metagenomic data can be performed using short-reads, assembled contigs (assembled overlapping reads), or reconstructed draft genomes through a self-accelerating data mining circle. The first strategy for metagenomic analysis is to directly use unassembled clean reads for the quantitative analysis of microbial community composition and function. The assembly-free approach is advantageous when studying rare organisms (with low sequencing depth or coverage) to avoid bias, such as their removal due to their inability to be assembled. In an assembly-free (read-based) approach, reads are directly compared against a reference database for taxonomic profiling and functional analyses. Generally, unassembled short reads retain the original abundance information and enable quantitative comparisons of microbial taxa, functional genes, and metabolic profiles (Yang *et al.*, 2013). However, they have a large amount of data and lack resolution for taxonomic and functional annotations.

14.3.4 Metagenomics, assembly-based approach

The assembly is a computational process that connects the clean short-reads sequences to form long contigs (>1,000 bp). This is especially relevant to 1) Recover genomic sequences (via binning as discussed in sections 15.3.5 and 15.3.6; 2) Analyze full-length protein-coding genes; 3) Identify strain-specific genomes (Langille *et al.*, 2010); 4) Analyze the genetic content (e.g., at the strain or species level), especially for uncultured microorganisms. Moreover, a metagenomic assembly reduces the data size to be analyzed in the downstream processing, though assembly requires substantial computational resources (Howe *et al.*, 2014). Two assembly strategies are used depending on whether a reference database is used: reference-based and de novo. Reference-based assembly aligns short-read fragments with reference genomes. However, this analysis cannot capture the differences between the genomes of novel species, resulting in an underestimation of microbial diversity in an open microbial system. (e.g., activated sludge, soil) (Howe *et al.*, 2014). As a result, a de novo strategy should be used when genetic novelty and diversity are high.

14.3.5 Metagenome-assembled Genome (MAG) Binning

The continuously decreasing sequencing cost has allowed researchers to access environmental metagenomes at increasing sequencing depths (e.g., > 50 Gbp). As a result, sufficient resolution can be obtained to retrieve partial or near-complete genomes of rare (1%), novel, and/or uncultured microorganisms from complex communities (Albertsen *et al.*, 2013). Binning is the computational process of clustering the assembled contigs or the short reads into groups representing an individual genome/taxon or genomes/taxa of closely related microorganisms (Albertsen *et al.*, 2013; Wrighton *et al.*, 2012). Binning can be based on genomic signatures such as 1) sequence composition, 2) homology, 3) coverage (abundance), or 4) a combination of these. Composition-based binning algorithms typically group DNA sequences based on their conserved nucleotide compositions, such as the tetranucleotide frequencies and GC content (McHardy *et al.*, 2007). These methods can be improved by providing sample-specific training data sets (e.g., a long DNA fragment with marker genes). The homology-based binning methods are based on a similarity search against existing genomes. However, this method is limited by the quality and representativeness of reference databases, the poor taxonomic resolution of short reads, and the accuracy and/or sensitivity of alignment tools. As a result, they are unreliable for assigning short reads. They often require longer assembled contigs (e.g., > 1 kb for the expert-trained PhyloPythiaS package) and manual efforts to ensure high assignment accuracies. Homology-based tools include MetaPhlAn2, MetaPhyler, and CARMA3. These are commonly applied for the taxonomic classification of shotgun metagenomic reads based on similarity comparisons with reference marker genes such as 16S rRNA, rpoB, or clade-specific markers. Besides, some packages consider the composition and homology of sequences for taxonomic classification and clustering, such as MEGAHIT (Li *et al.*, 2016), Metabat (Kang *et al.*, 2015) and MetaCluster (Wang *et al.*, 2012b).

14.3.6 Supervised and unsupervised binning

Two unsupervised approaches have been widely applied to reconstruct high-quality genomes of uncultured organisms from metagenomes: Tetra-ESOM (Dick *et al.*, 2009) and 'differential coverage binning' (Albertsen *et al.*, 2013). Tetra-ESOM analyzes the composition of DNA signatures by clustering tetranucleotide frequencies using emergent self-organizing maps (ESOM). While 'differential coverage binning' categorizes contigs based on their differential coverage profiles across multiple related metagenomes with the assumption that contigs from the same microorganisms will have similar abundance (coverage) profiles in a single metagenome. These draft genomes can be refined further using composition-based paired-end tracking, reassembly, and manual curation techniques. Several automated pipelines for platforms like CONCOCT (Alneberg *et al.*, 2014), MaxBin (Wu *et al.*, 2016), and Metabat (Kang *et al.*, 2015) allow genome reconstruction based on the coverage profiles and composition (tetranucleotide frequency patterns). The completeness and contamination in reconstructed genomes have been estimated by the presence/absence of marker genes, such as essential single copy marker genes conserved in 95% of bacteria (Dupont *et al.*, 2012), conserved phylogenetic marker genes (Wrighton *et al.*, 2012), or clusters of orthologous groups (COGs) (Raes *et al.*, 2007). Currently, CheckM is the only automated tool to assess the quality of a genome recovered from isolates, single cells, and metagenomes based on conserved marker genes (Parks *et al.*, 2015).

14.3.7 Functional annotation

Annotation is the process of analysing the structure and functions of assembled metagenomic contigs. Compared to unassembled short reads, assembled contigs are longer and more compact, allowing for a more robust and rapid analysis of specific species and their functional genes. Functional annotation of assembled contigs allows the prediction of the functional capacities of a microbial community. After genome assembly, binning, and gene calling are finished, several tools enable functional annotations. Gene function is commonly determined through similarity searches while using established tools like BLAST. However, it is computationally expensive and time-consuming to conduct a similarity search using BLASTX or PSI-BLAST, particularly for large query data sets and reference databases (such NCBI's RefSeq). Tools such as USEARCH (Edgar, 2010) and DIAMOND (Buchfink *et al.*, 2015) have been developed to overcome these computational challenges faster.

14.3.8 Genome-resolved Metatranscriptomics

Genome-resolved metatranscriptomics approach combines data from metagenomics and metatranscriptomics from the same environment to characterize 'active' and 'non-active' microorganisms and to compare gene expression patterns under different conditions such as treatment v/s control (Frias-Lopez *et al.*, 2008; Haroon *et al.*, 2013; Yu and Zhang,

2012). MAGs recovered from metagenomics are used to facilitate genome-resolved metatranscriptomics analyses for profiling the gene expression across the recovered population genomes (Ali *et al.*, 2020b).

14.4 DATA SHARING AND STORAGE

Sharing sample metadata, sequence data, and computational results is a widespread and effective method for exchanging knowledge. Data exchange enables comparative studies and avoids the needless repetition of processing the same data sets or sequencing similar microbial ecosystems. Several public access databases have been established to promote sequencing data sharing and storage, such as the National Center for Biotechnology Information (NCBI) and Genomes OnLine Database (GOLD).

14.5 BIOINFORMATICS ANALYSIS WORKFLOW EXAMPLES

The following sections describe the simple bioinformatics workflow examples for process amplicon, metagenomics and metatranscriptomics datasets.

14.5.1 Amplicon Sequences Processing Workflow

The amplicon sequences analysis workflow is shown in Fig. 3A. The first step of the process is to re-label the sequences in the raw FASTQ files with the corresponding sample name/ID. This enables the identification of the sequences from a specific sample in the later processing stages. The re-labelling can be performed using USEARCH (Edgar, 2013) with the command fastx_relabel. Subsequently, re-labelled forward and reserve sequences of different samples are concatenated separately into a single file. Then, the concatenated forward and reserve sequences are quality-filtered using trimomatic (Bolger *et al.*, 2014) with the settings SLIDINGWINDOW:5:3 and MINLEN:275, these settings should be adjusted based on the raw quality of the dataset determined through FASTQC package. The quality filtered sequences are hits to the PhiX genome USEARCH command filter_phix. PhiX genome-based library is an ideal sequencing control (typically with > 1% spike-in) for run quality monitoring (cluster generation, sequencing, and alignment). Later, the trimmed and filtered forward and reverse reads were merged using FLASH (Fast Length Adjustment of Short Reads) (Magoc and Salzberg, 2011) with the settings –m 25 –M 200, these parameters should be adjusted based on the input data quality. FLASH merges paired-end reads to create consensus sequences. Following that, the merged reads are compared letter by letter to the set of unique sequences, a process known as dereplication, and carried out with the USEARCH command fastx uniques. Then, the cluster_otus command performs 97% Operational Taxonomic Unit (OTU) clustering of unique sequences using the UPARSE-OTU algorithm (Edgar, 2013). Taxonomy is assigned to OTUs using sintax command available in USEARCH package and MiDAS database (Dueholm *et al.*, 2022). OTU abundances table is generated by mapping merged reads against the OTUs using otutab command available in USEARCH package. The OTU count and corresponding taxonomy table are imported into the Rstudio IDE environment using the ampvis2 package by the amp_load command (Andersen *et al.*, 2018). The ordination plot (Fig 3B) and heatmap (Fig. 3C) were generated using ampvis2 package. All the scripts, metadata and exercise files are available at this link (https://drive.google.com/drive/folders/1swJIv9Z1pyD52Jo630d-ivBTOhxt7dHu?usp=share_link).

14.5.2 Genome-resolved Metagenomics

In general, metagenomics is used to extract the genomes of a microbial population in the sample. The recovered MAGs from metagenomics analysis facilitate genome-resolved metatranscriptomics analyses and gene expression profiling across the recovered population genomes. Genome-resolved metagenomics workflow is presented in Fig. 4. In this exercise, two independent biomass samples for metagenomics were collected from the MBR in a previous study (Ali *et al.*, 2020b). Raw reads are processed for quality filtering using Cutadapt (Martin, 2011). Trimmed forward and reserve sequences of different samples are concatenated separately into a single file. Concatenated forward and reserve sequences were assembled using MEGAHIT (Li *et al.*, 2016). The assembled contigs were reformatted to simplify deflines and to remove short contigs (<1000 bp) using anvi'o command anvi-script-reformat-fasta (Eren *et al.*, 2021). The filtered reads were mapped back to the reformatted assembly file using Burrows-Wheeler Aligner (Li and Durbin, 2010) to generate coverage files for metagenomics binning. These files were converted to the sequence alignment/map (SAM) format using samtools (Li and Durbin, 2009). The SAM files are converted into a sorted and indexed BAM-file using samtools. Later, anvi'o uses sorted and indexed BAM-file in anvi-profile command to create a single profile that reports properties for each contig in a single sample based on mapping results. These profiles are merged into one anvi'o profile using anvi'o command anvi-merge. MAGs were obtained from assembled scaffolds by binning based on sequence composition, differential coverage, and read-pair linkage using metabat2 program (Kang *et al.*, 2019). Generated MAGs collections are imported into the merged profile database using anvi'o command anvi-import-collection. The command anvi-interactive provides an interactive interface that allows to visualize the results of unsupervised binning, perform supervised binning, or refine existing bins. The MAGs are manually refined if needed by the command anvi-refine provided in anvi'o package. Once the binning collection is ready, the anvi-summarize command provides a summary. The obtained summary, as shown in Table 1, includes details about the MAG completion as well as statistics like mean coverage, variability, etc. Subsequently, MAGs are functionally annotated using Prokka (Seemann, 2014). Finally, taxonomic classifications are assigned to MAGs based on the Genome Database Taxonomy (GTDB) using the GTDB-Tk program (Chaumeil *et al.*, 2022). Relative abundance values of metagenomics read that mapped to each MAG generated from 'bins_percent_recruitment.txt'. All the scripts and exercise files are available at this link (https://drive.google.com/drive/folders/1DQVaqD2VpSx6YbZUjOMHI97XcNP4Go_I?usp=share_link).

Table 1 Files obtained from the anvi-summarize command

File Name	Description
bins_summary.txt	Basic statistics of the recovered MAGs.
bins_across_samples/bins_percent_recruitment.txt	Coverage and detection of the MAGs.
bins_across_samples/mean_coverage.txt	The average number of sequencing reads that map to each MAG.
bin_by_bin	A folder with all the MAGs extracted from the metagenomics processing.

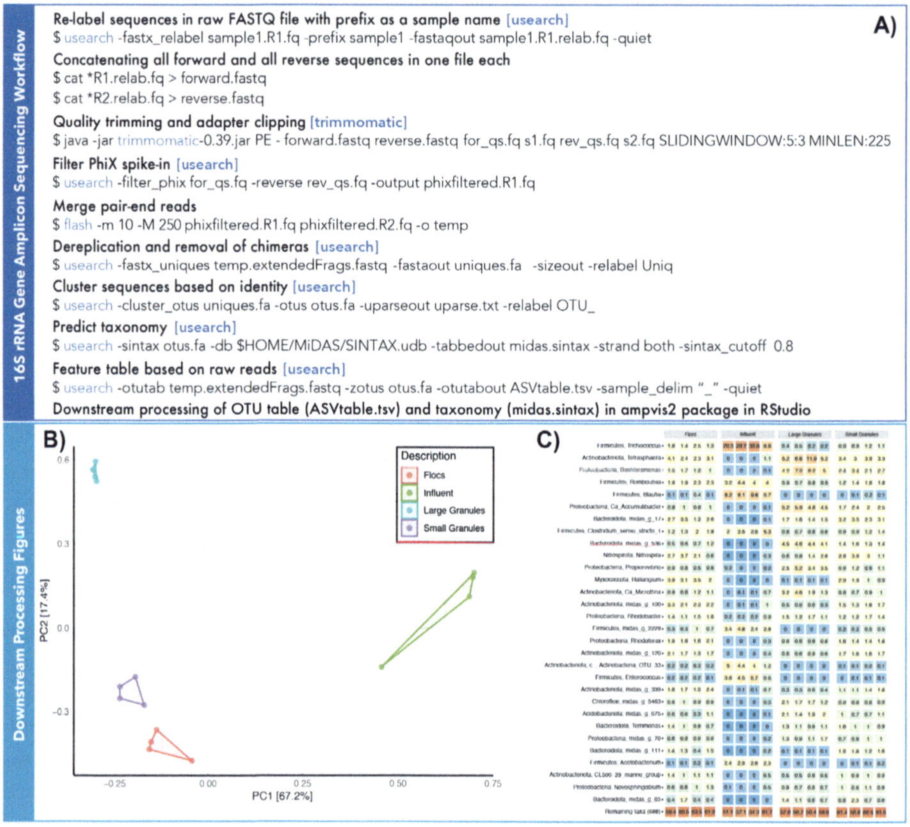

Figure 3 A) 16S rRNA gene Amplicon Sequencing Workflow.
B) Ordination plot of influent wastewater, flocs, small granules, and large granules based on Principal Components Analysis (PCA). Hellinger transformation is used to perform PCA as it will produce a more ecologically meaningful result.
C) Heatmap distribution of the dominant OTUs classified down to the genus level (f, o, c, and p represent family, order, class, and phylum, respectively) for the influent wastewater, flocs, small granules, and large granules. Figures in Panel B and C are adapted with permission from (Ali *et al.* Environmental Science & Technology, 2019, 53, 8291–8301). Copyright (2019) American Chemical Society.

Genome-resolved Metagenomics Workflow

```
Quality trimming and adapter clipping [cutadapt]
$ cutadapt -a $ADAPTER_FWD -A $ADAPTER_REV -m100 -q 25 -o sampl.R1.trim.fq -p sampl.R2.trim.fq  sampl.R1.fq  sampl.R2.fq
Concatenate all forward and all reverse sequences one file each
$ cat *.R1.trim.fq > forward.fastq
Prepare assembly [MEGAHIT]
$ megahit -o contigs -1 forward.fastq -2 reverse.fastq
Reformat contigs FASTA file [anvio-7.1]
$ anvi-script-reformat-fasta final.contigs.fa -o contigs-fixed.fa -l 1000 --simplify-names
Creating an Anvi'o contigs database [anvio-7.1]
$ anvi-gen-contigs-database -f contigs.fa -o contigs.db -n 'Project Marine Anammox'
Creating index for further mapping [bwa]
$ bwa index contigs.fa
Map raw sample reads against the assembly [samtools and bwa]
$ bwa mem -t 4 contigs.fa sampl.R1.trim.fq sampl.R2.trim.fq | samtools view -S -b - > sampl.bam
Sort BAM file of the sample [samtools]
$ samtools sort -o sampl.sort.bam sampl.bam
Create an index file for the sample [samtools]
$ samtools index sampl.sort.bam
Genome binning with contigs and sorted BAM files [metabat2]
$ metabat2 -i contigs.fa *sort.bam
Create an Anvi'o profiles [anvio-7.1]
$ anvi-profile -i sampl.sort.bam -c contigs.db --cluster-contigs
Merge Anvi'o profiles [anvio-7.1]
$ anvi-merge */PROFILE.db -o SAMPLES-MERGED -c contigs.db ---enforce-hierarchical-clustering
Downstream processing in Anvi'o and other packages [anvio-7.1]
```

Figure 4 Genome-resolved Metagenomics Workflow

14.5.3 Genome-resolved metatranscriptomics

In this exercise, metatranscriptomics samples are collected from the same MBR where metagenomics samples were collected (Ali *et al.*, 2020b). Genome-resolved metatranscriptomics workflow is presented in Fig. 5A. The reads are mapped to the predicted protein-coding genes generated from Prokka for each MAG using USEARCH command usearch_global. Reads with a sequence identity below 0.98 were discarded. The count tables are used to generate gene expression of microorganisms represented by the MAGs recovered from the MBR. The gene expression is based on transcription per million (TPM) metatranscriptomics reads that are mapped to each MAG (Fig. 5B&C). TPM values are calculated by dividing the read counts by the MAG size (kilobases), which gives reads per kilobase (RPK) value. The RPK values in a sample are then added, and the resultant value is divided by 1,000,000 to obtain a 'per million' scaling factor. Finally, the individual RPK values were divided by the 'per million' scaling factor, yielding the TPM value. All the scripts and exercise files are available at this link (https://drive.google.com/drive/folders/1I_wbxtSBsYaL_OIy1wPlWqnkGIhin4Zg?usp=share_link). For further comparative analysis, the count tables can be imported to RStudio, processed using the default DESeq2 workflow (Love *et al.*, 2014), and visualized using ggplot2. Principle component analysis (PCA) and hierarchical clustering analysis (HCA) of overall sample similarity can be performed using DESeq2 normalized count in the R platform.

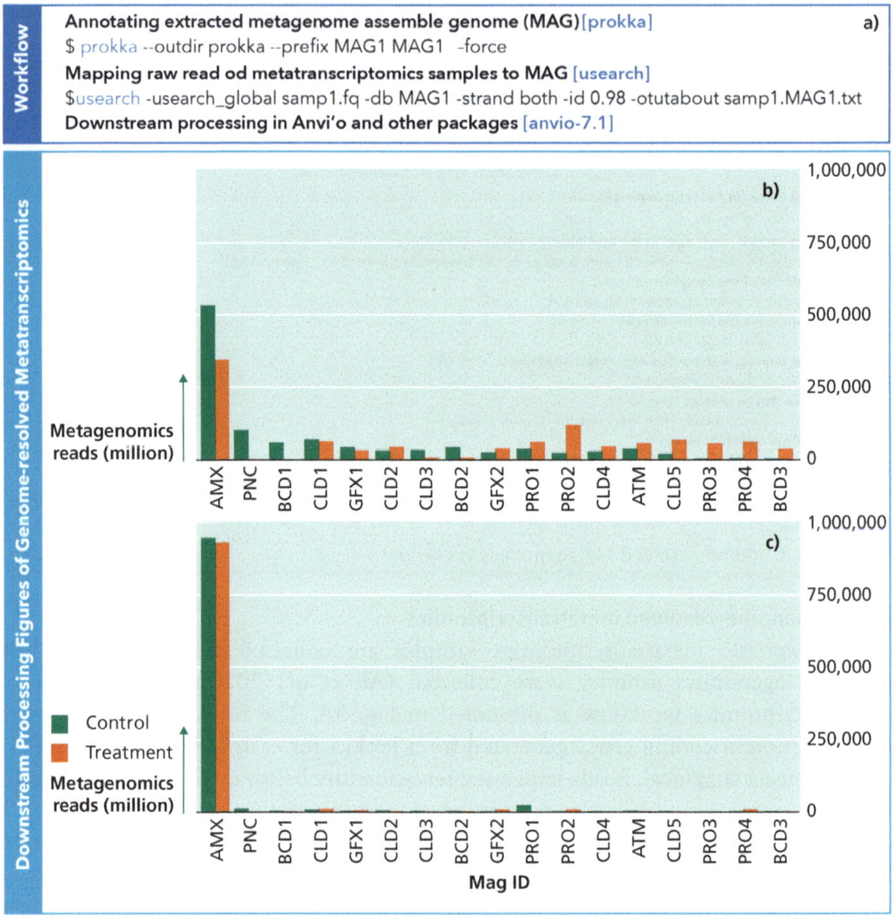

Figure 5 A) Genome-resolved Metatranscriptomics Workflow.
B) Relative abundance and
C) gene expression based on transcription per million (TPM) values of metagenomics and metatranscriptomics reads that mapped to each MAG recovered from the MBR. Figures in Panel B and C are adapted with permission from (Ali *et al*. Water Research, 2020, 170, 115345). Copyright (2019) Elsevier Ltd.

14.6 APPLICATIONS OF GENOMICS IN MEMBRANE FILTRATION RESEARCH

The use of 16S rRNA gene amplicon sequencing has been valuable for understanding the diversity and abundance of microorganisms in membrane-based water systems (Figure 3C) (Vries, 2022). For instance, it was used to create a microbiological database of bacteria detected in biofilms on rubber-coated valves in drinking water systems and to identify the specific bacterial phylum present in contaminated reverse osmosis (RO) membranes (Bereschenko, 2010; Beyer, 2019; Vries, 2022). Furthermore, by obtaining a thorough understanding of the bacterial communities within membranes, scientists have gained valuable insights regarding the types of microorganisms that are prone to causing fouling. These insights have been instrumental in refining membrane-cleaning strategies, specifically by targeting particular microorganisms or proactively preventing the formation of bacterial biofilms (Møllebjerg *et al.*, 2023). Additionally, 16S rRNA and other omics data can be used for ordination analysis to reveal similarities and clustering patterns among samples in membrane processes (Figure 3B). The findings from such studies can provide valuable information about the microbial ecology within membranes, helping researchers to gain insights into the microbial communities present in membranes and understand how they relate to one another.

Genome-centric metagenomics processing provides a more comprehensive and refined approach than 16S rRNA analysis. For example, MAGs enabled the identification of microorganisms at the species level (Rehman *et al.*, 2019). In Figure 6, two main aspects of genome-centric metagenomics analysis are presented. Firstly, it enables the construction of phylogenetic trees (Figure 6A). This allows us to understand the evolutionary relationships between different organisms and their role in the ecosystem. Secondly, genome-centric metagenomics processing allows the reconstruction of metabolic pathways using metagenome-assembled genomes (MAGs) based on gene presence or absence (Figure 6B). By analyzing the genetic content of MAGs, we can identify the presence or absence of specific genes involved in metabolic pathways. This information helps us understand the functional potential of microbial communities and their contributions to various biogeochemical processes in membrane-based systems. Overall, genome-centric metagenomics processing is a powerful tool that provides insights into the taxonomic and functional composition of microbial communities. It helps us understand the complex interactions between different organisms and their roles in ecosystem functioning. It also enables the discovery of genes that are enriched in biofilm development as well as the investigation of specific genes within microbial communities. These insights have played a crucial role in developing effective strategies to prevent biofouling. For example, antifouling strategies for seawater reverse osmosis (RO) membranes involve targeting the enriched Planctomycetes bacteria or employing inhibitory compounds such as azide, chlorate, cyanide, and thiocyanide to specifically target nitrate-reducing enzymes, which have been identified as enriched in the biofilm metagenome (Rehman *et al.*, 2019).

a) Phylogenetic tree **b) Metabolic pathways modelling**

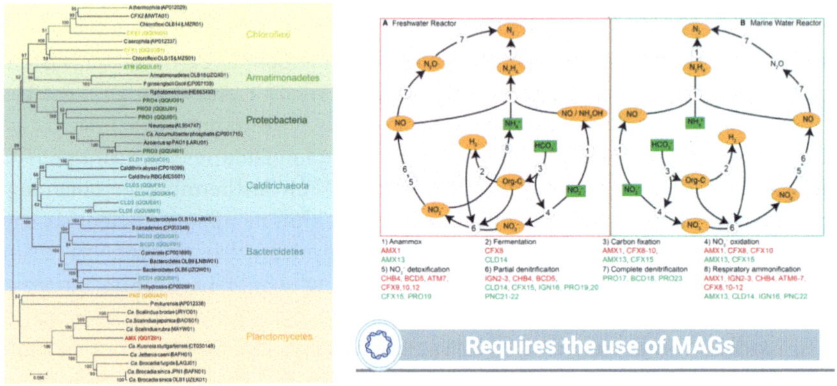

Figure 6 Genome-centric metagenomics processing enables the construction of A) Phylogenetic trees and B) Reconstruction of metabolic pathways using metagenome-assembled genomes (MAGs) based on gene presence or absence. Figures in Panel A is adapted with permission from (Ali *et al.* Water Research, 2020, 170, 115345). Copyright (2019) Elsevier Ltd. Figures in Panel B is adapted with permission from (Ali *et al.* Frontiers in Microbiology, 2020, 11, 1637.).

14.7 OUTLOOK

Recently, metagenomics has been successfully employed to discover novel microorganisms. However, the following points should be considered in future omics-based studies, including membrane-based filtration systems.

1. The omics-based study reports a relative abundance of microbial species as an actual abundance in the system. However, the relative abundance derived from such analysis should not be considered or interpreted as the system's actual abundance.
2. Assembled genomes (MAGs) have clear advantages for further functional analyses. However, obtaining correct assemblies is still challenging. The presence of genomic repeats, short overlap lengths, and phylogenetically close organisms can result in false-positive assembly outputs, such as joining non-overlapping fragments from different parts of the genome or producing chimeric contigs from different organisms. Besides the difficulty in assembling low-abundance species and closely related strains (micro diversity), and the exclusion of significant amounts of unassembled data, an assembly-based annotation strategy can introduce biases for quantitative analysis.
3. NGS-based sequencing is still expensive because the cost rises with sequencing depth. Most omics-based studies use quantitative analysis due to the high cost of technical/biological replicates. This makes it challenging to determine whether observed differences (particularly small ones) within or between experimental groups are statistically significant and robust (after considering replicate variance). Therefore, future research should consider the reproducibility of omics data before delving into differentially expressed features.

4. The completeness of database resources inevitably affects the taxonomic and functional annotation of sequence data. However, the databases improve as researchers continue to investigate microbial dark matter.

14.8 DATA AVAILABILITY

All raw sequencing data associated with amplicon sequencing analysis taken from the previous study (Ali *et al.*, 2019) are available at NCBI Sequence Read Archive under accession number SRP115069. All sequencing data associated with genome-resolve metagenomics analysis taken from a previous study (Ali *et al.*, 2020b) is available at NCBI under BioProject PRJNA482223.

14.9 REFERENCES

Albertsen, M., Hugenholtz, P., Skarshewski, A., Nielsen, K.L., Tyson, G.W. and Nielsen, P.H. 2013. Genome sequences of rare, uncultured bacteria obtained by differential coverage binning of multiple metagenomes. Nat Biotechnol 31(6), 533-538.

Albertsen, M., Karst, S.M., Ziegler, A.S., Kirkegaard, R.H. and Nielsen, P.H. 2015. Back to Basics--The Influence of DNA Extraction and Primer Choice on Phylogenetic Analysis of Activated Sludge Communities. PLoS One 10(7), e0132783.

Ali, M., Shaw, D.R. and Saikaly, P.E. 2020b. Application of an enrichment culture of the marine anammox bacterium "Ca. Scalindua sp. AMX11" for nitrogen removal under moderate salinity and in the presence of organic carbon. Water Res 170, 115345.

Ali, M., Shaw, D.R., Albertsen, M. and Saikaly, P.E. 2020a. Comparative Genome-Centric Analysis of Freshwater and Marine ANAMMOX Cultures Suggests Functional Redundancy in Nitrogen Removal Processes. Front Microbiol 11, 1637.

Ali, M., Wang, Z., Salam, K.W., Hari, A.R., Pronk, M., van Loosdrecht, M.C.M. and Saikaly, P.E. 2019. Importance of Species Sorting and Immigration on the Bacterial Assembly of Different-Sized Aggregates in a Full-Scale Aerobic Granular Sludge Plant. Environ Sci Technol 53(14), 8291-8301.

Alneberg, J., Bjarnason, B.S., de Bruijn, I., Schirmer, M., Quick, J., Ijaz, U.Z., Lahti, L., Loman, N.J., Andersson, A.F. and Quince, C. 2014. Binning metagenomic contigs by coverage and composition. Nat Methods 11(11), 1144-1146.

Alqahtani, M.F., Bajracharya, S., Katuri, K.P., Ali, M., Ragab, A., Michoud, G., Daffonchio, D. and Saikaly, P.E. 2019. Enrichment of Marinobacter sp. and Halophilic Homoacetogens at the Biocathode of Microbial Electrosynthesis System Inoculated With Red Sea Brine Pool. Front Microbiol 10, 2563.

Andersen, K.S., Kirkegaard, R.H., Karst, S.M. and Albertsen, M. 2018. ampvis2: an R package to analyse and visualise 16S rRNA amplicon data. bioRxiv.

Bereschenko, L.A. (2010) Biofilm development on new and cleaned membrane surfaces, Wageningen University.

Beyer, F. (2019) Microbiological and process technological aspects of nanofiltration and reverse osmosis membrane biofouling, Wageningen University.

Bodilis, J., Nsigue-Meilo, S., Besaury, L. and Quillet, L. 2012. Variable copy number, intra-genomic heterogeneities and lateral transfers of the 16S rRNA gene in Pseudomonas. PLoS One 7(4), e35647.

Bolger, A.M., Lohse, M. and Usadel, B. 2014. Trimmomatic: a flexible trimmer for Illumina sequence data. Bioinformatics 30(15), 2114-2120.

Bougouffa, S., Yang, J.K., Lee, O.O., Wang, Y., Batang, Z., Al-Suwailem, A. and Qian, P.Y. 2013. Distinctive microbial community structure in highly stratified deep-sea brine water columns. Appl Environ Microbiol 79(11), 3425-3437.

Caporaso, J.G., Kuczynski, J., Stombaugh, J., Bittinger, K., Bushman, F.D., Costello, E.K., Fierer, N., Peña, A.G., Goodrich, J.K., Gordon, J.I., Huttley, G.A., Kelley, S.T., Knights, D., Koenig, J.E., Ley, R.E., Lozupone, C.A., McDonald, D., Muegge, B.D., Pirrung, M., Reeder, J., Sevinsky, J.R., Turnbaugh, P.J., Walters, W.A., Widmann, J., Yatsunenko, T., Zaneveld, J. and Knight, R. 2010. QIIME allows analysis of high-throughput community sequencing data. Nature Methods 7, 335–336.

Chaumeil, P.A., Mussig, A.J., Hugenholtz, P. and Parks, D.H. 2022. GTDB-Tk v2: memory friendly classification with the Genome Taxonomy Database. Bioinformatics.

Desai, N., Antonopoulos, D., Gilbert, J.A., Glass, E.M. and Meyer, F. 2012. From genomics to metagenomics. Curr Opin Biotechnol 23(1), 72-76.

Dick, G.J., Andersson, A.F., Baker, B.J., Simmons, S.L., Thomas, B.C., Yelton, A.P. and Banfield, J.F. 2009. Community-wide analysis of microbial genome sequence signatures. Genome Biol 10(8), R85.

Dueholm, M.K.D., Nierychlo, M., Andersen, K.S., Rudkjobing, V., Knutsson, S., Mi, D.A.S.G.C., Albertsen, M. and Nielsen, P.H. 2022. MiDAS 4: A global catalogue of full-length 16S rRNA gene sequences and taxonomy for studies of bacterial communities in wastewater treatment plants. Nat Commun 13(1), 1908.

Dupont, C.L., Rusch, D.B., Yooseph, S., Lombardo, M.J., Richter, R.A., Valas, R., Novotny, M., Yee-Greenbaum, J., Selengut, J.D., Haft, D.H., Halpern, A.L., Lasken, R.S., Nealson, K., Friedman, R. and Venter, J.C. 2012. Genomic insights to SAR86, an abundant and uncultivated marine bacterial lineage. ISME J 6(6), 1186-1199.

Edgar, R.C. 2013. UPARSE: highly accurate OTU sequences from microbial amplicon reads. Nat Methods 10(10), 996-998.

Eren, A.M., Kiefl, E., Shaiber, A., Veseli, I., Miller, S.E., Schechter, M.S., Fink, I., Pan, J.N., Yousef, M., Fogarty, E.C., Trigodet, F., Watson, A.R., Esen, O.C., Moore, R.M., Clayssen, Q., Lee, M.D., Kivenson, V., Graham, E.D., Merrill, B.D., Karkman, A., Blankenberg, D., Eppley, J.M., Sjodin, A., Scott, J.J., Vazquez-Campos, X., McKay, L.J., McDaniel, E.A., Stevens, S.L.R., Anderson, R.E., Fuessel, J., Fernandez-Guerra, A., Maignien, L., Delmont, T.O. and Willis, A.D. 2021. Community-led, integrated, reproducible multi-omics with anvi'o. Nat Microbiol 6(1), 3-6.

Frias-Lopez, J., Shi, Y., Tyson, G.W., Coleman, M.L., Schuster, S.C., Chisholm, S.W. and Delong, E.F. 2008. Microbial community gene expression in ocean surface waters. Proc Natl Acad Sci U S A 105(10), 3805-3810.

Garcia Martin, H., Ivanova, N., Kunin, V., Warnecke, F., Barry, K.W., McHardy, A.C., Yeates, C., He, S., Salamov, A.A., Szeto, E., Dalin, E., Putnam, N.H., Shapiro, H.J., Pangilinan, J.L., Rigoutsos, I., Kyrpides, N.C., Blackall, L.L., McMahon, K.D. and Hugenholtz, P. 2006. Metagenomic analysis of two enhanced biological phosphorus removal (EBPR) sludge communities. Nat Biotechnol 24(10), 1263-1269.

Garcias-Bonet, N., Fusi, M., Ali, M., Shaw, D.R., Saikaly, P.E., Daffonchio, D. and Duarte, C.M. 2018. High denitrification and anaerobic ammonium oxidation contributes to net nitrogen loss in a seagrass ecosystem in the central Red Sea. Biogeosciences 15(23), 7333-7346.

Goodrich, J.K., Di Rienzi, S.C., Poole, A.C., Koren, O., Walters, W.A., Caporaso, J.G., Knight, R. and Ley, R.E. 2014. Conducting a microbiome study. Cell 158(2), 250-262.

Haroon, M.F., Hu, S., Shi, Y., Imelfort, M., Keller, J., Hugenholtz, P., Yuan, Z. and Tyson, G.W. 2013. Anaerobic oxidation of methane coupled to nitrate reduction in a novel archaeal lineage. Nature 500(7464), 567-570.

Head, S.R., Komori, H.K., LaMere, S.A., Whisenant, T., Van Nieuwerburgh, F., Salomon, D.R. and Ordoukhanian, P. 2014. Library construction for next-generation sequencing: overviews and challenges. Biotechniques 56(2), 61-64, 66, 68, passim.

Howe, A.C., Jansson, J.K., Malfatti, S.A., Tringe, S.G., Tiedje, J.M. and Brown, C.T. 2014. Tackling soil diversity with the assembly of large, complex metagenomes. Proc Natl Acad Sci U S A 111(13), 4904-4909.

Hugenholtz, P. 2002. Exploring prokaryotic diversity in the genomic era. Genome Biology 3(2), 1-8.

Imelfort, M., Parks, D., Woodcroft, B.J., Dennis, P., Hugenholtz, P. and Tyson, G.W. 2014. GroopM: an automated tool for the recovery of population genomes from related metagenomes. PeerJ 2, e603.

Kang, D.D., Froula, J., Egan, R. and Wang, Z. 2015. MetaBAT, an efficient tool for accurately reconstructing single genomes from complex microbial communities. PeerJ 3, e1165.

Kang, D.D., Li, F., Kirton, E., Thomas, A., Egan, R., An, H. and Wang, Z. 2019. MetaBAT 2: an adaptive binning algorithm for robust and efficient genome reconstruction from metagenome assemblies. PeerJ 7, e7359.

Knight, R., Jansson, J., Field, D., Fierer, N., Desai, N., Fuhrman, J.A., Hugenholtz, P., van der Lelie, D., Meyer, F., Stevens, R., Bailey, M.J., Gordon, J.I., Kowalchuk, G.A. and Gilbert, J.A. 2012. Unlocking the potential of metagenomics through replicated experimental design. Nat Biotechnol 30(6), 513-520.

Koren, S., Schatz, M.C., Walenz, B.P., Martin, J., Howard, J.T., Ganapathy, G., Wang, Z., Rasko, D.A., McCombie, W.R., Jarvis, E.D. and Adam, M.P. 2012. Hybrid error correction and de novo assembly of single-molecule sequencing reads. Nat Biotechnol 30(7), 693-700.

Kunin, V., Copeland, A., Lapidus, A., Mavromatis, K. and Hugenholtz, P. 2008. A bioinformatician's guide to metagenomics. Microbiol Mol Biol Rev 72(4), 557-578, Table of Contents.

Li, D., Luo, R., Liu, C.M., Leung, C.M., Ting, H.F., Sadakane, K., Yamashita, H. and Lam, T.W. 2016. MEGAHIT v1.0: A fast and scalable metagenome assembler driven by advanced methodologies and community practices. Methods 102, 3-11.

Li, H. and Durbin, R. 2009. Fast and accurate short read alignment with Burrows-Wheeler transform. Bioinformatics 25(14), 1754-1760.

Li, H. and Durbin, R. 2010. Fast and accurate long-read alignment with Burrows-Wheeler transform. Bioinformatics 26(5), 589-595.

Love, M.I., Huber, W. and Anders, S. 2014. Moderated estimation of fold change and dispersion for RNA-seq data with DESeq2. Genome Biol 15(12), 550.

Lovley, D.R. 2003. Cleaning up with genomics: applying molecular biology to bioremediation. Nat Rev Microbiol 1(1), 35-44.

Magoc, T. and Salzberg, S.L. 2011. FLASH: fast length adjustment of short reads to improve genome assemblies. Bioinformatics 27(21), 2957-2963.

Martin, M. 2011. Cutadapt removes adapter sequences from high-throughput sequencing reads. EMBNet Journal, 10-12.

Mason, O.U., Hazen, T.C., Borglin, S., Chain, P.S., Dubinsky, E.A., Fortney, J.L., Han, J., Holman, H.Y., Hultman, J., Lamendella, R., Mackelprang, R., Malfatti, S., Tom, L.M., Tringe, S.G., Woyke, T., Zhou, J., Rubin, E.M. and Jansson, J.K. 2012. Metagenome, metatranscriptome and single-cell sequencing reveal microbial response to Deepwater Horizon oil spill. ISME J 6(9), 1715-1727.

Matar, G.K., Ali, M., Bagchi, S., Nunes, S., Liu, W.T. and Saikaly, P.E. 2021. Relative Importance of Stochastic Assembly Process of Membrane Biofilm Increased as Biofilm Aged. Front Microbiol 12, 708531.

McHardy, A.C., Martin, H.G., Tsirigos, A., Hugenholtz, P. and Rigoutsos, I. 2007. Accurate phylogenetic classification of variable-length DNA fragments. Nat Methods 4(1), 63-72.

Møllebjerg, A., Zarebska, A., Nielsen, H.B., Hansen, L.B.S., Sørensen, S.R., Seredynska-Sobecka, B., Villacorte, L.O., Gori, K., Palmén, L.G. and Meyer, R.L. 2023. Novel high-throughput screening platform identifies enzymes to tackle biofouling on reverse osmosis membranes. Desalination 554.

Nielsen, P.H., Daims, H., Lemmer, H., Arslan-Alaton, I. and Olmez-Hanci, T. (2009) FISH Handbook for Biological Wastewater Treatment, IWA Publishing, London, UK.

Parks, D.H., Imelfort, M., Skennerton, C.T., Hugenholtz, P. and Tyson, G.W. 2015. CheckM: assessing the quality of microbial genomes recovered from isolates, single cells, and metagenomes. Genome Res 25(7), 1043-1055.

Peng, Y., Leung, H.C., Yiu, S.M. and Chin, F.Y. 2012. IDBA-UD: a de novo assembler for single-cell and metagenomic sequencing data with highly uneven depth. Bioinformatics 28(11), 1420-1428.

Pruesse, E., Peplies, J. and Glockner, F.O. 2012. SINA: accurate high-throughput multiple sequence alignment of ribosomal RNA genes. Bioinformatics 28(14), 1823-1829.

Raes, J., Korbel, J.O., Lercher, M.J., von Mering, C. and Bork, P. 2007. Prediction of effective genome size in metagenomic samples. Genome Biol 8(1), R10.

Rehman, Z.U., Ali, M., Iftikhar, H. and Leiknes, T. 2019. Genome-resolved metagenomic analysis reveals roles of microbial community members in full-scale seawater reverse osmosis plant. Water Res 149, 263-271.

Riesgo, A., Perez-Porro, A.R., Carmona, S., Leys, S.P. and Giribet, G. 2012. Optimization of preservation and storage time of sponge tissues to obtain quality mRNA for next-generation sequencing. Mol Ecol Resour 12(2), 312-322.

Roesch, L.F., Fulthorpe, R.R., Riva, A., Casella, G., Hadwin, A.K., Kent, A.D., Daroub, S.H., Camargo, F.A., Farmerie, W.G. and Triplett, E.W. 2007. Pyrosequencing enumerates and contrasts soil microbial diversity. ISME J 1(4), 283-290.

Schloss, P.D., Westcott, S.L., Ryabin, T., Hall, J.R., Hartmann, M., Hollister, E.B., Lesniewski, R.A., Oakley, B.B., Parks, D.H., Robinson, C.J., Sahl, J.W., Stres, B., Thallinger, G.G., Van Horn, D.J. and Weber, C.F. 2009. Introducing mothur: open-source, platform-independent, community-supported software for describing and comparing microbial communities. Appl Environ Microbiol 75(23), 7537-7541.

Scholz, M.B., Lo, C.C. and Chain, P.S. 2012. Next generation sequencing and bioinformatic bottlenecks: the current state of metagenomic data analysis. Curr Opin Biotechnol 23(1), 9-15.

Seemann, T. 2014. Prokka: rapid prokaryotic genome annotation. Bioinformatics 30(14), 2068-2069.

Shaw, D.R., Ali, M., Katuri, K.P., Gralnick, J.A., Reimann, J., Mesman, R., van Niftrik, L., Jetten, M.S.M. and Saikaly, P.E. 2020. Extracellular electron transfer-dependent anaerobic oxidation of ammonium by anammox bacteria. Nat Commun 11(1), 2058.

Siegwald, L., Touzet, H., Lemoine, Y., Hot, D., Audebert, C. and Caboche, S. 2017. Assessment of Common and Emerging Bioinformatics Pipelines for Targeted Metagenomics. PLoS One 12(1), e0169563.

Speth, D.R., Lagkouvardos, I., Wang, Y., Qian, P.Y., Dutilh, B.E. and Jetten, M.S.M. 2017. Draft Genome of Scalindua rubra, Obtained from the Interface Above the Discovery Deep Brine in the Red Sea, Sheds Light on Potential Salt Adaptation Strategies in Anammox Bacteria. Microb Ecol 74(1), 1-5.

Thomas, T., Gilbert, J. and Meyer, F. 2012. Metagenomics - a guide from sampling to data analysis. Microb Inform Exp 2(1), 3.

Trego, A.C., O'Sullivan, S., Quince, C., Mills, S., Ijaz, U.Z. and Collins, G. 2020. Size Shapes the Active Microbiome of Methanogenic Granules, Corroborating a Biofilm Life Cycle. mSystems 5(5).

van Dijk, E.L., Auger, H., Jaszczyszyn, Y. and Thermes, C. 2014. Ten years of next-generation sequencing technology. Trends Genet 30(9), 418-426.

Vries, H.J.d. (2022) Characterizing bacteria involved in fouling of spiral wound membranes, Wageningen University.

Wang, X., Hu, M., Xia, Y., Wen, X. and Ding, K. 2012a. Pyrosequencing analysis of bacterial diversity in 14 wastewater treatment systems in China. Appl Environ Microbiol 78(19), 7042-7047.

Wang, Y., Cao, H., Zhang, G., Bougouffa, S., Lee, O.O., Al-Suwailem, A. and Qian, P.Y. 2013. Autotrophic microbe metagenomes and metabolic pathways differentiate adjacent Red Sea brine pools. Sci Rep 3, 1748.

Wang, Y., Leung, H.C., Yiu, S.M. and Chin, F.Y. 2012b. MetaCluster 4.0: a novel binning algorithm for NGS reads and huge number of species. J Comput Biol 19(2), 241-249.

Wrighton, K.C., THOMAS, B.C., SHARON, I., MILLER, C.S., CASTELLE, C.J., VERBERKMOES, N.C., WILKINS, M.J., HETTICH, R.L., LIPTON, M.S., WILLIAMS, K.H., LONG, P.E. and BANFIELD, J.F. 2012. Fermentation, Hydrogen, and Sulfur Metabolism in Multiple Uncultivated Bacterial Phyla. Science 337(6102), 1661-1665.

Wu, L., Ning, D., Zhang, B., Li, Y., Zhang, P., Shan, X., Zhang, Q., Brown, M.R., Li, Z., Van Nostrand, J.D., Ling, F., Xiao, N., Zhang, Y., Vierheilig, J., Wells, G.F., Yang, Y., Deng, Y., Tu, Q., Wang, A., Global Water Microbiome, C., Zhang, T., He, Z., Keller, J., Nielsen, P.H., Alvarez, P.J.J., Criddle, C.S., Wagner, M., Tiedje, J.M., He, Q., Curtis, T.P., Stahl, D.A., Alvarez-Cohen, L., Rittmann, B.E., Wen, X. and Zhou, J. 2019. Global diversity and biogeography of bacterial communities in wastewater treatment plants. Nat Microbiol 4(7), 1183-1195.

Wu, Y.-W., Tang, Y.-H., Tringe, S.G., Simmons, B.A. and Singer, S.W. 2016. MaxBin: an automated binning method to recover individual genomes from metagenomes using an expectation-maximization algorithm. Microbiome 2, 26.

Yang, Y., Li, B., Ju, F. and Zhang, T. 2013. Exploring variation of antibiotic resistance genes in activated sludge over a four-year period through a metagenomic approach. Environ Sci Technol 47(18), 10197-10205.

Yu, K. and Zhang, T. 2012. Metagenomic and metatranscriptomic analysis of microbial community structure and gene expression of activated sludge. PLoS One 7(5), e38183.

Zhang, T., Shao, M.F. and Ye, L. 2012. 454 pyrosequencing reveals bacterial diversity of activated sludge from 14 sewage treatment plants. ISME J 6(6), 1137-1147.

doi: 10.2166/9781789062977_0337

Chapter 15

Biofouling Potential Measurement using Bacterial Growth Potential (BGP)

Almotasembellah Abushaban, UM6P, Morocco

Karima Bakkali, UM6P, Morocco

Léonie Le Bouille, IHE Delft / CIRSEE-Suez, France

Mohamed ChakerNecibi, UM6P, Morocco

Sergio G. Salinas-Rodriguez, IHE Delft, The Netherlands

The learning objectives of this chapter are the following:

- Define biological fouling in membrane systems
- Apply bacterial growth potential measurements as a biofouling indicator
- Define and apply microbial adenosine triphosphate for assessing biomass activity and bacterial growth
- Present and discuss the latest regulation to limit biofouling in SWRO systems
- Present several cases studies on assessing biofouling potential along SWRO pre-treatment and feedwater.

15.1 INTRODUCTION

Numerous seawater reverse osmosis (SWRO) desalination plants still struggle to control biological and organic fouling because there are no standard methods to monitor these types of fouling in desalination plants. Biological fouling results from microbial growth in membrane systems, which may lead to operational problems such as increased head loss across feed spacers in spiral wound elements and decreased permeability of SWRO membranes. Biofilm formation in SWRO is inevitable if the feed water supports significant bacterial growth due to the presence of easily biodegradable (dissolved) nutrients.

While there is currently no standard test method for measuring the progression of biofilm formation in the plant, it is still possible to identify the formation of biofilms by testing and monitoring water quality, membrane fouling, and a variety of other factors.

Several approaches are being studied to monitor and control biofouling in SWRO. A good approach for biofouling control is to lower biological fouling potential in SWRO feed water through the pre-treatment. By pre-treating feed water, it can be reduced to a condition where bacteria and other contaminants cannot grow as well or kill off the type of microorganism that causes bio fouling. This approach is attractive because it can be used as an early warning system allowing adjustment of the operational conditions of the pre-treatment processes to meet the required quality in SWRO feed water and consequently better control of biofouling in SWRO systems. For this purpose, few methods were developed to measure biological/organic fouling potential such as assimilable organic carbon (AOC), biodegradable dissolved organic carbon (BDOC) and bacterial growth potential (BGP). The use of AOC and BGP methods has gained interest as high levels of AOC/BGP are directly linked to biofilm formation and thus more severe biofouling is expected at higher AOC/BGP in the SWRO feed water.

In general, to measure AOC or BGP in any water sample, four steps should be followed as following (Abushaban *et al.*, 2022; Salinas-Rodriguez *et al.*, 2021):

- Bacterial inactivation by pasteurization, filtration or sterilization. This step allows the standardization of the initial microbial population by adding a constant inoculum concentration.
- Bacterial inoculum by adding constant concentration of specific bacteria or indigenous bacteria.
- Incubation at a constant temperature.
- Bacterial enumeration: Different enumeration methods can be used depending on the bacterial culture such as total direct cell count, bioluminescence, turbidity, microbial electrolysis, flow cytometry (FCM) and ATP.

15.2 MATERIALS

15.2.1 Laboratory equipment

- Eppendorf single channel pipette (adjustable volume) 10-100 μL, 100-1000 μL, 0.5-5 mL
- Eppendorf Biopur® pipette tips (sterile, pyrogen-free, DNA-free and ATP-free) for 10-100 μL, 100-1000 μL, 0.5-5 mL
- Eppendorf Biopur® 1.5 ml Safe-Lock micro test tubes (sterile, pyrogen-free, human-DNA free, bacterial-DNA free, RNase-free, DNase-free, PCR-inhibitor free and ATP-free), individually sealed
- 50 mL centrifuges tubes with screw caps, Roth (sterile)
- 0.1 μm PVDF membrane Filters, Millex-GP, Merck Millipore (sterile)
- Syringes 10 mL, 5 mL, BD (sterile)
- Tube holders

Remark:

All materials and consumables that is used for microbial ATP measurement **should be sterile** including pipette, pipette tips, centrifuge tubes, Eppendorf tubes, etc. The consumable should be used for one single use.

15.2.2 CHEMICALS

Chemicals for ATP measurement

- **Bacteria lysis** is used to extract ATP from the microbial cell by destroying the microbial cell membrane. It can be stored at room temperature.
- **ATP Water-Glo**: is the light generating reagent containing luciferin and luciferase. ATP Water-Glo is stored at room temperature and after reconstitution with buffer the reagent is stored at 4 °C.
- **ATP standard** is a liquid stable of 100 nM (50,000 ng/L) (Biothema or Promega Corp.) of ATP concentration.

Staining chemicals for flow cytometry:

- **SYBR Green I (SG)**: It is used to measure the total cell counts (TCC) in a water sample. It needs to be stored at 4 °C.
- **Propidium iodide (PI)**: It is used to measure the intact cell counts in a water sample. It needs to be stored at 4 °C.

15.2.3 Instrumental equipment

- AccuBlock™ Digital Dry Baths, Block
- Vortex mixer (Heidolph Reax 2000 shaker)
- Incubator (30 °C)
- Autoclave
- Muffle furnace (up to 550 °C)

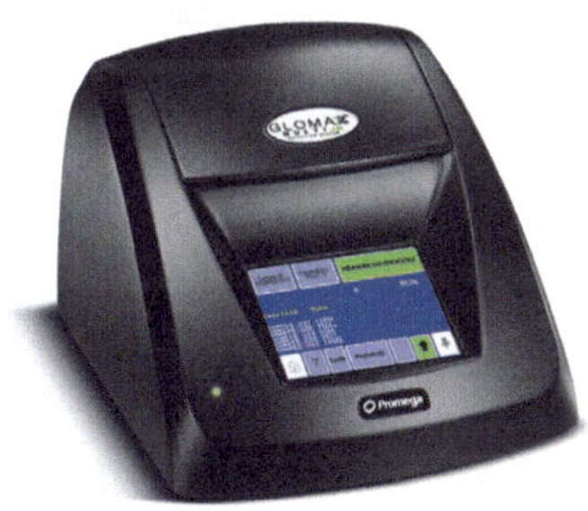

Figure 1 The GloMax® 20/20 Single Tube Luminometer.

- Promega GloMax®-20/20 single tube Luminometer: Different companies have different Luminometer. GloMax® 20/20 single tube Luminometer (Figure 1) developed by the Promega Company measures the light emitted during Luciferin-Luciferase reaction. To measure luminescence of a sample, the sample should be transferred into a 1.5-Micro centrifuge tube before inserting it inside the Luminometer. The cover of the equipment should be closed during the measurement to avoid any interference of any other external light. The result of the measurements is displayed within 10 seconds on the computer screen in relative light units (RLU).

- BD Accuri C6 or C6Plus Flow Cytometer system: Flow cytometer (FCM) is a device used to measure particles or cells which are present in the fluid and pass the laser intercept of the equipment and results in light forwarding and scattering. The flow cytometer can also measure the size of the cell which can be determined by the amount of light scattered by individual cell (Prest *et al.*, 2016). Since the flow cytometer is based on light forwarding and scattering, which can be due to cells or any other particle present in the sample, a standard flow cytometric staining protocol was developed at Eawag (Gatza *et al.*, 2013) to distinguish between the bacterial cells and other particles present in sample. The staining protocol is employed by BD AccuriTM C6 and C6 Plus (Figure 2) software analysis template which can be used to measure not only the total cells present in water but also the viable cells. The template consists a single, fixed gate that can determine total bacterial cells when stained with SYBR® Green I, or only viable bacteria cells when co-stained with SYBR® Green I and propidium iodide (Abushaban, 2020; Gatza *et al.*, 2013).

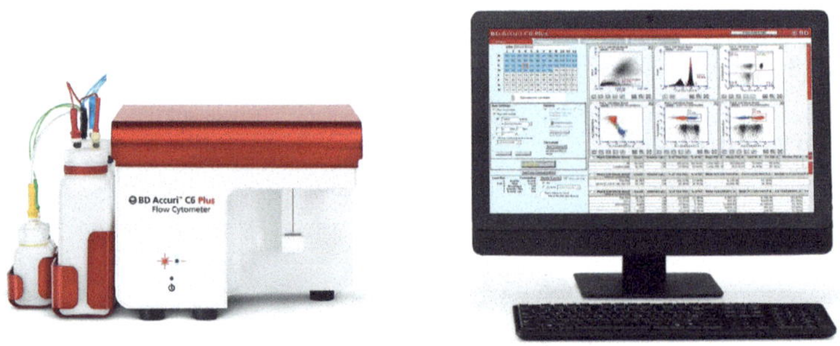

Figure 2 The BD Accuri C6 Flow Cytometer system.

15.3 METHODS AND EXPERIMENTAL PROCEDURE

15.3.1 Sample collection and storage

Seawater sample needs to be collected in sterile glass and stored in cooler box (4 °C) during transportation. Amber color glass sampling bottles with tight sealing screw caps are preferred. The volume of the collected sample should be at least 100 mL.

15.3.2 Cleaning glassware

Vials, bottles, caps and all glassware materials must be washed before any usage. Glassware may contain some external sticky contaminants such as particles and bacteria from inside and outside, which can lead to inaccurate results if not cleaned.

1. Wash all glassware and caps using lab detergent such as Alconox Ultrasonic Cleaner.
2. Rinsed all glassware with Milli-Q water (Milli-Q® water Optimized purification, 18.2 MΩ.cm at 25 °C, EC < 10 µS/cm, TOC < 30 µg/L, Millipore, USA).
3. Submerge all glassware and caps overnight in 0.2 M HCl.
4. Once again, rinse all glassware and caps three times with Milli-Q and let them dry in the air.
5. After draying, vials need to be covered with aluminum foil to avoid any potential organic contaminants.
6. Heat vials and bottles in oven furnace at 550 °C for 6 hours (effective time to expose glassware to heat).
7. On the other hand, vials caps are bathed in sodium persulfate solution (100 g/L) for 1 hour at 60 °C. Then, rinse all caps with Milli-Q water and let them air dried. Powder free gloves were used during the handling process.
8. Finally, take out the aluminum foil and close all vials and bottles with caps.

15.3.3 Preparation of artificial seawater

Artificial seawater (ASW) needs to be prepared with similar characteristic of the real seawater in terms of ion compositions, electrical conductivity and pH. Table 1 shows the ion composition of the average global seawater conditions (Villacorte, 2014). For preparation of the ASW, all glassware needs to be cleaned before using them as mentioned in the previous section. The salts used in the preparation of ASW may be contaminated. Therefore, the salts with high enough melting points must be burned at 550 °C for 6 hours to make carbon volatile. (Table 1). The amount of each salt is then dissolved in Milli-Q water (Milli-Q® water Optimized purification, 18.2 MΩ.cm at 25 °C, EC < 10 µS/cm, TOC < 30 µg/L, Millipore, USA) and stirred for at least 24 hours. Moreover, once the ASW is prepared, it needs to be autoclaved again to ensure that the prepared ASW is bacteria free.

Table 1 Ion concentrations of artificial seawater to mimic the average global seawater (Abushaban, 2020)

Component name	Chemical formula	Concentration (g/L)	Heating temperature (°C)
Sodium carbonate	Na_2CO_3	0.002	100 °C
Sodium hydrogen carbonate	$NaHCO_3$	0.213	NA
Potassium chloride	KCl	0.739	550 °C
Calcium chloride di hydrate	$CaCl_2.2H_2O$	1.540	100 °C
Sodium sulfate	Na_2SO_4	3.993	550 °C
Magnesium chloride hexahydrate	$MgCl_2.6H_2O$	10.873	100 °C
Sodium chloride	NaCl	22.3	550 °C

The autoclaving temperature of ASW should be set to 100 °C for 15 minutes, as the melting point of magnesium chloride hexahydrate ($MgCl_2.6H_2O$) is around 117 °C. It is worth mentioning that Sodium hydrogen carbonate ($NaHCO_3$) should not be heated because its melting point is 50 °C and should therefore be added after the autoclaving step.

15.3.4 Intact cell count by flow cytometry

To measure the number of intact cells using FCM in a water sample, two staining solutions can be used such as SYBR Green I (SG) and Propidium Iodide (PI). The former is used to distinguish the total bacterial cells from debris and the latter is used to differentiate the intact cells from damaged bacterial cells. The staining should be only done for samples containing bacterial cells lower than 10^7 cells/mL. For samples containing bacterial cells more than 10^7 cells/mL, they should be diluted with filtered ASW before staining. The procedure to measure intact cell concentration of a water sample is as follows:

- Preheat the samples before staining for 5 min at 35 °C.
- Add 5µL of the stock solution (SG/PI) in 500 µL sample.
- Place the stained samples in dark place and at 35 °C for exactly 10 minutes.
- From computer screen, choose 50 µL run limits, with medium speed and threshold 700 on F1 was selected.
- Insert the sample into the FCM device and start the measurement.

Remark:

The maximum number of cells should not exceed 2×10^7 cells/mL. In case the sample contains higher number of cells, but lower than 10^7 cells/mL, the sample can be diluted after staining, as mentioned earlier and measured again.

15.3.5 Measurement of Bacterial growth potential

To measure BGP of a seawater sample, four steps should be followed as including, bacterial inactivation, bacterial inoculation, sample incubation, and bacterial growth monitoring/ bacterial enumeration.

I. **Bacterial inactivation**: to inactivate bacteria in the seawater sample, the collected sample in glass bottle needs to be placed in an autoclave at a temperature between 70 and 100 °C and for 15-20 min. The bottle should be closed properly with a screw cap. It is important not to increase the temperature to more than 100 °C because higher temperature could negatively affect the Free ATP concentration in the seawater sample. After autoclaving, take the sample out of the autoclave and let it cool down (at room temperature).

II. **Bacterial inoculation:**

1. Transfer 20 mL of the autoclaved sample to a small cleaned vial (40 mL) in triplicate (3 vials). The vials should be cleaned following the pre-mentioned protocol for cleaning. The vial should not be filled completely to avoid any contamination form the cap. Sterilized syringe or pipette can be used to transfer the sample. It is recommended to do all steps on a clean surface of closed room (to avoid contamination from air).
2. Close the vial immediately after sample transfer using the cleaned vial cap.
3. Count the intact cell concentration of the same non-autoclaved sample (inoculum concentration) using flow cytometry (FCM).
4. Calculate the volume of inoculum to be added to the autoclaved sample (20 mL) using the following formula (equation 1). The final concentration of the inoculum should be 10,000 cells/mL.

$$V_1 = \frac{C_2 \cdot V_2}{C_1} \qquad \text{Eq. 1}$$

Where,
C_1 is the concentration of the sample measured by FCM minus the concentration of the inoculum (10,000 cells/mL),
V_1 is the volume of the inoculum to be added to the sample,
C_2 is the final concentration of the inoculum (10,000 cells/mL), and
V_2 is the final volume of the sample (20 mL).
Example: Assume the concentration of intact cells in the sample is 1.5 M cells/mL. The volume of the inoculum is (10,000 cells/mL × 20 mL ÷ 1,490,000 cells/mL) = 134.2 µL.

5. Transfer the calculated inoculum volume to the vials which have 20 mL autoclaved samples using a sterilized pipette tip.
   Note: The sample should be vortexed before pipetting the inoculum. Moreover, open the vial just before transferring the inoculum.
6. Once the inoculum is transferred in triplicate, the vials should be closed with a cleaned cap and vortexed.

III. **Sample incubation**: Place the inoculated sample at the incubator at 30 °C. Make sure that each vial is labelled including the date of sample collection and date of inoculation. The number of replications of each sample should be mentioned on the label to avoid mixing samples.

IV. **Sample enumeration**: The bacterial growth of inoculum needs to be monitored daily for at least 5 to 7 days. Microbial ATP or intact cell concentration using FCM can be used to monitor bacterial growth. The detailed protocol of these methods is presented in section 15.7. The microbial concentration on the day of inoculation should be measured as well as a reference concentration (Day 0). The concentration of day 0 (day of inoculation) should be around 10,000 cells/mL (intact cells measured by FCM) or less than 10 ng/L (Microbial ATP concentration).

15.3.6 Bacterial yield and calibration line

Establishing a calibration line of BGP is necessary to convert microbial growth to carbon concentration. Calibration curve is prepared by monitoring the bacterial growth of ASW of different known glucose concentrations (0, 10, 25, 50, 75 and 100 µg-C/L). To establish the curve, a correlation is made between added glucose concentration and its corresponding maximum bacterial growth of each concentration (Figure 3). It is worth mentioning that the bacterial growth may vary depending on the microorganism present in inoculum of indigenous microbial consortium (Wang *et al.*, 2014; Weinrich *et al.*, 2010), therefore the bacterial yield needs to be determined for each location.

The bacterial growth can be also inhibited due to absence of essential nutrients such as nitrogen and phosphorus. Therefore, nitrogen and phosphorous concentrations are added according to C: N: P proportional weight of 20:4:1 in artificial seawater.
The conversion of microbial growth to carbon concentrations is only possible if the bacterial yield is known. The bacterial yield is determined by calculating the slope of the established calibration curve which equals to the average bacterial growth that corresponds to 1 µg/L of glucose.

To prepare BGP calibration curve, the following procedure should be followed:

1. Prepare 1 L of artificial seawater following the above-mentioned procedure.
2. Prepare 20 mL of different glucose concentrations (for example; 0, 10, 25, 50, 75 and 100 µg-C/L) in triplicate (3×6 concentrations) using the sterilized artificial seawater (point 1). To prepare these concentrations, one stock solution of glucose with 2 mg/L concentration can be firstly prepared (using ASW of point 1) before preparing the low glucose concentration samples.
3. To allow bacterial growth, nitrogen and phosphorous concentrations should be added similar to point 2 and according to C: N: P proportional weight of 20:4:1.
4. Vortex the prepared samples containing C, N, and P before adding the inoculum concentration. To calculate the inoculum concentration, follow step II.4 under section 15.3.4.
5. Follow the same steps of BGP measurement to incubate and enumerate the samples.
6. After 7 days of bacterial growth monitoring, draw the relationship between the added glucose concentration and the corresponded maximum bacterial growth. The relationship should be a linear between them.
7. Bacterial yield can be calculated by identifying the slope of the correlation between maximum growth and the added glucose concentration.
8. Use the linear correlation to convert the maximum bacterial growth in ng/L to carbon concentration µg-C/L.

Note: A correlation needs to be established between bacterial growth and carbon concentration for a specific location.

Remark: To make sure that the calibration line is valid for your specific seawater sample, a correlation can be investigated using a real seawater (RSW). The slope using RSW should be similar or close to the slope prepared using ASW.

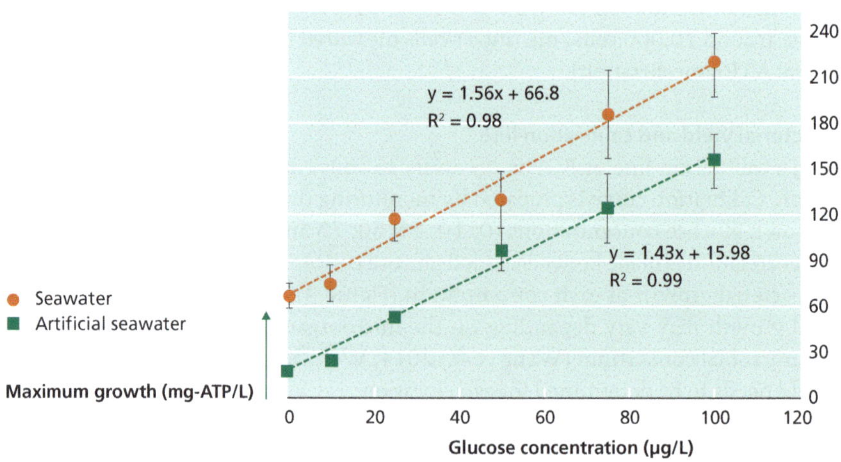

Figure 3 The correlation between added glucose concentration and the BGP in seawater and artificial seawater (Abushaban *et al.*, 2019a).

15.4 APPLICATIONS

15.4.1 Example A: BGP monitoring of an SWRO pre-treatment

BGP concentration were measured in a full-scale membrane-based desalination plant in Australia. The pre-treatment processes of SWRO system includes a drum screen, coagulation and flocculation, dual media filter (DMF), and cartridge filter. Four samples were collected in October 2016 through the pre-treatment as following; just before coagulation (raw seawater), after coagulation and flocculation, after DMF, and after cartridge filter.

Results showed that bacteria started to grow immediately in seawater and reached a maximum growth within 2 days (Figure 4a). Afterwards, microbial ATP gradually decreased due to the partial decay of bacteria or because bacterial activity decreased due to the depletion of nutrients. The maximum bacterial growth was observed (305 ng-ATP/L) in raw seawater (Figure 4b), indicating the highest potential of bacterial growth. Slightly lower potential of bacterial growth (262 ng-ATP/L) was observed after coagulation and flocculation, while a significant reduction (> 55 %) of the bacterial growth potential was found after DMF. The high reduction in BGP through DMF could indicate that the DMF is a biologically active media filter. The maximum bacterial growth decreased modestly through the cartridge filter to 86 ng-ATP/L. This result shows that the seawater after pre-treatment still supports further bacterial growth when it is compared with the BGP of the blank.

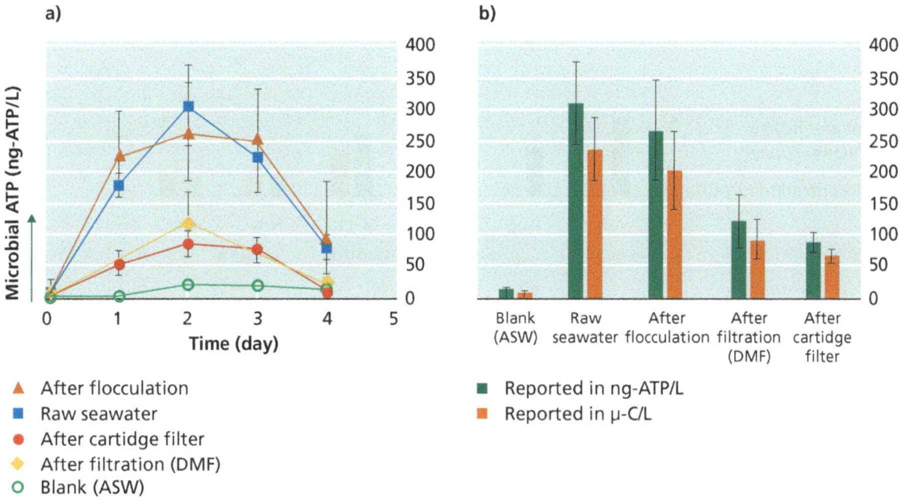

Figure 4 (a) bacterial growth over time and (b) maximum growth of different seawater samples collected through the pre-treatment processes of an SWRO desalination plant in Australia. (Abushaban *et al.*, 2021, 2018).

15.4.2 Example B: BGP in the intake and SWRO feed water

BGP was monitored at the intake and feedwater of an SWRO desalination plant in the Middle East. High BGP variations were observed in the seawater intake, in which the average monthly BGP ranged between 200 and 2,500 µg-C/L as glucose (Figure 5). Low BGPs were measured in the seawater intake during the Autumn, whereas extremely high BGPs were observed in September due to algal blooms which is widely reported in the Arabian Sea in September and October. The high BGP during summer might be attributed to carbon release from the algal cells present in seawater.

On average, the removal of BGP along the SWRO pre-treatment is 62 %, in which the maximum BGP removal (85%) was observed in July. However, low BGP removal or even negative removal were noted from October to December which could be attributed to the addition of antiscalant or the make-up water used for diluting antiscalant. The higher organic concentration after antiscalant addition has been observed in several SWRO and RO plants(Abushaban *et al.*, 2021; Vrouwenvelder *et al.*, 2000). Although reasonable concentration of organic and biological fouling potential was removed through the pre-treatment, still considerable concentration remains in the SWRO feed water (> 100 μg-C/L) which could lead to biofouling operational problem in the SWRO.

The monitored BGP based on microbial ATP illustrate its applicability to evaluate the pre-treatment processes of desalination systems.

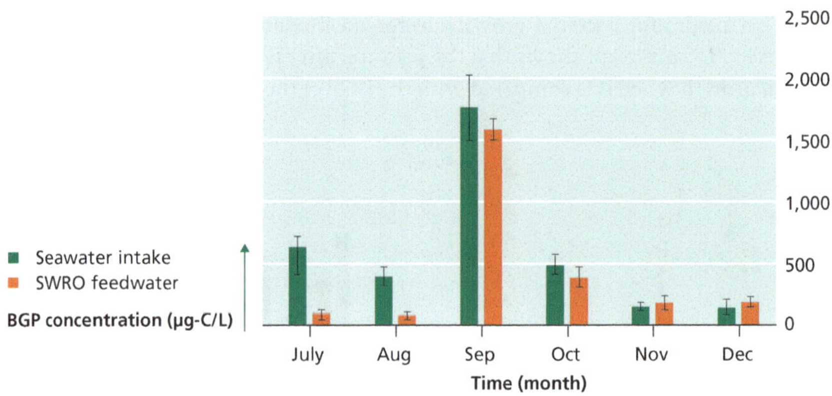

Figure 5 Monitored bacterial growth potential in the intake and SWRO feed water of an SWRO desalination plant located in the Middle East (Abushaban *et al.*, 2020).

15.5 DATA DISCUSSION AND INTERPRETATION

The correlation between BGP in the SWRO feed water and the normalized pressure drop/ permeability in the SWRO membrane system has been demonstrated (Abushaban, 2020). It was reported that the higher BGP in SWRO feed water corresponds to higher normalized pressure drop. In general, biofouling was observed in SWRO membrane systems when the measured BGP in the SWRO feed water is more than 100 μg-C/L.

Abushaban *et al.* (2020) preliminary investigated the correlation between the measured BGP in SWRO feed water and the RO membrane cleaning in place (CIP) frequency in the four SWRO plants. The CIP frequency (CIPs per year) was used as a surrogate parameter for biofouling. It is reported that a higher CIP frequency corresponded to a higher BGP of SWRO feed water (Figure 6). Accordingly, a tentative threshold concentration of BGP (< 70 μg/L) was proposed for SWRO feed water in order to ensure a chemical cleaning frequency of once/year or lower. However, to establish a real correlation, more data needs to be collected and many more SWRO plants need to be monitored for longer periods of time with different operating conditions.

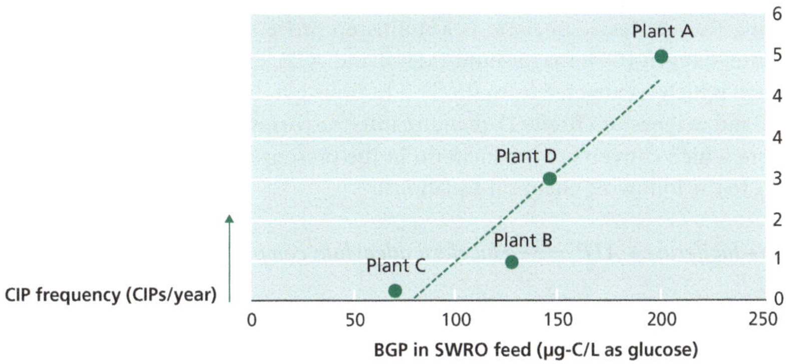

Figure 6 The correlation between bacterial growth potential in the feed water and cleaning frequency of SWRO membrane systems (Abushaban, 2019; Dhakal *et al.*, 2020).

15.6 ATP MEASURMENT

15.6.1 Introduction

ATP is a substance present in all living cells (including bacteria) that provides energy for many metabolic processes. In particular, it is used as a coenzyme in living cells and it is often called the 'molecular unit of currency' of intracellular energy transfer (Knowles, 1980). The structure of ATP has an ordered carbon compound as shown in Figure 7. ATP consists of adenosine and three phosphate groups. Adenosine composes of an adenine ring and a ribose sugar. The critical part of ATP is the phosphorous part - the triphosphate. Three phosphorous groups are connected by oxygen, and also there are side oxygen connected to the phosphorous atoms which leads to high repulsion between the negative charges of oxygen in the normal condition. Therefore, ATP has a lot of potential energy. ATP converts to adenosine diphosphate (ADP) if one of the three phosphates is broken down. This conversion is an extremely crucial reaction due to the realization of the energy after the reaction. The reaction and realized energy of one Mole ATP is shown in eq (2).

$$ATP + H_2O \rightarrow ADP + P_i \Delta G^{\circ} = -30.5 kJ/mol\,(-7.3 kcal/mol) \quad \text{Eq. 2}$$

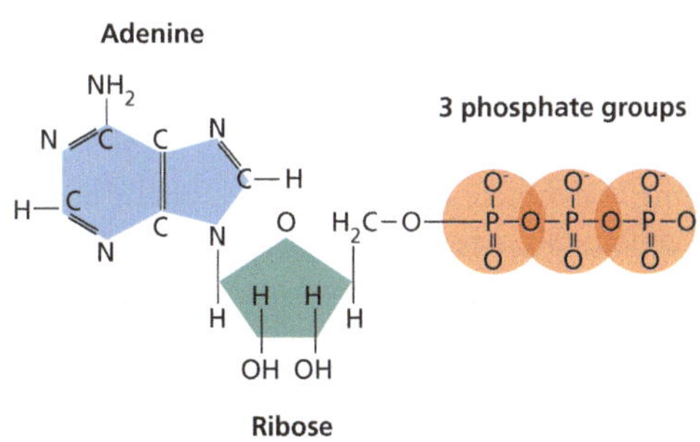

Figure 7 Schematic diagram showing the structure of ATP.

Several methodologies are used for ATP determination, but so far, the most successful technique is the bioluminescent method, because of its sensitivity and the wide range concentration that can be measured (Van der Kooij *et al.*, 2003). ATP bioluminescence has been used for determining levels of ATP in many different cell types. The bioluminescence method involving the Luciferase enzyme is a multistep process which mainly requires Luciferin substrate, oxygen (O_2), magnesium (Mg^{2+}) and ATP. Enzyme luciferase uses as catalyst of reaction which is extracted from firefly (Photinus pyralis). Luciferase converts in presence of ATP and magnesium firefly D-luciferin into the corresponding enzyme-bound luciferil adenylate which converts to oxyluciferin in the presence of oxygen. This process occurs according to the following chemical equations:

$$D-luceferin+luciferase+ATP\xrightarrow{Mg}luceferil\ adenylate\ complex+PP_i \qquad \text{Eq. 3}$$

$$Luceferin\ adenylate\ complex\xrightarrow{O_2}Oxyluceferin+AMP^{*}+CO_2+light \qquad \text{Eq. 4}$$

*AMP = Adenosine monophosphate

Light is emitted from a rapid loss of energy of the oxyluciferine molecule. The light emission is in the range between 500 to 700 nm wavelengths (Riemann, 1979). Under optimum conditions, light intensity is linearly related to ATP concentration. Cellular ATP can be measured by direct lysis of the cells with suitable detergent. In general, the determination of ATP concentration includes the following procedures (Van der Kooij *et al.*, 2003) which also presented in Figure 8:

- Collection of a representative sample.
- Extraction of ATP from the microorganisms.
- Addition of reagents for luciferine – luciferase assay.
- Recording the light emitted.
- Calculation of the ATP concentration from calibrated data.

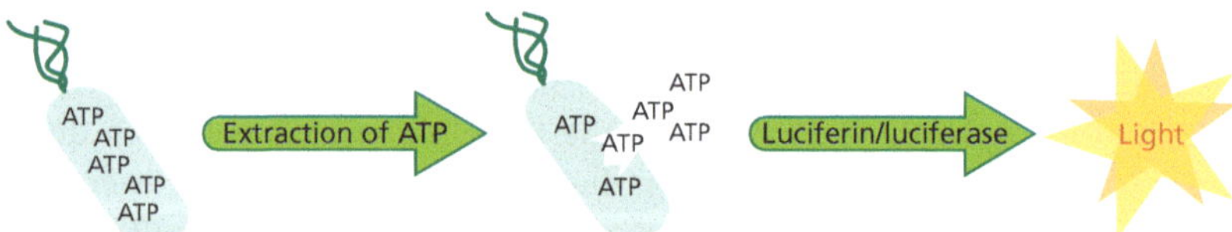

Figure 8 The simple protocol of bioluminescent method.

15.6.2 Material and methods

Material

The laboratory equipment, instrumental equipment and chemicals are listed in section 15.2
Note: All materials and consumables to be used for microbial ATP measurement should *be sterile* including pipette, pipette tips, centrifuge tubes, Eppendorf tubes, etc. The consumable should be used for one single use.

Methods

Preparation of Water-Glo detection reagent

- Clean the bench, pipette and the cover of the pipette tips with Ethanol (70 %),
- Carefully open the WaterGlo™ Substrate vial and do not touch the inside part of the stopper,
- Transfer the contents of the WaterGlo™ Buffer vial to the WaterGlo ™ Substrate vial,
- Replace the stopper and slowly invert the vial several times to dissolve the contents, and
- Allow reconstituted detection reagent to stand at room temperature for 1 hour before use.

Note: The prepared Water-Glo detection reagent should be Stored at the fridge at 5 °C and it can be taken out of the fridge 15 min before the measurement.

Direct ATP method

a. Calibration curve

Since the concentration of ATP is unknown in the sample, a calibration curve needs to be established by measuring the relative light unit (RLU) of different ATP concentration. To prepare different ATP concentrations, ATP standard (100 nM, 50,000 ng-ATP/L, Biothema) needs to be diluted in the water medium. To prepare the calibration curve, the following procedure is followed:

1. Biothema ATP Standard (100 nM) is taken out of the fridge 5 min before using it.
2. Use sterilized artificial seawater as water medium to dilute standard ATP.
3. Use the table 2 below to prepare different ATP standard solutions.
4. Switch on the heating block and adjust the heating temperature at 38 °C.
5. Distribute the prepared concentration to 6 Eppendorf tubes (3 for total ATP and 3 for free ATP) in which 100 µL is pipette into each tube.
6. Add 100 µL of Lysis reagents on each Eppendorf tubes prepared for total ATP measurements.
7. Place your prepared samples into the heating block and start measuring the samples as per described in the protocol below.

Table 2 The required volume of ATP standard and medium to prepare a calibration curve using ATP direct method.

Final Concentrate (ng ATP/L)	Volume of the ATP standard (µL)	Original concentrate (ng ATP/L)	Volume of medium (µL)
5,000	60	50,000	540
500	100	5,000	900
250	45	5,000	855
100	18	5,000	882
50	100	500	900
25	18	500	882
5	100	50	900
0.5	100	5	900
0	0	0	1,000

b. Measurement protocol

The ATP measurement to be used is the method developed by Abushaban *et al.*, (2018) for seawater samples. The protocol of ATP measurement using the direct method is illustrated in Figure 9.

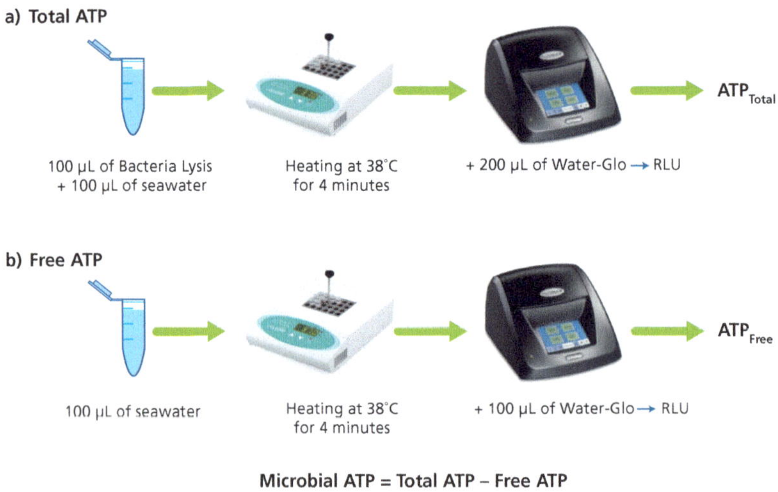

Figure 9 Scheme of Direct Method Set up (Abushaban *et al.*, 2018).

Total ATP

- Transfer 100µL of seawater sample into 1.5mL Eppendorf tubes in triplicates.
- Use pipette to transfer 100µL of Bacteria Lysis into the Eppendorf tubes containing the sample and mix with the pipette.
- Heat the sample in a heating block for 4 minutes and at a temperature of 38 °C together with the Water-Glo reagents in separate Eppendorf tubes.
- Add 200µL of heated Water-Glo reagent to the heated sample using a pipette.
- Mix the sample by using the sucking and release the mixture once and then immediately place the tube into the GloMax Luminometer (measurement to be conducted within 20 second of contact time).
- The measured RLU of the sample will be recorded directly in the excel sheet connected with the luminometer.

Free ATP Measurement:

- Transfer 100µL of seawater sample into 1.5mL Eppendorf tubes in triplicates.
- Heat the sample in a heating block for 4minutes at 38 °C together with the WaterGlo reagents in separate Eppendorf tubes.
- Add 100µL of heated Water-Glo reagent to the heated sample using a pipette.
- Mix the sample by using the sucking and release the mixture once and then immediately place the tube into the GloMax Luminometer (measurement to be conducted within 20 second of contact time).
- The measured RLU of the sample will be recorded directly in the excel sheet connected with the luminometer

- Calculate the ATP concentration based on the calibration curve. Bacterial ATP = Total ATP – Free ATP

Note: The lid cover of the luminometer should be immediately closed after inserting the Eppendorf tube to avoid any external light influence.

ATP Filtration method

a. Preparation of the Calibration Curve:

ATP calibration line is needed to convert the measured relative light unit to a concentration of ATP. For this purpose, different solutions of different ATP concentrations need to be prepared and measured using Luminometer. For the ATP filtration method, ATP standard of 2 nM (1,000 ng-ATP/L, Promega Corp.) is used. Moreover, bacterial lysis is used as a medium solution to dilute ATP standard. To prepare the calibration curve of ATP filtration method, the following procedure is followed:

1. Promega ATP Standard (2 nM) is taken out of the fridge 5 min before using it.
2. Use the sterile bacterial lysis reagent as water medium to dilute standard ATP.
3. Use the table 3 below to prepare different ATP standard solutions.
4. Switch on the heating block and adjust the heating temperature at 38 °C.
5. Distribute the prepared concentration to 3 Eppendorf tubes in which 100 µL is pipette into each tube.
6. Place your prepared samples into the heating block and start measure the samples as per described protocol below.

Table 3 Dilution of ATP standard in bacterial lysis for ATP filtration method calibration curve (Abushaban, 2020)

Final Concentrate (ng-ATP/L)	Volume of bacterial lysis reagent (µL)	ATP standard (ng ATP/L)	Volume of the ATP standard (µL)
500	250	1,000	250
250	375	1,000	125
100	450	1,000	50
50	950	1,000	50
25	250	50	250
5	450	50	50
0	1,000	0	0

b. Protocol of the ATP filtration method

The followings are the steps to measure microbial ATP using ATP filtration method which is also presented in Figure 10:

1. Filter 5 mL of seawater sample through 0.1 µm filter in order to accumulate bacterial cells on the filter surface. The filtrate is discarded.
2. Filter 2 mL of rinsing solution (sterilized ASW) through the same filter to eliminate free ATP which could be retained in the filter holder from the 1st step. The filtrate is discarded.

3. Filter 5 mL of bacteria lysis through the same filter to extract ATP from bacterial cells. Additionally, inject air to the filter holder in order to push the remaining bacteria lysis solution.
4. Transfer 100 μL of collected sample into 1.5 mL Eppendorf tubes (triplicate).
5. Heat all the samples and ATP Water-Glo detection reagent at 38 °C using a heating block.
6. Transfer 100 μL of the heated ATP Water-Glo detection reagent to the heated sample and mix the two solutions (reagent and sample).
7. Immediately insert the mixed sample in the luminometer to measure RLU value of the sample.
8. Calculate the ATP concentration from the calibration curve using the below formula:

   a. $$ATP = \frac{1}{m} \cdot (RLU - b)$$

   Where:
   RLU: the relative light unit obtained from Luminometer
   b: The intercept with y-axis of calibration curve.
   m: The slope of calibration curve.
   RLU: Relative light units value recorded by the luminometer
   ATP: The calculated ATP concentration based on calibration curve

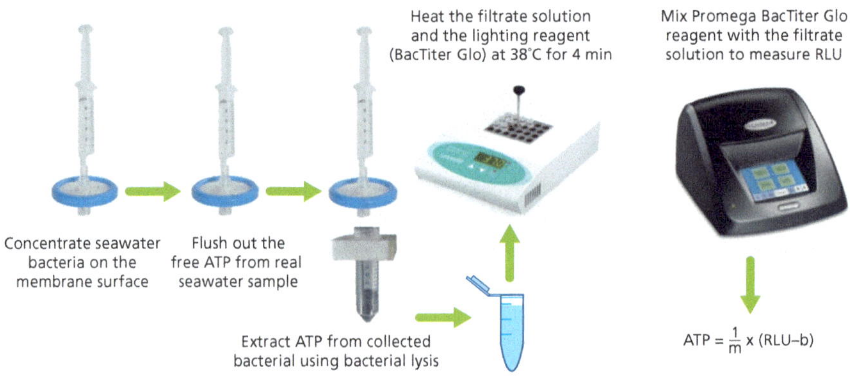

Figure10 The protocol of microbial ATP measurement in seawater (Abushaban *et al.*, 2019b).

15.7 REFERENCES

Abushaban, A., 2019. Assessing Bacterial Growth Potential in Seawater Reverse Osmosis Pretreatment: Method Development and Applications. CRC Press.

Abushaban, A., Mangal, M.N., Salinas-Rodriguez, S.G., Nnebuo, C., Mondal, S., Goueli, S.A., Schippers, J.C., Kennedy, M.D., 2018. Direct measurement of ATP in seawater and application of ATP to monitor bacterial growth potential in SWRO pre-treatment systems. Desalination and Water Treatment 99, 91–101.

Abushaban, A., Salinas-Rodriguez, S.G., Dhakal, N., Schippers, J.C., Kennedy, M.D., 2019a. Assessing pretreatment and seawater reverse osmosis performance using an ATP-based bacterial growth potential method. Desalination 467, 210–218. https://doi.org/10.1016/j.desal.2019.06.001

Abushaban, A., Salinas-Rodriguez, S.G., Kapala, M., Pastorelli, D., Schippers, J.C., Mondal, S., Goueli, S., Kennedy, M.D., 2020. Monitoring Biofouling Potential Using ATP-Based Bacterial Growth Potential in SWRO Pre-Treatment of a Full-Scale Plant. Membranes 10. https://doi.org/10.3390/membranes10110360

Abushaban, A., Salinas-Rodriguez, S.G., Mangal, M.N., Mondal, S., Goueli, S.A., Knezev, A., Vrouwenvelder, J.S., Schippers, J.C., Kennedy, M.D., 2019b. ATP measurement in seawater reverse osmosis systems: Eliminating seawater matrix effects using a filtration-based method. Desalination 453, 1–9.

Abushaban, A., Salinas-Rodriguez, S.G., Pastorelli, D., Schippers, J.C., Mondal, S., Goueli, S., Kennedy, M.D., 2021. Assessing Pretreatment Effectiveness for Particulate, Organic and Biological Fouling in a Full-Scale SWRO Desalination Plant. Membranes 11, 167–167.

Abushaban, A., Salinas-Rodriguez, S.G., Philibert, M., Le Bouille, L., Necibi, M.C., Chehbouni, A., 2022. Biofouling potential indicators to assess pretreatment and mitigate biofouling in SWRO membranes: A short review. Desalination 527, 115543. https://doi.org/10.1016/j.desal.2021.115543

Dhakal, N., Abushaban, A., Mangal, N., Abunada, M., Schippers, J.C., Kennedy, M.D., 2020. Membrane Fouling and Scaling in Reverse Osmosis, in: Membrane Desalination. CRC Press, pp. 325–344.

Gatza, E., Hammes, F., Prest, E., 2013. Rapid and accurate quantitation of bacteria in drinking water is essential to monitor, control, and optimize water treatment processes, and to illuminate the biology of low nutrient water systems. Historically, laboratories have relied on heterotrophic pl.

Knowles, J.R., 1980. Enzyme-catalyzed phosphoryl transfer reactions. Annual Review of Biochemistry 49, 877–919. https://doi.org/10.1146/annurev.bi.49.070180.004305

Prest, E.I., Hammes, F., van Loosdrecht, M.C.M., Vrouwenvelder, J.S., 2016. Biological stability of drinking water: controlling factors, methods, and challenges. Frontiers in Microbiology 7, 45–45. https://doi.org/10.3389/fmicb.2016.00045

Riemann, B.O., 1979. The occurrence and ecological importance of dissolved ATP in fresh water. Freshwater Biology 9, 481–490. https://doi.org/10.1111/j.1365-2427.1979.tb01532.x

Salinas-Rodriguez, S.G., Mangal, M.N., Villacorte, L.O., Abushaban, A., 2021. Methods for Assessing Fouling and Scaling of Saline Water in Membrane-Based Desalination, in: Removal of Pollutants from Saline Water. CRC Press.

Van der Kooij, Albrechtsen H., D., Corfitzen, C., Ashworth, J., Parry, I., Enkiri, F., Hambsch, B., Hametner, C., Kloiber, R., H., Veenendaal., 2003. Assessment of the microbial growth support potential of products in contact with drinking water. CPDW Project: European Commission Joint Research Centre.

Villacorte, L.O., 2014. Algal blooms and membrane based desalination technology. Environmental engineering and water technology department. Delft University of Technolog, CRC Press/ Balkema.

Vrouwenvelder, J.S., Manolarakis, S.A., Veenendaal, H.R., Van der Kooij, D., 2000. Biofouling potential of chemicals used for scale control in RO and NF membranes. Desalination 132, 1–10.

Wang, Q., Tao, T., Xin, K., Li, S., Zhang, W., 2014. A review research of assimilable organic carbon bioassay. Desalination and Water Treatment 52, 2734–2740. https://doi.org/10.1080/19443994.2013.830683

Weinrich, L.A., Jjemba, P.K., Giraldo, E., LeChevallier, M.W., 2010. Implications of organic carbon in the deterioration of water quality in reclaimed water distribution systems. Water Research 44, 5367–5375. https://doi.org/10.1016/j.watres.2010.06.035

doi: 10.2166/9781789062977_0355

Chapter 16

Assessing Biological Stability of Ultra-low Nutrient Water by Measuring Bacterial Growth Potential

Mohaned Sousi, IHE Delft, The Netherlands

The learning objectives of this chapter are the following:

- Define biological stability of drinking water and the factors influencing it
- Present and discuss the method for measuring bacterial growth potential of ultra-low nutrient drinking water
- Present a case study for applying the developed bacterial growth potential method.

16.1 INTRODUCTION

Water utilities do not only aim at producing high quality drinking water that is safe for human consumption, but also maintaining this quality during distribution until water reaches the consumer's tap. Water distribution systems provide a complex environment for bacterial growth either in the form of planktonic bacteria or biofilm (Liu *et al.*, 2013), where such systems usually contain pipes of a wide range of diameters and made of various materials, in addition to the fluctuating water flow throughout the day. Several studies have shown that bacterial growth could take place in water distribution systems whether residual chlorine was added or not (Prest *et al.*, 2016b).

The concept of biological stability of drinking water appeared by the end of the twentieth century, where the issue of bacterial growth in water distribution networks has gained increasing attention. Biologically stable drinking water does not promote excessive bacterial growth in distribution systems and until water reaches the consumption point (van der Kooij and Veenendaal, 2014; Liu *et al.*, 2013; Huck, 1990). Several serious problems are associated with biologically unstable water, including threats posed to human health due to the growth of (opportunistic) pathogens, deterioration of the aesthetic aspects of water (taste, odour, and colour), and operational problems related to bio-corrosion of pipes and fittings (Volk and LeChevallier, 1999; Berry *et al.*, 2006). Biologically stable drinking water can be achieved by applying a multi-barrier treatment strategy to reduce the concentration of nutrients that promote bacterial growth in water (Smeets *et al.*, 2009; van der Kooij and Veenendaal, 2014).

Biological stability of drinking water is traditionally assessed with the presumption that a small fraction of organic carbon is promoting bacterial growth. Several laboratory-based methods were developed to measure this fraction of organic carbon, namely assimilable organic carbon (AOC) method (van der Kooij *et al.*, 1982; van der Kooij and Hijnen, 1984) and biodegradable dissolved organic carbon (BDOC) method (Servais *et al.*, 1987). One of the main disadvantages of these traditional methods is the use of pure bacterial strains that might not completely consume the available organic carbon present in water, resulting in underestimation of bacterial growth that could take place in distribution systems where diverse bacterial strains are present.

Major developments in the field of microbiological methods have occurred in the past years, allowing for rapid, less laborious, and more accurate measurements of bacteria in water samples compared with the traditional HPC method (van Nevel *et al.*, 2017). One of these major developments is flow cytometry (FCM) which can be coupled with DNA staining to enable complete enumeration of bacterial cells in a water sample (Prest *et al.*, 2016a). FCM has been applied for biological stability assessment to measure AOC using natural bacterial consortium (Hammes and Egli, 2005), allowing for more accurate estimation of AOC. Additionally, FCM has been used for direct measurement of bacterial growth potential (BGP) of water expressed as cell count (cells/mL) (Sousi *et al.*, 2020a; Nescerecka *et al.*, 2018; Farhat *et al.*, 2018; Prest *et al.*, 2016a), without converting the obtained growth into AOC. BGP of water can also be measured with adenosine triphosphate (ATP) as an alternative bacterial parameter (Abushaban *et al.*, 2019; Vital *et al.*, 2012; Farhat *et al.*, 2018). In addition, combining BGP measurements with 16S rRNA gene sequencing enables in-depth understanding of bacterial growth characteristics of water during treatment and distribution (Liu *et al.*, 2020; Liu *et al.*, 2018; Li *et al.*, 2017).

Current methods to assess bacterial growth potential are suitable for drinking water that is produced using conventional water treatment technologies such as coagulation, flocculation, sedimentation, (rapid or slow) sand filtration and activated carbon filtration. However, this chapter focuses on assessing bacterial growth potential of water with ultra-low nutrient content produced by advan0ced water purification systems, such as reverse osmosis (RO). Post-treatment is a key process to make RO permeate potable and suitable for distribution. The process of re-adding the essential minerals to RO permeate is called remineralisation, which can be conducted using several methods, including: blending RO permeate with source water, direct dosing of chemicals, calcite contactors, and micronized calcite dosing. Remineralisation by calcite contactors is widely applied in practice since it is a simple and cost-efficient method, where RO permeate percolates through a calcite bed to dissolve calcium carbonate into calcium (Ca^{2+}) ions and hydrogen carbonate (HCO^{3-}) (Ruggieri *et al.*, 2008; Hasson and Bendrihem, 2006).

Research has shown that bacterial growth-promoting nutrients could be considerably removed by RO filtration, resulting in a very low level of bacterial growth in RO permeate (Escobar *et al.*, 2000; Thayanukul *et al.*, 2013; Park and Hu, 2010; Dixon *et al.*, 2012). However, this may be influenced by post-treatment, more specifically by remineralisation since it involves the addition of substances to RO permeate.

16.2 MATERIALS AND EXPERIMENTAL SET-UP

Bacterial regrowth potential (BGP) method is applied to assess biological stability in drinking water. This method implies monitoring of total and/or intact bacterial counts in water samples over time using flow cytometry concept (FCM). To perform BGP measurements, the following materials and equipment are required:

16.2.1 Equipment

A. Flow cytometer (FCM)

Flow cytometry (FCM) coupled with fluorescent staining has emerged as a leading tool for single-cell analysis in microbiology. It is used as a rapid and accurate enumeration tool for total bacteria or specific bacterial groups in water samples. Figure 1 describes the concept of flow cytometry.

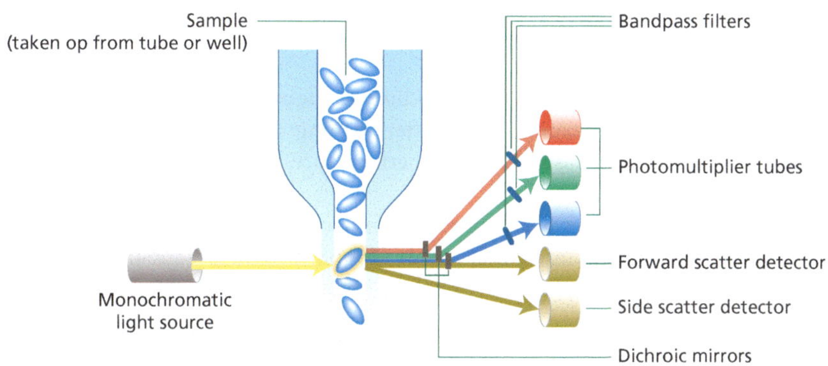

Figure 1 Flow cytometry principle (Safford and Bischel, 2019)

The labelled cells with a fluorochrome are transported through a flow cell where they are subjected to a laser beam of a specific wavelength. The flow cell is designed to allow only one fluorescently labelled cell to pass through the laser beam at a time where it gets excited and emits light which is captured by the detectors. The data gathered can be analysed statistically by flow cytometry software.

FCM measurements are performed using a BD Accuri C6® flow cytometer equipped with a 50 mW laser emitting at a fixed wavelength of 488 nm. Green fluorescence intensity is collected in FL1 channel (533 ± 30 nm) and red fluorescence is collected in FL3 channel (> 670 nm), while sideward and forward scattered light intensities were collected as well. All parameters are collected in logarithmic signals. The FCM is equipped with volumetric counting hardware, calibrated to measure the number of particles in 50 μL of a 500 μL sample. Measurements were performed at pre-set flow rate of 35 μL/min. The BD Accuri CFlow® software is used to process all data. An electronic gate on green/red fluorescence density plots was used to distinguish stained microbial cells from instrument noise or water sample background.

The following chemicals are used for FCM calibration and cells staining.

1- *Partec calibration beads, 3 μm (for daily calibration).*
Calibration beads for laser position with blue 488 nm emission, which is used for daily control before starting with the actual measurements to check the performance of the FCM and to give an indication if the maintenance is needed. According to the product supplier, 5 times diluted Partec calibration bead should give a count of 55,000 Events/mL on a specific provided gate with 10% standard deviation (the recommended range is between 50,000 - 60,000 Events/mL):

a) Shake Partec calibration beads bottle gently before pipetting 100 μL into 400 μL of filtered Evian water in an eppendorf 1.5 mL plastic vial (500 μL total sample volume of 5 times diluted beads sample).
b) Vortex gently the vial to mix its content and then fix it under the flow cytometry SIP.
c) Run the 500 μL of the bead control solution on medium speed and threshold of 500 on Fl1 using the gate template provided by the producer.

2- *Spherotech 8-peak validation beads (for calibration after maintenance works):*
Concentration: 1×10^7 particles/mL
Storage buffer: deionized water with 0.02% sodium Azide and 0.01% NP40
Storage temperature: 2-8 °C, expirers 1 year after opening
Use: routine validation of the flow cytometer performance.

Description: 8-peak validation beads product contains Rainbow particles (3.0-3.4 μm) with 8 different fluorescent intensities. Each Rainbow particle contains a mixture of fluorophores that enable their excitation with the blue laser (488 nm) to validate FL1, FL2 and FL3 channels of the flow cytometer.

Procedures: the beads sample should be prepared and measured as follows:

a) Prepare the beads solution by diluting 4 drops of 8-peaks bead in 1 mL of 0.22 μm filtered Evian water and diluting 4 drops of 6-peaks/peak 1 beads and 4 drops of the 6-peaks/peak 2-6 beads in 1 mL of 0.22 μm filtered Evian water.
b) Pipette 1 mL of the 8-peaks calibration bead solution in an eppendorf 1.5 mL plastic vial and vortex it properly before fixing it under the flow cytometry SIP and starting the measurement.
c) Run the beads solution with the settings of 50,000 Events run limit and slow speed.
d) Wipe the SIP and proceed the same way with the prepared 6-peaks beads solution.
e) Run 0. 22 μm filtered Evian water at the end of the procedure.
f) Check the results by counting the peaks numbers.

Recommended range: see Figure 2.

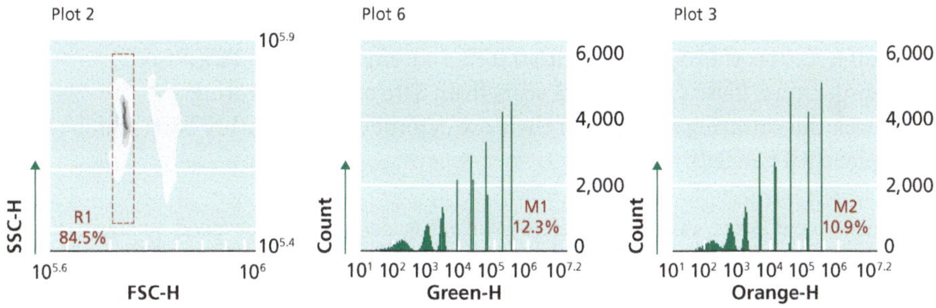

Figure 2 recommended range for 8-peak validation beads (own data)

3- Spherotech 6-peak validation beads (for calibration after maintenance works):
Concentration: 5×10^6 particles/mL
Storage buffer: deionized water with 0.02% sodium Azide and 0.01% NP40
Storage temperature: 2-8 °C, expirers 1 year after opening
Use: routine validation of the flow cytometer performance.

Description: 6-peak validation beads product contains a mixed population of 3.2 µm particles in six different fluorescent intensities. The particles can be excited at wavelengths in the 600-650 nm range.

Procedures: see 8-peak validation beads.
Recommended range: see Figure 3.

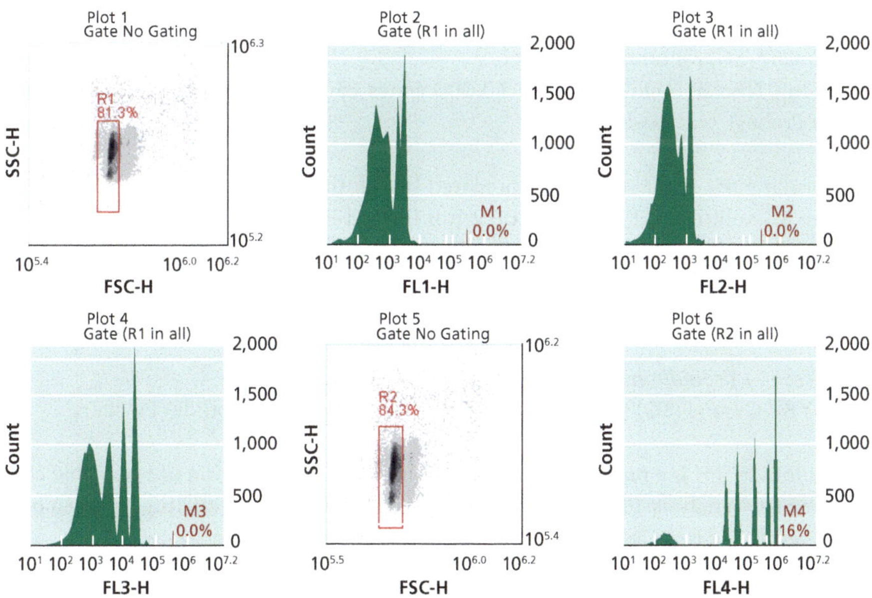

Figure 3 recommended range for 6-peak validation beads (own data)

4- *Spherotech AccuCount fluorescent particles, 7.0-7.9 μm (for weekly calibration).*
Concentration: 102,000 beads/mL
Storage buffer: 0.016 M PBS, pH 7.4 with 0.02% Sodium Azide and 0.2% BSA
Storage temperature: 2-8 °C, expirers 3 years from date of manufacturing
Use: to check the counting accuracy of the flow cytometer on monthly basis (can be used more frequently if needed).

Description: AccuCount validation beads product is designed to be used as stand-alone, quality control reagent to validate accurate volume measurement by the flow cytometer running with CFlow Plus software.

Procedures: the beads sample should be prepared and measured as follows:

a) Prepare 10 times diluted spherotech beads solution in an eppendorf 1.5 mL plastic vial (500 μL total sample volume of 450 μL filtered Evan water and 50 μL spherotech beads stock), the sample should be at room temperature.
b) Vortex the eppendorf plastic tube gently and then invert it a few times before putting it under the machine SIP.
c) Run the analysis of the sample with the following settings: run limits = 50 μL on medium speed.
d) Repeat steps b and c two more times with the same eppendorf plastic tube (the same sample will be analyzed in triplicate).
e) Check the count on Q1-UR gate and perform the calculations.

Recommended range: the difference between the obtained average count (the beads to be counted in triplicate every time) and the expected count of 102,000 events/mL should not exceed 10 %.

5- *SYBR Green I (SG) for total cells counts.*
SYBR Green I Nucleic Acid Gel Stain, 10,000 times concentrated in DMSO Invitrogen (Molecular Probes), was used.

A 1 mL working SG stock solution is prepared by diluting the Invitrogen 10,000 times concentrate SG solution 100 times in 0.22 μm filtered DMSO (IC Millex – LG, 0.2 μm, Millipore), where 10 μL of SG concentrate solution was added to 990 μL of filtered DMSO. The prepared stock solutions can be kept at -20 °C for future measurements. The working SG stock solution can be used for total cell count (TCC) using FCM.

6- *SYBR Green I + Propidium iodide (SG+PI) for intact cells counts*
A mix of SYBR Green I (SG) stain (described above) and Propidium iodide (PI) stain.

Propidium Iodide (PI) is a ready-to-use stock solution for the exclusion of nonviable cells in flow cytometry analysis. PI binds to double stranded DNA by intercalating between base pairs, but is excluded from cells with intact plasma membranes.

A working SG + PI stock solution is prepared by adding 10 μL of Invitrogen 10,000 times concentrate SG solution and 10 μL of PI ready-to-use stock solution in 980 μL of 0.22 μm filtered DMSO (IC Millex – LG, 0.2 μm, Millipore), so that the working SG + PI stock solution contains 1:100 diluted SG and 0.3 mM of PI. The prepared stock solutions can be kept at -20 °C for future measurements. The working SG + PI stock solution can be used for intact cell count (ICC) using FCM.

B. Autoclave

An autoclave is a pressure chamber which is used to sterilize water samples and supplies by subjecting them to a high pressure saturated steam at 121 °C for 15 to 20 minutes depending on the size of the load and the contents.

C. Water bath

Water bath is a system to control water temperature in where bottles containing water samples are placed for wet-pasteurization.

16.2.2 Materials and Methods

A. Glassware (Figure 4):
- Duran® graduated clear glass bottles with screw solid plastic cap for sampling.
- Clear glass vials with screw silicone septum cap for incubating.

B. Eppendorf plastic tubes of 1.5 mL.
C. Plastic syringes.
D. Pipettes and plastic tips of 1 mL, 200 µL and 5 µL.
E. Stopwatch.

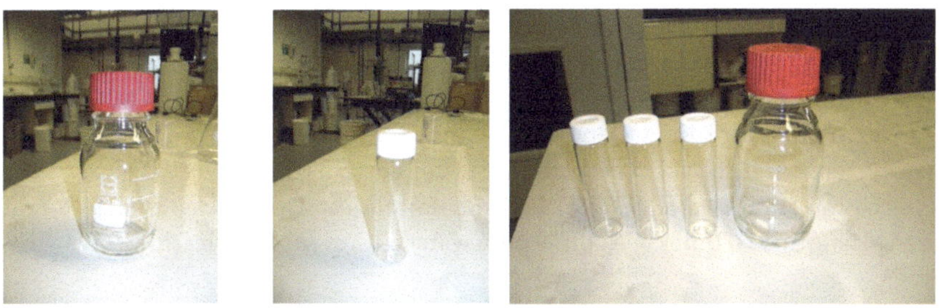

Figure 4 Glassware used for bacterial growth potential measurement (Own photos)

16.2.3 Method

1. **Preparation of glassware:** Glassware should be made AOC-free by applying the following cleaning protocol in a clean laboratory environment to avoid air-borne contamination (Hammes *et al.*, 2006).

A. Rinse the glassware and the caps with a cleaning solution (Alconox, 10 g/L in demi water) with brushing the internal side thoroughly.
B. Rinse it 3 times with Milli-Q water and leave it to air dry.
C. Put all the glassware (excluding the plastic caps) inside the muffle furnace. Glass bottles and vials should be wrapped with aluminium foil to avoid contamination.
D. Set the muffle furnace at 550 °C, the glassware stays inside for 24 hours as follows:

- 1.5 hours for the furnace to warm up and to reach the targeted temperature.
- 6 hours of effective exposure to heat at the targeted temperature.
- The rest of the time is for cooling down the temperature.

E. Clean the plastic caps by soaking them in heated (60 °C) sodium persulfate ($Na_2S_2O_8$, 100g/L) in water bath for 1 hour, thereafter rinse them 3 times with Milli-Q water and leave them to air dry.

F. After taking the cleaned glassware out of the muffle furnace, close them with the cleaned caps and keep them in a clean and closed place away from contamination.

2. **Preparation of stock solutions**: For blank preparation, four different inorganic stock solutions are prepared in AOC-free bottles using ultrapure water (i.e., Milli-Q water) with final concentrations of 67.2 g/L $NaHCO_3$ (for pH adjustment and buffer addition), combined 294 g/L $CaCl_2.2H_2O$ and 67 g/L $MgCl_2 \cdot 6H_2O$ (for calcium and magnesium addition), 0.219 g/L KH_2PO_4 (for phosphate addition), and 3.607 g/L KNO_3 (for nitrogen addition). Additionally, organic stock solutions of 1,000 ± 50 mg-C/L are prepared using Milli-Q water in AOC-free bottles for researching bacterial inocula, as follows: sodium acetate and glucose representing readily-available organic carbon, and laminarin (from Laminaria digitata) and gelatin (type B, from bovine skin) representing complex organic carbon. The prepared stock solutions were kept in the fridge at 4 °C and were used for multiple experiments. Reagent grade chemicals (>99% purity) were used throughout this study (J.T.Baker® Reagents Salts, ACS Grade, the USA).

3. **Preparation of blanks**: An ultra-pure blank was prepared by adjusting the pH and mineral content of RO permeate at the laboratory to 122 mg/L HCO_3^- (final pH of 7.8 ± 0.2), 40 mg/L Ca , 4 mg/L Mg^{2+}, 5 μg/L PO_4-P, and 50 μg-N/L. These concentrations were obtained by the addition of 2.5 μL/mL of $NaHCO_3$, 0.5 μL/mL of $CaCl_2$ and $MgCl_2$ and 0.1 μL/mL of both KH_2PO_4 and KNO_3 stock solutions. The BGP of ultra-pure blank (laboratory-remineralised RO permeate) is in the range of 50 ± 20 × 10^3 ICC/mL (ICC: intact cell counts as measured by FCM). For chemical addition, pipettes are used after rinsing sterilised plastic tips 10 times with ultra-pure water to prevent AOC leaching into the water samples.

Moreover, a broth of trace elements can be used for growth limitation experiments, where two stock solutions should be prepared (pH ~7): stock solution A containing 5 mg/L $CoCl_2.6H_2O$ and 10 mg/L H_3BO_3; and stock solution B containing 500 mg/L $MnSO_4.7H_2O$, 10 mg/L $ZnSO_4.7H_2O$, and 300 mg/L $FeSO_4.7H_2O$. The stock solutions should be kept in the dark at room temperature. Aliquots of 4 and 3.7 mL/L from stock solutions A and B, respectively, were added in water samples, resulting in final concentrations of 5 μg/L Co, 6.5 μg/L B, 359 μg/L Mn, 8.5 μg/L Zn, 215 μg/L Fe, and 345 μg/L S. Moreover, adding phosphate and nitrogen was accompanied with the addition of 29.2 μg/L K.

The ultra-pure blank is made carbon-limited to ensure detecting any potential carbon contaminations during the handling of samples, which comes in different forms such as: (i) carbon attached to the glassware and caps, (ii) volatile carbon present in the laboratory environment, and (iii) carbon contamination present in reagent grade chemicals used in the laboratory, in addition to the original carbon content of the blanks.

4. **Choice of inoculum**: the following steps are applied to test the suitability of a certain bacterial inoculum for BGP measurement, where a natural bacterial inoculum is recommended to ensure the consumption of available organic compounds to a large extent:

A. Prepare sample for BGP measurement as shown in part 6 of this section.
B. After inoculating with the target bacterial consortium, add the different organic stock solutions mentioned in part 2 of this section in separate test bottles, considering the following final concentrations per organic stock solution: 0, 10, 20, 50, 100, and 200 µg-C/L.
C. Perform the above-mentioned test procedure using different bacterial consortium, if available.
D. Select the bacterial inoculum that yields the maximum BGP in all organic carbon type.

5. **Sampling procedures**: water samples are collected as follows:

E. Open sampling tap for at least 10 minutes to flush the piping before sampling.
F. Collect 200 mL of the water sample in an AOC-free Duran® graduated clear glass bottle.
G. Keep the sampling bottle in a cooling box away from potential contaminations (Note: prevent any contact between water sample and cap by handling the bottle gently).
H. Samples are transported to the laboratory within 3 hours.

6. **Bacterial growth potential (BGP) test procedure**: water samples are pre-treated, inoculated, and measured at the laboratory, as follows:

A. Pre-treat water samples by pasteurisation at 70 °C using a water bath to inactivate indigenous bacteria. Water level inside sampling bottle and in water bath should be comparable to ensure effective pasteurisation. Sampling bottle should be air-tight to avoid contamination with water vapour. Effective pasteurization time is 30 min excluding the time needed for water sample to reach the target temperature (the warming up time is determined using a reference sample with similar volume to actual water sample).
B. Put them in an ice bath to cool down to room temperature.
C. Inoculated them with ~10^4 ICC/mL of a natural bacteria consortium originating from water that contains diverse bacterial species, e.g., tap water. For inoculation, pipettes are used after rinsing sterilised plastic tips (at 121 °C for 15 minutes) 10 times with ultra-pure water to prevent AOC leaching into the water samples.
D. Distribute each water sample in three individual AOC-free vials (i.e., triplicate measurements per sample) by direct pouring from sampling bottle into the vials. Use a reference vial for volume measurement (20 ± 2 mL sample per vial).
E. Incubate the vials in the dark at 30 °C under static conditions during a test period of 20 days.
F. Aliquots of ~1 mL are directly poured from the incubated vials into 1.5 mL Eppendorf tubes to perform FCM measurement, either intact cell count (ICC), total cell count (TCC), or both. FCM measurements are performed daily in the first test week, and then every two days. The FCM measurements are performed as follows:
   - Pipette 500 µL of the sample in 1.5 mL eppendorf plastic tube. Samples have to be diluted using 0.1 µm filtered bottle water (e.g., Evian) at >200,000 TCC or ICC/mL (dilution rates are 5, 10 or 20). For instance, to achieve 5 times dilution, 400 µL of filtered bottle water should be added to 100 µL of sample.
   - Preheat the sample in dark to 35 °C for 5 min.

- Stain the sample by adding 5 µL of SG (for TCC) or SG + PI (for ICC). Dilution ratio is 1:100.
- Incubate the stained sample in dark at 35 °C for exactly 10 minutes.
- Run the FCM at the following settings: run limit: 50 µL (identical settings to volumetric calibration of FCM); fluidics: medium speed; and threshold on FL1-H channel of green fluorescence to 700.
- Data acquisition was performed using BD Accuri CFlow® software where a digital gate was set on FL1/FL3 density plot to distinguish the stained bacterial cells from inorganic particles and instrument noise. The FCM detection limit is 10^3 TCC or ICC/mL. Figure 5 shows an example of data acquisition using FL1/FL3 density plot.

G. Measuring each incubating vial for TCC and/or ICC implies that triplicate measurements per sample are performed. Statistical analysis based on Dixon's Q-test is conducted to define the outlier. Thereafter, average TCC and/or ICC is reported per measurement day and standard deviation is presented as an error bar. Student's t-test and one-way analysis of variance (ANOVA) were applied to compare BGP of different sample types. A confidence level of 95 % was considered (alpha of 0.05).

H. BGP results are expressed as the maximum TCC or ICC obtained during the whole test period of 20 days.

Figure 6 represents the complete sample handling and BGP test procedure.

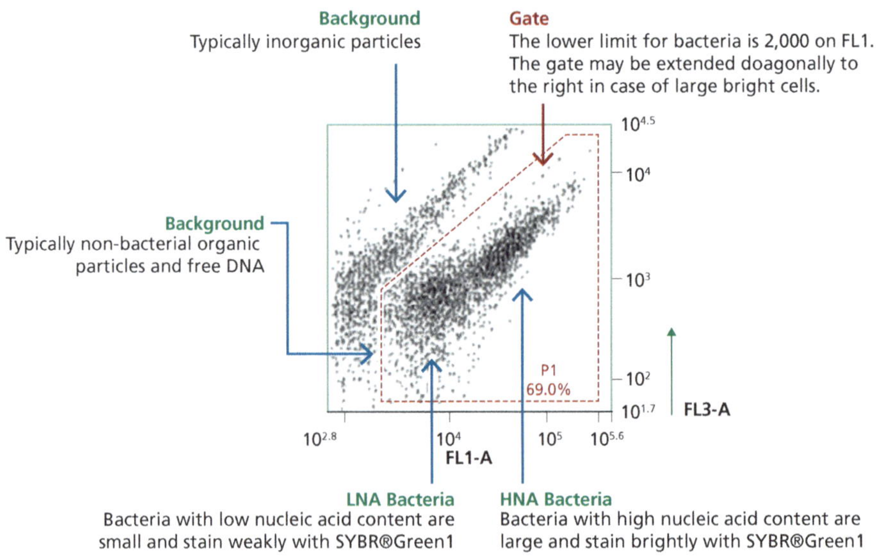

Figure 5 An example of data acquisition using FCM BD Accuri CFlow® software (FL1/FL3 density plot) (own data)

16.3 EXAMPLES OF APPLICATION

This method can be applied in studies about water treatment plants that are based on advanced technologies to produce ultra-low nutrient water, such as, membrane technology (e.g., RO filtration) and distillation. RO filtration is a superior barrier for bacteria and growth-promoting nutrients in water, as a result of which RO-treated has a high degree of biological stability, which was the main application while developing the proposed method in this chapter. The following paragraph shows an example of using this method in a case-study in the Netherlands.

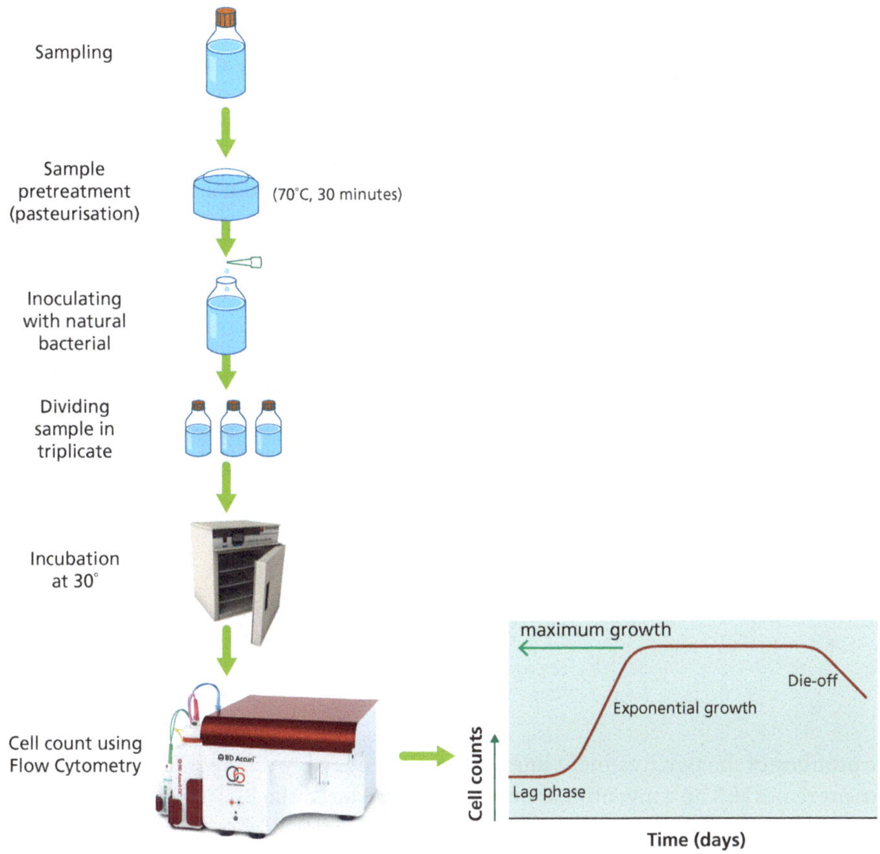

Figure 6 illustration of the steps to follow for measuring BGP of ultra-pure water

The Kamerik drinking water treatment plant (Oasen Drinking Water Company, Gouda, Netherlands) currently produces 340 m^3/h of drinking water from anaerobic groundwater (AGW) by conventional water treatment processes, which are given in Figure 8A in the following order: spray aeration on the surface of rapid sand filters (so-called dry sand filtration, DSF), tower aeration, pellet softening (SOF), carry-over submerged rapid sand filtration (RSF), granular activated carbon filtration (ACF), and UV disinfection (UVD) before storing the conventionally treated water (CTW) in the clean water reservoir.

Installed in parallel for research purposes, a pilot-scale advanced treatment line with a capacity of 7 m^3/h treats the same source water with the following processes (Figure 8B): anaerobic RO filtration (RO) with a total recovery of 75%, followed by post-treatment comprising anaerobic ion exchange (IEX) to remove residual ammonium, remineralisation using anaerobic calcite contactors (CC) to correct the calcium and bicarbonate concentrations to the required level (40 mg/L Ca^{2+}, 122 mg/L HCO_3^-), magnesium dosing ($MgCl_2$, 4 mg/L Mg^{2+}), and tower aeration for the introduction of oxygen and the removal of methane and excess carbon dioxide. The finished drinking water after RO filtration and all post-treatment processes is denoted as site-Remin and has a final pH of 7.8 ± 0.2. Water samples were collected after each treatment step in both the conventional and RO-based treatment lines. To identify bacterial growth-limiting nutrients, BGP of water samples was measured with the addition of different combinations of nutrients as previously described by Prest *et al.* (2016a), and shown in Table 1. The used nutrient stocks were carbon (1.07 g/L $C_2H_3NaO_2$), phosphate (0.219 g/L KH_2PO_4), nitrogen (3.61 g/L KNO_3), and a broth of trace elements (5 mg/L $CoCl_2.6H_2O$, 10 mg/L H_3BO_3, 10 mg/L $CaSO_4.5H_2O$, 500 mg/L $MnSO_4.7H_2O$, 10 mg/L $ZnSO_4.7H_2O$, 300 mg/L $FeSO_4.7H_2O$). Nutrients were added according to the ratio of C:N:P = 100:10:1 (Hammes and Egli, 2005). The blank (lab-Remin) and samples of the finished drinking water produced by the RO-based and conventional treatment lines (site-Remin and CTW, respectively) were tested.

Table 1 BGP test matrix to identify the bacterial growth-limiting nutrient in water samples

Test #	C ($C_2H_3NaO_2$)	P (KH_2PO_4)	N (KNO_3)	TE*	Investigation
1	–	–	–	–	actual BGP
2	–	+	+	+	C-limited BGP
3	+	–	+	+	P-limited BGP
4	+	+	–	+	N-limited BGP
5	+	+	+	–	TE-limited BGP
6	+	+	+	+	positive control

The profiling of the two treatment lines showed considerably different degrees of BGP and nutrient removal. The conventional treatment line reduced the BGP by ~60% (from 1,250 ± 100 ×10³ in AGW to the range of 450 ×10³ – 550 ×10³ ICC/mL across the different treatment steps), where the BGP of conventionally treated water (CTW) was 515 ± 5 ×10³ ICC/mL (Figure 8a). Meanwhile, DOC decreased from 7.2 mg/L in AGW to 6.0 mg/L in CTW (Table 2). Notably, the humic substances, which accounted for >70% of DOC in AGW, showed the highest removal in the conventional treatment line (from 5.2 mg/L to 4.3 mg/L).

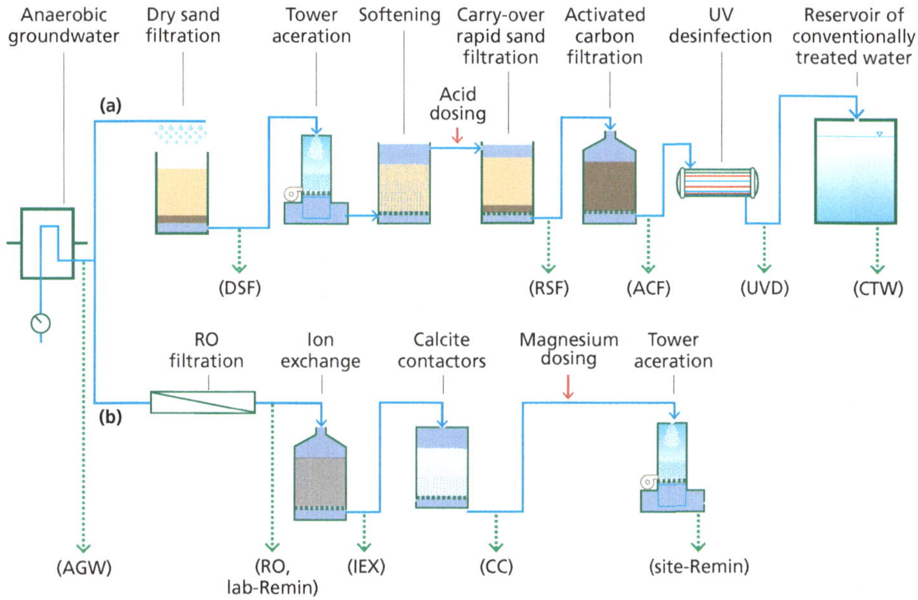

Figure 7 Full-scale conventional (a) and pilot-scale RO-based (b) water treatment lines at the drinking water treatment plant. Sampling locations (dashed arrows) and codes (between brackets) are indicated.

Phosphate was also considerably reduced, mainly during DSF (>98%, from 553 µg/L PO_4-P in AGW to 7 µg/L PO_4-P in DSF), reaching down to 1 µg/L PO_4-P in CTW (Table 2). Similarly, ammonium was also reduced below 0.02 mg/L NH_4-N (limit of detection) by the conventional treatment (Table 2). The results showed that nitrification was the main mechanism for ammonium removal, where ammonium (NH_4^+) in AGW (2.90 ± 0.10 mg/L NH_4-N) was completely converted into nitrate (NO_3^-) in CTW (2.77 ± 0.40 mg/L NO_3-N). Methane, which was present at 2,000–4,000 µg-CH_4/L in AGW, was reduced to 10–20 µg-CH_4/L in CTW.

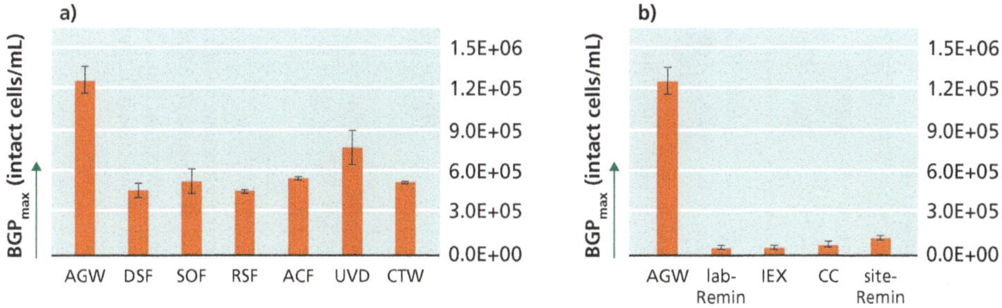

Figure 8 Bacterial growth potential (BGP) at each step of the conventional (A) and RO-based (B) treatment lines. BGPs of RO permeate and ion exchange effluent were measured after remineralisation at the laboratory (i.e., lab-Remin and IEX respectively). Error bars represent the standard deviation of triplicate measurements.

Table 2 The concentration of carbon (LC-OCD fractions), phosphate, nitrogen (ammonium, nitrite, and nitrate), and methane at each step of the conventional and RO-based treatment lines

Treatment		Source water	Conventional line				RO-based line			
Sample type		**AGW**	**DSF**	**RSF**	**ACF**	**CTW**	**RO**	**IEX**	**CC**	**site-Remin**
Carbon (μg-C/L)[a]	DOC	7,242	7,237	6,636	6,105	5,987	36	32	20	27
	Bio-polymers	3	16	8	6	10	4	0	3	13
	Humic Substances	5,202	5,170	4,610	4,486	4,323	0	0	0	0
	Building Blocks	1,110	1,095	1,151	1,027	994	2	1	5	6
	Neutrals	869	809	801	652	636	12	66	7	22
	Acids	0	0	0	0	0	3	2	2	3
Nitrogen (mg-N/L)[b]	Ammonium (NH_4^+)	2.90 ± 0.10	0.82 ± 0.15	<0.02	n.a.	<0.02	0.17 ± 0.02	<0.02	<0.02	<0.02
	Nitrite (NO_2^-)	<0.003	0.022 ± 0.004	<0.003	n.a.	<0.003	n.a.	n.a.	n.a.	<0.003
	Nitrate (NO_3^-)	<0.23	n.a.	n.a.	n.a.	2.77 ± 0.40	n.a.	n.a.	n.a.	0.23 ± 0.05
Methane (μg-CH_4/L)[c]		2,000–4,000	n.a.	n.a.	n.a.	10–20	n.a.	n.a.	n.a.	<5–14
Phosphate (μg/L PO_4-P)[d]		553 ±17	6.8 ± 0.6	1.1 ± 0.1	0.7 ± 0.1	1.1 ± 0.1	0.9 ± 0.1	0.9 ± 0.1	7.0 ± 0.5	7.3 ± 0.1

[a] The reporting limit is 100 μg-C/L for biopolymers and 200 μg-C/L for the other LC-OCD fractions

[b] Limit of detection: 0.02 mg-N/L for ammonium, 0.003 mg-N/L for nitrite, and 0.23 mg-N/L for nitrate

[c] Limit of detection: 5 μg-CH_4/L

[d] Limit of detection is 0.3 μg/L PO_4^-P. The concentrations after SOF and UVD are 3.8 ± 0.1 and 1.1 + 0.1, respectively. n.a., not measured

The RO-based treatment showed a substantial BGP reduction (>96%) from ~1,250 ± 100 $\times 10^3$ ICC/mL in AGW to ~50 ± 12 $\times 10^3$ ICC/mL in lab-Remin (i.e., RO permeate after remineralisation at the laboratory, Figure 8b). However, the BGP increased by 160% after remineralisation using calcite contactors (CC) and tower aeration (site-Remin), reaching 130 ± 10 $\times 10^3$ ICC/mL. The LC-OCD analysis revealed that all organic matter fractions were considerably retained by RO filtration to levels below the reporting limit (Table 2).

Despite the increase in BGP after post-treatment, there was no detectable increase in any DOC fraction by the LC-OCD. For phosphate, a sharp decrease from 553 to 1 µg/L PO_4-P was observed after RO filtration, followed by an increase across the post-treatment to 7 µg/L PO_4-P (Table 2). In contrast to conventional treatment, nitrification was insignificant within the RO-based treatment line, where ammonium in AGW was mostly retained by the RO membrane (0.17 ± 0.02 mg/L NH_4-N in RO permeate), and was further removed by absorption in ion exchange resins (<0.02 mg/L NH_4-N). This resulted in a low concentration of nitrate in RO-treated water (0.23 ± 0.05 mg/L NH_4-N) (Table 2). Methane in RO-treated water was at similar concentrations as in CTW.

The investigation of the growth-limiting nutrient (Figure 9) revealed that the growth in the examined water types was limited by organic carbon. For all samples, the difference between the actual BGP (i.e., without nutrient addition to the sample) and the C-limited BGP (i.e., samples spiked with all nutrients except for carbon) was insignificant ($p > 0.05$). Contrarily, the BGP of samples with limited phosphate, nitrogen, and trace elements (Fe, Mn, Zn, Co, and B) was significantly ($p < 0.05$) higher than the actual BGP of the corresponding sample. Interestingly, the P-limited BGP of site-Remin was 50% higher than that of CTW and lab-Remin.

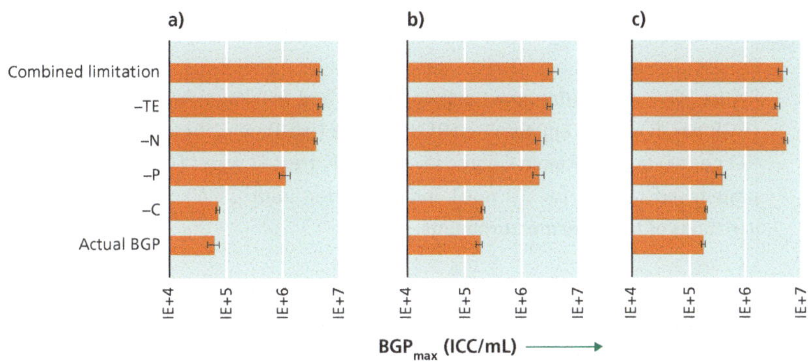

Figure 9 BGP of lab-Remin (the blank, A), site-Remin (B), and CTW (C) with the addition of different nutrients as given in Table 1 (Lower nutrient concentrations were added in the case of lab-Remin and site-Remin). Actual BGP: no nutrients added, –C: no carbon added, –P: no phosphate added, –N: no nitrogen added, –TE: no trace elements added, Combined limitation: all nutrients added. Error bars represent the standard deviation of triplicate tests.

The initial intact cell count and BGP of lab-Remin, site-Remin, and CTW were monitored for a period of 2 years (Figure 10). The results demonstrated superior performance of the RO-based treatment line, where the initial cell count of lab-Remin (<10^3 ICC/mL) and site-Remin ($25 \times 10^3 - 200 \times 10^3$ ICC/mL) were systematically lower than that of CTW ($400 \times 10^3 - 600 \times 10^3$ ICC/mL). Similarly, the BGP was subsequently reduced by >75% with the RO-based treatment line compared with the conventional one, where no pronounced seasonal variations were observed and the BGP was stable around $35 \times 10^3 - 60 \times 10^3$, $90 \times 10^3 - 150 \times 10^3$, and $500 \times 10^3 - 700 \times 10^3$ ICC/mL for lab-Remin, site-Remin, and CTW, respectively.

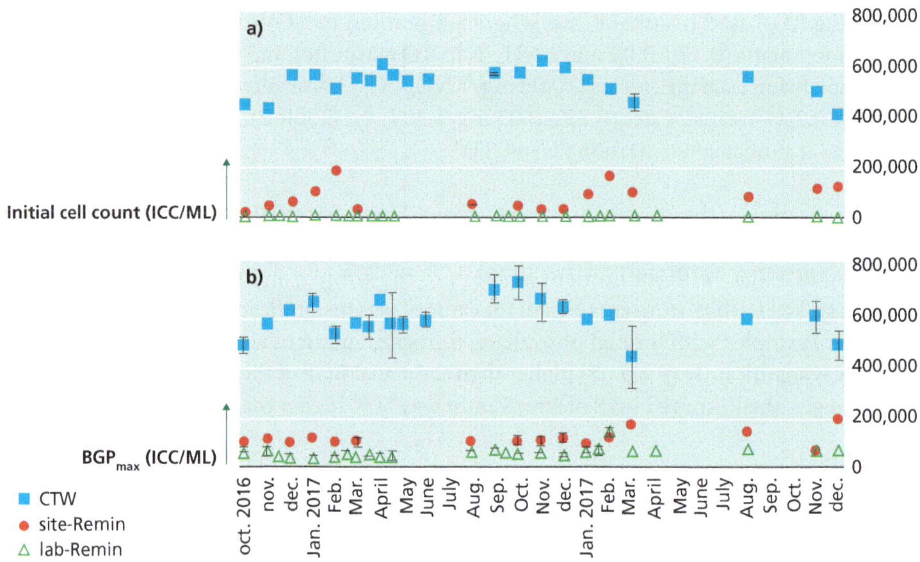

Figure 10 Initial intact cell count (A) and maximum bacterial growth potential (BGPmax, B) of lab-remineralised RO permeate (lab-Remin, the blank), site-remineralised RO permeate (site-Remin), and conventionally treated water (CTW). All samples were pasteurised (70 °C for 30 min) and inoculated with CTW. Error bars represent the variations of triplicate measurements.

16.4 ADDITIONAL CONSIDERATIONS

This chapter is based on previously published data by Mohaned Sousi and co-authors in his Ph.D. dissertation. The findings of his dissertation have also been published in peer-reviewed journals (Sousi *et al.*, 2018; Sousi *et al.*, 2020a; Sousi *et al.*, 2020b; Sousi *et al.*, 2021).

The proposed method in this chapter can be further developed, e.g., the limit of detection of the BGP method might be further lowered by testing different types of water as a blank other than RO permeate, for instance, distilled water produced at the laboratory under controlled conditions. Moreover, further investigation on the impact of various steps in the procedure can be carried out, such as glassware materials and methods of collecting aliquots from incubated glassware. Another factor that could be critical for other types of ultra-low nutrient water is the contamination caused by inoculation, even though this effect was negligible for the inoculum concentration used in this dissertation. Preparing AOC-free inoculum could be, therefore, considered, where regular validation is required by testing the ability of this inoculum to consume readily available as well as complex organic carbon in addition to performing 16S rRNA gene sequencing analysis for diversity control.

16.5 REFERENCES

Abushaban, A., Salinas-Rodriguez, S.G., Mangal, M.N., Mondal, S., Goueli, S.A., Knezev, A., *et al.* (2019). ATP measurement in seawater reverse osmosis systems: Eliminating seawater matrix effects using a filtration-based method. Desalination 453. https://doi.org/10.1016/j.desal.2018.11.020

Berry, D., Xi, C., Raskin, L. (2006). Microbial ecology of drinking water distribution systems. Curr. Opin. Biotechnol. 17:3. https://doi.org/10.1016/j.copbio.2006.05.007

Dixon, M.B., Qiu, T., Blaikie, M., Pelekani, C. (2012). The application of the bacterial regrowth potential method and fow cytometry for biofouling detection at the Penneshaw desalination plant in south Australia. Desalination 284. https://doi.org/10.1016/j.desal.2011.09.006

Escobar, I.C., Hong, S., Randall, A.A. (2000). Removal of assimilable organic carbon and biodegradable dissolved organic carbon by reverse osmosis and nanofiltration membranes. Journal of Membrane Science 175:1. https://doi.org/10.1016/S0376-7388(00)00398-7

Farhat, N., Hammes, F., Prest, E., Vrouwenvelder, J. (2018). A uniform bacterial growth potential assay for different water types. Water Res. 142. https://doi.org/10.1016/j.watres.2018.06.010

Hammes, F., Salhi, E., Köster, O., Kaiser, H.-P., Egli, T., von Gunten, U. (2006). Mechanistic and kinetic evaluation of organic disinfection by-product and assimilable organic carbon (AOC) formation during the ozonation of drinking water. Water Res. 40:12. http://dx.doi.org/10.1016/j.watres.2006.04.029

Hammes, F.A., Egli, T. (2005). New method for assimilable organic carbon determination using flow-cytometric enumeration and a natural microbial consortium as inoculum. Environ. Sci. Technol. 39:9. https://doi.org/10.1021/es048277c

Hasson, D., Bendrihem, O. (2006). Modeling remineralization of desalinated water by limestone dissolution. Desalination 190:1–3. http://dx.doi.org/10.1016/j.desal.2005.09.003

Huck, P.M. (1990). Measurement of biodegradable organic matter and bacterial growth potential in drinking water. J. Am. Water Works Assoc. 82:7. https://doi.org/10.1002/j.1551-8833.1990.tb06995.x

Li, Q., Yu, S., Li, L., Liu, G., Gu, Z., Liu, M., *et al.* (2017). Microbial communities shaped by treatment processes in a drinking water treatment plant and their contribution and threat to drinking water safety. Front. Microbiol. 8:2465. https://doi.org/10.3389/fmicb.2017.02465

Liu, G., Verberk, J.Q.J.C., van Dijk, J.C. (2013). Bacteriology of drinking water distribution systems: An integral and multidimensional review. Appl Microbiol Biotechnol 97:21. https://doi.org/10.1007/s00253-013-5217-y

Liu, G., Zhang, Y., Liu, X., Hammes, F., Liu, W.-T., Medema, G., *et al.* (2020). 360-degree distribution of biofilm quantity and community in an operational unchlorinated drinking water distribution pipe. Environ. Sci. Technol. 54:9. https://doi.org/10.1021/acs.est.9b06603

Liu, G., Zhang, Y., van der Mark, E., Magic-Knezev, A., Pinto, A., van den Bogert, B., *et al.* (2018). Assessing the origin of bacteria in tap water and distribution system in an unchlorinated drinking water system by SourceTracker using microbial community fingerprints. Water Res. 138. https://doi.org/10.1016/j.watres.2018.03.043

Nescerecka, A., Juhna, T., Hammes, F. (2018). Identifying the underlying causes of biological instability in a full-scale drinking water supply system. Water Res. 135. https://doi.org/10.1016/j.watres.2018.02.006

Park, S.K., Hu, J.Y. (2010). Assessment of the extent of bacterial growth in reverse osmosis system for improving drinking water quality. J. Environ. Sci. Health, Part A: Tox. Hazard. Subst. Environ. Eng. 45:8. https://doi.org/10.1080/10934521003772386

Prest, E.I., Hammes, F., Kötzsch, S., van Loosdrecht, M.C.M., Vrouwenvelder, J.S. (2016a). A systematic approach for the assessment of bacterial growth-controlling factors linked to biological stability of drinking water in distribution systems. Water Sci. Technol. Water Supply 16:4. https://doi.org/10.2166/ws.2016.001

Prest, E.I., Hammes, F., van Loosdrecht, M.C.M., Vrouwenvelder, J.S. (2016b). Biological stability of drinking water: Controlling factors, methods, and challenges. Front. Microbiol. 7:45. https://doi.org/10.3389/fmicb.2016.00045

Ruggieri, F., Fernandez-Turiel, J.L., Gimeno, D., Valero, F., García, J.C., Medina, M.E. (2008). Limestone selection criteria for EDR water remineralization. Desalination 227:1–3. http://dx.doi.org/10.1016/j.desal.2007.07.020

Safford, H.R., Bischel, H.N. (2019). Flow cytometry applications in water treatment, distribution, and reuse: A review. Water Res. 151. https://doi.org/10.1016/j.watres.2018.12.016

Servais, P., Billen, G., Hascoët, M.C. (1987). Determination of the biodegradable fraction of dissolved organic matter in waters. Water Res. 21:4. https://doi.org/10.1016/0043-1354(87)90192-8

Smeets, P.W.M.H., Medema, G.J., van Dijk, J.C. (2009). The Dutch secret: How to provide safe drinking water without chlorine in the Netherlands. Drink. Water Eng. Sci. 2:1. https://doi.org/10.5194/dwes-2-1-2009

Sousi, M., Liu, G., Salinas-Rodriguez, S.G., Chen, L., Dusseldorp, J., Wessels, P., *et al.* (2020a). Multi-parametric assessment of biological stability of drinking water produced from groundwater: Reverse osmosis vs. conventional treatment. Water Res. 186:116317. https://doi.org/10.1016/j.watres.2020.116317

Sousi, M., Liu, G., Salinas-Rodriguez, S.G., Knezev, A., Blankert, B., Schippers, J.C., *et al.* (2018). Further developing the bacterial growth potential method for ultra-pure drinking water produced by remineralization of reverse osmosis permeate. Water Res. 145. https://doi.org/10.1016/j.watres.2018.09.002

Sousi, M., Salinas-Rodriguez, S.G., Liu, G., Dusseldorp, J., Kemperman, A.J.B., Schippers, J.C., *et al.* (2021). Comparing the bacterial growth potential of ultra-low nutrient drinking water assessed by growth tests based on flow cytometric intact cell count versus adenosine triphosphate. Water Res. https://doi.org/10.1016/j.watres.2021.117506

Sousi, M., Salinas-Rodriguez, S.G., Liu, G., Schippers, J.C., Kennedy, M.D., van der Meer, W. (2020b). Measuring bacterial growth potential of ultra-low nutrient drinking water produced by reverse osmosis: Effect of sample pre-treatment and bacterial inoculum. Front. Microbiol. 11:791. https://doi.org/10.3389/fmicb.2020.00791

Thayanukul, P., Kurisu, F., Kasuga, I., Furumai, H. (2013). Evaluation of microbial regrowth potential by assimilable organic carbon in various reclaimed water and distribution systems. Water Res. 47:1. https://doi.org/10.1016/j.watres.2012.09.051

van der Kooij, D., Hijnen, W.A.M. (1984). Substrate utilization by an oxalate-consuming Spirillum species in relation to its growth in ozonated water. Appl. Environ. Microbiol. 47:3.

van der Kooij, D., Veenendaal, H.R. (2014). "Regrowth problems and biostability assessment in the Netherlands" in Microbial growth in drinking-water supplies: Problems, causes, control and research needs, ed. D. van der Kooij, P.W.J.J. van der Wielen (London: IWA Publishing.), 291–337.

van der Kooij, D., Visser, A., Hijnen, W.A.M. (1982). Determining the concentration of easily assimilable organic carbon in drinking water. J. Am. Water Works Assoc. 74:10. https://doi.org/10.1002/j.1551-8833.1982.tb05000.x

van Nevel, S., Koetzsch, S., Proctor, C.R., Besmer, M.D., Prest, E.I., Vrouwenvelder, J.S., *et al.* (2017). Flow cytometric bacterial cell counts challenge conventional heterotrophic plate counts for routine microbiological drinking water monitoring. Water Res. 113. https://doi.org/10.1016/j.watres.2017.01.065

Vital, M., Dignum, M., Magic-Knezev, A., Ross, P., Rietveld, L., Hammes, F. (2012). Flow cytometry and adenosine tri-phosphate analysis: Alternative possibilities to evaluate major bacteriological changes in drinking water treatment and distribution systems. Water Res. 46:15. https://doi.org/10.1016/j.watres.2012.06.010

Volk, C.J., LeChevallier, M.W. (1999). Impacts of the reduction of nutrient levels on bacterial water quality in distribution systems. Appl. Environ. Microbiol. 65:11.

doi: 10.2166/9781789062977_0375

Chapter 17

Optical Coherence Tomography (OCT) as a Tool for (Bio)-fouling Assessment in Desalination Systems

Johannes S. Vrouwenvelder, KAUST, Saudi Arabia

Luca Fortunato, KAUST, Saudi Arabia

The learning objectives of this chapter are the following:

- Present and discuss the use of Optical Coherence Tomography in membrane-based desalination systems
- Monitor biofilm formation non-invasively *in-situ* in spacer filled channels
- Define and apply fouling descriptors to quantify biofouling development by using Optical Coherence Tomography
- Understand the impact of biofouling on performance decline in spacer filled channels.

This chapter is based, with permission from copyright holder, on two papers previously published in *Journal of Membrane Science* Volume 524, 15 February 2017, Pages 673 doi: 10.1016/j.memsci.2016.11.052 and in *Bioresource Technology* Volume 229, April 2017, Pages 231-235 doi: 10.1016/j.biortech.2017.01.021

17.1 INTRODUCTION

In the last decades the use of membrane filtration to produce high quality drinking water has increased. One of the major problems of membrane filtration systems is biofouling (Ridgway and Flemming, 1996; Vrouwenvelder *et al.*, 2008a). Biofilm formation is caused by the accumulation of microorganisms, including extracellular polymeric substances

(EPS) produced by microorganisms, on a surface due to either deposition and/or growth. A biofilm causing an unacceptable decline in membrane performance is defined as biofouling. Performance losses are caused by increase in feed channel pressure drop, permeate flux reduction and/or salt passage (Matin *et al.*, 2011).

The complex configuration of the spiral-wound membrane modules makes it difficult to study biofouling *in-situ*. Lab-scale monitors have been developed to allow easier access and better analyses of biofilm development in spiral wound membrane modules (Flemming, 2003; Hemming *et al.*, 1998). Membrane fouling simulator (MFS) was proved to be a suitable tool to study biofouling in spiral wound membrane systems (Vrouwenvelder *et al.*, 2006).

A key aspect of biomass studies involves the analysis of biomass structure (Halan *et al.*, 2012), which can predict the biomass behavior, and thus, the impact on membrane filtration performance. Several approaches are reported in literature to study biomass, most often involving destructive methods (Flemming and Wingender, 2010; Herzberg and Elimelech, 2007). Microscopic techniques are considered an important tool for biomass structure investigation. However, these techniques involve sample preparation, and are less suitable to study the biomass development *in-situ*.

To better understand the biomass development in membrane systems, *in-situ* qualitative and quantitative analyses of the biomass under operational conditions are needed). Several techniques are currently available to study the biomass formation under membrane operational conditions, such as nuclear magnetic resonance spectroscopy (NMR), planar optodes and optical coherence tomography (OCT) (Valladares Linares *et al.*, 2016).

Optical coherence tomography (OCT) can investigate biomass formation and 3D structure *in-situ*, without any biomass staining procedures. The OCT has been used to study biofouling in membrane filtration systems (Derlon *et al.*, 2012; Wibisono *et al.*, 2015). The biofilm time-resolved deformation was calculated in real-time from OCT cross sectional scans (Blauert *et al.*, 2015). Fortunato *et al.* (2017b) monitored in real-time the fouling layer evolution in a submerged membrane bioreactor. West *et al* (2015a) correlated the biomass accumulation to the feed channel pressure drop increase in time using OCT. Yang *et al.* (2000) demonstrated the importance of 3D structural analyses for biofilms grown on a membrane surface. The 3D image analysis offers several advantages with respect to the 2D analysis, such as quantification of biomass growth defined by biovolume, porosity, heterogeneity, thickness, and spatial distribution. In the last years, the use of OCT has been extended to several membrane processes and configurations (Fortunato *et al.*, 2019; Im *et al.*, 2021; Jang *et al.*, 2022; Pathak *et al.*, 2018; Ranieri *et al.*, 2022; Ricceri *et al.*, 2022; Scarascia *et al.*, 2021).

The objective of this study was to assess the biomass formation in a spacer filled flow channel under representative conditions for spiral wound membrane filtration systems. A novel approach is proposed to process 3D datasets acquired with OCT and to visualize and quantify the biomass distribution over the feed spacer and membrane surfaces. The proposed approach allows to evaluate the impact of accumulated biomass on membrane filtration performance measured by feed channel pressure drop and permeate flux.

17.2 MATERIALS, EXPERIMENTAL SET-UP

For all the experiments the biomass was grown on sheet of membrane and spacer in membrane fouling simulator (MFS) (Vrouwenvelder *et al.*, 2007). To enable *in-situ* non-destructive observation of the biomass formation by OCT, the MFS cover contained five millimeter thick glass window. For each experiment a 20 cm × 10 cm ultrafiltration (PAN UF, with a molecular cut-off of 150 kDa) membrane coupon and 31 mil (787 μm, Trisep, USA) thick feed spacer was inserted into the MFS. The ultrafiltration (UF) membrane was necessary to allow water permeation at one bar through the membrane due to the low hydraulic pressure thereby mimicking the flux through the system and resulting hydraulic resistance. Moreover, the use of this membrane enabled the investigation of the biofouling in spacer filled channel without any influence of concentration polarization or other types of fouling.

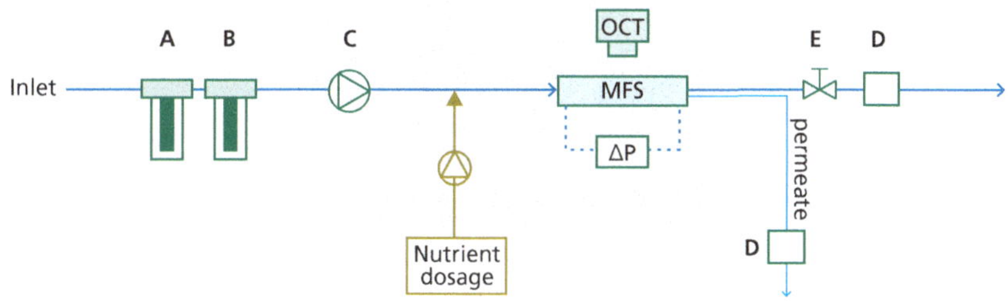

Figure 1 Schematic representation of the experimental setup consisting of carbon (A) and cartridge filters (B), a tank containing nutrient solution, pump (C), dosing pump, flow meter (D), pressure reducing valve (E), differential pressure transmitter, membrane fouling simulator (MFS), and optical coherence tomography (OCT) device.

The MFS was operated under constant hydraulic pressure of one bar at ambient temperature (20 °C). The MFS was fed with tap water by a gear pump (Cole Palmer, USA) at a flow rate of 45.5 $L \cdot h^{-1}$, resulting in a 0.16 $m \cdot s^{-1}$ linear flow velocity at the inlet side of the flow channel, representative for practice (Vrouwenvelder *et al.*, 2009). The tap water was filtered through carbon and cartridge filters (5 μm pore size) to remove residual chlorine and to avoid larger particles entering the MFS (Figure 1). Water permeation though the UF membrane was accomplished with one bar pressure. The hydraulic pressure was regulated by a back-pressure valve (Hydra cell, Wanner Engineering Inc., USA) located on the outflow of the MFS. During the five days experimental period the biomass development was monitored by OCT imaging and its impact on performance was evaluated by the feed channel pressure drop (Deltabar, Endress + Hauser PMD75, Germany) (Bucs *et al.*, 2015), and permeate flux (Sensirion, Switzerland) measurements.

17.3 METHODS

17.3.1 Imaging with Optical Coherence Tomography

An OCT (Thorlabs GANYMEDE GmbH, Dachau, Germany) with a central wavelength of 930 nm equipped with a 5× telecentric scan lens (Thorlabs LSM 03BB) was used to investigate the biomass growth in the MFS flow channel containing membrane and feed spacer sheets. The MFS was mounted on a stage under the OCT probe in order to monitor the biomass development over time in a fixed area (one feed spacer square element) positioned at 5 cm from the feed inlet over time (Figure 2). The monitored area corresponds to 5.3 mm × 5.3 mm with 2.7 μm axial resolution. The OCT lens depth of field was adjusted to 950 μm (slightly higher than the total flow channel height of 787 μm) to allow capturing a part of the membrane and cover glass window. The resulting image stack resolution was (545 × 545 × 482) pixels, with a lateral resolution of 11 μm.

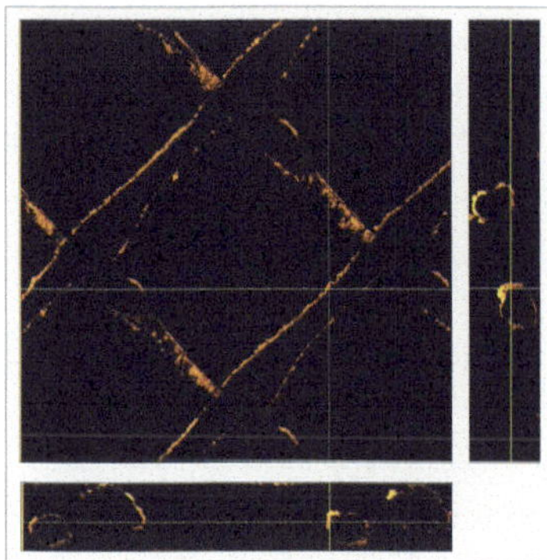

Figure 2 Orthogonal view of OCT images of accumulated biomass (orange color) on the feed spacer, membrane, and cover glass window (5.3 mm × 5.3 mm × 0.95 mm) in the MFS after one day of operation. The yellow lines show the location of the orthoslices.

17.4 DATA ANALYSIS

17.4.1 Biovolume calculation

The OCT images were processed using ImageJ software (Version 1.48). A multi-step processing sequence was applied, consisting of (1) subtraction the initial image t_0 from the image taken at any given time (t_X), (2) adjustment of contrast and brightness of the resulting image (3) application of a median filter and (4) binarization of the image with Otsu algorithms (Otsu, 1979). This approach allows the elimination of the cover glass, membrane, and feed spacer from the OCT image stack, and allowing the quantification of the accumulated biomass (Figure 3).

The initial scan was subtracted from the successive scans (step 1) in order to eliminate the over or under estimation of the accumulated biomass in the scanned area the feed spacer geometry and other structures present in the flow channel need to be eliminated from the scans. The resulting stack was then processed with a customized MATLAB code to obtain the thickness map.

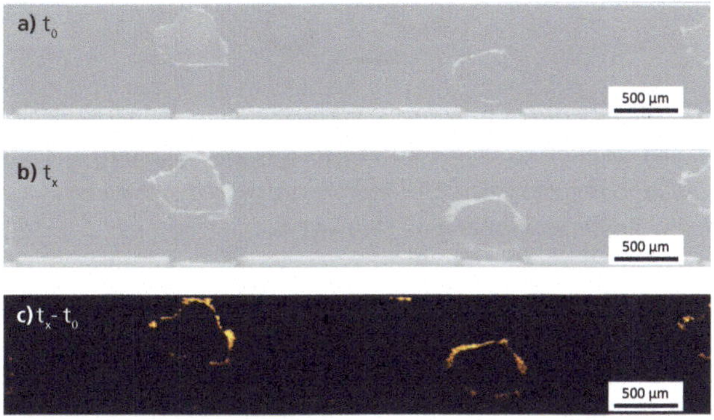

Figure 3 OCT scans at different times at the same position: (a) image before biomass formation at t_0, (b) image with accumulated biomass after certain time tX and spatially-resolved biomass quantification (c) after subtracting the image at time 0 from the image taken after a certain time period ($t_X - t_0$). The final image shows only the biomass (orange color) without the background signals (glass, membrane, and feed spacer).

The binarized datasets were then further analyzed to assess the accumulated biomass volume (V_{Tot}) using the ImageJ plug-in voxel counter. Two different biomass descriptors were used to quantify the biomass development in the flow channel. The total biovolume (mm^3/cm^2) for the scanned (monitored) area was calculated with the following equation:

$$V_{Scanned} = \frac{V_{Tot}}{A_{Scanned}} \quad \text{Eq. 1}$$

where V_{Tot} is the total biomass volume and $A_{Scanned}$ is the scanned area (in this case 0.53 cm × 0.53 cm). The specific biovolume ($V_{Specific}$) was calculated using the following equation:

$$V_{Specific} = \frac{\sum V^i_{biomass}}{\sum A^i} = \frac{V_{Tot}}{\sum A^i} \quad \text{Eq. 2}$$

where $V_{biomass}$ is the biomass volume, A^i the covered area of the investigated element (i) of the flow channel (membrane, feed spacer, cover glass). The total biomass V_{Tot} is the sum of biomass accumulated on the membrane, spacer, and cover glass surface. The specific biovolume ($V^i_{Specific}$) for each element was calculated using the following equation:

$$V^i_{Specific} = \frac{V^i_{biomass}}{A^i} \quad \text{Eq. 3}$$

where $V^i_{Specific}$ is the specific biovolume of each individual flow cell element (i.e., membrane, feed spacer, cover glass). The developed approach allows to separately evaluate the accumulated biomass on the membrane, feed spacer and cover glass surface respectively.

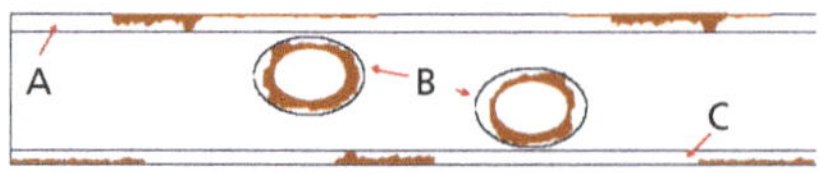

Figure 4 2-D view of the spatial distribution of the biomass on the three elements (membrane, feed spacer and cover glass). Three masked areas A, B and C (the boundaries of the masked area are represented by dashed lines) are distinguished in correspondence of the three elements. Biofilm is represented by brown color.

Three different masks (A, B, C) were created for the three elements one for the spacer (B) and two for the glass (A) and membranes (C) (Figure 4). The size of masks was determined according to the maximum thickness of the biomass observed on the surface of the elements. For the cases where the biomass is attached simultaneously to two elements (Figure 5), the biomass volume is calculated by equally distributing the biomass over the two elements. First the voxels are counted in the areas where the masks belonging to two different elements intersect (A∩B and B∩C) and the total number of voxels are divided by two and subtracted from the total number of voxels counted in each mask (Eqs. 4 -11).

Figure 5 2-D representation of areas where the biomass is simultaneously attached to two elements (membrane, feed spacer and cover glass). The hatched regions represents the areas where the biomass is attached to two elements. The dashed lines represent the boundaries of the three different masked areas and the brown color symbolizes the biomass.

Voxel counting in the three different masks: glass (A), spacer (B) and membrane (C). For the cases where the biomass is attached simultaneously to two elements (Figure 5), the biomass volume is calculated by equally distributing over the two elements. First the voxels are counted in the areas where the masks belonging to two different elements intersects (A∩B and B∩C) and the total number of voxel are divided by two and subtracted to the total number of voxels counted in each mask (Eq. 4 – 11).

$$A = \sum Voxel_A \quad \text{Eq. 4}$$

$$B = \sum Voxel_B \quad \text{Eq. 5}$$

$$C = \sum Voxel_C \quad \text{Eq. 6}$$

$$A \cap B = \sum Voxel_{A \cap B} \quad \text{Eq. 7}$$

$$B \cap C = \sum Voxel_{B \cap C} \quad \text{Eq. 8}$$

$$Voxel_{SPACER} = B - \frac{A \cap B}{2} - \frac{B \cap C}{2} \quad \text{Eq. 9}$$

$$Voxel_{GLASS} = A - \frac{A \cap B}{2} \quad \text{Eq. 10}$$

$$Voxel_{MEMBRANE} = C - \frac{B \cap C}{2} \quad \text{Eq. 11}$$

17.4.2 Image Processing

The OCT was used to monitor the biomass formation at a fixed position in the spacer filled channel two times per day throughout the five days experimental period. To quantify the biomass development the accumulated biomass volume was calculated from the OCT scans. The feed spacer was not transparent for the OCT. When the feed spacer was present a shift of the location of the membrane and possible biomass below the feed spacer filaments were observed (Figure 3a,b). The applied image processing method allows visualization of the biomass only and thus excludes the membrane, feed spacer and cover glass structure from the collected scans (Figure 6). The rendered volume development over time shown in Figure 6 represents only the biomass.

Figure 6 Three-dimensional (3D) rendered OCT image with biomass (brown color), the spacer, membrane and cover glass were eliminated by using the scan at time zero as baseline.

17.5 DATA DISCUSSION AND INTERPRETATION

17.5.1 Biomass Quantification

The OCT scans confirmed the presence of biomass after one day of operation with nutrient dosage (Figure 8a). As reported in the material and methods, the biomass grown in a specific area can be quantified with different descriptors as biomass volume (V_{tot}), scanned biovolume (V) and specific biovolume ($V_{Specific}$). The scanned biovolume normalizes the biomass volume for the scanned area, allowing comparison of data obtained with the same feed spacer (and flow channel height). However, the specific biovolume is the only descriptor that allows comparing the biomass volume with different feed spacers, normalizing the biomass volume for the available surfaces (membrane, feed spacer and glass window) in the flow cell. In Table 1 are reported the biomass values over the time according to different descriptors.

The OCT scans taken periodically during the experimental period confirm the exponential biomass growth ($r^2 = 0.97$). In the first two days of nutrient dosage only a small amount of biomass was detected. A specific biovolume of 0.22 $mm^3 \cdot cm^{-2}$ was detected in the position monitored after one day, corresponding to 0.9 percent of the available volume. From the third day a steep increase in biomass volume was observed (Figure 8a). Towards the end of the study the rate of increase in biomass volume started to decrease. At the end of the experimental period, the final biomass volume occupied 24.9% of the monitored area reaching a specific biovolume of 6.29 $mm^3 \cdot cm^{-2}$.

Table 1 Biomass development in the flow cell in time with the four descriptors

Time (hours)	Biomass Volume (mm^3)	Scanned Biovolume ($mm^3 \cdot cm^{-2}$)	Specific Biovolume ($mm^3 \cdot cm^{-2}$)	Feed channel void volume %*
27	0.17	0.61	0.22	0.9
39	0.22	0.78	0.28	1.1
45	0.45	1.60	0.58	2.3
54	0.93	3.31	1.19	4.7
63	1.69	6.02	2.17	8.6
72	2.7	9.61	3.47	13.7
81	3.19	11.36	4.09	16.2
93	3.96	14.10	5.08	20.1
102	4.50	16.02	5.78	22.9
114	4.90	17.44	6.29	24.9

*Percentage of the occupied volume occupied by the biomass from the total available volume.
The fixed area = 5.3 mm x 5.3 mm; flow channel height = 0.787 μm; feed channel volume = 22.1 mm^3; feed spacer porosity = 0.89; available feed channel volume = 19.7 mm^3.

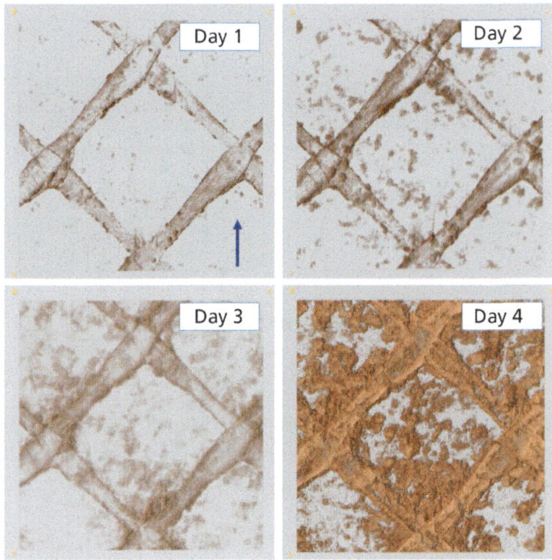

Figure 7 Biomass development over time. Flow direction from bottom to top (arrow).

17.5.2 Membrane Performance

Pressure drop over the feed channel and permeate flux through the membrane were monitored throughout the experimental period. Additionally, biomass volume was calculated from the OCT scans.

Biomass accumulation was confirmed by the feed channel pressure drop. A rapid increase in feed channel pressure drop was observed after two days of operation with nutrient dosage (Figure 8b). By the end of the experimental period the normalized feed channel pressure drop reached a value of 980 mbar/m due to biomass accumulation.

Feed channel pressure drop was measured over the whole flow channel length (20 cm) while the OCT data is from a fixed position. In the present study, the OCT scans covered a much smaller area than pressure drop measurements, 5.3 mm × 5.3 mm in our case with a 2.7 µm resolution positioned at 5 cm from the feed inlet. This gives the possibility to detect biomass deposition and growth at an early stage with micrometers resolution.

Because of the use of a UF membrane, the initial permeate flux of the clean membrane was 105 $L{\cdot}m^{-2}{\cdot}h^{-1}$. With nutrient dosage a small flux decline was observed at the first day of the experimental period, followed by a rapid decrease (1.27 $L{\cdot}m^{-2}{\cdot}h^{-1}$) on the second and third days (Figure 8c). On the fourth day the rate of flux decline was slowed down (0.21 $L{\cdot}m^{-2}{\cdot}h^{-1}$) and reached a final permeate flux of 30 $L{\cdot}m^{-2}{\cdot}h^{-1}$ at the end of the experimental period.

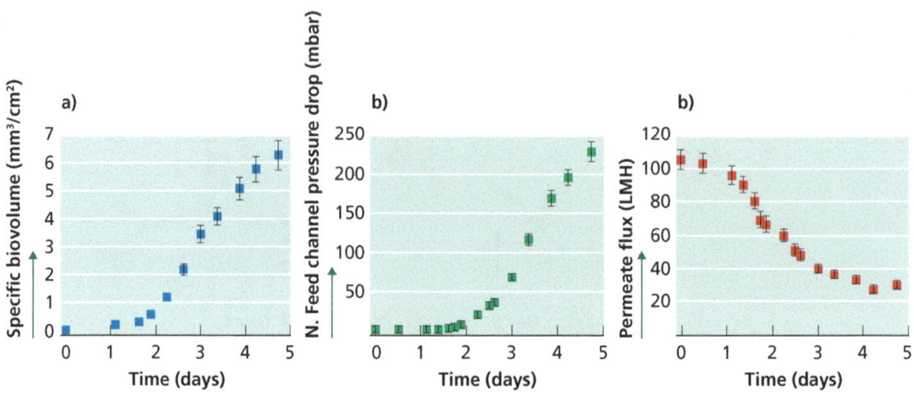

Figure 8 Development of biomass and membrane performances over time. (a) Specific biovolume calculated from the OCT scans. (b) Normalized pressure drop over the MFS feed channel due to biomass development. (c) Permeate flux.

17.6 APPLICATIONS, EXAMPLES

17.6.1 Biomass Distribution

Pressure drop over the feed channel and permeate flux through the membrane was monitored throughout the experimental period. Additionally, biomass volume was calculated from the OCT scans.

Biomass accumulation was confirmed by the feed channel pressure drop. A rapid increase in feed channel pressure drop was observed after two days of operation with nutrient dosage (Figure 8b). By the end of the experimental period the normalized feed channel pressure drop reached a value of 980 mbar/m due to biomass accumulation.

Feed channel pressure drop was measured over the whole flow channel length (20 cm) while the OCT data is from a fixed position. In the present study, the OCT scans covered a much smaller area than pressure drop measurements, 5.3 mm × 5.3 mm in our case with a 2.7 μm resolution positioned at 5 cm from the feed inlet. This gives the possibility to detect biomass deposition and growth at an early stage with micrometers resolution.

Because of the use of a UF membrane, the initial permeate flux of the clean membrane was 105 $L{\cdot}m^{-2}{\cdot}h^{-1}$. With nutrient dosage a small flux decline was observed at the first day of the experimental period, followed by a rapid decrease (1.27 $L{\cdot}m^{-2}{\cdot}h^{-1}$) on the second and third days (Figure 8c). On the fourth day the rate of flux decline was slowed down (0.21 $L{\cdot}m^{-2}{\cdot}h^{-1}$) and reached a final permeate flux of 30 $L{\cdot}m^{-2}{\cdot}h^{-1}$ at the end of the experimental period.

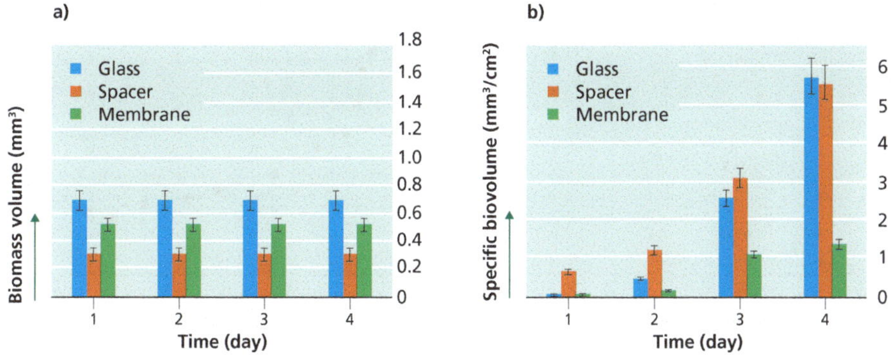

Figure 9 Biomass volume (a) and specific biovolume (b) in time on the feed spacer, and membrane surface in the MFS. Specific biovolume is the biomass volume over the available surface area (area of both membrane and cover glass was each 28 mm^2 and of feed spacer was 21.9 mm^2).

17.6.2 Biomass and performance decline

Based on the OCT images, the accumulated biomass volume was calculated for each measurement time thus enabling to quantify changes in biomass volume. As the biomass volume increased the feed channel pressure drop increased (Figure 10a) and the permeate flux decreased (Figure 10b). The two performance indicators feed channel pressure drop and permeate flux, were seen to respond differently by the increasing biomass volume. During the biomass accumulation in the flow cell two phases were observed in the rate of permeate flux decline (a sharp decrease followed by less sharp decrease), while the feed channel pressure drop increased with increasing biomass.

Increase in the channel pressure drop can be explained by the biomass distribution in the flow channel (Figure 10a). Quantification of the accumulated biomass volume on the membrane and feed spacer surfaces showed more biomass accumulation on the feed spacer than on the membrane surface (Figure 9a).

The impact of the accumulated biomass on the different flow channel elements (membrane, feed spacer and cover glass) on pressure drop increase is shown in Figure 11. The biomass accumulated on the feed spacer and on the cover glass had a higher impact on pressure drop increase than the biomass accumulated on the membrane surface.

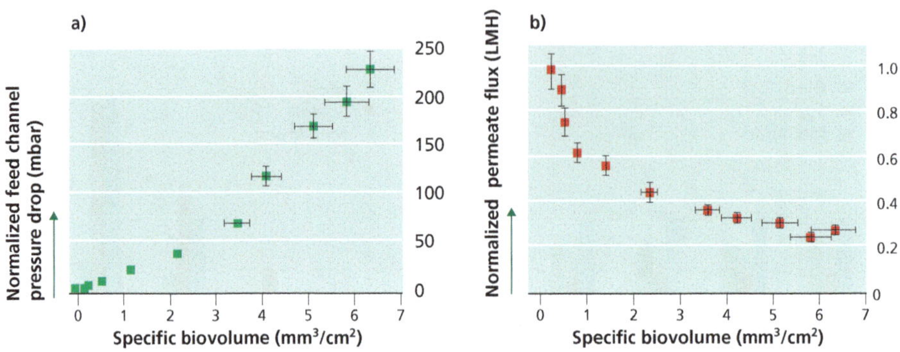

Figure 10 Normalized feed channel pressure drop (a), and permeate flux (b) as function of the accumulated biovolume during the 5 day experimental period

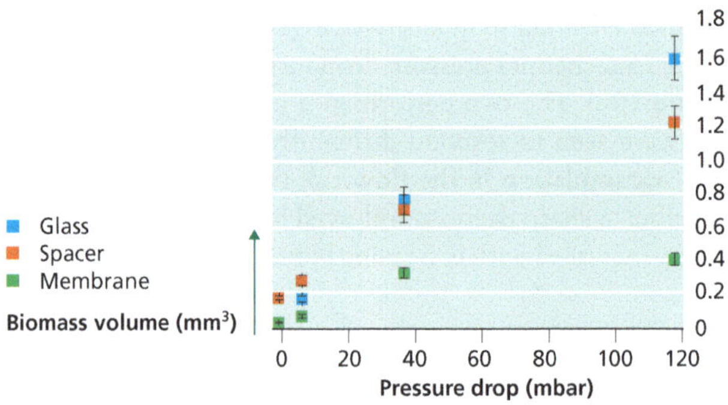

Figure 11 Accumulated biomass volume on the three different elements (membrane, feed spacer and cover glass) in function of feed channel pressure drop increase.

17.6.3 Biomass Thickness Map

Biofilm thickness maps are presented in this study as a new tool to assess the biofilm spatial distribution on a surface. It is a similar approach as the classical distribution map that depicts the distribution of a phenomenon on a surface, where different colors are used to show and evaluate the distribution in a physical map.

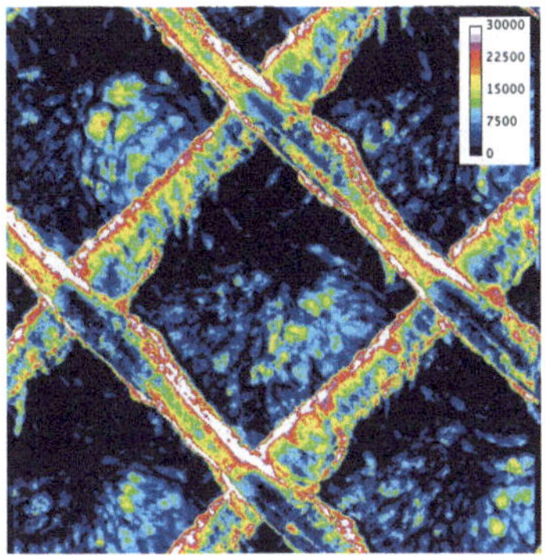

Figure 12 Biofilm thickness map generated from a 3D OCT dataset. The calibration bar allows estimating the biofilm thickness deposited in the spacer filled channel.

In Figure 12 the biofilm thickness map generated from a 3D OCT dataset is shown for a spacer filled channel representative of a spiral-wound membrane. Applying 3D OCT image analysis to a specific area enables the analysis of information related to the spatial distribution and to the homogeneity of the biofilm. In Figure 12, areas shown in red represent higher amounts of biofilm, and are more easily and rapidly detected. In this study, the approach was used to assess the biomass thickness and the spatial distribution over a single frame (Figure 12), and to evaluate the biofouling development over the time (Figure 13). The same approach can be extended to the evaluation of the biofilm distribution over any surfaces.

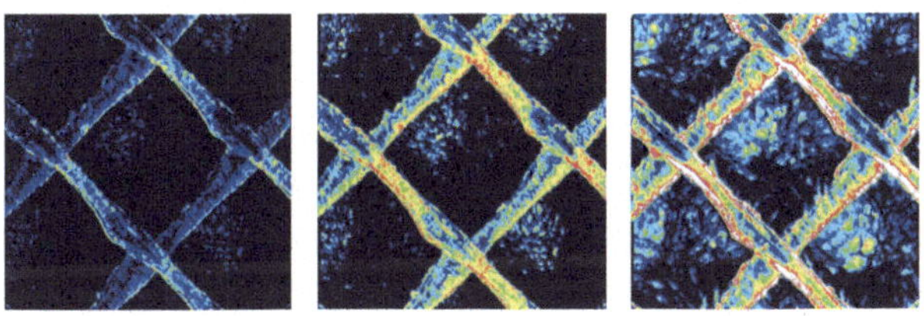

Figure 13 Biofilm development in spacer filled channel (20 hours, 30 hours, 40 hours). Biofilm thickness map. Flow direction is from left to right.

17.7 ADDITIONAL CONSIDERATIONS

In this study a novel approach for 3D reconstruction, assessment and visualization of OCT images was presented. The method presented enables monitoring and quantification of biomass growth during operation. The approach was used to evaluate the effect of the (i) biomass on membrane performance and evaluate the (ii) biomass spatial distribution in the flow channel.

17.7.1 OCT Image Analysis

The novel image processing method presented in this study (i) eliminates the background signal (feed spacer, membrane, and cover glass) from the images and (ii) enables reduction of the noise of the OCT scans. By applying the scan at time zero as a baseline, all changes in the subsequent images can be normalized to time zero. Besides subtracting the signals due to the three elements (spacer, cover glass and membrane) it also removes the signal due to the water present in the flow cell. Therefore, the proposed approach reduces the background noise and facilitates the binarization.

West *et al.* (2015b) used an image masking process on OCT scans to avoid the structures not corresponding to the biomass. The data presented in this study are similar with the results shown by West *et al.* (2015), a different OCT scan processing method has been applied. The method presented in this study enabled the detailed visualization of the biomass deposition in the monitored area.

Other imaging techniques used to study biofilms such as confocal laser scanning microscopy (CLSM) and scanning electron microscopy (SEM) generate images with a higher resolution, however, OCT enables studying larger areas necessary to gain knowledge on biofouling behavior and how it may influence the performance of membrane filtration systems. As reported by Wagner *et al.* (2010), the structural information at micro-scale and nano-scale level might be of minor relevance to characterize the behavior of macro-scale biofilm processes as they occur in membrane filtration systems.

Meso-scale investigation of the biomass by OCT gives insight to the biofouling distribution in a spiral wound membrane module. Biomass formation under different conditions, like various spacer geometry, hydrodynamic conditions or cleaning strategies can be evaluated at a meso-scale range, due to the repetitive geometry of the feed spacer (Bucs *et al.*, 2015; Radu *et al.*, 2014). The possibility to evaluate biomass development under operational conditions, *in-situ*, at a meso-scale range (mm^3) is one of the advantages of OCT compared with other imaging techniques.

Obtaining 3D biomass structures formed under representative conditions for spiral wound membrane systems may be used as additional tool to better understand the impact of different operational conditions on the biomass formation and to evaluate the effect of control strategies on the biomass structure. In-situ real-time detailed image analysis on the acquired biomass morphology could be used to evaluate how the biomass structure responds to the operational conditions (i.e., feed pressure).

The proposed approach for analyzing OCT scans can be used to evaluate biomass development: (i) under various operating conditions, (ii) on different membranes and spacers (e.g., coatings/modifications) and (iii) in the presence of biocides.

17.7.2 Biomass accumulation and membrane performance

The delay in increasing feed channel pressure drop with respect to the biomass increase as detected by OCT scans can be explained by the higher sensitivity of the OCT and the position of the scanned area. When biofilm starts to form and grow in the feed channel the pressure drop starts to increase. However, in an early stage of biomass accumulation the biomass may not have an immediate impact on pressure drop. Bucs *et al.* (2014) demonstrated that a 5 µm thin biofilm and small biofilm patches in the flow channel may not be detected by pressure drop measurements. Conversely, OCT imaging allows to capture and visualize these thin biofilms.

The higher impact on pressure drop increase of biomass accumulation on the feed spacer has been observed in other studies as well (Bucs *et al.*, 2014). As shown in Figure 9 the biomass accumulated on the membrane has lower impact on the feed channel pressure drop in respect to the biomass accumulated on the other elements.

The effect of biomass on permeate flux in spiral wound elements (reverse osmosis, nanofiltration) depends on the membrane surface coverage, flow channeling, biofilm hydraulic resistance, biofilm porosity and thickness (Dreszer *et al.*, 2013a). At the initial phase of biofilm formation, studies showed that a thin biofilm layer is deposited on the surface (Flemming *et al.*, 1997). At this phase the biofilm is a thin and porous structure with low hydraulic resistance, meaning that the membrane surface coverage will be the main factor which impacts water flux (West *et al.*, 2015b). As the biofilm grows (i.e. more biomass volume), thickness, porosity and hydraulic resistance change. Studies have shown that young biofilms are less porous and tend to have a low hydraulic resistance compared to a mature biofilm (Dreszer *et al.*, 2013b; Martin *et al.*, 2014). The rapid flux decline observed in the early stage of biomass accumulation may be attributed to the pore blocking fouling mechanism in UF membranes (Wang and Tarabara, 2008). Once the biomass layer is formed on the membrane surface the flux depends mainly on its properties and the flux decline rate decreases (Figure 8c, days 3 and 4). As shown in figure 9a, on the third and fourth day the biomass volume only slightly increases on the membrane surface while sharply increases on the other two elements (feed spacer and glass).

The biomass accumulation in the flow channel had different impact on the membrane performance parameters. While the pressure drop increases as the biomass increases, the permeate flux decrease is significantly affected in the initial phase of biomass accumulation.

17.7.3 Biomass location in the flow channel

The biomass accumulation occurred mainly on the feed spacer in the early stages may be an indication of either a higher affinity of bacteria to attach to the feed spacer material (polypropylene) or preferential deposition due to the hydrodynamics of the system. Other studies have also reported that at initial stages of biomass formation, more biomass accumulates on the feed spacer than on the membrane surface (Baker *et al.*, 1995; Van Paassen *et al.*, 1998; Vrouwenvelder *et al.*, 2008b). As reported in Vrouwenvelder *et al.* (2008) feed spacers play an important role in biofouling development and in membrane cleanability (Creber *et al.*, 2010).

A lower biomass volume was measured on the membrane surface compared to the cover glass surfaces (Figure 9). The difference in the biomass volume can be attributed to the water flux through the membrane. For this study a UF membrane was used, resulting in a high-water flux (105 LMH). It was shown previously that the biomass compacts, decreasing in thickness and thus in biomass volume under high flux conditions (Dreszer *et al.*, 2014; Valladares Linares *et al.*, 2016b). This may have affected the measured biomass volume and underestimated the amount on the membrane surface. However, in a spiral wound membrane system the flow channel is delimited by membranes on both sides therefore the biomass accumulated on the membrane surface may have a lower impact on pressure drop.

17.7.4 Use of OCT in biofouling studies

The main advantage of OCT is that it allows observation and monitoring of biomass development during MFS operation without sample preparation such as the use of stains or contrast agents. The effect of various cleaning strategies (e.g., chemical cleaning, air flushing, back washing etc.) on biomass developed can also be evaluated. Moreover, the reconstructed 3D biomass structures can be further imported into modeling software for mathematical modeling to increase the understanding of biofouling processes. The 3D biomass analysis and mapping presented in this study shows that OCT is a promising tool to study biofouling in membrane systems.

17.7.5 Mapping The Biofouling

In biofouling monitoring, there is a need to quickly assess and evaluate the spatial development of the biofilm on the membrane over time. Compared to the method used for direct observation through the membrane (DOTM) (Chen *et al.*, 2004), 3D visualization and thickness maps allow a better understanding of the biofilm deposition, providing a depth-resolved biofilm structure. As matter of fact, cross-sectional analysis is necessary to enable the distinction of biofilm accumulation in different elements (i.e., membrane, feed spacer and glass for the spacer filled channel) and for quantifying the thickness. With respect to the 3D volume rendering image analysis (Fortunato *et al.*, 2017a), thickness mapping images require less imaging skills, less computational resources and automated data handling is therefore more feasible. Furthermore, interpretation of fouled systems applying 3D rendered volume images is more difficult compared to the biomass thickness maps. In the 3D rendering images, biofilm deposited on the glass may obscure visualization of lower areas and thus impact the image results. Thickness maps can be obtained directly from raw images without any correction (e.g., time zero as baseline) or data segmentation, where the image data is not only due to the biofilm but also the other elements (e.g., feed spacer, glass, and membrane).

Another advantage of mapping images is the calibration of the color scales, making it possible to relate each color with a corresponding thickness value. In this way, quantifying the amount of biofilm deposited over a specific area can be done while evaluating the spatial distribution. An example of calibrated maps for the system studied in this paper is shown in Figure 13.

The approach proposed can easily be applied to any membrane configuration. It is possible to evaluate the biofilm thickness distribution on the membrane surface and distinguish between different fouled areas. The colored scale allows identifying zones with lower biofilm deposition, which can further be related to the water flow (i.e., lower hydraulic resistance).

17.8 SUMMARY

The use of optical coherence tomography (OCT) to investigate biomass in membrane systems has increased with time. OCT enables characterizing the (bio)-fouling in-situ and non-destructively. In this study, a novel approach to process three-dimensional (3D) OCT scans is proposed. The approach allows obtaining spatially-resolved detailed structural biomass information. The 3D biomass reconstruction enables analysis of the biomass only, obtained by subtracting the time zero scan to all images. A 3D time series analysis of biomass development in a spacer filled channel under representative conditions (cross-flow velocity) for a spiral wound membrane element was performed. The flow cell was operated with monitoring of ultrafiltration membrane performance: feed channel pressure drop and permeate flux. The biomass development in the flow cell was detected by OCT before a performance decline was observed. Feed channel pressure drop continuously increased with increasing biomass volume, while flux decline was mainly affected in the initial phase of biomass accumulation. The novel OCT imaging approach enabled the assessment of spatial biomass distribution in the flow cell, discriminating the total biomass volume between the membrane, feed spacer and glass window. Biomass accumulation was stronger on the feed spacer during the early stage of biofouling, impacting the feed channel pressure drop stronger than permeate flux.

17.9 REFERENCES

Baker, J., Stephenson, T., Dard, S., Cote, P., 1995. Characterisation of fouling of nanofiltration membranes used to treat surface waters. Environmental Technology 16, 977–985.

Blauert, F., Horn, H., Wagner, M., 2015. Time-resolved biofilm deformation measurements using optical coherence tomography 112, 1893–1905. https://doi.org/10.1002/bit.25590

Bucs, Szilárd S., Farhat, N., Siddiqui, A., Valladares Linares, R., Radu, A., Kruithof, J.C., Vrouwenvelder, J.S., 2015. Development of a setup to enable stable and accurate flow conditions for membrane biofouling studies. Desalination and Water Treatment 57, 12893–12901. https://doi.org/10.1080/19443994.2015.1057037

Bucs, Szilard S., Linares, R.V., Marston, J.O., Radu, A.I., Vrouwenvelder, J.S., Picioreanu, C., 2015. Experimental and numerical characterization of the water flow in spacer-filled channels of spiral-wound membranes. Water Research 87, 299–310. https://doi.org/10.1016/j.watres.2015.09.036

Bucs, Sz.S., Radu, A.I., Lavric, V., Vrouwenvelder, J.S., Picioreanu, C., 2014. Effect of different commercial feed spacers on biofouling of reverse osmosis membrane systems: A numerical study. Desalination 343, 26–37. https://doi.org/10.1016/j.desal.2013.11.007

Chen, V., Li, H., Fane, a. G., 2004. Non-invasive observation of synthetic membrane processes - A review of methods. Journal of Membrane Science 241, 23–44. https://doi.org/10.1016/j.memsci.2004.04.029

Creber, S.A., Vrouwenvelder, J.S., van Loosdrecht, M.C.M., Johns, M.L., 2010. Chemical cleaning of biofouling in reverse osmosis membranes evaluated using magnetic resonance imaging. Journal of Membrane Science 362, 202–210. https://doi.org/10.1016/j.memsci.2010.06.052

Derlon, N., Peter-Varbanets, M., Scheidegger, A., Pronk, W., Morgenroth, E., 2012. Predation influences the structure of biofilm developed on ultrafiltration membranes. Water Research 46, 3323–3333. https://doi.org/http://dx.doi.org/10.1016/j.watres.2012.03.031

Dreszer, C., Vrouwenvelder, J.S., Paulitsch-Fuchs, A.H., Zwijnenburg, A., Kruithof, J.C., Flemming, H.C., 2013a. Hydraulic resistance of biofilms. Journal of Membrane Science 429, 436–447. https://doi.org/10.1016/j.memsci.2012.11.030

Dreszer, C., Vrouwenvelder, J.S., Paulitsch-Fuchs, A.H., Zwijnenburg, A., Kruithof, J.C., Flemming, H.C., 2013b. Hydraulic resistance of biofilms. Journal of Membrane Science 429, 436–447. https://doi.org/10.1016/j.memsci.2012.11.030

Dreszer, C., Wexler, A.D., Drusová, S., Overdijk, T., Zwijnenburg, A., Flemming, H.-C., Kruithof, J.C., Vrouwenvelder, J.S., 2014. In-situ biofilm characterization in membrane systems using Optical Coherence Tomography: formation, structure, detachment and impact of flux change. Water Res 67, 243–54. https://doi.org/10.1016/j.watres.2014.09.006

Flemming, H.C., 2003. Role and levels of real-time monitoring for successful anti-fouling strategies - An overview, in: Water Science and Technology. pp. 1–8.

Flemming, H.-C., Schaule, G., Griebe, T., Schmitt, J., Tamachkiarowa, A., 1997. Biofouling—the Achilles heel of membrane processes. Desalination 113, 215–225. https://doi.org/10.1016/S0011-9164(97)00132-X

Flemming, H.-C., Wingender, J., 2010. The biofilm matrix. Nat Rev Micro 8, 623–633.

Fortunato, L., Bucs, S., Linares, R.V., Cali, C., Vrouwenvelder, J.S., Leiknes, T., 2017a. Spatially-resolved in-situ quantification of biofouling using optical coherence tomography (OCT) and 3D image analysis in a spacer filled channel. Journal of Membrane Science 524, 673–681. https://doi.org/10.1016/j.memsci.2016.11.052

Fortunato, L., Li, M., Cheng, T., Rehman, Z.U.Z.U., Heidrich, W., Leiknes, T., 2019. Cake layer characterization in Activated Sludge Membrane Bioreactors: Real-time analysis. Journal of Membrane Science 578, 163–171. https://doi.org/10.1016/J.MEMSCI.2019.02.026

Fortunato, L., Qamar, A., Wang, Y., Jeong, S., Leiknes, T., 2017b. In-situ assessment of biofilm formation in submerged membrane system using optical coherence tomography and computational fluid dynamics. Journal of Membrane Science 521, 84–94. https://doi.org/10.1016/j.memsci.2016.09.004

Halan, B., Buehler, K., Schmid, A., 2012. Biofilms as living catalysts in continuous chemical syntheses. Trends in Biotechnology 30, 453–465. https://doi.org/10.1016/j.tibtech.2012.05.003

Hemming, H., Tamachkiarowa, A., Klahre, J., Schmitt, J., 1998. Monitoring of fouling and biofouling in technical systems. Water Science and Technology 38, 291–298. https://doi.org/10.1016/S0273-1223(98)00704-5

Herzberg, M., Elimelech, M., 2007. Biofouling of reverse osmosis membranes: Role of biofilm-enhanced osmotic pressure. Journal of Membrane Science 295, 11–20. https://doi.org/10.1016/j.memsci.2007.02.024

Im, S.J., Fortunato, L., Jang, A., 2021. Real-time fouling monitoring and membrane autopsy analysis in forward osmosis for wastewater reuse. Water Research 197, 117098. https://doi.org/https://doi.org/10.1016/j.watres.2021.117098

Jang, Y., Lee, J.G., Fortunato, L., Lee, J., Lee, Y., An, A.K., Ghaffour, N., Lee, S., Jeong, S., 2022. Colloidal silica fouling mechanism in direct-contact membrane distillation. Desalination 527. https://doi.org/10.1016/j.desal.2022.115554

Martin, K.J., Bolster, D., Derlon, N., Morgenroth, E., Nerenberg, R., 2014. Effect of fouling layer spatial distribution on permeate flux: A theoretical and experimental study. Journal of Membrane Science 471, 130–137. https://doi.org/10.1016/j.memsci.2014.07.045

Matin, A., Khan, Z., Zaidi, S.M.J., Boyce, M.C., 2011. Biofouling in reverse osmosis membranes for seawater desalination: Phenomena and prevention. https://doi.org/10.1016/j.desal.2011.06.063

Otsu, N., 1979. Threshold selection method from gray-level histograms. IEEE Trans Syst Man Cybern.

Pathak, N., Fortunato, L., Li, S., Chekli, L., Phuntsho, S., Ghaffour, N., Leiknes, T.O., Shon, H.K.H.K., 2018. Evaluating the effect of different draw solutes in a baffled osmotic membrane bioreactor-microfiltration using optical coherence tomography with real wastewater. Bioresource Technology 263, 306–316. https://doi.org/10.1016/j.biortech.2018.04.123

Radu, A.I., van Steen, M.S.H., Vrouwenvelder, J.S., van Loosdrecht, M.C.M., Picioreanu, C., 2014. Spacer geometry and particle deposition in spiral wound membrane feed channels. Water Res 64, 160–76. https://doi.org/10.1016/j.watres.2014.06.040

Ranieri, L., Vrouwenvelder, J.S., Fortunato, L., 2022. Periodic fouling control strategies in gravity-driven membrane bioreactors (GD-MBRs): Impact on treatment performance and membrane fouling properties. Science of The Total Environment 838, 156340. https://doi.org/https://doi.org/10.1016/j.scitotenv.2022.156340

Ricceri, F., Blankert, B., Ghaffour, N., Vrouwenvelder, J.S., Tiraferri, A., Fortunato, L., 2022. Unraveling the role of feed temperature and cross-flow velocity on organic fouling in membrane distillation using response surface methodology. Desalination 540, 115971. https://doi.org/https://doi.org/10.1016/j.desal.2022.115971

Ridgway, H.F., Flemming, H.-C., 1996. Membrane biofouling in water treatment membrane processes. Water Treatment Membrane Processes.

Scarascia, G., Fortunato, L., Myshkevych, Y., Cheng, H., Leiknes, T., Hong, P.-Y., 2021. UV and bacteriophages as a chemical-free approach for cleaning membranes from anaerobic bioreactors. Proceedings of the National Academy of Sciences 118, e2016529118. https://doi.org/10.1073/PNAS.2016529118

Valladares Linares, R., Fortunato, L., Farhat, N.M., Bucs, S.S., Staal, M., Fridjonsson, E.O., Johns, M.L., Vrouwenvelder, J.S., Leiknes, T., 2016a. Mini-review: novel non-destructive in situ biofilm characterization techniques in membrane systems. Desalination and Water Treatment 1–8. https://doi.org/10.1080/19443994.2016.1180483

Valladares Linares, R., Wexler, A.D., Bucs, S.S., Dreszer, C., Zwijnenburg, A., Flemming, H.C., Kruithof, J.C., Vrouwenvelder, J.S., 2016b. Compaction and relaxation of biofilms. Desalination and Water Treatment 57, 12902–12914. https://doi.org/10.1080/19443994.2015.1057036

Van Paassen, J.A.M., Kruithof, J.C., Bakker, S.M., Kegel, F.S., 1998. Integrated multi-objective membrane systems for surface water treatment: pre-treatment of nanofiltration by riverbank filtration and conventional ground water treatment. Desalination 118, 239–248. https://doi.org/10.1016/S0011-9164(98)00137-4

Vrouwenvelder, J., Van Paassen, J., Wessels, L., Van Dam, A., Bakker, S., 2006. The Membrane Fouling Simulator: A practical tool for fouling prediction and control. Journal of Membrane Science 281, 316–324. https://doi.org/10.1016/j.memsci.2006.03.046

Vrouwenvelder, J.S., Bakker, S.M., Wessels, L.P., van Paassen, J.A.M., 2007. The Membrane Fouling Simulator as a new tool for biofouling control of spiral-wound membranes. Desalination 204, 170–174. https://doi.org/10.1016/j.desal.2006.04.028

Vrouwenvelder, J.S., Hinrichs, C., Van der Meer, W.G.J., Van Loosdrecht, M.C.M., Kruithof, J.C., 2009. Pressure drop increase by biofilm accumulation in spiral wound RO and NF membrane systems: role of substrate concentration, flow velocity, substrate load and flow direction. Biofouling 25, 543–55. https://doi.org/10.1080/08927010902972225

Vrouwenvelder, J.S., Manolarakis, S.A., van der Hoek, J.P., van Paassen, J.A.M., van der Meer, W.G.J., van Agtmaal, J.M.C., Prummel, H.D.M., Kruithof, J.C., van Loosdrecht, M.C.M., 2008a. Quantitative biofouling diagnosis in full scale nanofiltration and reverse osmosis installations. Water Research 42, 4856–4868. https://doi.org/10.1016/j.watres.2008.09.002

Vrouwenvelder, J.S., Manolarakis, S.A., van der Hoek, J.P., van Paassen, J.A.M., van der Meer, W.G.J., van Agtmaal, J.M.C., Prummel, H.D.M., Kruithof, J.C., van Loosdrecht, M.C.M., 2008b. Quantitative biofouling diagnosis in full scale nanofiltration and reverse osmosis installations. Water Research 42, 4856–4868. https://doi.org/10.1016/j.watres.2008.09.002

Wagner, M., Taherzadeh, D., Haisch, C., Horn, H., 2010. Investigation of the mesoscale structure and volumetric features of biofilms using optical coherence tomography. Biotechnology and Bioengineering 107, 844–853. https://doi.org/10.1002/bit.22864

Wang, F., Tarabara, V. V, 2008. Pore blocking mechanisms during early stages of membrane fouling by colloids. J Colloid Interface Sci 328, 464–9. https://doi.org/10.1016/j.jcis.2008.09.028

West, S., Wagner, M., Engelke, C., Horn, H., 2015a. Optical coherence tomography for the in situ three-dimensional visualization and quantification of feed spacer channel fouling in reverse osmosis membrane modules. Journal of Membrane Science 498, 345–352. https://doi.org/10.1016/j.memsci.2015.09.047

West, S., Wagner, M., Engelke, C., Horn, H., 2015b. Optical coherence tomography for the in situ three-dimensional visualization and quantification of feed spacer channel fouling in reverse osmosis membrane modules. Journal of Membrane Science 498, 345–352. https://doi.org/10.1016/j.memsci.2015.09.047

Wibisono, Y., El Obied, K.E., Cornelissen, E.R., Kemperman, a. J.B., Nijmeijer, K., 2015. Biofouling removal in spiral-wound nanofiltration elements using two-phase flow cleaning. Journal of Membrane Science 475, 131–146. https://doi.org/10.1016/j.memsci.2014.10.016

Yang, X., Beyenal, H., Harkin, G., Lewandowski, Z., 2000. Quantifying biofilm structure using image analysis. Journal of Microbiological Methods 39, 109–119. https://doi.org/10.1016/S0167-7012(99)00097-4

Part 6

General applications

doi: 10.2166/9781789062977_0399

Chapter 18

Membrane Autopsy

Javier Rodriguez Gómez, Genesys - PWT, Spain

Nuria Peña García, Genesys - PWT, Spain

The learning objectives of this chapter are the following:

- Describe the membrane autopsy procedure
- Present the experimental set-up for membrane autopsies
- Present the membrane autopsy protocol
- Mention methods for fouled membrane characterisation
- Describe tests for membrane damage characterisation
- Illustrate membrane performance characterisation
- Present tests for chemical cleaning efficiency

18.1 INTRODUCTION

Membrane systems suffer foulant in a higher or lower extent during their performance life. This foulant is related to feed water quality, pre-treatment efficiency and other factors related to design and plant operation. On the other hand, membranes may suffer irreversible damage which is necessary to identify.

Membrane autopsy is the main tool for identifying fouling nature and membrane failures. It is based on a complex group of analyses and tests which are used to find out the cause of membrane failure or performance decline. In this process there are some analyses and tests that are basic to understand the status of the membrane, but there are additional tests which can be used depending on the case.

Table 1 Main causes of membrane failures and their impact

Foulant	Damage: Physical / mechanical issues | Chemical issues
Reversible	Irreversible
↓ Permeability ↓ Salt rejection	↑ Permeability ↓ Salt rejection
↑ Differential pressure (dp) in some cases	

Considering the main issues that membranes may suffer, autopsies should involve those analyses and tests which allow to determine which is the main cause of membrane failure, how it affects membrane performance and if it is possible to recover it.

Membrane autopsies can provide critical information needed to help troubleshoot and/ or to optimize membrane systems performance, but considering that they include many results, it is necessary that an expert is reading the data to get a more reliable and useful diagnosis.

Identification of foulant can be carried out on any membrane type. But to check the impact of that failure on membrane performance, it is necessary to reproduce its performance. There are systems available to check membranes performance, but those mainly used are for NF and RO spiral bound membranes and UF hollow fibers. In this chapter NF and RO spiral wound membranes are mainly covered.

To give a good diagnosis about membrane failure, it is very important to have information about the problem detected in plant and some details about the site. This will also help to achieve the final diagnosis and to select the membranes to study.

Main details to consider would be:

- Water type
- Plant stages/pass
- Elements per pressure vessel
- Recirculation Yes/No
- Recovery
- Feed system (storage tanks) opened / covered
- Pretreatment details: Settings, flotation, UF, filtration,
- Microfilters: Type (wound/expanded/pleated), replacement, *dp* design/current, micron, etc.
- Reagents dosing: Coagulant, flocculant, pH adjustment, Chorine, reducer (eg., SBS), antiscalant
- SDI values
- CIP frequency and protocols

A complete study of the membrane would include the characterization of the full element first and the study and characterization of the membrane once unrolled. Through this way, it will be possible to distinguish if some issues are due to membrane configuration or other components different than membrane. The analysis and characterization of membrane samples will mainly give information about the membrane itself.

For a useful autopsy, membrane selection is essential. For example, for RO systems, it is recommended to autopsy one membrane from lead position of first stage and one from tail position from last stage to get a complete understanding of the plant. Lead element in the system will contain highest concentration of suspended solids and organics. Biofouling is also worst at the first stage where the bacteria attach to the first available hydrophobic surface that they find. Last membrane in the system receives water with a higher concentration of salts so it is the most likely to scale.

18.2 MATERIALS, EXPERIMENTAL SET-UP

Spiral wound membranes are built with an outer glass fiber covering which must be removed to unroll the element. For this purpose, an autopsy table and cutting machine are the basic tools (Figure 1).

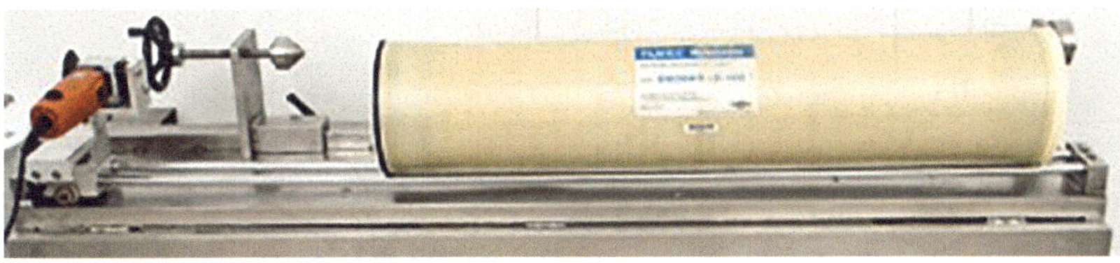

Figure 1 Full membrane element ready for autopsy.

The best surface to perform an autopsy is a level table, accessible from all sides and with a size slightly wider than the axial length of the element and at least the leaf/membrane envelopes length (Franz Leitz, 1996).

Considering the cylindrical shape of the membranes, it is necessary to fix the membrane to the table in such a way that it keeps blocked during the external housing cutting, unrolling and membrane sampling. For it, different shape cradles or systems to fix membrane to the table can be used.

Cutting of the external housing produces a significant amount of glass fiber dust, so it is very important to carry out this process considering personnel safety. Then, it is mandatory to follow some cautions:

Table 2 Safety during membrane autopsy

Cutting machine	It is advisable to use an electric cast saw better than grinder machines
Personnel protection	- Wear ample eye and face protection. - Wear gloves. - Wear long-sleeved garments to avoid skin contamination with fiberglass particles.

For sampling of the different components that can be analysed for the autopsy, it is necessary to arrange some simple tools which allow taking membrane samples, foulant and particles. Some of these tools are plastic bags, plastic bottles, scalpel, tweezers, spatula, scissors, etc.

18.3 MEMBRANE AUTOPSY PROTOCOL

The diagram in Figure 2 show main steps involved in a membrane autopsy:

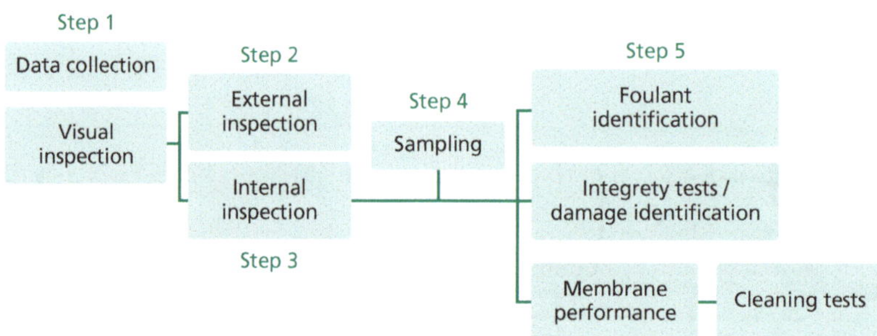

Figure 2 Main steps for implementing a membrane autopsy.

Before the autopsy starts it is important to have the following information (data collection):

- **Plant failure** to know the objective of the study. That information is essential also to determine the kind of tests and analyses to be carried out and the necessary samples.
- **Membrane position.** It will help to understand the results and to give more useful recommendations. For this it is essential also that the autopsied membrane is the correct for the purpose of the study. This point is essential for RO systems.
- **Manufacturer, model and type of membrane.** It is necessary to know the reference values of membrane performance, pH limits for cleaning tests, etc.
- **Membrane serial number.** It will help in some cases to verify in plant membrane position or to check performance parameter values with membrane manufacturer.

Since there are many details to consider during the autopsy, it is advisable to use a check list to be sure that all the items are reviewed, especially during the visual inspection of the membrane.

As summary, there are two main procedures during membrane autopsy:
- Steps 2 and 3: Visual inspection which reviews macroscopic issues.
- Step 5: Analyses and tests which give information about microscopic issues, foulant composition, structural changes, recovery of membrane performance, etc.

All the analysis and tests carried out during the autopsy should be selected to get the most accurate diagnosis.

18.4 METHODS

18.4.1 Visual inspection

External inspection

Before proceeding to open the membrane, some details of the membrane must be considered. Table 3 includes the main membrane details that must be checked.

Table 3 Details to consider during external inspection

Detail	Comment
Weight	By comparing to manufacturer reference, it gives an idea about the amount of foulant. Element must be drained to assure that retained water is removed. It is mainly useful when scaling is suspected
External housing integrity.	In some cases affected by overpressure issues.
Presence of particles/deposits on external housing.	Failures on element seal (O-ring).
Antitelescoping devices (ATD) condition.	In some cases damage is produced during transport.
Presence of particles/deposits on element ends.	Related to pre-treatment issues.
Permeate tube condition.	If there is a damage, there will be salt rejection issues.
Telescoping	
Spacer protrusion	

The best way to document these details is taking photographs. Some photographs are included in Figure 3, showing examples of main membrane failures that can be observed during visual inspection (*Images credit by Genesys Membrane Products S.L.U*).

Depending on the membrane manufacturer, it will be necessary or not to remove ATD to make element ends inspection.

If deposits and particles are observed during external inspection, samples should be taken just in case it is necessary to carry out additional analyses. These samples would provide information about plant performance.

During external inspection, it is recommendable to carry out also an integrity test of the full element. This integrity test can be carried out in two ways:
- **Bubble test**: During this test, a small pressure (3-5 psi) is put into the permeate tube while the element is submerged vertically under water. If the element continually emits bubbles and cannot hold the air pressure, then the element exhibits compromised mechanical integrity (Technical Service Bulletin Nitto-Hydranautics, 2017).

- **Vacuum decay test**: In this test, after element drain to remove water, permeate tube of the element is evacuated and isolated. A vacuum decay higher than 100 mbar/minute indicates mechanical integrity or a leak on the membrane element (Dupont, 2020).

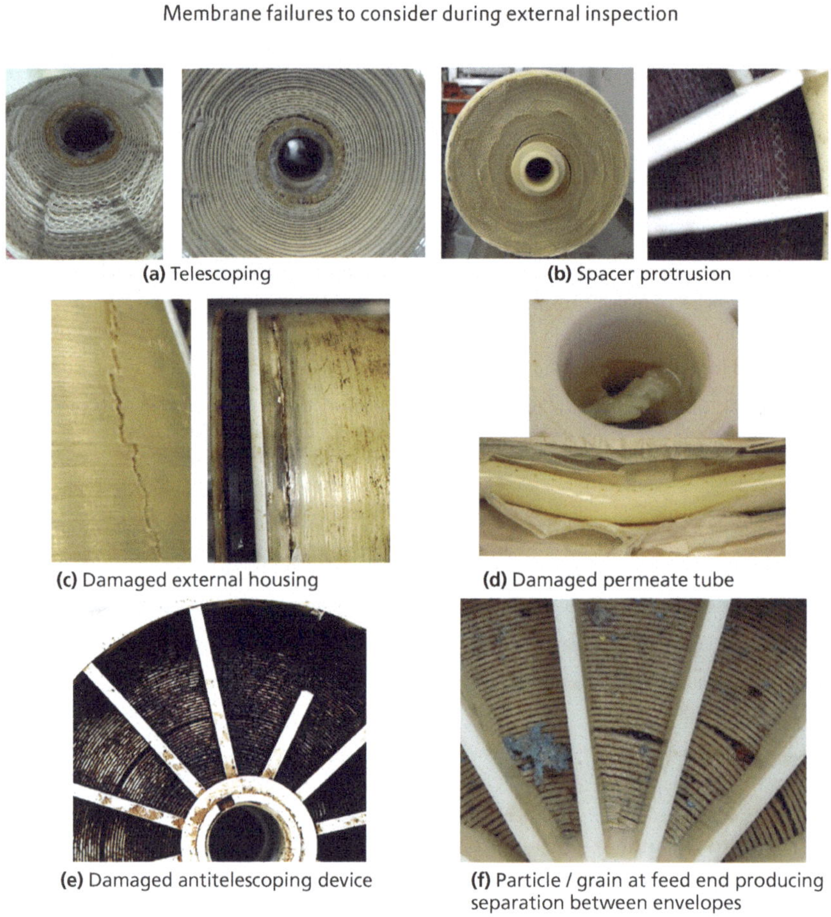

Figure 3 Examples of different issues that can be observed during external inspection of membranes during autopsy. (Images credit: Genesys Membrane Products S.L.)

Internal inspection

To proceed with the internal inspection of the membrane, it is necessary to remove the outer casing (Figure 4). Before unrolling the membrane, it is necessary to verify the feed side of the element and it is recommendable to mark it. It is the best way to verify the feed side once the membrane is unrolled.

For outer casing removal, at least four cuts are necessary to make easier the procedure: two cross sections to remove ATDs (cuts 1 and 2 at the following photograph) and two longitudinal sections (cuts 3 and 4). It is mandatory to follow the safety recommendations described in Section 18.2.

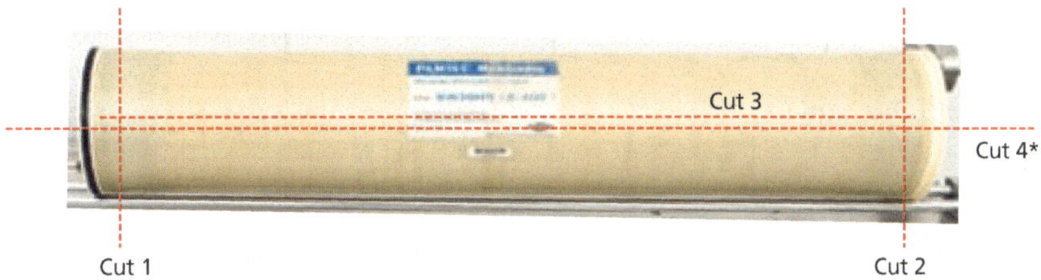

Figure 4 Recommendable cuts for external housing removal during autopsy.
*Note: cut 4 should be made approximately on the opposite side of cut 3.

With this procedure, two pieces of external housing with a quite similar size must be removed before unrolling the membrane (Figure 5). Once the membrane is unrolled, inspection can be carried out.

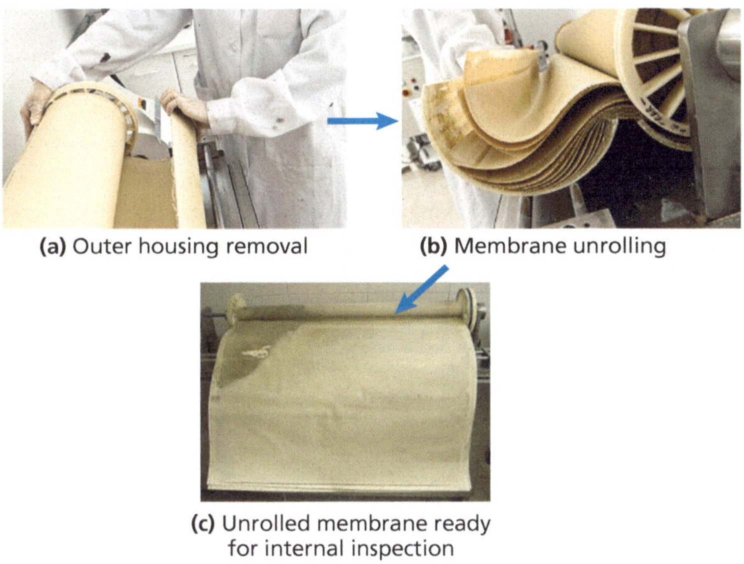

Figure 5 Membrane unrolling during autopsy. (Images credit: Genesys Membrane Products S.L.)

As during external inspection, some details must be checked during this internal inspection. Table 4 includes these details, which should be checked in the number of leaves/envelopes that assure a representative inspection of the membrane element.

Table 4 Details to consider during internal inspection

<table>
<tr><td>Odour</td><td>Odour description is important to distinguish some foulant components.</td></tr>
<tr><td>Presence of foulant /particles</td><td>During visual inspection it is possible to distinguish deposits from scaling, organics from inorganics, and if it will be necessary to make specific analyses to identify them.</td></tr>
<tr><td>Distribution of foulant/particles</td><td>It is very important to check if there is a higher concentration of foulant on feed or reject side, if there are different components and if there is presence of particles.</td></tr>
<tr><td>Failures on membrane surface</td><td>Creases, delamination, and other failures.</td></tr>
<tr><td>Spacer prints</td><td>If spacer prints are visually detected a significant damage from increases in dp can be expected.</td></tr>
<tr><td>Spacer material condition</td><td>Presence of foulant, wrinkles or deformation are important details to consider if detected.</td></tr>
<tr><td>Permeate carrier material condition</td><td>If presence of foulant is observed on permeate carrier it will be evidence of damage on membrane material.</td></tr>
<tr><td>Membrane colour</td><td>It is important to know in advance the colour of a new membrane to determine if there is a failure/problem or not.</td></tr>
<tr><td>Membrane leaves insertion</td><td rowspan="2">Failures on these components can lead on salt rejection reduction and permeate flow increase.</td></tr>
<tr><td>Glue lines condition</td></tr>
</table>

Similar to external inspection, the best evidence of the details and failures observed during membrane internal inspection is by taking photographs. The images in Figure 6 correspond to some of the main membrane failures that should be considered during this inspection.

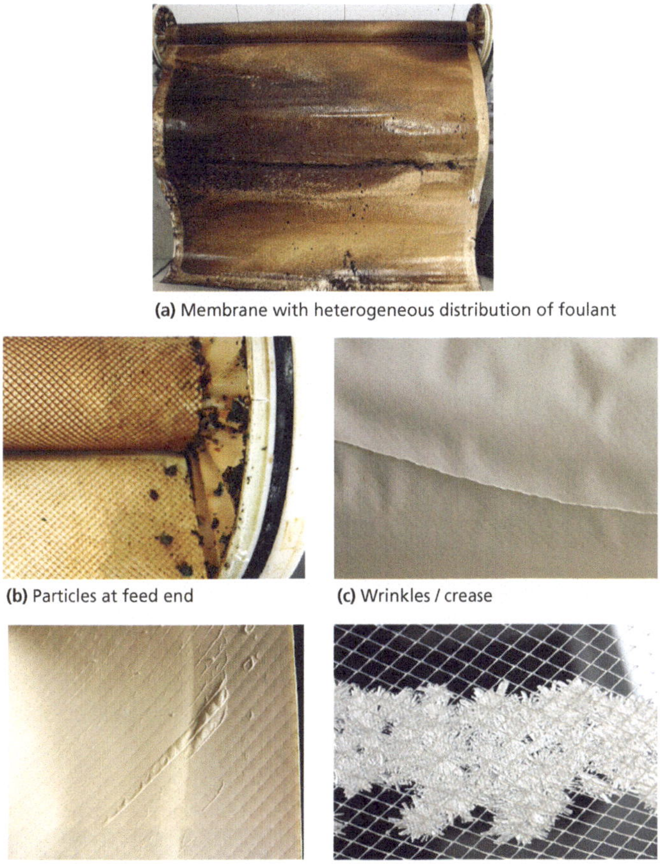

(a) Membrane with heterogeneous distribution of foulant

(b) Particles at feed end

(c) Wrinkles / crease

(d) Delamination on glue lines and membrane active layer

(e) Scaled spacer material

Figure 6 Examples of different issues that can be observed during internal inspection of membranes during autopsy. (Images credit: Genesys Membrane Products S.L.)

During internal inspection of the membrane element, the different samples needed for the rest of analyses and tests included in the autopsy procedure should be taken. Each autopsy may involve different samples depending on the objective of the study, but in general terms the minimum samples to consider are listed in Table 5.

Table 5 Recommended sampling during autopsy

Sample		Analysis / test
Particles / deposits from element ends		They will allow distinguishing elements from suspended matter reaching the membranes from foulant components developed on membrane surface.
Foulant		By analysing foulant removed from membrane surface, foulant components will be concentrated and it will be possible to distinguish some foulant components from membrane components during analyses.
Membrane	Raw sample	Microbiology
		Foulant identification
		Performance parameters
	Clean surface	Cleaning tests
		Presence of halogens/oxidant agents
		Polyamide bands conditions

18.4.2 Analytical methods for foulant and damage identification

Some of the analytical techniques commonly used for foulant identification are described in other chapters. The aim of this chapter is to guide during membrane autopsy proceeding the analytical techniques that can be used for membrane failure diagnosis. Those considered more simple, accessible, and more commonly used are described in Table 6.

Table 6 Analytical techniques for the identification of fouling components and membrane damage

Autopsy step	Method	Items to check	
Foulant identification	Analytical techniques	LOI / TGA	Quantification organic/inorganic based in the loss of weight after thermal degradation.
		SEM-EDX	Elementary identification of foulant components. It is possible to distinguish different foulant components.
		ATR-FTIR	Identification of functional groups and compounds from data base.
		Microbiological counts	It gives quantification of specific microorganisms.
		LC-OCD	Separation and identification of foulant components, mainly from NOM.

Autopsy step	Method		Items to check
Damage	Analytical techniques	SEM-EDX	Abrasion marks from particles and spacer material.
		ATR-FTIR	Polyamide bands changes indicate structural changes/degradation.
		XPS/ESCA	Halogens detection Elements oxidation state.
	Tests	Dye tes	Damage, mainly physical.
		Fujiwara test	Contact with halogens if they are on high concentration.

Table 6 includes some of the most common techniques used during autopsies but any technique which could be applied on solids and surfaces to provide information about both foulant identification and membrane conditions could be used.

Other techniques to be considered: optical microscopy, confocal microscopy, chemical solubility of foulant, humic and acids testing, contact angle testing, Zeta potential, FRX, DRX, ICP-MS, GC-MS, etc.

During autopsies, it is also common to calculate foulant density (mg of foulant/cm^2). Reference for this parameter may change depending on the characteristics of foulant, but it gives quite useful information about the distribution of foulant on a system if different membranes from same pressure vessel (PV) are analysed.

LOI and TGA

- *Loss on ignition (LOI):*

LOI is a common and widely used method to estimate the organic content of membrane foulant. Organic matter is oxidised at 500–550 °C to carbon dioxide and ash. The weight loss during this reaction is easily measured by weighing the dry sample (LOI 105 °C) before and after heating and is closely correlated to the organic matter.

- *Thermogravimetric analysis (TGA)*

For TGA, the mass of a substance is monitored as a function of temperature or time as the sample specimen is subjected to a controlled temperature program in a controlled atmosphere (Perkin Elmer, 2010-2015).

Both techniques provide the content of organic/inorganic component in percentage. This information is very valuable to determine which is the main cause in failure and also about the performance of the plant comparing results from different membrane position.

SEM-EDX

Scanning Electronic Microscopy with X-ray dispersive energy analysis (SEM-EDX) is used to study the membrane surface and to verify the elemental composition of its fouling. Elemental determination with the SEM-EDX system is based on analysis of X-rays produced via electron beam excitation of a sample area. This technique allows analysis of a sample in selective areas. The limited depth of analysis (typically a few microns), and the possibility to select the area of interest under the electron beam, allows for local analysis to reveal differences in composition. The identification and measurement of individual peak intensities in the X-ray spectrum is done with a computerized multichannel analyser.

SEM-EDX is one of the most powerful and basic tools used during membrane autopsies. It is useful not only for foulant identification, but also for studying foulant distribution on membrane surface and to distinguish different foulant components. Some damage can also be identified, mainly abrasion marks and marks from spacer. Concerning chemical damage, it is very difficult to detect them by this technique (Peña *et al.*, 2013).

The microphotographs in Figures 7a and 7b show characteristic membrane structure studied by SEM-EDX at high magnifications and different types of foulant and damage.

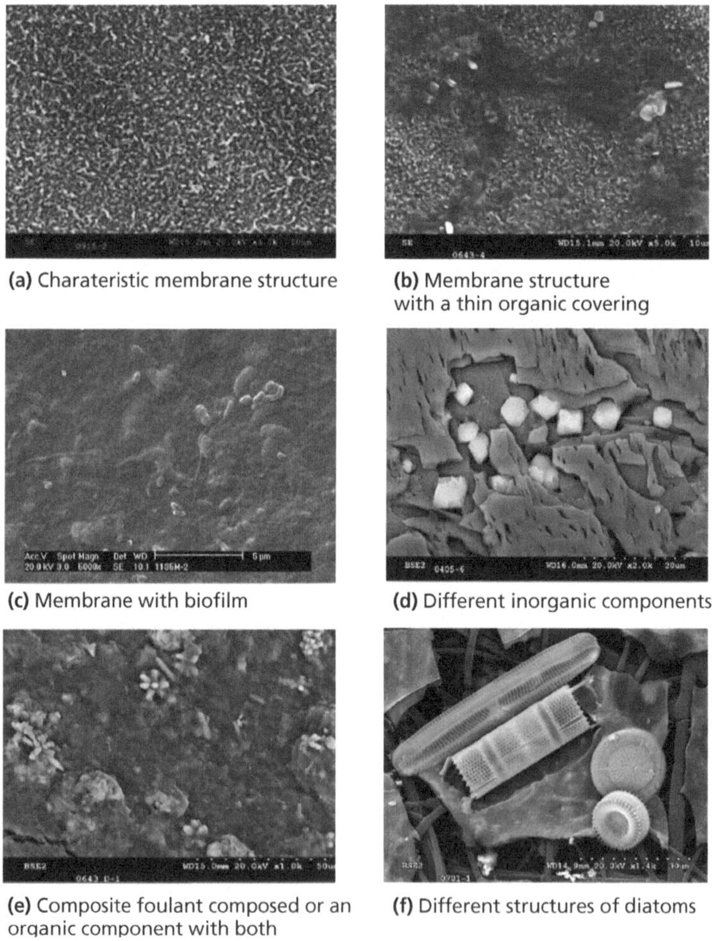

(a) Charateristic membrane structure

(b) Membrane structure with a thin organic covering

(c) Membrane with biofilm

(d) Different inorganic components

(e) Composite foulant composed or an organic component with both

(f) Different structures of diatoms

Figure 7a Examples of different observations and details obtained during analysis of membranes by SEM-EDX. (Images credit: Genesys Membrane Products S.L.)

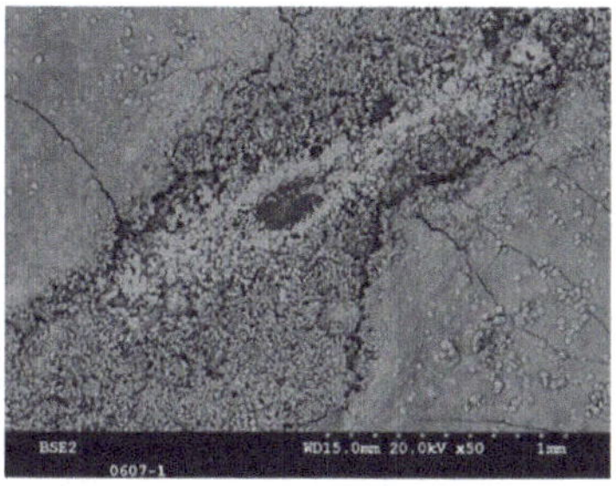

(a) Example of a membrane with different components on areas of contact with spacer material and the rest of the membrane surface

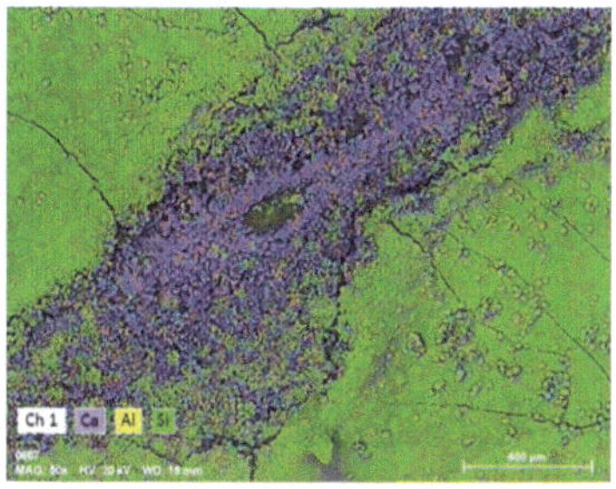

(b) Mapping allows to apply color to different elements detected during EDX analysis, wich makes much easier to study element distribution on membrane surface

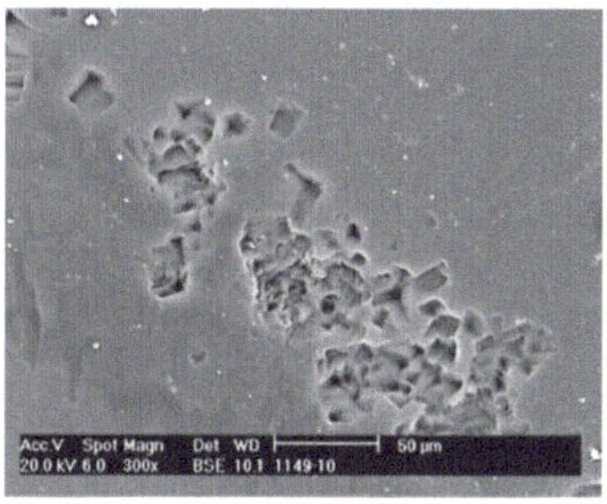

(c) Marks from scaling on a clean membrane surface

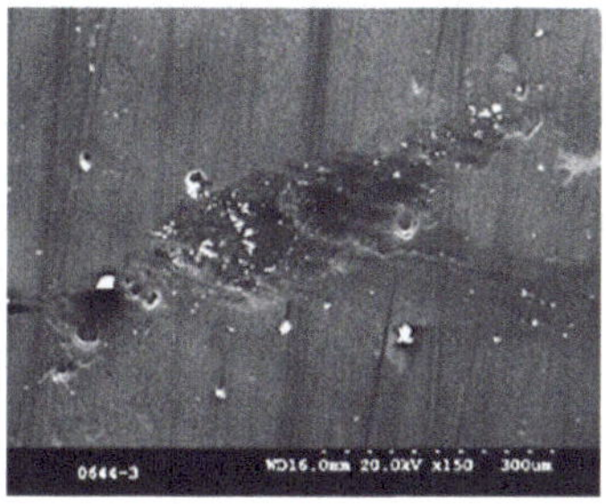

(d) Marks from spacer material on a clean membrane surface

Figure 7b Examples of different observations and details obtained during analysis of membranes by SEM-EDX. (Images credit: Genesys Membrane Products S.L.)

ATR-FTIR

In the mid-infrared, absorption of radiation is related to fundamental vibrations of the chemical bonds. Then, fourier transform infrared spectroscopy with attenuated total reflectance (ATR-FTIR) Spectrometry can provide valuable information related to the chemical structure of membrane or characterize the fouling layer that may be present on the membrane surface. IR spectra can be studied both in absorbance or transmittance.

As examples, Figures 8 and 9 include IR spectra in absorbance showing the characteristic spectrum of a clean polyamide-polysulphone membrane in good chemical conditions (blue lines) compared to spectra of membranes with thick covering and thin covering (red line). Membranes with thin covering will show most of the membrane bands and membranes with thick covering will mainly show bands from foulant composition. Identification of foulant will be obtained both by the identification of functional groups or by using IR data base. It is important to remind that the composite nature of the foulants may be difficult in some cases the identification and that commonly only the main component of foulant will be identified.

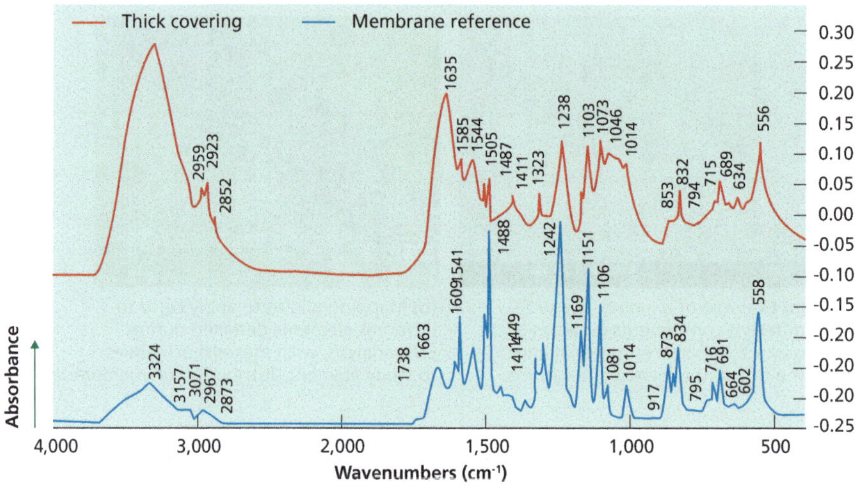

Figure 8 Example of membrane with thick foulant (red line) compared to a reference membrane (blue line): only some of the membrane bands are detected since there is a main presence of foulant.

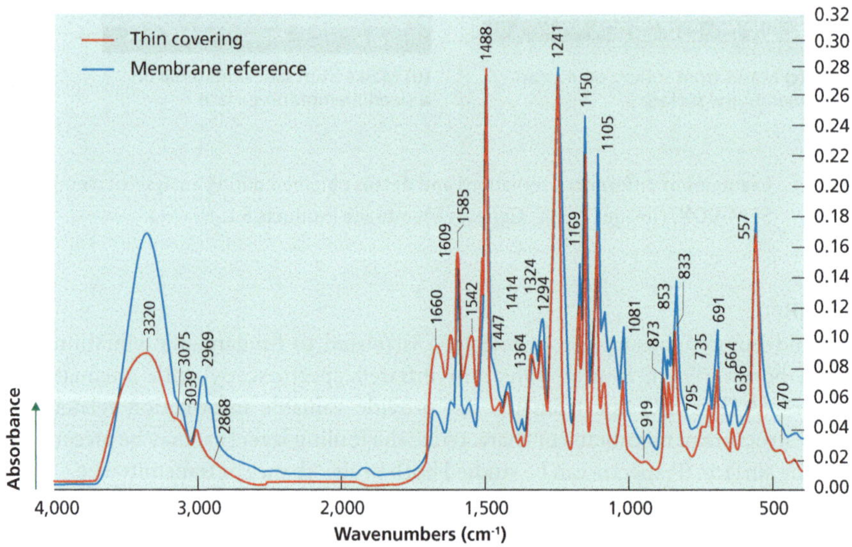

Figure 9 Example of membrane with thin foulant (red line) compared to a reference membrane (blue line): most of the membrane bands are detected.

Since IR spectrometry provides information related to the presence or absence of specific functional groups, shifts in the frequency of absorption bands and changes in relative band intensities indicate changes in the chemical structure or changes on the membrane surface.

By this technique it is possible to check the presence of polyamide layer bands in an RO membrane and to determine if there are any structural changes on it (Figure 10).

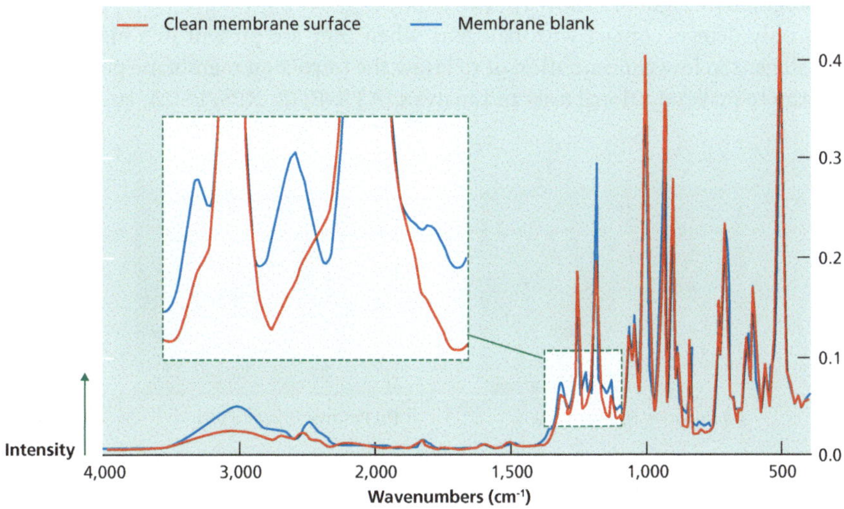

Figure 10 Example of membrane with significant chemical damage: Comparison of used membrane IR spectrum after a mechanical cleaning (red line) compared to membrane reference (green line).

There is an absence of polyamide bands.

In some case in which the decrease is not significant and it is necessary to check membrane conditions, quantification of the polyamide bands can be carried out to achieve additional information (Sandin *et al.*, 2013).

Fujiwara test

Fujiwara test (FJ) test detects significant levels of polyhalogen compounds so it will detect if the membrane is oxidised. This is a colorimetric test in which a pink colour in the analytical solution, indicates organically bound halogens. FJ test must be carried out on RO water rinsed membrane samples and without deposit (physically removed).

It is a qualitative test that can be easily carried out by the following procedure:

1. Introduce a sample of membrane in a test tube.
2. Add pyridine, add approximately the same volume of sodium hydroxide. Shake to mix the two layers.
3. Heat the mixture in a boiling water bath.
4. If a reddish-pink colour appears in the organic layer, it indicates that the membrane has been in contact with a halogen (normally chlorine)*.

*It is also advisable to carry out a blank test (without membrane) to verify that the mixture of pyridine and sodium hydroxide does not generate any colour without the sample.

The photographs in Figure 11 show a positive and a negative Fujiwara test. This is a qualitative test that only detects contact with halogens when they are present in a high extent. To detect halogens in low concentration or to know the impact on membrane performance, it is necessary to make additional tests and analysis (ATR-FTIR, XPS/ESCA, etc.).

(a) Negative Fujiwara test (b) Positive Fujiwara test

Figure 11 Examples of a negative (picture a) and positive results (picture b) obtained during Fujiwara test. (Images credit: Genesys Membrane Products S.L.)

XPS/ESCA

X-ray photoelectron spectroscopy (XPS), also known as electron spectroscopy for chemical analysis (ESCA) is a quantitative spectroscopic technique that measures the empirical formula, chemical state and electronic state of the elements that exist within a material. XPS spectra are obtained by irradiating a material with a beam of X-rays while simultaneously measuring the kinetic energy and number of electrons that escape from the top 1 to 10 nm of the material being analysed. This technique is able to detect and identify different halogens (chlorine, bromine, iodine) and it can distinguish the oxidation state of the detected elements.

Then, for example if the oxidant agent is sodium hypochlorite, this technique allows distinguishing between chloride (198,7 eV) and chlorine related to carbon by a quasi-covalent bound (200 eV) (Beverly *et al.*, 2000, Hiraoka *et al.*, 2011).

For seawater membranes, even if the disinfection process is based in NaOCl, the oxidant agent is bromine. Figure 12 includes a characteristic spectra of a clean membrane surface with a significant presence of bromine.

Depending on the detected halogen, different impact on membrane performance can be expected (Maugin, 2013).

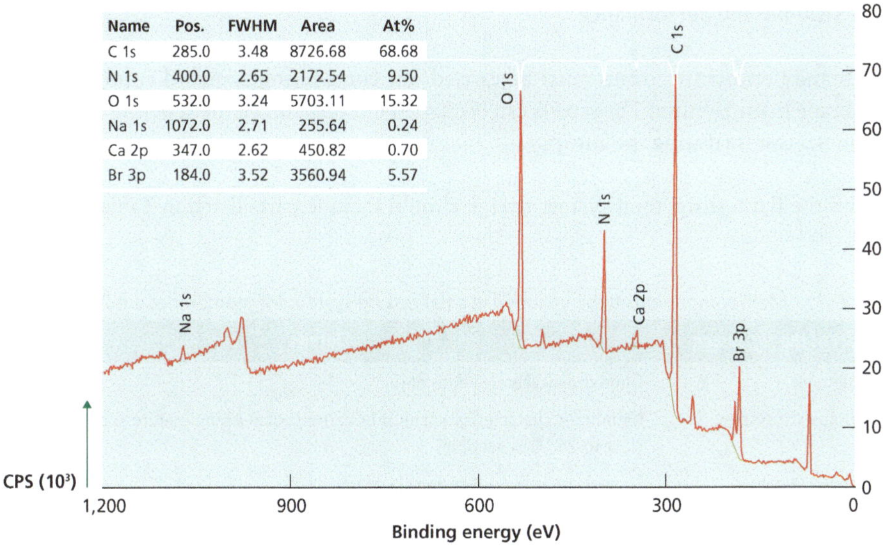

Figure 12 XPS spectrum (clean surface) of a membrane with presence of bromine.

Dye test

This test is based on the passage of a high molecular weight compound solution through the membrane. A high molecular weight compound should be retained by the polyamide layer, so when presence of this compound is detected on the permeate side of the membrane, it means that the polyamide layer is damaged on that area. This test indicates damage but not the source, although it is more sensitive to physical damage (Peña *et al.*, 2013).Different dyes can be used for this test, but one of the most common is Methylene blue. This test can be done on the full element or on membrane samples.

For this test, it is necessary to circulate a solution of aprox. 0.01% dye, applying the minimum pressure at which it permeates the membrane during 20-30 minutes.

If the surface of the membrane is physically damaged, the dye will pass through the membrane, showing colour on the permeate side (Figure 13).

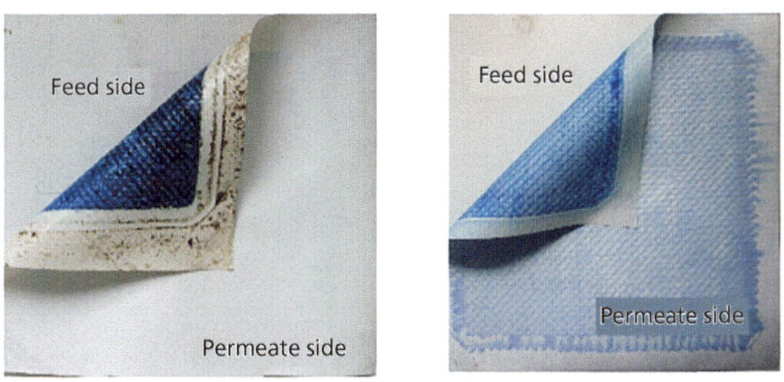

(a) Negative methylene blue test **(b)** Massive positive methylene blue test

Figure 13 Examples of a negative (picture a) and positive results (picture b) obtained during Methylene Blue test. (Images credit: Genesys Membrane Products S.L.)

18.4.3 Membrane performance

Membrane performance tests must be carried out considering standard conditions from membrane manufacturer. These tests can be carried out on a full element before the autopsy and on flat test rig during the autopsy.

Besides the flat test rig, total system design should include parts listed in Table 7 (weebly.com):

Table 7 Main components to be included in a flat test rig system for membranes autopsy

Part of the rig design	Details
Water tank	Good chemical and temperature.
Thermostatic system	Membrane characterization must be carried out at a temperature as constant and close to 25 °C as possible.
Heating element	To apply temperature during cleaning tests.
Pump	Both pressure and flow rates must be variable.
Piping	It must be pressure and corrosive resistant.
Pressure	gauge 0 to 70-80 bar.
Collection tank	A measuring cylinder is the most appropriate since it will provide volume reading.

There are various flat test rig systems available in the market. Main detail to consider for a correct flat test rig selection is to choose the one that allows with brackish and/or sea water samples depending on the needs.

The diagrams in Figure 14 obtained from references correspond to some examples of rig and system configuration that can be used for this purpose:

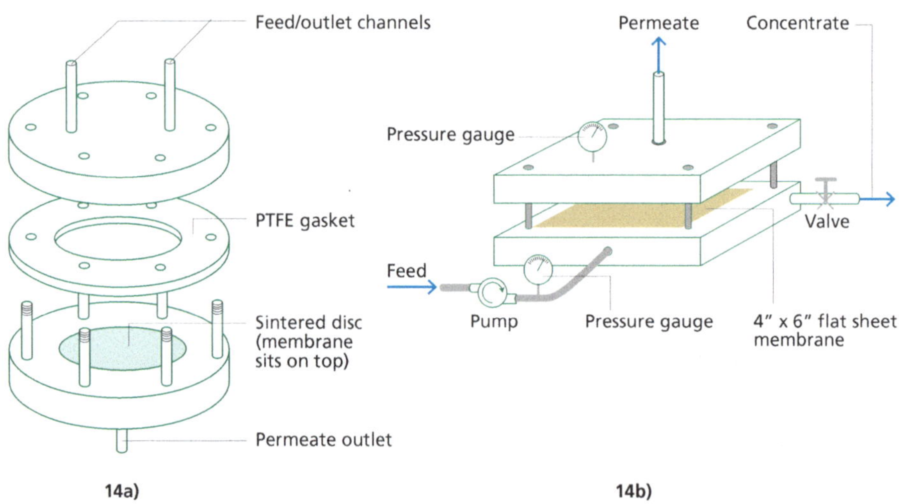

Figure 14 Examples of flat test rigs by (14a) Robinson *et al*, (2004) and (14b) Cartwright (2012).

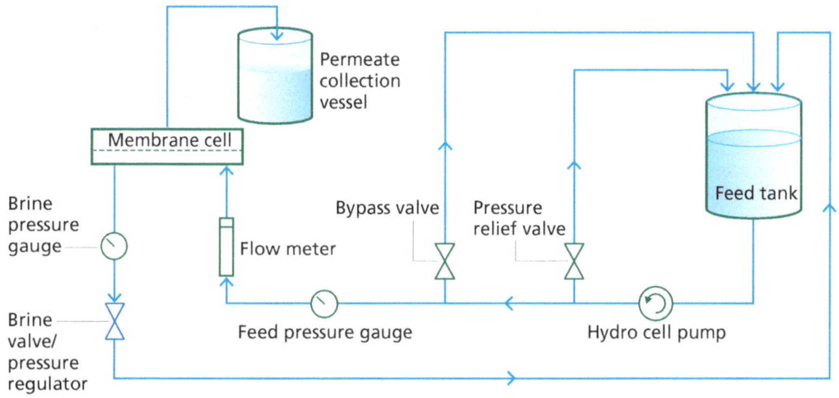

Figure 15 Example of system for membrane characterization (Farhat *et al.*, 2013).).

To characterize membrane coupons during autopsy, it is necessary to select representative samples. It is recommended also to use spacer material and permeate carrier from same membrane model.

Feed water must be prepared with sodium chloride (reverse osmosis membranes) or magnesium sulphate (nanofiltration membranes) dissolved in deionized water. The concentration of each salt will be specified by the manufacturer for the design values.

Salt solutions must be recirculated through membrane coupons by applying specified pressure. Most common characterization conditions are shown in Table 8.

Table 8 Most common conditions for membrane characterization (this info is available from membrane manufacturers data sheets for each specific type and membrane model)

	NF membrane BW	RO membrane SW	RO membrane
Salt concentration	2,000 mg/L $MgSO_4$	1,500 mg/L NaCl 2,000 mg/L NaCl	32,000 mg/L NaCl
Pressure, psi	70	150 / 225	600 / 800

The characterization procedure is implemented as follows:

1. Once solution is circulating through the membrane by applying the specified pressure and it is producing permeate, let it get stabilized during 20-30 minutes.
2. Measure permeate flow and register temperature to normalize the data to 25ºC. For this normalization, membrane manufacturers commonly include tables or equations to make needed calculations.
3. Take both feed water sample and permeate sample to measure conductivity or a selected ion to check salt rejection.
4. Take at least three measurements to get a representative average.

Comparing the results obtained during these tests with reference values established by membrane manufacturer. different conclusions will be obtained. The following figures show some examples of characteristic results of fouled (Figure 16) and damaged membranes (Figure 17).

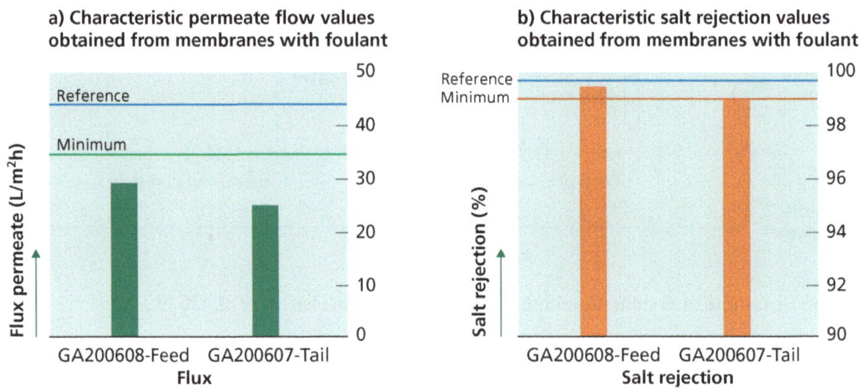

Figure 16 Characteristic results of membranes with foulant with (a) permeate flow lower than reference and (b) salt rejection lower but close to reference Both parameters can be recovered during cleaning tests / foulant removal.

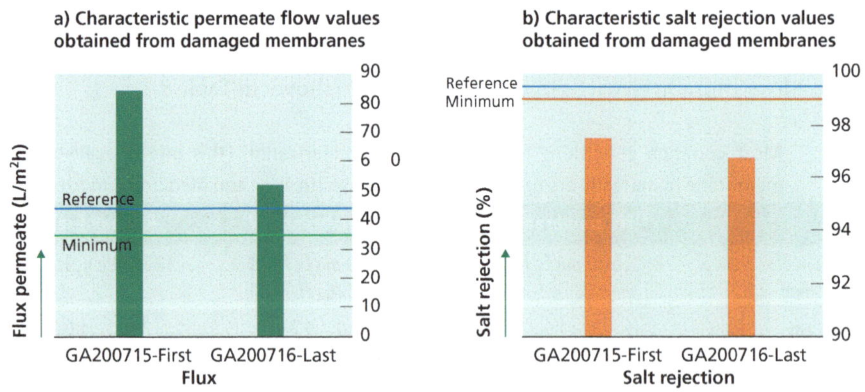

Figure 17 Characteristic results of damaged membranes where (a) permeate flux higher than reference and (b) salt rejection lower than reference. These parameters cannot be recovered during cleaning tests / foulant removal.

18.4.4 Cleaning tests

In the event of a reduction in productivity, it may be interesting to carry out individual cleaning tests in pilot plants/test rigs to find out the cleaning chemical products and the ideal cleaning conditions.

The choice of the most suitable chemical product for cleaning depends on the type of membrane fouling. For this reason, it is advisable to start the cleaning tests when foulant composition is already identified.

To carry out cleaning tests, performance data obtained from the characterization of the membrane are the baseline, but it is necessary to apply the different cleaners on different samples to obtain information about flux and salt rejection before and after the application of each product. Then, the procedure should be:

- Characterization of permeate flux and salt rejection as described in section 18.4.3. If cleaning tests are carried out on a full element, *dp* should be considered also.
- Application of the cleaner considering cleaner and/or membrane manufacturer recommendations.
- Rinsing with water until the complete removal of the cleaning solution. pH control can be used to be sure about cleaning solution removal.
- Characterization of permeate flux and salt rejection as described in section 18.4.3. If cleaning tests are carried out on a full element, *dp* measurement should also be considered.

By using this procedure with different cleaners it will be possible to select the most suitable product, which will be the one achieving the best results in terms of parameters improvement.

When working with samples at test rigs, if membrane initially gives a flux lower than reference, the aim of the cleaning test will be to reach that value.

If membrane is damaged, it is probable that in some cases the flux is already equal or higher than reference. Then, best cleaner will be the one achieving the highest flux increase in terms of percentage. In this case, it won't be possible to recover salt rejection and it may happen that the value gets even worse after foulant removal.

Considering the composite nature of foulants, it is very common to need a multi-step cleaning procedure based on alkaline-acid cleaners, alkaline-biocide-acid cleaners, etc. In these cases, once the best cleaners are selected, additional tests must be carried out to check their performance applied on a full cleaning protocol.

The application of each cleaner must take into account the recommendations of the product manufacturer and the membrane manufacturer, to follow the guidelines about pH, temperature, contact time, etc. In some case it will be necessary to improve contact time, concentration, etc., but always working within conditions that preserve membrane integrity.

18.5 REFERENCES

Beverly, Seal, Hong (2000), Identification of surface chemical functional groups correlated to failure of reverse osmosis polymeric membranes, J. Vac. Sci. Technol. 18 1107–1113.

Cartwright (2012) Reverse Osmosis Technology-Not just for Water Purification. Water conditioning & Purification International Magazine, 2012.

Farhat A., Ahmad F., Hilal N., Arafat H.A. (2013) Boron removal in new generation reverse osmosis (RO) membranes using two-pass RO without adjustment. Desalination 310, 50-59.

FilmtecTM (2020) Reverse Osmosis Membranes. Technical Manual. Page 143. From No.609-00071-0416

Leitz (1996) Membrane element autopsy manual. Water Treatment Technology Program Report No. 17. Water Treatment Engineering and Research Group, Environmental Resources Services. U.S. Department of the Interior.

Hiraoka, Iijima, Sakai (2011) Surface analysis of polyvinylchloride bombarded by Ar+ and charged water droplets, Surf. Interface Anal. 43, 236–240.

List of sourced parts for the final rig design. - Design of a Desalination by Reverse Osmosis Laboratory Test Facility (weebly.com)

Maugin, (2013) Fate of Reverse Osmosis (RO) membranes during oxidation by disinfectants used in water treatment: Impact on membrane structure and performances. Doctoral Thesis. King Abdullah University of Science and Technology, Thuwal, Kingdom of Saudi Arabia.

Peña García, del Vigo, Chesters, Armstrong, Wilson, Fazel (2013), A study of the physical and chemical damage on reverse osmosis membranes detected by autopsies. IDA World Congress on Desalination and Water Reuse, Tianjing, China. IDAWC/TIAN13-184

Perkin Elmer 2010-2015, A Beginner's Guide.

Robinson, J.P. *et al.*, 2004. Evidence for swelling-induced pore structure in dense PDMS nanofiltration membranes. Filtration, 4 (1), pp. 50-56.

Sandin *et al.,* (2013) Reverse osmosis membranes oxidation by hypochlorite and chlorine dioxide: spectroscopic techniques vs Fujiwara test. Desalination and Water Treatment. 51(1-3), p. 318-327.

Technical Service Bulletin, Nitto-Hidranautics. February 2017. TSB 101.03

doi: 10.2166/9781789062977_0421

Chapter 19

CFD as a Tool for Modelling Membrane Systems

Gustavo A. Fimbres Weihs, The University of Sydney, Australia

Francisco Javier García Picazo, City of San Diego, USA

Yie Kai Chong, Wen Yew Lam, Jia Xin Tan, Kathleen Foo, Weng Fung Twong, Yong Yeow Liang,

Universiti Malaysia Pahang Al-Sultan Abdullah, Malaysia

The learning objectives of this chapter are the following:

- Define basic principles of CFD
- Define and apply CFD for modelling membrane systems
- Present and discuss the equations involved and steps for building a CFD model of a membrane system
- Understand the theoretical background of mass transport and permeate flux and how they interplay to result in concentration polarisation and fouling.

19.1 INTRODUCTION

Computational fluid dynamics (CFD) is a computer-based numerical method used to analyse systems that involve fluid flow and/or heat and mass transfer (Versteeg & Malalasekera, 2007). CFD bridges the two different approaches for solving engineering problems before the computer era, theoretical and experimental; it relies on mathematical models while being easy to adapt to almost any realistic condition (Anderson & Wendt, 1995). Another feature of CFD is its versatility, as it allows the analysis of systems for a variety of applications such as chemical reactions (Salehi *et al.*, 2016), aerodynamics (Snel, 2003), dispersion of pollutants (Chu *et al.*, 2005), blood flows (Byun & Rhee, 2004), among many others.

The popularity of CFD has increased as the processing power of computers has become more capable in the last decades. Some of the advantages of CFD over experiment-based approaches for analysing and understanding systems are (Versteeg & Malalasekera, 2007):

- significantly higher resolution,
- excellent reproducibility and repeatability,
- ability to analyse complex systems,
- capacity to extract multiple characteristics of the fluid (velocity, concentration, vorticity, etc.),
- non-intrusiveness, and
- set-up cost and time reduction.

Theoretical analysis of fluids is typically intractable for realistic conditions, so typically several assumptions are considered in order to be able to solve these models (Wiley & Fletcher, 2003). In addition, these simplified analytical models are very difficult to validate in the laboratory due to the presence of external sources of noise (Liang *et al.*, 2020b). In those cases, CFD stands out a useful technique as it allows the study of complex models that are difficult to analyse mathematically or to set up experimentally, albeit at the expense of computational power.

There are multiple software options available to perform CFD analysis, including commercial offerings such as ANSYS-Fluent®, ANSYS-CFX® and COMSOL-Multiphysics®, as well as freely available open-source options such as OpenCFD-OpenFOAM® (Table 1). The most general differences between them are the user interface, the flexibility to edit the simulation parameters and the numerical approach used to obtain the results. In general, they all make use of the computer CPU (ANSYS-CFX®) to perform calculations, while some may use the GPU or both (ANSYS-Fluent®). The selection of software should be based on the type of analysis being performed.

Table 1 Commercially available software for CFD analysis and their characteristics

Software	Developer	Type	Method	Reference	Used by
Fluent®	ANSYS	Proprietary	Cell-centred finite volume or finite element	ANSYS Inc. (2020)	Gavelli *et al.* (2008), Liu *et al.* (2013), Su *et al.* (2019)
CFX®	ANSYS	Proprietary	Cell-vertex finite volume	ANSYS Inc. (2012)	Liang *et al.* (2018b), Toh *et al.* (2020b)
Multiphysics®	COMSOL	Proprietary	Finite element	COMSOL AB (2008)	Baghel *et al.* (2020), Brunner *et al.* (2018)
OpenFOAM®	OpenCFD	Open source	Cell-centred finite volume	OpenCFD Ltd. (2016)	Haddadi *et al.* (2018), Liu and Hinrichsen (2014), Kone *et al.* (2018)

19.1.1 What is NOT modelled

As any other analysis method, CFD presents some limitations that are worthy of note when considering it for the analysis of fluid flows. For instance, some simulations may require very large computational time to perform even using specialised hardware. The use of high-performance computing hardware can significantly reduce the computational time, but this is often expensive (Jamshed, 2015). In addition, CFD is not suitable for all cases or purposes; for example, simulation of the transport of individual molecules in a fluid domain can be extremely difficult (or even impossible) to model. Determining the effect of the physicochemical properties of a membrane on its transport performance is also not within the applications of CFD, as the relevant phenomena do not occur in the fluid phase but within the membrane matrix. Moreover, CFD is not useful to determine the effect of the membrane roughness on the boundary layer, as roughness is in order of 10 nm (Boussu *et al.*, 2005), while the concentration boundary layer typically ranges from 1 to 100 μm in thickness (Rodrigues *et al.*, 2013). Given the nature of CFD, variables cannot be explicitly computed at every possible point of the domain, but they are rather approximated. Major aspects of the numerical methods that form the basis of CFD are discussed in section 18.1.2.

19.1.2 How modelling can assist membrane systems

Membrane separation processes (MSP) are operations where a fluid is forced through a thin semipermeable barrier, called a 'membrane'. Transmembrane pressure and electrochemical potential are generally the driving forces used to operate membrane systems (Baker, 2012). One of the key components of MSPs is the membrane itself, as its properties determine the efficiency of separation of the components in the fluid. There are many applications for MSPs such as gas purification, crystallisation and membrane reactors; nonetheless, this chapter focuses on applications in the field of water treatment.

Desalination and tertiary wastewater treatment stand out as the most widely spread applications of membrane systems. These processes are very effective to separate small solutes from water mainly using pressure as the driving force. Nevertheless, they require very high operating pressure (of the order of 10 to 70 bar), which comes with high operation costs. Another issue faced by MSPs is concentration polarisation (CP) which results in a significant decrease in efficiency. CP occurs when the solutes rejected by the membrane accumulate in the vicinity of its surface, forming a region of high concentration, i.e., the concentration boundary layer. Given the nature of MSPs, CP is impossible to prevent, although it can be mitigated so its effect on efficiency is minimised, most commonly via mass transfer enhancement strategies. Other than CP, particulate fouling is also one of the challenging issues that persist in reverse osmosis systems since the technology of RO was first introduced into the desalination process and other pressure-driven operations such as microfiltration, ultrafiltration and nanofiltration.

Over the last decades, CFD has been used to study mass transfer in MSPs and to find innovative approaches to enhance it. Modifying the geometry and location of spacers (Fimbres-Weihs & Wiley, 2008), imposing an unsteady flow at the inlet (Liang *et al.*, 2018b) and vibration-assisted modules (Su *et al.*, 2018), are among the approaches investigated using CFD for mass transfer enhancement in MSPs. Moreover, CFD has also been employed to study different types of fouling in membrane modules (Fimbres Weihs & Wiley, 2014; Radu *et al.*, 2010). The following sections take a closer look into some specific application cases of CFD analysis of membrane systems.

19.2 METHODS

Using CFD for the analysis of MSPs involves multiple steps that need to be performed to draw valuable conclusions on the model. A comprehensive CFD study comprises three main stages: set-up, solution and conclusion, with an additional stage for experimental validation often performed based on the scope of the study. Figure 1 summarises the typical workflow for CFD modelling. The steps required for setting up the model are in yellow, the ones involved in solving the model are in red and finally, in green, the postprocessing steps. The set-up stage involves the problem definition, generation of the geometry of the fluid domain, definition of the meshing strategy and the establishment of the boundary conditions. Despite seeming trivial, defining the goal and subsequent approach to model a membrane system is crucial, as it sets the foundation of the method.

The solution of the model comes in the next step. CFD uses a numerical approach to solve the Navier-Stokes equations for continuity, momentum and mass transfer. These equations are summarised as follows:

$$\nabla \cdot \vec{v} = 0 \qquad \text{Eq. 1}$$

$$\rho \frac{\partial \vec{v}}{\partial t} + \rho\left(\vec{v} \cdot \nabla\right)\vec{v} = \mu \nabla^2 \vec{v} - \nabla p \qquad \text{Eq. 2}$$

$$\rho \frac{\partial w}{\partial t} + \rho \nabla \cdot (w\vec{v}) = \rho D \nabla^2 w \qquad \text{Eq. 3}$$

where ρ, μ and D are the density (kg m^{-3}), dynamic viscosity (kg m^{-1} s^{-1}) and diffusivity (m^2 s^{-1}), respectively. These physicochemical properties of the fluid are typically considered to remain constant, although, some variations may be caused by temperature and concentration fluctuations. CFD methods solve (numerically) these equations for the velocity vector ($\vec{V}$), the pressure field (p) and the solute mass fraction field (w).

It is worthy of note that there are currently no generally applicable fully-analytical solutions for the general case of the Navier-Stokes equations, hence the importance of computational methods. Given the numerical nature of CFD methods, they require boundary and initial conditions as inputs to the solution algorithm. Determining a solution to the model is probably the most time-consuming step of modelling, as it involves solving equations - iteratively until the solution error is below a specified tolerance. It is at this point, when the error is sufficiently small and the solution variables are not changing significantly after further iterations, that the solution is said to have converged.

Once the model solution has converged, it should be compared against previous results for the same or a similar model, in a step known as verification. Data processing and assessment come after the results are verified. These two steps are key parts of modelling as they are used to draw conclusion about the model and, ultimately, for decision making. The assessment step is crucial to suggest improvements and correct deficiencies in a real-world system.

Validation is not necessarily part of modelling, although, it is useful to prove that the model is valid for real-world applications. The set-up starts with the problem definition, this is, conceptualising the model that is to be solved. The steps followed for CFD modelling are detailed in the following section.

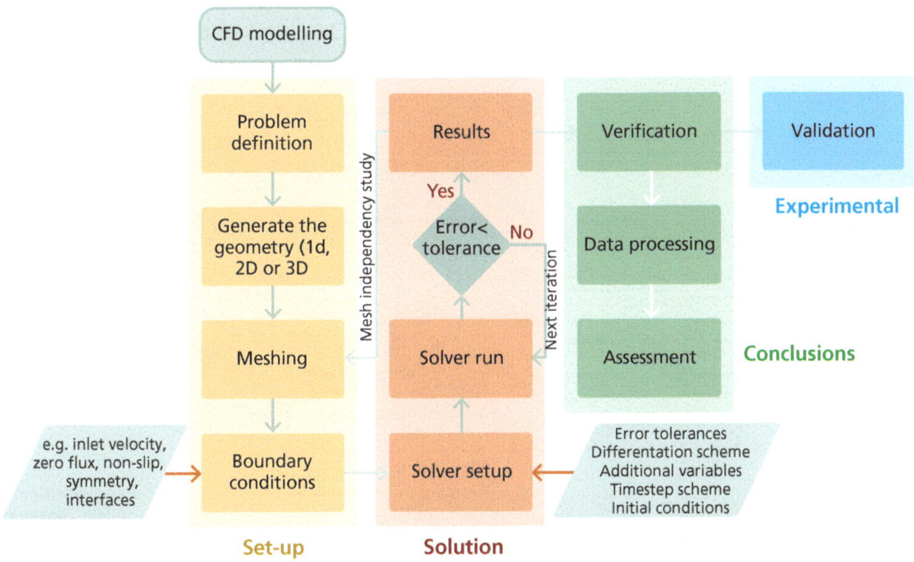

Figure 1 Overall process for CFD modelling of membrane systems.

19.2.1 Geometry

One of the first steps for a CFD simulation of an MSP consists in defining the domain. The geometry of the domain is important as it should represent the system to be modelled. Spirally wound membrane (SWM) modules are the most industrially spread, therefore, many of the CFD models are based in this type of modules. In SWM modules the membrane layers are separated using a spacer mesh with a small width (typically ~1 mm) to create a path for the flow (Scott, 1995). Thus, most CFD analysis of SWM modules consist of domain modelling a spacer-filled narrow channel in order to approximate this geometry. The length of commercial SWM modules is around 1 m which can be difficult to model due to large computational requirements. Nevertheless, the geometry of a SWM module consists of multiple repetitions of small units with the same geometry, called unitary cells. Thus, a common approach in CFD studies is to reduce the length of the channel to be modelled by considering a small number of unitary cells.

Membrane modules can be divided into four categories, namely flat sheet (FS) membrane, spiral-wound membrane (SWM), hollow fibre (HF) and tubular membrane as shown in Figure 2. Among these modules, HF (6000 to 8000 m^2/m^3) is the most compacted (highest surface area per volume), followed by SWM (600 to 800 m^2/m^3), FS (50 to 100 m^2/m^3) and tubular membrane (50 to 70 m^2/m^3) (Martín, 2016). Nevertheless, HF is susceptible to fouling and clogging, thus it can only be used to treat low viscous water (Berk, 2009).

On the other hand, tubular membrane modules have better antifouling properties compared to HF and SWM modules because of their relatively larger diameter (10 to 25 mm), so it can maintain high tangential velocity in the feed stream (Berk, 2009). Thus, it is widely used to treat wastewater with a high content of suspended solids, or viscous oil and water mixtures (Xue *et al.*, 2021). In terms of robustness, the SWM is stronger against the membrane breakage compared to HF (Lu & Chung, 2019). Moreover, FS membranes offer simplicity in design, but they cannot withstand higher pressures, thus they are only suitable for MF and UF (Berk, 2009).

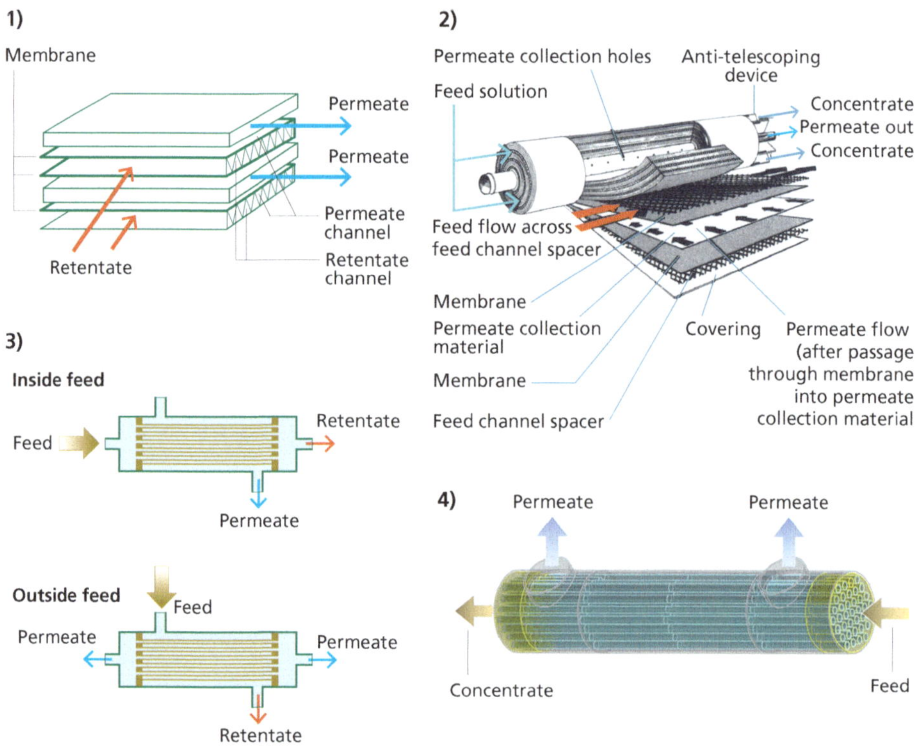

Figure 2 Schematic diagram of (1) flat sheet membrane module (Berk, 2009), (2) spiral-wound membrane module (Sparks & Chase, 2016), (3) hollow fibre module (Balster, 2016) and (4) tubular membrane module (Xue *et al.*, 2021).

19.2.2 Flow types

19.2.2.1 1D, 2D and 3D

As fluid flow is a 3D phenomenon, the most representative way to model it is using a 3D model. However, this may be computationally intensive. In some cases, it is useful to reduce the dimension of the model sacrificing information in order to reduce the computational requirements of a simulation. CFD allows to analyse a system as a 1D, 2D or 3D model and the choice between them depends on the applications intended for a study. For example, a 1D analysis may be useful to analyse a laminar flow between two steady plates, but it overlooks the variations of the flow properties along the other two dimensions. Furthermore, 2D models can be used for flow pattern identification disregarding other phenomena like vortex stretching. In the early stages of CFD, 2D models were the most widely used since the requirements to simulate in 3D were almost inaccessible. Nevertheless, as the capability of computers has improved over the last decades, 3D studies are gaining ground against their 2D analogues (Fimbres-Weihs & Wiley, 2010).

19.2.2.2 Laminar, transient, turbulent

A fluid flow can be classified in different ways according to its characteristics, such as laminar, transient or turbulent. In laminar flow, fluid particles move along a smooth path in parallel trajectories or layers with very low energy losses. On the other hand, turbulent flow is characterised by irregular paths for the flow particles and by large energy losses (Streeter *et al.*, 1985). A fluid flow must have two features in order to be considered turbulent, randomness and auto-similarity (Landahl *et al.*, 1989). Transient flow is an intermediate between laminar and turbulent flow, therefore, it shares some characteristics of these types of flow. Irregular flow trajectory and intermediate energy loss are characteristics of a transient flow.

The hydraulic Reynolds number (Re_h) is almost ubiquitous when characterising a fluid flow. It is the ratio of inertial forces and viscous forces in a fluid flow. The Re_h is described by:

$$Re_h = \frac{u_{eff} d_h \rho}{\mu} \qquad \text{Eq. 4}$$

where u_{eff} is the effective velocity of the flow, d_h is the hydraulic diameter, ρ is the density of the fluid and μ is its dynamic viscosity. A flow with $Re_h > 2100$ is typically considered as turbulent under specific conditions (Rajaratnam, 1976). The hydraulic diameter of the channel is described by:

$$d_h = \frac{4V_{ch}}{a_{ws}} \qquad \text{Eq. 5}$$

The volume of the channel (V_{ch}) and the area of the wetted surface (a_{ws}) depend on the geometry of the channel. Typical values for the d_h are in the order of 1×10^{-3} m.

Operating an MSP under turbulent flow conditions may enhance mass transfer, but implies excessive pressure drop caused by energy dissipation (Burn & Gray, 2015). Conversely, laminar flow may improve mass transfer at relatively low pressure drop. For this reason, most SWM modules operate at a Re_h between 300 and 400 (for reverse osmosis) (Liang *et al.*, 2020b).

19.2.2.3 Single phase

CFD has been used extensively to demonstrate the CP phenomena in a membrane channel for single phase applications. Common membrane applications that involve only liquid phase include desalination and oily-waste water treatment. Since the 2000s, CFD has been used to simulate the hydrodynamic and concentration profiles in the feed channel of SWMs (Fimbres Weihs & Wiley, 2007; Li *et al.*, 2016; Liang *et al.*, 2019) and HFs (Cancilla *et al.*, 2021; Junker *et al.*, 2021; Kaya *et al.*, 2014) for desalination processes. The concentration profile can be visualised in the channel by fixing a solute concentration on an impermeable wall or incorporating Darcy's law on a permeable wall (Pak *et al.*, 2008; Wardeh & Morvan, 2008). As the variations of density and viscosity are insignificant in a horizontal RO membrane system, a Newtonian fluid with constant properties is normally assumed in a SWM feed channel for the purposes of CFD simulation (Foo *et al.*, 2021).

The simulation of oil-in-water MF through CFD approaches has also been presented in the literature (Behroozi *et al.*, 2019; Lotfiyan *et al.*, 2014; Zare *et al.*, 2013). The Eulerian model (described in the following Section) is commonly used to solve the governing equations for the oil and water portions separately. Zare *et al.*, (2013) and Lotfiyan *et al.*, (2014) conducted 2D CFD studies of oil-in-water MF and the simulation could predict the CP profile accurately. However, the permeate flux was underpredicted due to the simplification of the model (e.g., neglecting pore blockage, size distribution of oil droplets, and interaction between oil droplets and the membrane surface). Later, Behroozi *et al.*, (2019) coupled the pore blocking phenomena into a 2D CFD model. The permeate flux predicted by the CFD simulation was comparable to the experiment with 4.62% error. In addition, it was found that the pore-blocking model reduced the relative error by 15% compared to the non-pore blocking model. Thus, considering the pore-blocking phenomena in the simulation of oil-in-water MF can improve the precision of the result.

19.2.2.4 Multiphase

Air sparging in liquid solutions has been extensively used to induce shear stress and mitigate fouling on the membrane surface (Martinelli *et al.*, 2010). This two-phase flow can be modelled using CFD with fluid models such as the Eulerian two-fluid model, or the volume-of-fluid (VOF) model incorporating into the governing equations. In the case of Eulerian two-fluid method, the gas and liquid phases are treated with their respective velocity fields, but both share a common pressure field. Hence, the governing equations are solved separately. VOF model on the other hand, is capable of tracking the gas/liquid interface throughout the computational domain by assuming a no-slip condition between the phases, and that all fluid properties can be calculated as weight-averaged volume fractions (Ndinisa *et al.*, 2005). The governing equations for Eulerian two-fluid (Asefi *et al.*, 2019) and VOF (Javid *et al.*, 2017) methods have been reported elsewhere.

The two-phase flow regime can be either slug or bubbly (Figure 3), which is determined by the superficial velocities of gas phase and liquid phase according to the channel geometry (Golrokh Sani *et al.*, 2021; Gupta *et al.*, 2009). Since the 2000s, modelling two-phase flow in membrane channels through CFD has been extensively studied to understand its impacts on shear stress and flux enhancement (Gupta *et al.*, 2009; Ndinisa *et al.*, 2005; Taha & Cui, 2002, 2006a, 2006b). From the simulation, several approaches were found to be effective for minimizing fouling while enhancing flux such as maintaining high gas flow rate and low liquid rate (Ratkovich *et al.*, 2009), maintaining bigger bubble size (Radaei *et al.*, 2018), and controlled pulse injection (Taha *et al.*, 2006; Taha & Cui, 2002).

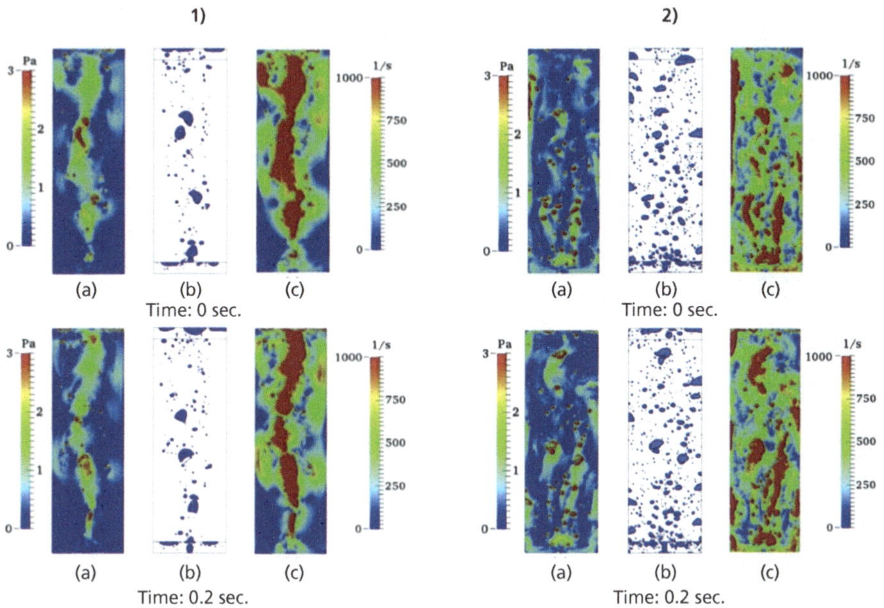

Figure 3 Surface profiles of (a) wall shear, (b) bubbles and (c) vorticity for (1) slug flow and (2) bubbly flow reported by (Javid *et al.*, 2017).

Table 2 Recent 5-years CFD multiphase simulation in membrane process for water treatment

Two phases flow patterns	Software used	Dimension	Modelling approaches	Membrane configuration	Membrane process	Application	Main findings of spacer performances	Ref
Bubbly and slug flows	In-house	2D	Eulerian two fluid	FS	NF	MgSO4 solution	The bubbly flow had marginal impact on the flux enhancement, but a change of the flow regime to slug flow resulted in a flux enhancement.	Asefi *et al.* (2019)
Bubbly and slug flows	OpenFOAM	3D	VOF	FS	UF	whey protein concentrate solution	The velocity gradient of fluid flow near bubbles resulted in the formation of vorticity, which increased shear stress in the membrane system. A significantly larger shear stress was reported in slug flow when compared with the bubbly flow.	Javid *et al.* (2017)
Bubbly and slug flows	ANSYS FLUENT	3D	Eulerian two fluid and VOF	FS	MF	Surface water	Slug flow causes a larger shear stress than the case in bubbly flow. However, the shear induced by bubbly flow was more effective for reducing fouling in pre-treated feed water.	Du *et al.* (2017)
Bubbly flow	ANSYS FLUENT	3D	VOF	HF	MF/MBR	Water and Xanthan gum solution	The fouling rate was found to be inversely proportional to the calculated RMS shear stress. Non-Newtonian fluid showed a reduced tendency in shear fluctuations for membrane cleaning than those observed in water. A larger pulse bubble size yielded more RMS shear stress, which is more effective in fouling control.	Radaei *et al.* (2018)

In recent years, the effects of slug and bubbly flow in a flat sheet channel on shear stress and flux enhancement have been compared through experimental and CFD studies. Air sparging was found to be effectively in reducing the CP by 73% for cross flow NF, leading to flux enhancement (Asefi *et al.*, 2019). However, only slug flow (0.8 min^{-1} to 1 min^{-1}) has significant effects to permeate flux enhancement but not for the bubbly flow (0.2 min^{-1} to 0.6 min^{-1}) (Asefi *et al.*, 2019). Furthermore, an experimental study showed that the slug flow and bubbly flow yield higher flux than single liquid phase flow by 78% and 30% respectively in UF (Javid *et al.*, 2017). This is because the slug flow and bubbly flow result in a higher velocity, which leads to a higher wall shear stress as shown in Figure 3 (Javid *et al.*, 2017). Further, the shear stress caused by the slug flow is higher than those reported in bubbly flow (Javid *et al.*, 2017). In addition, the peak shear stress was found higher for slug flow (15 Pa) than bubbly flow (1.4), although the shear stress is more even distributed for bubbly flow than slug flow (Du *et al.*, 2017). Table 2 summarises recent findings on multiphase for membrane proceses.

19.2.3 Boundary conditions

A set of boundary conditions is required to solve a CFD model due to its numerical nature. Boundary conditions are required at each interface of the model (i.e., inlet, outlet, walls, membrane surface and interfaces). Figure 4 shows a schematic of a membrane channel with the boundary conditions typically used for modelling of MSPs. Other boundary conditions may apply when analysing chemical reaction or heat transfer.

The boundary condition for the incoming flow can be set as inlet or periodic. Setting a flow profile is required for the inlet condition, while a periodic boundary wraps the outcoming flow velocity profile as the inflow. A fully developed profile giving a laminar flow is generally set as the boundary condition at the inlet. The solute concentration profile in the incoming flow can be adjusted according to the application of the model (e.g., constant or periodic). There are three options for boundary conditions at the exit. The outlet condition is the simplest but does not account for backward flow. The opening condition accounts for any recirculation re-entering the channel, but it requires value to be set for any transported variables (e.g., mass fraction and temperature) at the exit. In some cases, the area-averaged concentration at the exit can be calculated and assigned to the backward flow. Setting an opening at the exit usually involves a zero-pressure condition ($p = 0$).

Walls and spacer surfaces are considered impermeable walls with zero-flux, zero-concentration gradient and non-slip velocity conditions. Perhaps the most critical boundary condition is located at the membrane surface, where different approaches may be considered depending on the intended applications.

i. **Impermeable wall with constant concentration.** The concentration at the vicinity of the membrane surface increases due to the CP phenomenon. Under fully developed conditions for the concentration at the membrane surface is roughly twice the concentration at the bulk flow; hence a common assumption includes a constant concentration at the membrane surface. Other assumptions made include a zero flux throughout the membrane (impermeable wall). This assumption is not realistic for MSPs, but it has been shown that the flux across the membrane has very low impact on the characteristics of the boundary layer (Schwinge *et al.*, 2002b). In addition, the flux across the membrane may be estimated using a correlation developed by Geraldes and Alfonso (2006). These considerations are useful to analyse the mass transfer and hydrodynamics inside the channel while reducing computational time.

ii. **Semi-permeable wall.** This is a more realistic approach to membrane systems modelling as it accounts for the separation capacity of the membrane (Ahmad *et al.*, 2005). In this case, the flux throughout the membrane is computed using the physicochemical parameters of the membrane such as permeance and intrinsic solute rejection. This model is suitable for cases where the ratio between permeate flux and bulk flow velocity is higher. Calculating the flux is especially important for techno-economic assessment since it is used to determine the productivity of the system.

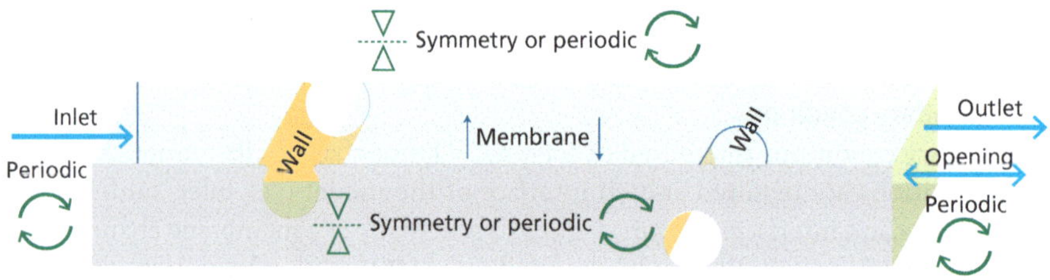

Figure 4 Schematic of a narrow spacer-filled membrane channel and possible boundary conditions for CFD analysis.

19.2.3.1 Steady-state and transient-state

There are two types of CFD simulations that can be used to model membrane systems, steady- and transient-state. The solution obtained in a steady-state simulation describes the behaviour of the system in the long term, when there is no variation over time ($dw/dt = 0$). In general, the variables of the system are recalculated over the domain until some convergence criterion is reached. Steady-state simulations require relatively low computational time; therefore, they are suitable for most of the engineering problems where the goal is to determine the final state of the system.

Transient-state simulations are used to determine the evolution of a system over time. These provide more information on the phenomena taking place in a membrane system and can be used for fundamental analysis (e.g., flow pattern analysis). In transient-state simulations a steady solution is computed at each timestep which involves large computational time.

19.2.4 Initial conditions

The finite volume method requires an initial condition for each element as any numerical method. The selection of the initial condition is normally based upon previous information on the model and using an arbitrary set of initial conditions may lead to excessive computational time. Using a good approximation of the transient-state as an initial condition can significantly speed up the simulations. This can be achieved by setting a previous steady-state solution as the initial condition for the transient case.

19.2.5 Meshing and algorithms

CFD is essentially a numerical technique that commonly uses a finite volume scheme to calculate the properties of the flow within the domain. The finite volume method is based on discretising the fluid domain in a finite number of smaller volumes, which collectively form the mesh. All assigned variables are calculated for these elements and some interpolation scheme is used to estimate their values at mid-points. The mesh can be structured or unstructured depending on requirements of the application. In a structure mesh the elements share the same pattern of construction, whereas in an unstructured mesh has multiple patterns. Unstructured meshes are typically suitable to model MSPs since mass transfer occurs differently at different locations of the domain. For MSPs, more elements are required at regions where the concentration gradient is higher, such as close to the membrane surface. The structured mesh is useful for regions where the flow is tangential to the boundary. Most CFD models combine both types of meshing to increase the resolution of the method at a low computational time. Figure 5 shows an example of the mesh used for CFD modelling including an unstructured pattern for the bulk flow region and a structured pattern for the region corresponding to the boundary layer (zoomed-in). A coarse mesh is shown as an example for better visualisation; however, it is recommended that meshes where the largest elements are about 5% of the channel height are used for CFD studies of membrane channels under typical operation conditions.

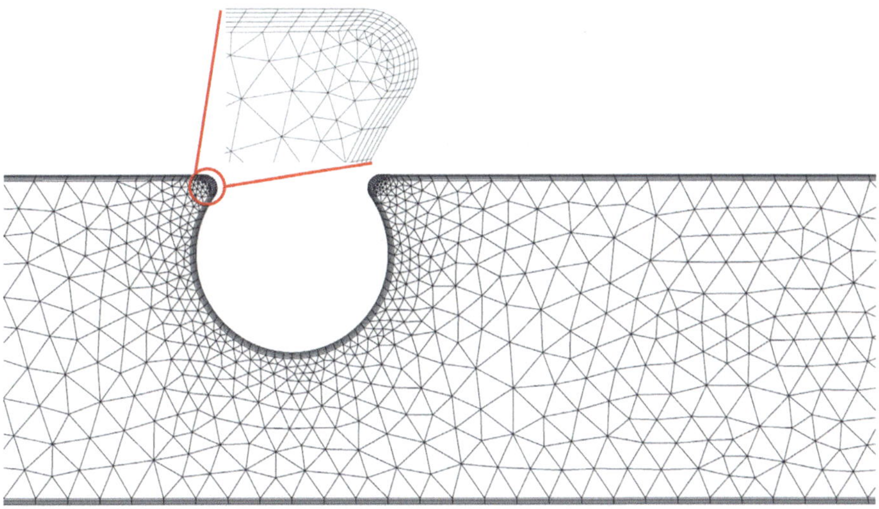

Figure 5 Typical meshing approach for mass transfer analysis on a narrow membrane channel using CFD.

Another meshing approach is using an adaptive mesh, which changes with every iteration on an established criterion. The solute concentration gradient may be used as a refinement criterion as it is advisable to have more elements in the regions with higher mass transfer. When using adaptive meshing the software will automatically refine those regions where the criterion is met at a certain number of iterations. Using adaptive refinement increases the computational time required to solve each iteration, however, it can reduce significantly the number of iterations required by the model. Both steady-state and transient-state cases can be solved using adaptive meshing on ANSYS FLUENT and ANSYS CFX.

19.2.6 Convergence

The results obtained via CFD simulation change with every iteration or timestep and a criterion is required to determine when to stop the simulation. The convergence criterion for steady-state simulations is the residual which measures the imbalance of a conserved variable for each control volume (ANSYS Inc., 2021). Given the numerical nature of the method, the residuals cannot be zero, but an acceptable tolerance is set arbitrarily depending on the precision required for the analysis. The residual tolerance can be set as maximum or root mean square (RMS) with the former being stricter. A solution is considered converged when the residual is equal or less than the tolerance setpoint. The residuals of the velocity and pressure fields are computed by default in most of the commercial software, however, the user can add additional criteria for other variables. It is important to note that using stringent criteria for the residual, can significantly increase the required computational time.

The velocity and pressure fields are typically sufficient to determine convergence in terms of hydrodynamics, but not for mass transfer. Additional variables such as solute concentration or mass transfer coefficient are suitable as a convergence criterion when analysing mass transfer phenomena.

Transient-state simulations require the solution to be converged in time, therefore, the relative error is computed between two different time-dependant states of the system, often referred as ‘snapshots’. The error between two snapshots depends on the timestep size; for example, using a very small timesteps will misleadingly reduce the relative error significantly. The ratio between the relative error and the timestep size, which is analogous to the time derivative, may be used to account for the effect of the timestep size. Another approach is to compute a time-averaged variable within a time window and use it as convergence criterion. The later approach is useful for oscillatory states where the variables are not converging to a single value over time, but rather they oscillate within a limit cycle.

19.3 DATA ANALYSIS

19.3.1 Verification

Another essential step of CFD modelling is the verification of the results. At this stage we already have precise results and now the main goal is to determine whether they are accurate or not. Verification is done by comparing the results obtained against some previously reported for a same (or similar) model or method. In most of the cases there is no true value to aim for, unless there are explicit analytical solutions available for the variables to be compared. Instead, there is a range within the results are considered acceptable. The results used for verification are commonly the area-averaged Sherwood number ($\overline{\mathrm{Sh}}$) and the global friction factor (f_{glob}), which are indicators of mass transfer and energy loss, respectively. The $\overline{\mathrm{Sh}}$ and the f_{glob} are described by the following equations:

$$\overline{Sh} = \frac{d_h}{w_w - \frac{1}{2}\left(w_{b,out} - w_{b,in}\right)} \left(\frac{\partial w}{\partial y}\right)_{w,AA} \qquad \text{Eq. 6}$$

$$f_{glob} = \frac{d_h}{2\rho u_{ave}^2} \frac{\Delta P}{L} \qquad \text{Eq. 7}$$

In Equation 6, w_w is the solute concentration at the membrane surface and, $w_{b,out}$ and $w_{b,in}$ are the solute concentration at the bulk at the outlet and the inlet, respectively. The partial derivative term is the area-average of the y-component of the solute concentration gradient vector at the membrane surface (m^{-1}). Moreover, in Equation 7, ρ is the density of the fluid ($kg\ m^{-3}$), u_{ave} is the average velocity of the bulk flow ($m\ s^{-1}$), ΔP is the pressure drop across the channel ($kg\ m^{-1}\ s^{-2}$) and L is the channel length (m). Despite being good indicators of the channel performance, using $\overline{\mathrm{Sh}}$ and f_{glob} for verification, overlooks the local aspects of the flow.

Another type of verification that is typically carried out in CFD modelling is mesh independency analysis. Mesh independency studies are used to quantify the error introduced by the meshing strategy by computing the flow variables using different meshes. The error for a variable is computed for meshes with different number of elements until certain tolerance is reached. The variables selected should be representative for the whole domain and for the phenomena that is being studied (e.g., hydrodynamics, mass transfer or heat transfer). In addition, the effect of the number of elements should be taken into account as, for example, the error between two meshes may be small because they have relatively the same number of elements. First developed by Roache (1997), the grid convergence index (GCI) is one of the first introduced criterion for mesh independency studies as it accounts for the entire domain and for the number of elements of the mesh. The GCI for a fine mesh (GCI_{fine}) is described by:

$$GCI_{fine} = \frac{3|e|R^{\eta}}{R^{\eta} - 1} \qquad \text{Eq. 8}$$

where η is the order of accuracy (or the dimension number), e is the relative error for an integral function between two meshes and R is the growth factor for the number of elements. Values for the GCI_{fine} less than 5% are typically considered acceptable but, of course, this depends on the application of the model. Although this criterion is very suitable for structured meshes, it presents some issues when used for unstructured meshes, as increasing the number of elements does not necessarily decrease the error. Refining in regions where the gradients (either velocity or concentration) are very small may cause an 'artificial' decrease in the GCI, while increasing the number of elements in a relatively small amount can lead to a higher GCI.

The GCI as a verification metric has more recently been replaced by alternative criteria for mesh independency analysis, as it has been shown that unstructured meshes are more suitable for modelling membrane systems. Recent CFD studies focus on the behaviour of integral functions (e.g., $\overline{Sh}$ and f_{glob}) in an effort to circumvent the effect of the number of elements of the mesh on the GCI. A typical approach consists in computing these integral functions for multiple discretisation meshes with an increasing number of elements, while also computing the relative error between each other until an asymptotical behaviour of the integral function is observed. An arbitrary threshold for the relative error is set depending on the level of accuracy required for the study, with 5% being acceptable for most applications. It is important to highlight that this verification approach requires using the same meshing strategy for all the meshes compared.

19.3.2 Validation

Validation assessment determines if the computational results agree with physical reality (NASA, 2022). The results obtained throughout a CFD simulation are compared against experimental data in order to validate them. In another words, validation is used to determine if the computational model is actually representative of the physical system that was intended to analyse. Some physical systems may be very difficult to setup or to monitor, this is why in some cases CFD analysis comes before and its validation can become an issue. In these cases, the validation may be performed by comparing against experimental results from a different model with similar characteristics.

Existing techniques for experimental analysis of flows include particle image velocimetry (PIV), micro-particle image velocimetry (μPIV) and holographic particle image velocimetry (HPIV). These techniques use optical devices such as lasers, cameras and lenses to capture the flow pattern structures in the fluid. PIV is based on the light scattering capacity of small particles that often need to be added to the fluid. A high-speed camera captures snapshots of the tracer particles distribution to visualise the flow pattern (Lindken & Burgmann, 2012). μPIV refers to the PIV variation where the fluid motion can be determined by the resolved length scales, ranging from 10^{-4} m to 10^{-7} m. A specialised high-resolution camera is required to achieve microscopic length scales for μPIV. The application of μPIV include visualisation of fluid patterns in microchannels, which makes this technique suitable for flow analysis in membrane narrow channels. Figure 6a shows the typical components of the experimental set-up for PIV and μPIV. A flow field captured using PIV is shown in Figure 6b.

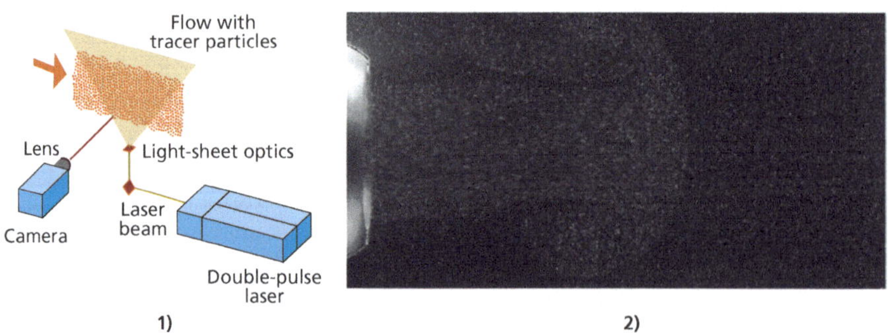

Figure 6 (1) Experimental set-up for PIV/μPIV (Lindken & Burgmann, 2012) and (2) jet flow image captured using PIV (Yu *et al.*, 2021).

HPIV stores the amplitude and phase of a light wave in a two-dimensional film, the hologram. Based on the wave interference phenomenon, the interference pattern of this scattered wave is compared against a second wave refer to as the reference wave (Hinsch, 2002). The interference pattern is then used to reconstruct the flow field by illuminating the hologram with a replica of the reference wave. Figure 7a depicts a basic set-up for HPIV with its components. In general, the laser beam is split into an object wave and a reference wave using an arrange of optical devices. The object wave is directed towards the flow and the reference wave to the holographic film (or device, for digital HPIV). Figure 7b shows an example of a HPIV hologram, whereas the corresponding particle distribution is shown in Figure 7c (Sun *et al.*, 2020).

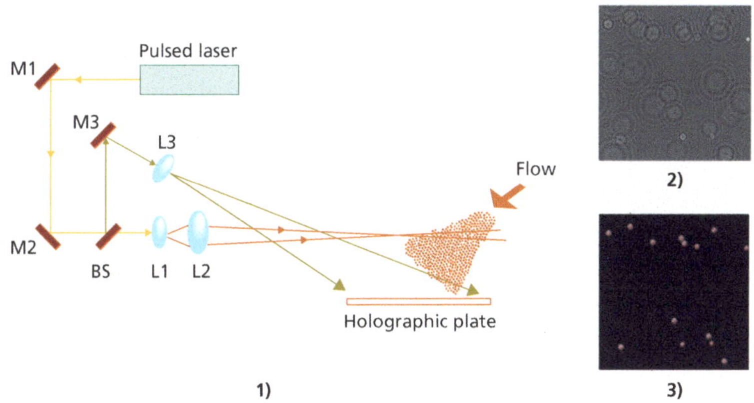

Figure 7 (1) Schematic of a HPIV set-up (Bryanston-Cross *et al.*, 1992), (2) hologram of an interference pattern and (3) corresponding particle distribution (Sun *et al.*, 2020).

Despite the excellent performance of the aforementioned experimental methods for flow characterisation, the applications for membrane systems are limited. For example, the small solutes (salts or organic molecules) common in membrane systems are difficult to emulate for PIV since, the latter requires particles to have certain characteristics to be able to scatter light.

19.4 DATA DISCUSSION AND INTERPRETATION

19.4.1.1 Data processing and assessment

Simulations may yield large amounts of data and their interpretation is often the cornerstone of CFD analysis. Numerical data has very little to no value if it is not processed and interpreted correctly to draw information about a physical system, as that is the purpose of CFD modelling. When presenting results, data is most often used to verify or validate the model, to explain physical phenomena taking place in the system or to propose improvements. CFD may also be coupled with different mathematic tools such as, calculus, Fourier analysis, statistics and linear algebra to produce valuable data for membrane system modelling. In addition, data science (Holemans *et al.*, 2022) and machine learning (Vinuesa & Brunton, 2022) are increasingly used to make CFD a more powerful tool for modelling.

19.5 APPLICATIONS, EXAMPLES

19.5.1 Flow stability

19.5.1.1 Laminar steady

The first uses of CFD for membrane modelling where focused on optimising the design of mesh spacers (Da Costa *et al.*, 1994). These studies used laminar flow in spacer-filled membrane channels to gather information about the CP phenomena. Cao *et al.* (2001) found that recirculation zones are formed before and after the spacers in a narrow membrane for laminar flow. The presence of recirculation zones significantly increases the mass transfer enhancement as they introduce low concentration flow to the boundary layer. Nevertheless, the effect of recirculation zones on mass transfer is limited as they remain attached to the boundary and eventually stabilise. Ahmad *et al.* (2005) went a step further by studying the mechanism by which the recirculation zones increase mass transfer, finding that increasing the shear rate at the membrane surface is critical to mitigate CP. As the role of shear rate (velocity in the normal direction) on the boundary layer development became more evident, the efforts were diverted towards finding ways to destabilise the boundary layer by increasing shear rate.

19.5.1.2 Laminar unsteady – oscillating vs vortex shedding

The fluid flow pattern may change over time under specific conditions without being considered turbulent. This flow regime is called laminar unsteady and it is characterised by unsteady recirculation zones. Recirculation zones may contract and expand or move while remaining attached to the wall affecting the boundary layer. Figure 8a shows a flow pattern for a recirculation zone. Destabilising recirculation zones may cause them to detach and get convected by the bulk flow in a phenomenon called vortex shedding. The efficiency of vortex shedding to mitigate CP is higher than that of recirculation zones as these only affect a small portion of the channel. Vortex shedding is characterised by two separate regions with swirl motion with one of them detaching from the boundary and moving downstream. The double swirl pattern for vortex shedding is shown in Figure 8b.

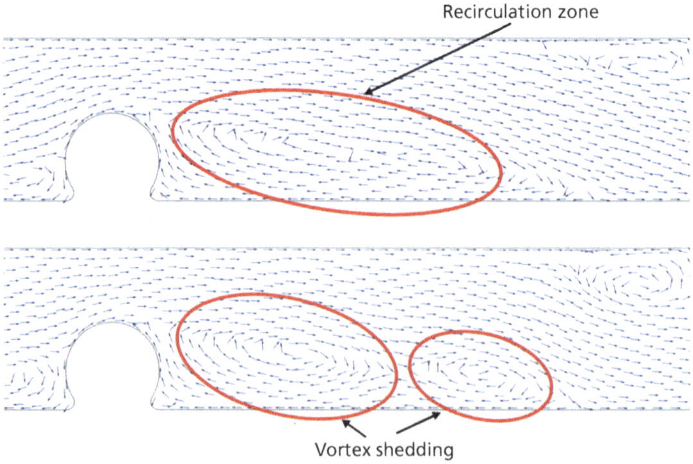

Figure 8 Fluid flow patterns for (top) recirculation zones and (bottom) vortex shedding in a spacer-filled membrane channel (Fimbres-Weihs *et al.*, 2006).

Vortex shedding occurs in flows at higher Reynolds number translating into more energy requirements. Because of this, some researchers have focused on finding the optimal Reynolds number that can cause vortex shedding, but at a lower energy requirement. Alexiadis *et al.* (2007) found that the critical Reynolds number (Re_{cr}) for the transition from recirculation zones to vortex shedding in spacer-filled membrane channels is within the range of 526-841.

Fimbres-Weihs *et al.* (2006) studied mass transfer under unsteady-flow conditions at hydraulic Reynolds numbers up to 1683, paying special attention to the causes of mass transfer enhancement. Wall shear was identified as having a significant impact on mass transfer enhancement; however, the inflow of lower concentration towards the membrane surface was found to dominate mass transfer phenomena. Thus, recent CFD studies on mass transfer enhancement are focused on maximising the inflow towards the membrane surface (i.e., normal velocity).

19.5.1.3 Quasiperiodic flow

The study of fluid flow instabilities is crucial as the effect of disturbance on the hydrodynamic stability and how the flow system responds to it remain unclear. Fluid periodicity is one of the indicators to measure the flow stability. Fluid periodicity in flow refers to a flow system that displays a recurring behaviour given at regular intervals. Quasiperiodic flow, on the other hand, refers to a recurrence behaviour with a component of uncertainty, such that it could be periodic on a small scale but unpredictable on a large scale, which would eventually result in imprecise measurements.

From fluid mechanic point of view, the flow can be classified into three states namely the laminar, transition and turbulent flow. The fluid flow states are usually determined by a dimensionless number known as Reynolds number. Reynolds number (*Re*) is defined as the ratio of inertial forces to viscous forces in a fluid flow. As *Re* increases, the inertial forces become larger, leading to flow instability (Masuda & Tagawa, 2019). Transition flow is a mixture of laminar and turbulent flow occurring simultaneously in a fluid channel with a range of *Re* between 2,000–4,000. During the transition state, aperiodic oscillatory flow or quasiperiodic flow usually occurs due to the unstable flow condition with mixed characteristics of laminar and turbulent flow characteristics. Nowadays, Computational Fluid Dynamics (CFD) has been widely used to study the hydrodynamics behaviour of quasiperiodic flow. One of the advantages of flow simulation by CFD is that it can help to generate different physical data such as vorticity or energy dissipation rate which cannot be measured easily in practical experiments.

Schwinge *et al.*, (2002a) investigated the effect of unsteady 2D flow in narrow spacer-filled channels for spiral wound membrane modules. Their simulation results show that when the membrane channel is filled with obstacles (spacer), an unsteady flow pattern is observed at *Re* as low as 200, depending on the geometry of the obstacles. Figure 9 shows the transition of flow from stable to unsteady conditions occurring at *Re* above 300 for a single filament located at the centre of a narrow channel. At this Reynolds number, the recirculation begins to form behind the filament and as *Re* increases, the extent of flow unsteadiness increases until it reaches a turbulent condition. For a single filament located close to the bottom wall, the flow is found to be stable for *Re* below 600 and it becomes unstable when Re increases above 600 as the flow disturbance induced by the recirculation behind the filament spreads the unsteadiness downstream the membrane channel (as shown in Figure 10). Nevertheless, their transient results also found that the channel walls close to the cylindrical filament tend to stabilise the fluid flow and slow down the transition of flow to the unsteady state.

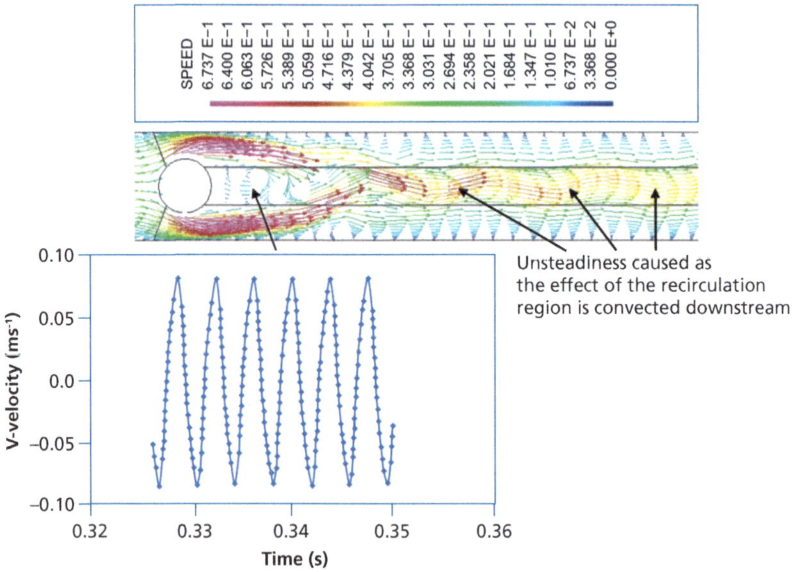

Figure 9 Unsteady flow caused by a cylinder located in the centre of the narrow channel for Re = 500 (Schwinge *et al.*, 2002a)

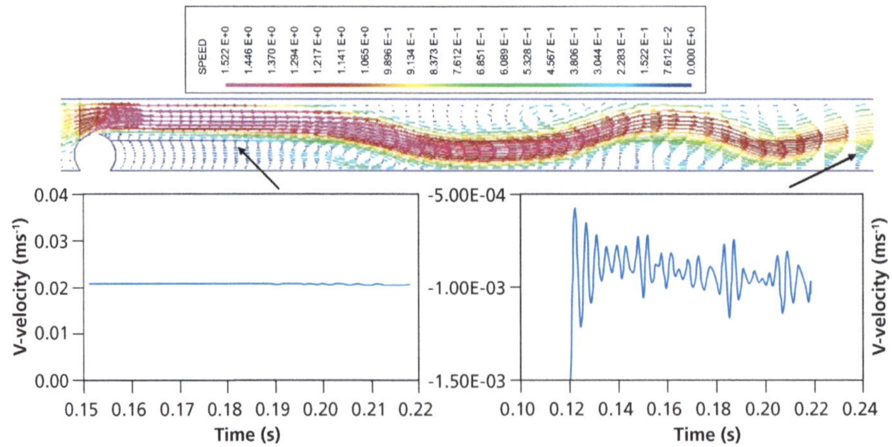

Figure 10 Unsteady flow caused by the cylinder close to the bottom wall for Re = 1,000 and small flow disturbances are convected downstream (Schwinge *et al.*, 2002a)

19.5.1.4 Turbulent flow

Turbulent flow generally occurs at a Reynolds number greater than 2,000. As Reynolds number increases, the fluid flow undergoes irregular fluctuations or mixing that eventually leads to an aperiodic oscillation. Not only that, the direction and magnitude of fluid flow are constantly changing at a turbulent condition. The characteristics of a turbulent flow include higher velocities, low viscosity and higher characteristic linear dimension compared to a laminar flow. Due to the random nature and irregularity of turbulent flow, the flow pattern is extremely difficult to understand. Hence, the governing equations for turbulent flow condition are not easy to develop due to the unsteady flow continuously changing with time, which increases the difficulty level for studying flow turbulence. Although the analysis of turbulent flow is very challenging, analysis of flow turbulence is important for industries as currently there are more membrane filtration processes are using turbulent flow to enhance mixing and reduce the extent of concentration polarisation in the membrane module.

In general, there are three main ways to simulate turbulent flow using CFD technique, including the Direct Numerical Simulation (DNS), Large Eddy Simulation (LES), and Reynolds-Averaged Navier-Stokes (RANS) equations. The computational cost associated with solving the flow simulation increases ascendingly with the order of RANS, LES, followed by the DNS technique in CFD modelling. For DNS and LES methods, CFD will take into account the majority of fluid scales in simulation and provide a comprehensive flow data. In contrast to DNS and LES, RANS do not require fine details of all the turbulent eddies, hence lower computational cost. However, the accuracy of flow simulation can be impacted for RANS method because the flow is only simply modelled and not fully (DNS) or partially (LES) resolved.

Jafarkhani *et al.* (2012) developed a 3D model with semi-circular baffles incorporated into the membrane tube to study the hydrodynamic behaviour of turbulent flow. Their results found that the intense fluctuations induced by the baffles increase the local wall shear stress and velocity in the membrane channel. The induced turbulence to the bulk flow at *Re* up to 7,500 results in a rapid change in the flow directions, which enhances the flow fluctuations and reduces the formation of concentration boundary layer on the membrane surface, consequently leading to a potential fouling reduction while improving the filtration flux performance.

19.5.2 Mass transfer and vortex shedding

Vortex shedding occurs as an oscillating flow at specific velocities when a fluid flow passed a bluff body depending on the size and shape of the body. The alternate formation and shedding of vortices produce alternating forces, which happen more frequently as flow velocity increases. This phenomenon plays a vital role in membrane technology due to the significant impact of vortex shedding on membrane efficiency, as well as the mass and heat transfer of a fluid flow passed a bluff body (Korinek *et al.*, 2017). Research has shown that vortex shedding is capable of enhancing membrane separation efficiency by increasing the transmembrane flow, while disrupting the formation of thermal and solute boundary layer on the membrane surfaces.

Su *et al.*, (2018) investigated the performance of a vibration-enhanced reverse osmosis membrane module on the membrane separation efficiency. Their CFD results showed that the vibration force results in more vortices along time at the downstream and upstream faces of the membrane, compared to that of non-vibration case (Figure 11). The vortex generation induced by the vibration increases the boundary shear rate in the membrane channel, which significantly reduces CP.

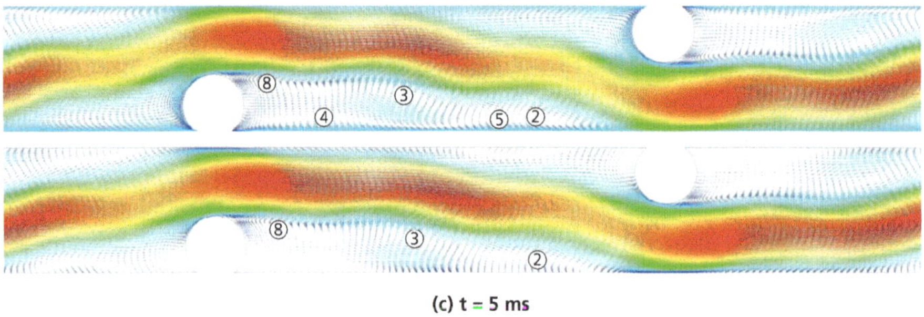

Figure 11 Velocity vector profile of vibration case (top) and non-vibration case (bottom) (Su *et al.*, 2018).

19.5.3 Spacer design

19.5.3.1 Two-Dimensional Feed spacer

The accumulation of solute due to membrane rejection leads to concentration polarisation (CP), which later causes the deterioration of membrane flux. Thus, intensive studies such as optimisation of feed spacer geometry in the membrane channel have been performed to mitigate CP by enhancing the mixing of fluid. The first two-dimensional (2D) models of spacer-filled membrane channels appeared in the early 2000s, aiming to understand its hydrodynamic (Chong *et al.*, 2022a) and concentration profiles. Later, the configurations of the spacer (e.g., cavity, submerged and zigzag) and other parameters (filament diameter, mesh length, etc.) have been studied thoroughly to understand their impacts to the flux enhancement and pressure loss.

Early 2D feed spacer study found that zig-zag spacer was the most efficient spacer for SWM module (Schwinge *et al.*, 2002b). Subsequent findings include 1) dependence of formation of recirculation region on spacer geometry and flow condition (Schwinge *et al.*, 2002a); 2) significance of vortex shedding for mass transfer enhancement. It was also concluded that the mass transfer enhancement depends on two important mechanism: 1) flow of low solute concentration to the membrane boundary layer and 2) increase of wall shear (Fimbres-Weihs *et al.*, 2006).

19.5.3.2 Three-Dimensional Feed Spacer

3D modelling of spacer-filled channel emerged rapidly around 2010 due to improvement of computer processing and storage of capacity. During the first decade of 21st century, the research direction regarding CFD analysis in spacer studies mainly focused on the hydrodynamic performance caused by the spacer. Later, the effect of various geometries such as the filament size, mesh length and flow attack angle on the mass transfer performance and pressure loss reduction were studied and investigated (Gu *et al.*, 2017). In the past decade, the research direction concerning CFD analysis of spacer-filled channels has shifted and focused more on the novel spacer geometries for further improvement on the mass transfer performance and reduction in the pressure loss (Park *et al.*, 2021).

Chong *et al.* (2022) studied the effects of submerged spacers with variations in the node geometries and sizes (as shown in Figure 12a) on the hydrodynamics and mass transfer performance through CFD. It is found that conventional spacers have similar or higher Sh than the submerged type spacers at lower Re_h (<100) because of the sideway flow. However, due to a greater vortex mixing effect, submerged type spacers performed better than the conventional spacers at higher Re_h (>200). In addition, the mass transfer performance of the spherical node spacer is inferior to the column node spacer as the flow can pass through the gap between the spherical nodes and membrane, thereby resulting in a lower local velocity at filament and mass transfer. In fact, as node size increases, pressure loss increases as well because of the larger nodes significantly impeding the flow, creating more form drag and skin friction in the membrane channel, thus increasing the pressure loss.

Recently, a honeycomb shape spacer (Figure 12b) was proposed in which it has the ability to generate high-magnitude turbulent kinetic energy in the area of spacer filaments, resulting in a smaller fouling deposit (Park *et al.*, 2021). Furthermore, optical coherence tomography (OCT) scans showed the fouling layer thickness could be decreased by 33% using the honeycomb spacer.

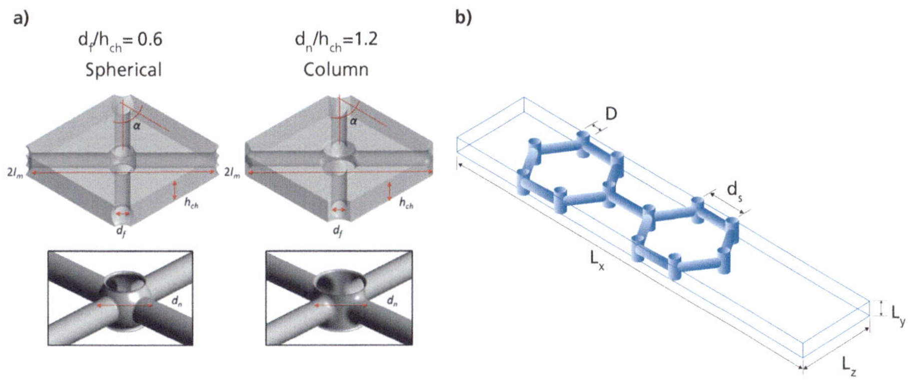

Figure 12 Schematic diagram of (a) spherical nodes submerged and column nodes submerged spacers and (b) honeycomb spacer (Park *et al.*, 2021)

19.5.4 Flow perturbation

One of the efforts to minimise concentration polarisation (and eventually fouling) in a membrane system is by introducing a disturbance into the fluid flow through an external flow perturbation technique. With the help of CFD, the technique is performed in an attempt to induce flow unsteadiness for promoting fluid mixing and therefore, reducing the fouling tendency in membrane modules (Fimbres-Weihs *et al.*, 2006; Schwinge *et al.*,

2002a). The flow perturbation technique generally involves creating unsteadiness to the bulk flow, which can be done by introducing an oscillating flow to the system; or causing a disturbance to the boundary layer adjacent to the membrane surface by using vibration or electro-osmosis approach. Further descriptions regarding the flow perturbation techniques as well as their effectiveness in enhancing membrane performance are elaborated in the following sections.

19.5.4.1 Electro-osmosis

Electro-osmosis is an electrokinetic phenomenon involving the movement of a thin fluid layer adjacent to a charged surface in response to an external electric field (Asadi *et al.*, 2013; Hu & Li, 2007; Jagannadh & Muralidhara, 1996; Ouyang *et al.*, 2013). The electrokinetic phenomenon occurs due to the electrostatic interaction between the similarly charged and oppositely charged ions in the vicinity of a solid/liquid surface, which then results in the motion of a thin fluid layer near the surface, as illustrated in Figure 13. Electro-osmosis has great potential to enhance mass transfer while minimising fouling tendency, particularly for membrane separation processes, such as reverse osmosis and nanofiltration because a disruption to the flow near the membrane surface tends to promote back-transport of solute and reduce polarisation effect (Liang *et al.*, 2014a). Further, the approach is suitable for water treatment and desalination processes as there are salts and charged species involved in the system that can respond to the applied electric field (Jagannadh & Muralidhara, 1996).

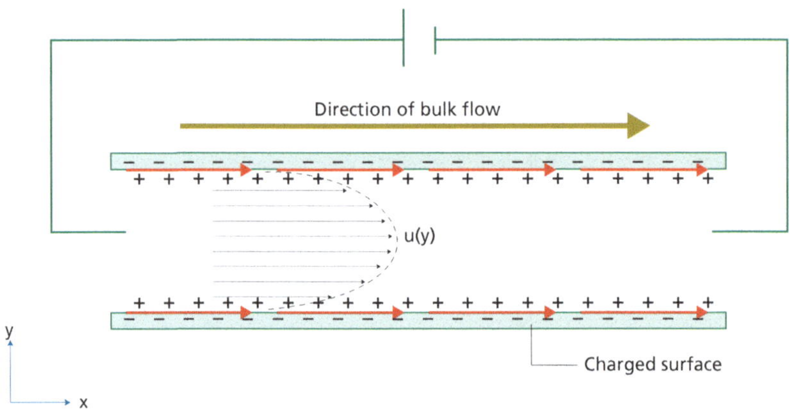

Figure 13 Schematic of electro-osmosis technique with red arrows representing the electro-osmotic flow (EOF) adjacent to a charged surface in an empty flow channel.

Several theories have been developed to describe electro-osmosis, including the Helmholtz-Smoluchowski (HS) theory, the Spiegler friction model, the Schmid theory, and ion hydration theory. Among those early theoretical approaches, an early study of Spiegler and Macleish (1981) investigated the electro-osmosis technique on a reverse osmosis desalination process by filling the feed stream with ferric hydroxide. Their results showed that the electro-osmotic backwashing of the membrane managed to recover a range of 30–100% of flux loss in the system. Despite the great potential of electro-osmosis, the technique is difficult to analyse experimentally in a membrane system due to various technical constraints. Hence, there is significant value in utilising CFD as a tool for understanding the mechanisms of electro-osmotic flow (EOF) in enhancing mass transfer and fluid flow in a membrane process. The following sections further discuss the modelling of electro-osmosis in membrane system studies and the many learnings obtained using CFD.

19.5.4.2 Modelling electro-osmosis in CFD

Electro-osmotic flow refers to the motion of a thin fluid layer that carries a net electric charge acting upon a fluid/solid interface (i.e., membrane surface) in response to an electric field. This layer of net charge is known as the electric double layer (EDL) and is often characterised by the Debye length (λ_D) (Cummings *et al.*, 2000), which is formed due to the charge separation occurred near the fluid/solid interface. The movement of fluid layer (EOF) is then driven by the excess charged ions in the double layer field via viscous effect under the influence of external electric field (Hu & Li, 2007; Probstein, 1989; Russel *et al.*, 1991). For an incompressible fluid flow in a membrane system, the electro-osmotic effects can be introduced into the Navier-Stokes momentum equation by means of an external force given as follows (Hu & Li, 2007):

$$\rho\frac{\partial\vec{v}}{\partial t}+\rho\left(\vec{v}\cdot\nabla\right)\vec{v}=\mu\nabla^2\vec{v}-\nabla p+\rho_e\vec{E}+\rho g \qquad \text{Eq. 9}$$

where ρ, $\vec{v}$, p, μ, ρ_e, $\vec{E}$, and g are the density, velocity vector, pressure, fluid viscosity, electric charge density, electric field, and gravitational acceleration, respectively. The electric field ($\vec{E}$) can be calculated as follows:

$$\vec{E}=-\nabla\left(\phi+\psi\right) \qquad \text{Eq. 10}$$

where ϕ and ψ are the potentials due to the double layer and external electric field, respectively (Patankar & Hu, 1998; Rawool & Mitra, 2006). The charge density (ρ_e) can be related to the electric potential by the Poisson equation (Hunter, 2013):

$$\nabla^2\phi=\frac{-\rho_e}{\varepsilon} \qquad \text{Eq. 11}$$

where ε is the fluid permittivity. Excluding any disturbance to the double layer, the charge density can be solved by the Boltzmann distribution or other similar correlations (Hunter, 2013; Probstein, 1989).

Nevertheless, the high computational requirement and cost is one of the challenges for solving the Poisson and Navier-Stokes equations at the scale of Debye length ranging up to several hundred nanometres in numerical simulation of electro-osmosis considering a typical membrane channel height of ~10^{-3} m. For the sake of simplicity, the thickness of double layer is generally assumed to be neglected in most EOF studies (Hu & Li, 2007). In fact, the coupled relations between the Poisson and Navier-Stokes equations can be simplified by dropping the EOF term $\rho_e\vec{E}$ from equation 9 and instead, replace the no-slip boundary condition at channel surfaces by using the Helmholtz-Smoluchowski (HS) slip boundary (Anderson, 1989; Hu & Li, 2007; Ren *et al.*, 2003; Santiago, 2001; Zhang *et al.*, 2006).

Using the simplified HS approximation, the electro-osmotic flow can be incorporated as an artificial forced slip velocity applied along the surface of the membrane channel in a separation module. The HS slip velocity is applied outside the edge of the double layer assuming the thickness of EDL much smaller compared to the channel height (Probstein, 1989). In addition, the slip velocity equation assumes a 1D charge distribution when the fluid velocity is small and/or the inertial terms in the momentum equation do not dominate (Patankar & Hu, 1998). Assuming time constant with negligible pressure gradient and gravitational acceleration, the coupled system of Poisson and Navier-Stokes equation from equation 9 can be simplified to the following HS equation, which corresponds linearly with the magnitude of electric field, expressed as follows (Probstein, 1989):

$$u_s = -\frac{\varepsilon \varsigma E_x}{\mu} \qquad \text{Eq. 12}$$

where u_s is the forced slip velocity, ζ is the zeta potential, and E_x is the magnitude of electric field in the x-direction. The permittivity (ε) is assumed to be uniform for the case of an RO system, such that the permittivity value of the membrane can be regarded as similar to that of water (Liang *et al.*, 2014a). This is because of the structure of RO membrane which is mostly comprised of the high porosity support layers (Antony *et al.*, 2013), along with an extremely fine selective membrane layer of 1×10^{-7} m or less (Baker, 2004), in order to assume a uniform permittivity value and to safely neglect any effect related with the nonuniformity in this case.

The HS approximation was compared against a more rigorous charge density distribution (CD) solution under the influence of uniform and non-uniform electric fields (Liang *et al.*, 2014b). The case study with uniform electric field showed that the HS approximation agrees well with the CD solution at increasing solute concentration in a 2D unobstructed membrane channel. Their results found that the HS approximation is more accurate at higher solute mass fractions of 0.001 or more, which are the typical salt profiles encountered in RO desalination process. Hence, the HS forced-slip model is suitable for modelling the electro-osmotic effect in an RO membrane system.

19.5.4.3 Significant learnings of EOF slip velocity in CFD studies

The HS slip velocity model has been extensively validated and is suitable to be applied for the typical flow conditions encountered in real RO membrane modules (Liang *et al.*, 2014a; Liang *et al.*, 2014b, 2016b). It is reported that the EOF perturbation induced near the membrane surface results in the shedding of vortices in a 2D spacer-filled membrane channel, which tends to enhance fluid mixing and reduce concentration polarisation (as shown in Figure 14). In addition, the EOF induced vortex shedding also increases wall shear along the membrane that could potentially reduce the fouling tendency in the membrane system (Liang *et al.*, 2016a; Liang *et al.*, 2016b). The effectiveness of EOF slip velocity in enhancing membrane performance and the significant findings in CFD membrane studies are further elaborated in Table 3.

Table 3 Significant findings of application of electro-osmosis in membrane system

Membrane model	Reference	Main findings
2D unobstructed RO rectangular channel	Chan *et al.* (2020b)	A reduced-order model was developed for fast predictions of concentration polarisation in an RO membrane system under permeation condition.
2D unobstructed RO rectangular channel	Chan *et al.* (2020a)	A reduced-order model was proposed to study the effect of permeation in an RO membrane system with EOF slip velocity considered.
2D spacer-filled channel for RO SWM modules	Liang *et al.* (2016b)	First CFD study incorporating steady and unsteady EOF for enhancing mass transfer in 2D spacer-filled channels, using HS forced-slip approximation.
2D unobstructed RO membrane channel	Ratnayake *et al.* (2016)	A spatio-temporal frequency response analysis was performed to investigate the effect of waves of different frequencies for an EOF forced slip velocity and the impact of changes on solute concentration gradients.
2D spacer-filled channel for RO SWM modules	Liang *et al.* (2018a)	CFD study of non-sinusoidal waveforms of EOF slip velocity reported a similar membrane performance as those obtained by sinusoidal slip velocity, in terms of both mass transfer and wall shear.
2D spacer-filled RO membrane channel	Liang *et al.* (2020a)	A comparison study between forced slip velocity and oscillating feed perturbation found that both approaches predict similarly in terms of hydrodynamics and flux performance.
2D spacer-filled RO membrane channel	Foo *et al.* (2020)	CFD study of the effect of varying feed spacer geometries on membrane performance revealed that the resonant slip frequency increases as spacer size is increased due to stronger shear layer interactions. An increased in distance between spacers leads to a greater flux due to forced-slip, albeit the actual flux is smaller.
2D spacer-filled channel for RO SWM modules	Foo *et al.* (2021)	CFD study of varying spacer configurations found that the submerged configuration results in the largest flux increase under forced-slip effect at the expense of a relatively larger pressure drop.

19.5.4.4 Oscillating flow

Inducing oscillations in the bulk flow is an approach to enhance mass transfer in membrane systems by generating instabilities. Li *et al.* (1998) studied the use of oscillating inflow to promote vortex shedding at lower *Re* in microfiltration systems. Their results suggested that using a time-dependant inflow reduces the cake layer resistance and, therefore, increases the flux. CFD modelling was used by Liang *et al.* (2018b) to show that an oscillating inflow may promote vortex shedding increasing the magnitude of the flow velocity towards the membrane surface which was later validated by Liang *et al.* (2020b).

The oscillatory inflow (OI) approach is implemented in CFD by changing the inlet velocity from time-independent to time-dependant, giving place to a waveform. Sine and cosine functions are typically used as waveforms for OI, though, others like square and sawtooth waves are also possible to use. The sine waveform for OI is described by:

$$u_{osc} = u_{ave}\left[1 + A\sin(2\pi ft)\right] \quad \text{Eq. 13}$$

where u_{osc} is the oscillatory inlet velocity (m s^{-1}), u_{ave} is the average inlet velocity, A is the oscillation normalised amplitude, f is the oscillation frequency (s^{-1}) and t is time (s). Figure 15 shows different waveforms previously used for the OI technique (plot as normalised velocity).

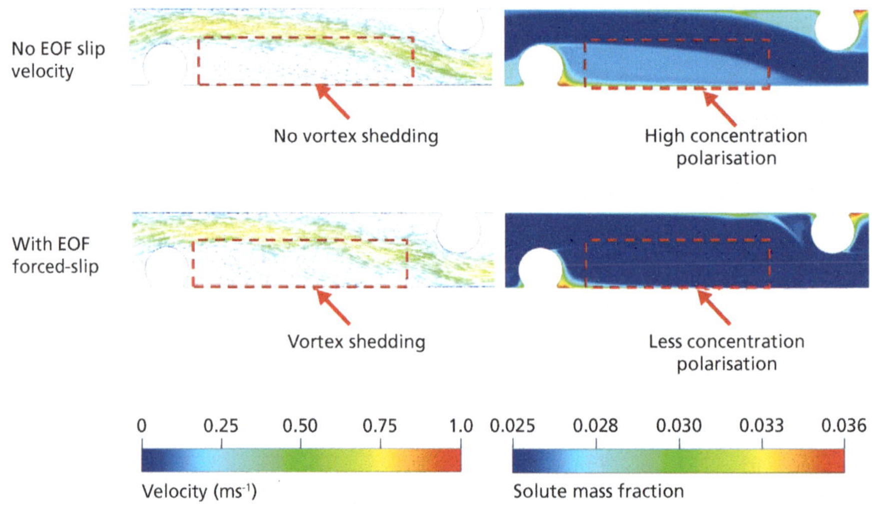

Figure 14 Effect of EOF slip velocity on velocity and solute concentration profiles in a unit cell of spacer-filled membrane channel (Foo *et al.*, 2020).

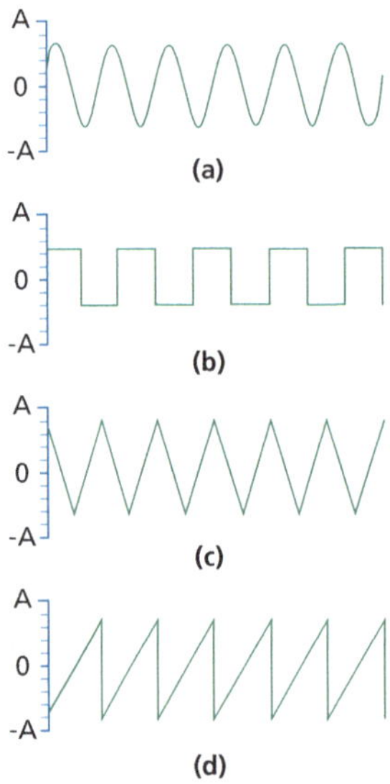

Figure 15 Waveforms for CFD analysis of OI technique: (a) sine, (b) square, (c) triangular and (d) sawtooth wave (Liang *et al.*, 2018b).

The optimal frequency of the oscillation to maximise mass transfer enhancement lies between 10 Hz and 1,000 Hz for typical spacer-filled membrane channels. A frequency response analysis can be used to identify the optimal response frequency for a membrane system. In a frequency response analysis, an input stimulus (commonly a short pulse) is introduced in the system and the response in velocity or concentration is recorded to find the maximum response. Recent approaches have undertaken the challenge of promoting vortex shedding at the minimum energy expenses by optimising the oscillating inflow amplitude and frequency.

19.5.4.5 Vibrations

Another approach to destabilise the boundary layer, instead of causing fluctuations in the flow, is to constantly move the module. Vibration-assisted modules use a mechanical device to induce oscillations on the channel enhancing mass transfer. Su *et al.* (2018) conducted a CFD study on concentration polarisation and permeate flux in a vibration enhanced system obtaining a reduction of up to 10% on the CP modulus when applying vibrations to the module. In addition, both the Sherwood number and permeate flux increased in about 15% and 5%, respectively, compared to the case without vibrations. In the same work, the results for permeate flux were validated experimentally. Studies on the effectiveness of vibration-assisted modules to mitigate CP are relatively recent and they are mainly focused on desalination processes (Li *et al.*, 2017; Su *et al.*, 2019). Thus, multiple aspects may be analysed to improve this technique such as optimal frequency and amplitude, waveform, etc.

19.5.5 Fouling modelling

Fouling phenomenon occurs in membranes due to solute deposited into the membrane pores or onto the surface. The phenomenon usually begins with a condition known as concentration polarisation (CP), which involves the solute accumulation near the membrane surface due to a greater applied pressure, compared to the osmotic pressure difference in the membrane process. There are two major types of fouling namely particulate fouling and biofouling phenomena. Particulate fouling occurs when foreign particles such as proteins, carbohydrates, oil, silt or clay are deposited into the membrane pores, leading to pore-blocking. The biofouling phenomenon, on the other hand, occurs due to the precipitation of microorganisms on the membrane surface, which leads to the formation of biofilm on the solid material. It is known that a prolonged fouling phenomenon would reduce the overall performance of a membrane module as permeate flux and membrane separation efficiency decrease (Fritzmann *et al.*, 2007).

19.5.5.1 Particulate fouling

Particulate fouling generally occurs when a solution containing particles with sizes ranging from few nanometres to micrometres dispersed evenly in the solution, eventually clogging the membrane pores. Particulate fouling condition can be reversed or becomes non-reversible depending on the types of foulant binding to the membrane surface. The phenomenon is said to be reversible if the particulate removal can be achieved via physical cleaning, and it is irreversible when foulant requires a chemical cleaning (Leam *et al.*, 2020). The particulates can be categorized into organic and inorganic substances, such that the organic particles are usually made up of proteins, carbohydrates, oils, etc., while the inorganic particles consist of silt, clay, silica sediments, etc (Qasim *et al.*, 2019). A continuous deposition of particles near the membrane surface over time results in the formation of a cake layer adjacent to the surface, which would impose a cake resistance onto the mass flow and eventually reduce the mass transport across the membrane, leading to a decrease in permeate flux (Jiang *et al.*, 2017).

For the past few decades, CFD has been used to analyse the effects of particulate fouling on membrane performance as the software is capable of providing analysis for numerous flow and heat transfer mechanism, mass transport of soluble substrate, and the hydrodynamics of fluid flowing through the membrane for optimisation of membrane design through model simulations. A study on the particle deposition in a spiral wound membrane module was conducted to identify the optimum spacer type for preventing particulate fouling (Li *et al.*, 2012). Flow that has lower velocity will lead to lower shear stress across membrane surface that promotes particulate fouling. Their study addressed the effects of curvature on the flow pattern for four different types of spacer configurations, namely zigzag, submerged, *i*-cavity, and *o*-cavity, respectively by changing the dimensionless radius of curvature, η given as follows:

$$\eta = \frac{R_o + R_i}{R_{ave}} \qquad \text{Eq. 14}$$

where R_o is the outer radius, R_i is the inner radius, and R_{ave} is their arithmetic mean (Li & Tung, 2008).

Based on the simulation results for empty channels and four spacer configurations of spacer-filled channels, a higher shear stress has the potential for a lower tendency for particle deposition as well as fouling. For submerged spacer-filled channels, it was found that the deposition ratio is greater at the outer membrane compared with that at the surface of the inner membrane. This can be explained by the lower shear stress observed at the surface of the outer membrane compared with that at the inner membrane due to a lower flow velocity near the outer membrane surface, which promotes the particulate fouling at the outer membrane surface. An illustration that depicts the deposition ratio at different positions of membrane in the submerged spacer-filled channels is shown in Figure 16.

In the case of asymmetric spacer-filled channels, it is reported that curve channels result in a higher particulate deposition ratio at the outer membrane surface compared with that of the flat channel because of a lower shear stress. This occurs because the curve channel tends to direct the fluid flow into the inner membrane surface, which restrains the recirculation size and decreases the velocity near the inner surface of the membrane. A summary on the significant findings of particulate fouling in membrane studies by CFD is listed in Table 4.

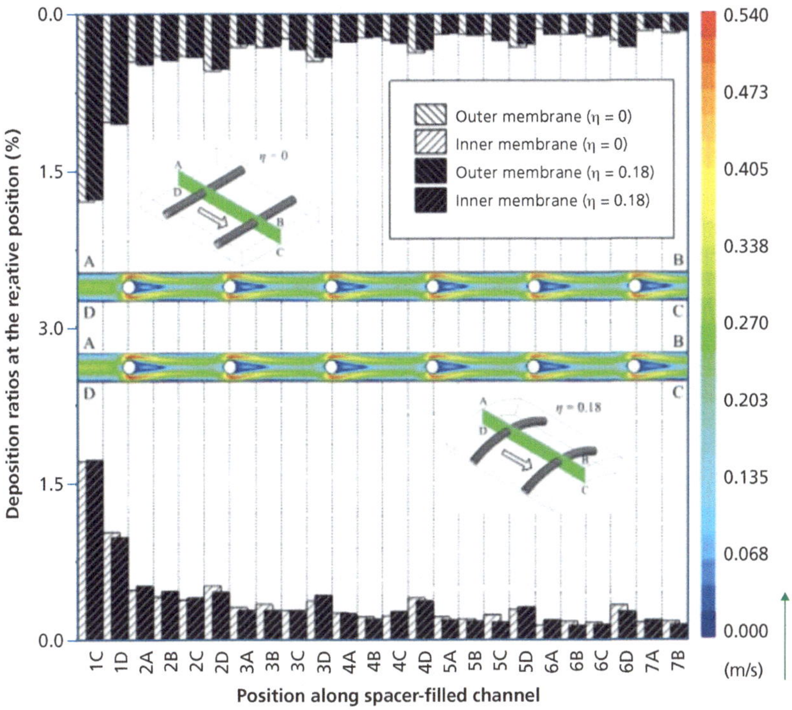

Figure 16 The deposition ratios on the inner and outer membranes in the flat and curved channel filled with submerged spacers at various positions (Li *et al.*, 2012).

Table 4 Significant findings of CFD studies on particulate fouling in membranes

Author	Research types	Main findings	Observations
Lin *et al.* (2022a)	CFD and Response surface methodology (RSM)	Diagonal-flow feed channel results in a higher salt rejection and water flux with an average crossflow velocity in the channel increased by ~50%, compared to that in the conventional feed channel.	The study is limited to feed channels without considering the effects of feed spacer.
Rahimi *et al.* (2009)	CFD and experimental	Fouling on membrane is not uniform and the possibilities of fouling to happen are higher in regions with lower shear stress.	The study is limited to an incompressible flow system.
Li *et al.* (2012)	CFD only	Recirculation occurred adjacent to the membrane and behind the filament causing a higher shear stress, which potentially reduces particulate fouling in the membrane system.	This study is limited to a low volume fraction of discrete phase that ranges less than 10-12%. It is also limited to a two-phase flow.

19.5.5.2 Tracer test

Tracer test or commonly known as the Salt Tracer Response Technique (STRT), is frequently used to evaluate the effects of cake-enhanced osmotic pressure (CEOP) on the development of concentration polarisation in a membrane system. It is reported that a larger extent of CEOP indicates an elevated solute concentration near the membrane surface due to the formation of a cake-layer, which obstructs the back-diffusion of solute into the bulk solution (Taheri *et al.*, 2015). This eventually causes an increase in the transmembrane pressure (TMP), in which more energy is required to push the fluid through the membrane.

CFD technique is used to assess and interpret the results from tracer test for estimating the fouling resistance and concentration polarisation. A study conducted by Fimbres Weihs & Wiley (2014) focuses on the effect of cake-enhanced osmotic pressure on particulate fouling. The tracer test technique is used to monitor the TMP, permeate flux, and solute concentration in the permeate by injecting sodium chloride as the tracer into the feed stream of a membrane separation unit (Chong *et al.*, 2007). A *CP* index is used to measure the extent of *CP*, which can be defined in a one-dimensional mass balance differential equation, given as follows:

$$CP = \frac{w_w - w_p}{w_b - w_p} = \exp\left(\frac{J_v}{k_{mt}}\right) \qquad \text{Eq. 15}$$

where *CP* is the local *CP* index based on local solute bulk concentration, k_{mt} is the mass transfer coefficient, and J_v is the volumetric permeate flux.

Results from the tracer test found that a step change in feed concentration does not affect the degree of concentration polarisation across the membrane. The *CP* index was found to be over-estimated, which can be explained by the entrance effects limited by spacers, such that the presence of spacers separating the membrane leaves would eventually shift the velocity profile of fluid towards the module inlet. The over-estimation in the fouled membrane is greater and the overestimation of *CP* index decreases with higher tracer concentration but increases with fouling layer mass, which reduces the membrane performance in terms of reducing particulate fouling. However, it is important to make assumptions to maintain a constant CPM index to identify precise error results on constant pressure tracer response tests because changing salt concentration may alter the specific cake resistance that will lead to deviation in over-estimation.

19.5.5.3 Biofouling

Biofouling phenomena occur due to the accumulation and adhesion of microorganisms on the membrane surface, leading to the formation of a biofilm on the membrane. The biofilm layer is commonly composed of an extracellular polymeric substance (EPS) matrix, which is a polymer-like material enclosed with microorganisms on the surface (Unal, 2022). The extent of biofouling can depend on several factors such as membrane characteristics, influent composition and microorganism types. Most biofouling instances occur as a result of a symbiotic interaction between bacteria, algae and fungus. Another fact to consider when dealing with biofouling, is that a significant amount of the cells attached to the membrane are dead cells, meaning that biofouling leads to organic fouling. Recent advances on membrane science have focused on producing membranes with high surface hydrophobicity and low surface roughness, as it has been found that membranes with these characteristics have a lower tendency for biofouling, due to smaller exposed areas and less active sites for microbial adhesion (Maddah & Chogle, 2016). Furthermore, the development of a biofilm depends, in general, on the substrate availability and the type of microorganisms on the feed stream.

A recent study by Lin *et al.* (2022b) numerically and experimentally investigated the effects of feed spacer geometries and channel porosity on the degree of biofouling in the membrane system. Feed spacer geometries were varied in terms of the filament length, diameter, mesh angle, as well as spacer thickness to obtain different channel porosities. The porosity of the feed spacer membrane channel is calculated using the following equation:

$$\delta_{channel} = 1 - \frac{V_{spacer}}{V_{channel}} = 1 - \frac{\left(\frac{\pi}{4}D^2\right)L}{HL^2 \sin\alpha} = 1 - \frac{\pi D^2}{2HL\sin\alpha} \quad \text{Eq. 16}$$

where D is filament diameter (m), L is filament length (m), H is spacer thickness (m), and α is mesh angle. The channel porosity is an important factor in determining the feed channel pressure (FCP) drop. A larger channel porosity will generally result in less FCP drop, as there are less flow obstructions per unit of channel volume. Furthermore, larger porosity also generally leads to less biomass accumulation, due to less stagnant flow regions and larger shear.

Lin *et al.* (2022b) noted that several experimental studies report that higher shear stress conditions tend to promote biofouling because of a higher nutrient load enhances biomass accumulation, although the biofilm forming on high shear stress regions would be thinner and more compacted. In addition, their analysis shows that the average cross-flow velocity decreases with increasing filament length and spacer thickness due to reduced turbulence caused by the number of spacer filament per unit channel length and the increment of space between spacer filament and membrane surface, respectively. However, the average velocity increases at a larger mesh angle or larger filament diameter. This is because a decrease in the distance between the neighbouring spacer meshes and a narrow space between the spacer filament and the membrane surface cause more turbulence in the fluid flow. It was also found that an increase in the filament diameter and/or a decrease in the spacer thickness promotes the accumulation of biomass, which favours the growth of bacteria on the membrane surface as the interspace between the membrane and spacer filaments is reduced. Therefore, a thicker feed spacer is recommended as higher feed channel porosity tends to reduce biofouling in the membrane system (Bucs *et al.*, 2014). A summary of significant findings related to biofouling in membrane systems by CFD is presented in Table 5.

Table 5 Significant findings of CFD studies on biofouling in membrane system

References	Research types	Main findings	Observations
Vrouwenvelder *et al.* (2010)	CFD only	The presence of spacer strongly affects the level of pressure drop as an increment of 10 times the pressure drop in the feed channel is reported, when compared with the experimental case without spacer. Fouling on the feed spacer is more important than fouling on the membrane.	This study is limited to one spacer geometry, specifically on the spacer thickness of 31 mil.
Gu *et al.* (2017)	CFD only	Fully woven spacers result in the highest water flux and have lower average concentration polarisation moduli.	The study is limited to hydrodynamics and ignores mass transfer along the channel.
Li *et al.* (2016)	CFD only	Regions with high concentration of biomass are isolated to zones near the spacer filaments in which the flow is relatively static. The fouling tendencies are higher in the stagnant regions because the particles can settle down more easily.	The study is limited to one type of spacer geometry without considering the effects of spacer geometry on CP.
Chen and Wu (2021)	CFD only	A larger average pore size of the membrane increases the nucleation frequency and growth rate of membrane biofouling. At lower velocity, the flow rate decreases more than 50%, partly contributed by the decreased low permeate flux caused by fouling.	This study lacks focus on the effects of pressure drop in determining the degree of concentration polarisation.

An approach to model biofouling development and its effect on mass transfer in membrane systems is to consider the biofilm as a different phase. The biofilm phase is considered to have different mass transfer properties for the models. The solute concentration and the biofilm are linked by introducing a mathematical model to describe the biomass growth as a function of the substrate concentration. The Monod equation is possibly the simplest model to describe the cell growth rate in terms of the substrate concentration. This differential equation states that the cell growth rate is directly proportional to the number of cells, but limited by the substrate availability. The Monod equation is described as:

$$\frac{dC_X}{dt} = \mu_{\max} \frac{C_S}{K_S + C_S} C_X \qquad \text{Eq. 17}$$

where C_X is the biomass concentration (kg m^{-3}), μ_{max} is maximum growth rate (s^{-1}), C_S (kgm^{-3}) is the substrate concentration and K_S is the half-velocity constant (kg m^{-3}).

More sophisticated models may take into account the cell growth phase, the inhibitory effect of the components of the media (e.g., sodium chloride, chlorine, reactive oxygen species, etc.). Radu *et al.* (2010) conducted a numerical study on the biofilm formation in a spacer-filled membrane channel coupling COMSOL and MATLAB. Figure 17 shows a graphic overview of the process for biofouling modelling used by Radu *et al.* (2010). The biomass growth rate is modelled using Monod equation with a random seeding on the filament wall or membrane surface. A finite element mesh is used to discretise the fluid and the biofilm sub-domains. The hydrodynamics, solute concentration and substrate concentration are solved via CFD. The biofilm growth is modelled according to the substrate distribution and then updated into the CFD model. Finally, an attachment/detachment step is added based on the mechanical stress. This process is repeated for each timestep to follow the time evolution of the biofilms.

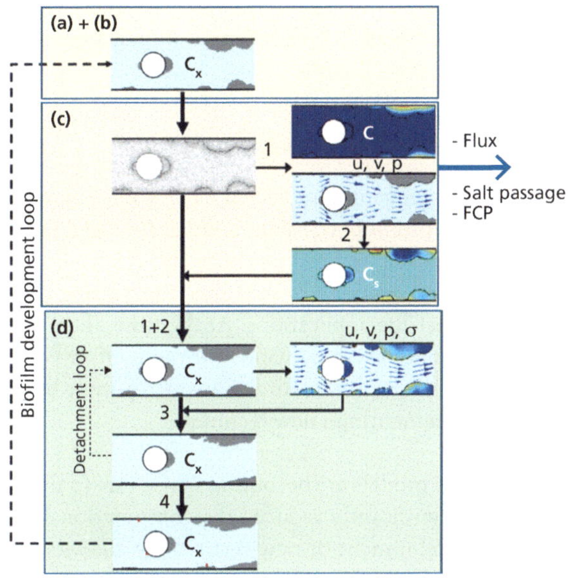

Figure 17 Adaptive algorithm for modelling of biofilm evolution using a hybrid CFD/numerical approach (Radu *et al.*, 2010).

19.6 ADDITIONAL CONSIDERATIONS

19.6.1 Multi-scale modelling

Despite of being a powerful tool for analysis, CFD studies typically focus only on a small-scale model in order to facilitate the modelling of an MSP. Multi-scale modelling refers to the analysis of fluids at different scales of space and/or time (Steinhauser, 2017). While analysing the operational advantages of a certain technique in a small-scale model may be useful to determine its effectivity, in some cases, it is worth to analyse the large-scale effects as they are directly linked to the feasibility of the approach. For example, using an instantaneous pulse as the inflow for a membrane system may enhance mass transfer over a short period of time, but this effect is negligible compared to the large operation periods of actual systems. Indicators such as pressure drop and permeate flux can be used to extrapolate the efficiency of a full-length membrane system. Furthermore, optimising the separation performance of the system can lead to energy savings during operation; although, these may be negligible when compared to the energy requirements of the whole process.

19.6.1.1 Techno-economics

Techno-economic analyses are used to determine the cost-effectiveness of a process or a modification made to it, by estimating the overall production cost of the product (Toh *et al.*, 2020a). A unitary cost per volume (e.g., dollars per m^3) is used to reflect the processing cost for producing treated water. A comprehensive techno-economic analysis would include both capital and operational costs which are difficult to determine as they depend on geographical and time factors (Toh *et al.*, 2020a). The results obtained via CFD analysis are useful to carry out a simplified techno-economic assessment considering only operational costs related to the membrane operation such as pre-treatment, operating pressure, pressure drop and permeate flux (Liang *et al.*, 2019).

19.7 OUTLOOK

Instead of focusing on the spacer geometry, recent CFD studies on membrane systems are focusing on new strategies to promote transient laminar flow and induce vortex shedding such as unsteady inflow, vibration-assisted modules and electroosmosis. In addition, the number of studies including experimental validation is increasing due to the necessity of assessing the feasibility for real-life applications. At the time, there is a gap between CFD studies and real-life applications which demands for more effective ways to scale-up the models. Comprehensive techno-economic studies are increasingly being included to assess the economic benefits of implementing a new technique.

Over the last decade, 2D CFD models are becoming sparse due to the increasing capability of modern computers that significantly reduced the computational time required for 3D simulations. The soaring development of new technology allows higher resolution with less computational time and has introduced new approaches for data analysis. New studies including advanced modelling techniques are frequently appearing in an effort to extract the most information from CFD analysis. This modelling techniques include tools like big data, machine learning and reduced-order modelling. The arrival of techniques such as micro- and holographic-particle image velocimetry is allowing the validation of micro-scale phenomena studied by CFD during the recent years.

19.8 REFERENCES

Ahmad, A. L., Lau, K. K., Bakar, M. A., & Shukor, S. A. (2005). Integrated CFD simulation of concentration polarization in narrow membrane channel. Computers & Chemical Engineering, 29(10), 2087-2095. doi:https://doi.org/10.1016/j.compchemeng.2005.06.001

Alexiadis, A., Wiley, D., Fletcher, D., & Bao, J. (2007). Laminar flow transitions in a 2D channel with circular spacers. Industrial & Engineering Chemistry Research, 46(16), 5387-5396. doi:https://doi.org/10.1021/ie0607797

Anderson, J. D., & Wendt, J. (1995). Computational fluid dynamics (Vol. 206): Springer.

Anderson, J. L. (1989). Colloid transport by interfacial forces. Annual review of fluid Mechanics, 21(1), 61-99.

ANSYS Inc. (2012). ANSYS CFX Solver theory guide. Canonsburg.

ANSYS Inc. (2020). ANSYS Fluent Users Guide. Canonsburg.

ANSYS Inc. (2021). ANSYS CFX Solver Manager User's Guide. Canonsburg.

Antony, A., Chilcott, T., Coster, H., & Leslie, G. (2013). In situ structural and functional characterization of reverse osmosis membranes using electrical impedance spectroscopy. Journal of Membrane science, 425-426, 89-97. doi:https://doi.org/10.1016/j.memsci.2012.09.028

Asadi, A., Huat, B. B., Nahazanan, H., & Keykhah, H. A. (2013). Theory of electroosmosis in soil. International Journal of Electrochemical Science, 8(1), 1016-1025.

Asefi, H., Alighardashi, A., Fazeli, M., & Fouladitajar, A. (2019). CFD modeling and simulation of concentration polarization reduction by gas sparging cross-flow nanofiltration. Journal of Environmental Chemical Engineering, 7(5), 103275. doi:https://doi.org/10.1016/j.jece.20r19.103275

Baghel, R., Kalla, S., Upadhyaya, S., Chaurasia, S., & Singh, K. (2020). CFD modeling of vacuum membrane distillation for removal of Naphthol blue black dye from aqueous solution using COMSOL multiphysics. Chemical Engineering Research and Design, 158, 77-88. doi:https://doi.org/10.1016/j.cherd.2020.03.016

Baker, R. W. (2004). Membrane Technology and Applications. England: John Wiley & Sons Ltd.

Baker, R. W. (2012). Membrane technology and applications: John Wiley & Sons.

Balster, J. (2016). Hollow Fiber Membrane Module. In E. Drioli & L. Giorno (Eds.), Encyclopedia of Membranes (pp. 955-957). Berlin, Heidelberg: Springer Berlin Heidelberg.

Behroozi, A. H., Kasiri, N., & Mohammadi, T. (2019). Multi-phenomenal macroscopic investigation of cross-flow membrane flux in microfiltration of oil-in-water emulsion, experimental & computational. Journal of Water Process Engineering, 32, 100962. doi:https://doi.org/10.1016/j.jwpe.2019.100962

Berk, Z. (2009). Chapter 10 - Membrane processes. In Z. Berk (Ed.), Food Process Engineering and Technology (pp. 233-257). San Diego: Academic Press.

Boussu, K., Van der Bruggen, B., Volodin, A., Snauwaert, J., Van Haesendonck, C., & Vandecasteele, C. (2005). Roughness and hydrophobicity studies of nanofiltration membranes using different modes of AFM. Journal of colloid and interface science, 286(2), 632-638. doi:https://doi.org/10.1016/j.jcis.2005.01.095

Brunner, D., Khawaja, H., Moatamedi, M., & Boiger, G. (2018). CFD modelling of pressure and shear rate in torsionally vibrating structures using ANSYS CFX and COMSOL Multiphysics. The International Journal of Multiphysics, 12(4), 349-358. doi:https://doi.org/10.21152/1750-9548.12.4.349

Bucs, S. S., Valladares Linares, R., van Loosdrecht, M. C., Kruithof, J. C., & Vrouwenvelder, J. S. (2014). Impact of organic nutrient load on biomass accumulation, feed channel pressure drop increase and permeate flux decline in membrane systems. Water Res, 67, 227-242. doi:10.1016/j.watres.2014.09.005

Burn, S., & Gray, S. (2015). Efficient Desalination by Reverse Osmosis: A guide to RO practice (Vol. 14). London, UK: IWA Publishing.

Byun, H. S., & Rhee, K. (2004). CFD modeling of blood flow following coil embolization of aneurysms. Medical engineering & physics, 26(9), 755-761. doi:https://doi.org/10.1016/j.medengphy.2004.06.008

Cancilla, N., Gurreri, L., Marotta, G., Ciofalo, M., Cipollina, A., Tamburini, A., & Micale, G. (2021). CFD prediction of shell-side flow and mass transfer in regular fiber arrays. International Journal of Heat and Mass Transfer, 168, 120855. doi:https://doi.org/10.1016/j.ijheatmasstransfer.2020.120855

Cao, Z., Wiley, D., & Fane, A. (2001). CFD simulations of net-type turbulence promoters in a narrow channel. Journal of Membrane science, 185(2), 157-176. doi:https://doi.org/10.1016/s0376-7388(00)00643-8

Chan, F. S., Tan, C. K., Ratnayake, P., Junaidi, M. U. M., & Liang, Y. Y. (2020a). Reduced-order modelling of concentration polarization with varying permeation: Analysis of electro-osmosis in membranes. Desalination, 495, 114677. doi:https://doi.org/10.1016/j.desal.2020.114677

Chan, F. S., Tan, C. K., Ratnayake, P., & Liang, Y. Y. (2020b). Reduced-order modeling of flow and concentration polarization in membrane systems with permeation. AIChE journal, 66(4), e16851. doi:https://doi.org/10.1002/aic.16851

Chen, L., & Wu, B. (2021). Research Progress in Computational Fluid Dynamics Simulations of Membrane Distillation Processes: A Review. Membranes (Basel), 11(7). doi:10.3390/membranes11070513

Chong, T. H., Wong, F. S., & Fane, A. G. (2007). Enhanced concentration polarization by unstirred fouling layers in reverse osmosis: Detection by sodium chloride tracer response technique. Journal of Membrane science, 287(2), 198-210. doi:10.1016/j.memsci.2006.10.035

Chong, Y. K., Liang, Y. Y., Lau, W. J., & Fimbres Weihs, G. A. (2022). 3D CFD study of hydrodynamics and mass transfer phenomena for spiral wound membrane submerged-type feed spacer with different node geometries and sizes. International Journal of Heat and Mass Transfer, 191, 122819. doi:https://doi.org/10.1016/j.ijheatmasstransfer.2022.122819

Chu, A., Kwok, R. C.-W., & Yu, K. (2005). Study of pollution dispersion in urban areas using Computational Fluid Dynamics (CFD) and Geographic Information System (GIS). Environmental Modelling & Software, 20(3), 273-277. doi:https://doi.org/10.1016/j.envsoft.2004.05.007

COMSOL AB. (2008). COMSOL Multiphysics User's Guide. Burlington.

Cummings, E. B., Griffiths, S. K., Nilson, R. H., & Paul, P. H. (2000). Conditions for Similitude between the Fluid Velocity and Electric Field in Electroosmotic Flow. Analytical Chemistry, 72(11), 2526-2532. doi:10.1021/ac991165x

Da Costa, A., Fane, A., & Wiley, D. (1994). Spacer characterization and pressure drop modelling in spacer-filled channels for ultrafiltration. Journal of Membrane science, 87(1-2), 79-98. doi:https://doi.org/10.1016/0376-7388(93)e0076-p

Du, X., Liu, X., Wang, Y., Radaei, E., Lian, B., Leslie, G., Li, G., & Liang, H. (2017). Particle deposition on flat sheet membranes under bubbly and slug flow aeration in coagulation-microfiltration process: Effects of particle characteristic and shear stress. Journal of Membrane science, 541, 668-676. doi:https://doi.org/10.1016/j.memsci.2017.07.023

Fimbres-Weihs, G., & Wiley, D. (2008). Numerical study of two-dimensional multi-layer spacer designs for minimum drag and maximum mass transfer. Journal of Membrane science, 325(2), 809-822. doi:https://doi.org/10.1016/j.memsci.2008.09.005

Fimbres-Weihs, G., & Wiley, D. (2010). Review of 3D CFD modeling of flow and mass transfer in narrow spacer-filled channels in membrane modules. Chemical Engineering and Processing: Process Intensification, 49(7), 759-781. doi:https://doi.org/10.1016/j.cep.2010.01.007

Fimbres-Weihs, G. A., Wiley, D. E., & Fletcher, D. F. (2006). Unsteady Flows with Mass Transfer in Narrow Zigzag Spacer-Filled Channels: A Numerical Study. Industrial & Engineering Chemistry Research, 45(19), 6594-6603. doi:10.1021/ie060243l

Fimbres Weihs, G., & Wiley, D. (2007). Numerical study of mass transfer in three-dimensional spacer-filled narrow channels with steady flow. Journal of Membrane science, 306, 228-243. doi:10.1016/j.memsci.2007.08.043

Fimbres Weihs, G. A., & Wiley, D. E. (2014). CFD analysis of tracer response technique under cake-enhanced osmotic pressure. Journal of Membrane science, 449, 38-49. doi:10.1016/j.memsci.2013.08.015

Foo, K., Liang, Y. Y., & Fimbres Weihs, G. A. (2020). CFD study of the effect of SWM feed spacer geometry on mass transfer enhancement driven by forced transient slip velocity. Journal of Membrane science, 597. doi:10.1016/j.memsci.2019.117643

Foo, K., Liang, Y. Y., Tan, C. K., & Fimbres Weihs, G. A. (2021). Coupled effects of circular and elliptical feed spacers under forced-slip on viscous dissipation and mass transfer enhancement based on CFD. Journal of Membrane science, 637, 119599. doi:https://doi.org/10.1016/j.memsci.2021.119599

Fritzmann, C., Löwenberg, J., Wintgens, T., & Melin, T. (2007). State-of-the-art of reverse osmosis desalination. Desalination, 216(1-3), 1-76. doi:10.1016/j.desal.2006.12.009

Gavelli, F., Bullister, E., & Kytomaa, H. (2008). Application of CFD (Fluent) to LNG spills into geometrically complex environments. Journal of hazardous materials, 159(1), 158-168. doi:https://doi.org/10.1016/j.jhazmat.2008.02.037

Geraldes, V., & Afonso, M. D. (2006). Generalized mass transfer correction factor for nanofiltration and reverse osmosis. AIChE journal, 52(10), 3353-3362. doi:https://doi.org/10.1002/aic.10968

Golrokh Sani, A., Najafi, H., & Azimi, S. S. (2021). CFD simulation of air-sparged slug flow in the flat-sheet membrane: A concentration polarization study. Separation and Purification Technology, 270, 118816. doi:https://doi.org/10.1016/j.seppur.2021.118816

Gu, B., Adjiman, C. S., & Xu, X. Y. (2017). The effect of feed spacer geometry on membrane performance and concentration polarisation based on 3D CFD simulations. Journal of Membrane science, 527, 78-91. doi:https://doi.org/10.1016/j.memsci.2016.12.058

Gupta, R., Fletcher, D. F., & Haynes, B. S. (2009). On the CFD modelling of Taylor flow in microchannels. Chemical Engineering Science, 64(12), 2941-2950. doi:https://doi.org/10.1016/j.ces.2009.03.018

Haddadi, B., Jordan, C., Miltner, M., & Harasek, M. (2018). Membrane modeling using CFD: Combined evaluation of mass transfer and geometrical influences in 1D and 3D. Journal of Membrane science, 563, 199-209. doi:https://doi.org/10.1016/j.memsci.2018.05.040

Hinsch, K. D. (2002). Holographic particle image velocimetry. Measurement Science and Technology, 13(7), R61-R72. doi:10.1088/0957-0233/13/7/201

Holemans, T., Yang, Z., & Vanierschot, M. (2022). Efficient Reduced Order Modeling of Large Data Sets Obtained from CFD Simulations. Fluids, 7(3), 110. doi:https://doi.org/10.3390/fluids7030110

Hu, G., & Li, D. (2007). Multiscale phenomena in microfluidics and nanofluidics. Chemical Engineering Science, 62(13), 3443-3454. doi:https://doi.org/10.1016/j.ces.2006.11.058

Hunter, R. J. (2013). Zeta potential in colloid science: principles and applications (Vol. 2): Academic press.

Jafarkhani, M., Moraveji, M. K., Davarnejad, R., Moztarzadeh, F., & Mozafari, M. (2012). Three-dimensional simulation of turbulent flow in a membrane tube filled with semi-circular baffles. Desalination, 294, 8-16. doi:https://doi.org/10.1016/j.desal.2012.02.031

Jagannadh, S. N., & Muralidhara, H. S. (1996). Electrokinetics Methods To Control Membrane Fouling. Industrial & Engineering Chemistry Research, 35(4), 1133-1140. doi:10.1021/ie9503712

Jamshed, S. (2015). Using HPC for Computational Fluid Dynamics: A Guide to High Performance Computing for CFD Engineers: Academic Press.

Javid, S. M., Passandideh-Fard, M., Faezian, A., & Goharimanesh, M. (2017). Slug and bubble flows in a flat sheet ultrafiltration module: Experiments and numerical simulation. International Journal of Multiphase Flow, 91, 39-50. doi:https://doi.org/10.1016/j.ijmultiphaseflow.2016.12.006

Jiang, S., Li, Y., & Ladewig, B. P. (2017). A review of reverse osmosis membrane fouling and control strategies. Sci Total Environ, 595, 567-583. doi:10.1016/j.scitotenv.2017.03.235

Junker, M. A., de Vos, W. M., Lammertink, R. G. H., & de Grooth, J. (2021). Bridging the gap between lab-scale and commercial dimensions of hollow fiber nanofiltration membranes. Journal of Membrane science, 624, 119100. doi:https://doi.org/10.1016/j.memsci.2021.119100

Kaya, R., Deveci, G., Turken, T., Sengur, R., Guclu, S., Koseoglu-Imer, D. Y., & Koyuncu, I. (2014). Analysis of wall shear stress on the outside-in type hollow fiber membrane modules by CFD simulation. Desalination, 351, 109-119. doi:https://doi.org/10.1016/j.desal.2014.07.033

Kone, J.-P., Zhang, X., Yan, Y., Hu, G., & Ahmadi, G. (2018). CFD modeling and simulation of PEM fuel cell using OpenFOAM. Energy Procedia, 145, 64-69. doi:https://doi.org/10.1016/j.egypro.2018.04.011

Landahl, M. T., Mollo Christensen, E., & Korman, M. S. (1989). Turbulence and random processes in fluid mechanics. In: Acoustical Society of America.

Leam, J. J., Bilad, M. R., Wibisono, Y., Hakim Wirzal, M. D., & Ahmed, I. (2020). Membrane Technology for Microalgae Harvesting. 97-110. doi:10.1016/b978-0-12-817536-1.00007-2

Li, H. y., Bertram, C. D., & Wiley, D. E. (1998). Mechanisms by which pulsatile flow affects cross flow microfiltration. AIChE journal, 44(9), 1950-1961. doi:https://doi.org/10.1002/aic.690440903

Li, M., Bui, T., & Chao, S. (2016). Three-dimensional CFD analysis of hydrodynamics and concentration polarization in an industrial RO feed channel. Desalination, 397, 194-204. doi:https://doi.org/10.1016/j.desal.2016.07.005

Li, W., Su, X., Palazzolo, A., Ahmed, S., & Thomas, E. (2017). Reverse osmosis membrane, seawater desalination with vibration assisted reduced inorganic fouling. Desalination, 417, 102-114. doi:. https://doi.org/10.1016/j.desal.2017.05.016

Li, Y.-L., & Tung, K.-L. (2008). The effect of curvature of a spacer-filled channel on fluid flow in spiral-wound membrane modules. Journal of Membrane science, 319(1-2), 286-297. doi:10.1016/j.memsci.2008.03.069

Li, Y.-L., Tung, K.-L., Chen, Y.-S., & Hwang, K.-J. (2012). CFD analysis of the initial stages of particle deposition in spiral-wound membrane modules. Desalination, 287, 200-208. doi:10.1016/j.desal.2011.10.001

Liang, Y. Y., Chapman, M. B., Fimbres-Weihs, G. A., & Wiley, D. E. (2014a). CFD modelling of electro-osmotic permeate flux enhancement on the feed side of a membrane module. Journal of Membrane science, 470, 378-388. doi:https://doi.org/10.1016/j.memsci.2014.07.039

Liang, Y. Y., Fimbres-Weihs, G. A., & Fletcher, D. F. (2018a). CFD study of the effect of unsteady slip velocity waveform on shear stress in membrane systems. Chemical Engineering Science, 192, 16-24. doi:https://doi.org/10.1016/j.ces.2018.07.009

Liang, Y. Y., Fimbres-Weihs, G. A., Setiawan, R., & Wiley, D. E. (2016a). CFD modelling of unsteady electro-osmotic permeate flux enhancement in membrane systems. Chemical Engineering Science, 146, 189-198. doi:https://doi.org/10.1016/j.ces.2016.02.028

Liang, Y. Y., Fimbres-Weihs, G. A., & Wiley, D. E. (2014b). Approximation for modelling electro-osmotic mixing in the boundary layer of membrane systems. Journal of Membrane science, 450, 18-27. doi:https://doi.org/10.1016/j.memsci.2013.08.031

Liang, Y. Y., Fimbres-Weihs, G. A., & Wiley, D. E. (2016b). CFD modelling of electro-osmotic permeate flux enhancement in spacer-filled membrane channels. Journal of Membrane science, 507, 107-118. doi:https://doi.org/10.1016/j.memsci.2016.02.012

Liang, Y. Y., Fimbres-Weihs, G. A., & Wiley, D. E. (2020a). Comparison of oscillating flow and slip velocity mass transfer enhancement in spacer-filled membrane channels: CFD analysis and validation. Journal of Membrane science, 593, 117433. doi:https://doi.org/10.1016/j.memsci.2019.117433

Liang, Y. Y., Fimbres Weihs, G. A., & Fletcher, D. F. (2018b). CFD study of the effect of unsteady slip velocity waveform on shear stress in membrane systems. Chemical Engineering Science, 192, 16-24. doi:https://doi.org/10.1016/j.ces.2018.07.009

Liang, Y. Y., Fimbres Weihs, G. A., & Wiley, D. E. (2020b). Comparison of oscillating flow and slip velocity mass transfer enhancement in spacer-filled membrane channels: CFD analysis and validation. Journal of Membrane science, 593, 117433. doi:https://doi.org/10.1016/j.memsci.2019.117433

Liang, Y. Y., Toh, K. Y., & Fimbres Weihs, G. A. (2019). 3D CFD study of the effect of multi-layer spacers on membrane performance under steady flow. Journal of Membrane science, 580, 256-267. doi:https://doi.org/10.1016/j.memsci.2019.02.015

Lin, W., Lei, J., Wang, Q., Wang, X.-m., & Huang, X. (2022a). Performance enhancement of spiral-wound reverse osmosis membrane elements with novel diagonal-flow feed channels. Desalination, 523, 115447.

Lin, W., Wang, Q., Sun, L., Wang, D., Cabrera, J., Li, D., Hu, L., Jiang, G., Wang, X.-m., & Huang, X. (2022b). The critical role of feed spacer channel porosity in membrane biofouling: Insights and implications. Journal of Membrane science, 649, 120395.

Lindken, R., & Burgmann, S. (2012). 14 - Laser-optical methods for transport studies in low temperature fuel cells. In C. Hartnig & C. Roth (Eds.), Polymer Electrolyte Membrane and Direct Methanol Fuel Cell Technology (Vol. 2, pp. 425-461): Woodhead Publishing.

Liu, D., Bu, C., & Chen, X. (2013). Development and test of CFD–DEM model for complex geometry: A coupling algorithm for Fluent and DEM. Computers & Chemical Engineering, 58, 260-268. doi:https://doi.org/10.1016/j.compchemeng.2013.07.006

Liu, Y., & Hinrichsen, O. (2014). CFD modeling of bubbling fluidized beds using OpenFOAM®: Model validation and comparison of TVD differencing schemes. Computers & Chemical Engineering, 69, 75-88. doi:https://doi.org/10.1016/j.compchemeng.2014.07.002

Lotfiyan, H., Zokaee Ashtiani, F., Fouladitajar, A., & Armand, S. B. (2014). Computational fluid dynamics modeling and experimental studies of oil-in-water emulsion microfiltration in a flat sheet membrane using Eulerian approach. Journal of Membrane science, 472, 1-9. doi:https://doi.org/10.1016/j.memsci.2014.08.036

Maddah, H., & Chogle, A. (2016). Biofouling in reverse osmosis: phenomena, monitoring, controlling and remediation. Applied Water Science, 7(6), 2637-2651. doi:10.1007/s13201-016-0493-1

Martín, M. M. (2016). Chapter 4 - Water. In M. M. Martín (Ed.), Industrial Chemical Process Analysis and Design (pp. 125-197). Boston: Elsevier.

Martinelli, L., Guigui, C., & Line, A. (2010). Characterisation of hydrodynamics induced by air injection related to membrane fouling behaviour. Desalination, 250(2), 587-591. doi:https://doi.org/10.1016/j.desal.2009.09.029

NASA. (2022). Overview of CFD Verification & Validation. Retrieved from https://www.grc.nasa.gov/www/wind/valid/tutorial/overview.html

Ndinisa, N. V., Wiley, D. E., & Fletcher, D. F. (2005). Computational Fluid Dynamics Simulations of Taylor Bubbles in Tubular Membranes: Model Validation and Application to Laminar Flow Systems. Chemical Engineering Research and Design, 83(1), 40-49. doi:https://doi.org/10.1205/cherd.03394

OpenCFD Ltd. (2016). OpenFOAM: User Guide. Retrieved from https://www.openfoam.com/documentation/guides/latest/doc/guide-schemes.html

Ouyang, H., Bao, J., Fimbres-Weihs, G. A., & Wiley, D. E. (2013). Control study on mixing enhancement in boundary layers of membrane systems. Journal of Process Control, 23(8), 1197-1204. doi:https://doi.org/10.1016/j.jprocont.2013.07.005

Pak, A., Mohammadi, T., Hosseinalipour, S. M., & Allahdini, V. (2008). CFD modeling of porous membranes. Desalination, 222(1), 482-488. doi:https://doi.org/10.1016/j.desal.2007.01.152

Park, S., Jeong, Y. D., Lee, J. H., Kim, J., Jeong, K., & Cho, K. H. (2021). 3D printed honeycomb-shaped feed channel spacer for membrane fouling mitigation in nanofiltration. Journal of Membrane science, 620, 118665. doi:https://doi.org/10.1016/j.memsci.2020.118665

Patankar, N. A., & Hu, H. H. (1998). Numerical Simulation of Electroosmotic Flow. Analytical Chemistry, 70(9), 1870-1881. doi:10.1021/ac970846u

Probstein, R. F. (1989). Physiochemical Hydrodynamics: An Introduction. Hoboken, New Jersey: John Wiley & Sons.

Qasim, M., Badrelzaman, M., Darwish, N. N., Darwish, N. A., & Hilal, N. (2019). Reverse osmosis desalination: A state-of-the-art review. Desalination, 459, 59-104. doi:10.1016/j.desal.2019.02.008

Radaei, E., Liu, X., Tng, K. H., Wang, Y., Trujillo, F. J., & Leslie, G. (2018). Insights on pulsed bubble control of membrane fouling: Effect of bubble size and frequency. Journal of Membrane science, 554, 59-70. doi:https://doi.org/10.1016/j.memsci.2018.02.058

Radu, A. I., Vrouwenvelder, J. S., van Loosdrecht, M. C. M., & Picioreanu, C. (2010). Modeling the effect of biofilm formation on reverse osmosis performance: Flux, feed channel pressure drop and solute passage. Journal of Membrane science, 365(1), 1-15. doi:https://doi.org/10.1016/j.memsci.2010.07.036

Rahimi, M., Madaeni, S. S., Abolhasani, M., & Alsairafi, A. A. (2009). CFD and experimental studies of fouling of a microfiltration membrane. Chemical Engineering and Processing: Process Intensification, 48(9), 1405-1413. doi:10.1016/j.cep.2009.07.008

Rajaratnam, N. (1976). Turbulent jets: Elsevier.

Ratkovich, N., Chan, C. C. V., Berube, P. R., & Nopens, I. (2009). Experimental study and CFD modelling of a two-phase slug flow for an airlift tubular membrane. Chemical Engineering Science, 64(16), 3576-3584. doi:https://doi.org/10.1016/j.ces.2009.04.048

Ratnayake, P., Setiawan, R., Bao, J., Fimbres-Weihs, G. A., & Wiley, D. E. (2016). Spatio-temporal frequency response analysis of forced slip velocity effect on solute concentration oscillations in a reverse osmosis membrane channel. Computers & Chemical Engineering, 84, 151-161. doi:https://doi.org/10.1016/j.compchemeng.2015.08.016

Rawool, A. S., & Mitra, S. K. (2006). Numerical simulation of electroosmotic effect in serpentine channels. Microfluidics and Nanofluidics, 2(3), 261-269. doi:10.1007/s10404-005-0076-1

Ren, L., Sinton, D., & Li, D. (2003). Numerical simulation of microfluidic injection processes in crossing microchannels. Journal of Micromechanics and Microengineering, 13(5), 739-747. doi:10.1088/0960-1317/13/5/329

Roache, P. J. (1997). Quantification of uncertainty in computational fluid dynamics. Annual review of fluid Mechanics, 29(1), 123-160. doi:https://doi.org/10.1146/annurev.fluid.29.1.123

Rodrigues, C., Garcia-Algado, P., Semião, V., de Pinho, M. N., & Geraldes, V. (2013). Concentration boundary layer visualization in nanofiltration by holographic interferometry with light deflection correction. Journal of Membrane science, 447, 306-314. doi:https://doi.org/10.1016/j.memsci.2013.07.035

Russel, W. B., Russel, W., Saville, D. A., & Schowalter, W. R. (1991). Colloidal dispersions: Cambridge university press.

Salehi, M.-S., Askarishahi, M., Godini, H. R., Görke, O., & Wozny, G. n. (2016). Sustainable process design for oxidative coupling of methane (OCM): comprehensive reactor engineering via computational fluid dynamics (CFD) analysis of OCM packed-bed membrane reactors. Industrial & Engineering Chemistry Research, 55(12), 3287-3299. doi:https://doi.org/10.1021/acs.iecr.5b03292

Santiago, J. G. (2001). Electroosmotic Flows in Microchannels with Finite Inertial and Pressure Forces. Analytical Chemistry, 73(10), 2353-2365. doi:10.1021/ac0101398

Schwinge, J., Wiley, D. E., & Fletcher, D. F. (2002a). A CFD study of unsteady flow in narrow spacer-filled channels for spiral-wound membrane modules. Desalination, 146(1), 195-201. doi:https://doi.org/10.1016/S0011-9164(02)00470-8

Schwinge, J., Wiley, D. E., & Fletcher, D. F. (2002b). Simulation of the Flow around Spacer Filaments between Channel Walls. 2. Mass-Transfer Enhancement. Industrial & Engineering Chemistry Research, 41(19), 4879-4888. doi:10.1021/ie011015o

Scott, K. (1995). Handbook of industrial membranes: Elsevier.

Snel, H. (2003). Review of aerodynamics for wind turbines. Wind Energy: An International Journal for Progress and Applications in Wind Power Conversion Technology, 6(3), 203-211. doi:https://doi.org/10.1002/WE.97

Sparks, T., & Chase, G. (2016). Section 2 - Filter Media. In T. Sparks & G. Chase (Eds.), Filters and Filtration Handbook (Sixth Edition) (pp. 55-115). Oxford: Butterworth-Heinemann.

Spiegler, K. S., & Macleish, J. H. (1981). Molecular (osmotic and electro-osmotic) backwash of cellulose acetate hyperfiltration membranes. Journal of Membrane science, 8(2), 173-192. doi:https://doi.org/10.1016/S0376-7388(00)82089-X

Steinhauser, M. O. (2017). Computational multiscale modeling of fluids and solids: Springer.

Streeter, V., Wylie, E., & Bedford, K. (1985). Fluid Mechanics (8th ed.): McGraw-Hill.

Su, X., Li, W., Palazzolo, A., & Ahmed, S. (2018). Concentration polarization and permeate flux variation in a vibration enhanced reverse osmosis membrane module. Desalination, 433, 75-88. doi:https://doi.org/10.1016/j.desal.2018.01.001

Su, X., Li, W., Palazzolo, A., & Ahmed, S. (2019). Permeate flux increase by colloidal fouling control in a vibration enhanced reverse osmosis membrane desalination system. Desalination, 453, 22-36. doi:https://doi.org/10.1016/j.desal.2018.12.003

Sun, B., Ahmed, A., Atkinson, C., & Soria, J. (2020). A novel 4D digital holographic PIV/PTV (4D-DHPIV/PTV) methodology using iterative predictive inverse reconstruction. Measurement Science and Technology, 31(10), 104002. doi:10.1088/1361-6501/ab8ee8

Taha, T., Cheong, W. L., Field, R. W., & Cui, Z. F. (2006). Gas-sparged ultrafiltration using horizontal and inclined tubular membranes—A CFD study. Journal of Membrane science, 279(1), 487-494. doi:https://doi.org/10.1016/j.memsci.2005.12.063

Taha, T., & Cui, Z. F. (2002). CFD modelling of gas-sparged ultrafiltration in tubular membranes. Journal of Membrane science, 210(1), 13-27. doi:https://doi.org/10.1016/S0376-7388(02)00360-5

Taha, T., & Cui, Z. F. (2006a). CFD modelling of slug flow in vertical tubes. Chemical Engineering Science, 61(2), 676-687. doi:https://doi.org/10.1016/j.ces.2005.07.022

Taha, T., & Cui, Z. F. (2006b). CFD modelling of slug flow inside square capillaries. Chemical Engineering Science, 61(2), 665-675. doi:https://doi.org/10.1016/j.ces.2005.07.023

Taheri, A. H., Sim, L. N., Chong, T. H., Krantz, W. B., & Fane, A. G. (2015). Prediction of reverse osmosis fouling using the feed fouling monitor and salt tracer response technique. Journal of Membrane science, 475, 433-444. doi:10.1016/j.memsci.2014.10.043

Toh, K., Liang, Y., Lau, W., & Weihs, G. F. (2020a). The techno-economic case for coupling advanced spacers to high-permeance RO membranes for desalination. Desalination, 491, 114534. doi:https://doi.org/10.1016/j.desal.2020.114534

Toh, K. Y., Liang, Y. Y., Lau, W. J., & Fimbres Weihs, G. A. (2020b). A review of CFD modelling and performance metrics for osmotic membrane processes. Membranes, 10(10), 285. doi:https://doi.org/10.3390/membranes10100285

Unal, B. O. (2022). Membrane autopsy study to characterize fouling type of RO membrane used in an industrial zone wastewater reuse plant. Desalination, 529, 115648.

Versteeg, H. K., & Malalasekera, W. (2007). An introduction to computational fluid dynamics: the finite volume method: Pearson education.

Vinuesa, R., & Brunton, S. L. (2022). Enhancing computational fluid dynamics with machine learning. Nature Computational Science, 2(6), 358-366. doi:https://doi.org/10.1038/s43588-022-00264-7

Vrouwenvelder, J. S., Picioreanu, C., Kruithof, J. C., & van Loosdrecht, M. C. M. (2010). Biofouling in spiral wound membrane systems: Three-dimensional CFD model based evaluation of experimental data. Journal of Membrane science, 346(1), 71-85. doi:10.1016/j.memsci.2009.09.025

Wardeh, S., & Morvan, H. P. (2008). CFD simulations of flow and concentration polarization in spacer-filled channels for application to water desalination. Chemical Engineering Research and Design, 86(10), 1107-1116. doi:https://doi.org/10.1016/j.cherd.2008.04.010

Wiley, D. E., & Fletcher, D. F. (2003). Techniques for computational fluid dynamics modelling of flow in membrane channels. Journal of Membrane science, 211(1), 127-137. doi:https://doi.org/10.1016/s0376-7388(02)00412-x

Xue, Y. L., Zhang, R., Cao, B., & Li, P. (2021). Chapter 20 - Tubular membranes and modules. In T.-S. Chung & Y. Feng (Eds.), Hollow Fiber Membranes (pp. 431-448): Elsevier.

Zare, M., Zokaee Ashtiani, F., & Fouladitajar, A. (2013). CFD modeling and simulation of concentration polarization in microfiltration of oil–water emulsions; Application of an Eulerian multiphase model. Desalination, 324, 37-47. doi:https://doi.org/10.1016/j.desal.2013.05.022

Zhang, J.-b., He, G.-w., & Liu, F. (2006). Electro-osmotic flow and mixing in heterogeneous microchannels. Physical Review E, 73(5), 056305. doi:10.1103/PhysRevE.73.056305

IWA Publishing's authorised EU representative for General Product Safety Regulations is Diane D'Arras, 15 rue Duret, 75116 Paris, France, e-mail: safety@iwap.co.uk.

Printed and bound by CPI Group (UK) Ltd, Croydon, CR0 4YY
14/05/2026
02110727-0001